Texts in Theoretical Computer Science
An EATCS Series

T0236825

Texts in Theoretical Computer Science
An EATCS Series

Editors: W. Brauer G. Rozenberg A. Salomaa
On behalf of the European Association
for Theoretical Computer Science (EATCS)

Advisory Board: G. Ausiello M. Broy C.S. Calude
S. Even J. Hartmanis N. Jones T. Leighton M. Nivat
C. Papadimitriou D. Scott

Springer
Berlin
Heidelberg
New York
Hong Kong
London
Milan
Paris
Tokyo

Juraj Hromkovič

Algorithmics for Hard Problems

Introduction
to Combinatorial Optimization,
Randomization, Approximation,
and Heuristics

Second Edition

With 71 Figures

 Springer

Author

Prof. Dr. Juraj Hromkovič
Swiss Federal Institute of Technology
Department of Computer Science
ETH Zürich, ETH Zentrum RZ F2
8092 Zürich, Switzerland

Series Editors

Prof. Dr. Wilfried Brauer
Institut für Informatik, Technische Universität München
Boltzmannstr. 3, 85748 Garching bei München, Germany
Brauer@informatik.tu-muenchen.de

Prof. Dr. Grzegorz Rozenberg
Leiden Institute of Advanced Computer Science
University of Leiden
Niels Bohrweg 1, 2333 CA Leiden, The Netherlands
rozenber@liacs.nl

Prof. Dr. Arto Salomaa
Turku Centre for Computer Science
Lemminkäisenkatu 14 A, 20520 Turku, Finland
asalomaa@utu.fi

Corrected printing 2004

Die Deutsche Bibliothek – CIP-Einheitsaufnahme

Hromkovič, Juraj: Algorithmics for hard problems / J. Hromkovič. –
Berlin; Heidelberg; New York; Hong Kong; London; Milan; Paris; Tokyo: Springer, 2003
(Texts in theoretical computer science)

ACM Computing Classification (1998):
F.2, F.1.2–3, I.1.2, G.1.2, G.1.6, G.2.1, G.3, I.2.8

ISBN 978-3-642-07909-2

Springer-Verlag is a part of Springer Science+Business Media

springeronline.com

© Springer-Verlag Berlin Heidelberg 2010

Printed in Germany

Illustrations: Ingrid Zámečniková
Cover Design: *KünkelLopka Werbeagentur*

To PETRA and PAULA

You have been told also that life is darkness,
and in your weariness you echo what was said by the weary.

And I say that life is indeed darkness
save when there is urge,

And all urge is blind save when there is knowledge,

And all knowledge is vain save where there is work,

And all work is empty save when there is love;

And when you work with love you bind yourself
to yourself, and to one another, and to God ...

Work is love made visible.

And if you cannot work with love but only with distances,
it is better that you should leave your work
and sit at the gate of the temple and take alms of those
who work with joy.

KAHLIL GIBRAN
The Prophet

Preface to the Second, Enlarged Edition

The term algorithm is the central notion of computer science and algorithmics is one of the few fundamental kernels of theoretical computer science. Recent developments confirm this claim. Hardly any other area of theoretical computer science has been more lively and has achieved comparably deep progress and breakthroughs so fascinating (such as the PCP-theorem and efficient algorithms for primality testing) in recent years. The most exciting development happened exactly in the field of algorithmics for hard problems, which is the topic of this book.

The goal of this textbook is to give a transparent, systematic introduction to the concepts and to the methods for designing algorithms for hard problems. Simplicity is the main educational characteristic of this textbook. All ideas, concepts, algorithms, analyses, and proofs are first explained in an informal way in order to develop the right intuition, and then carefully specified in detail. Following this strategy we preferred to illustrate the algorithm design methods using the most transparent examples rather than to present the best, but too technical, results. The consequence is that there are sections where the first edition of this book does not go deep enough for advanced courses.

To smooth this drawback in the second edition, we extended the materials for some of the topics of central interest – randomized algorithms for primality testing and applications of linear programming in the design of approximation algorithms. This second edition contains both the Solovay-Strassen algorithm and the Miller-Rabin algorithm for primality testing with a selfcontained analysis of their behaviour (error probability). In order to give all related details, we extended the section about algebra and number theory in an appropriate way. To explain the power of the method of relaxation to linear programming, we added the concept of LP-duality and the presentation of the primal-dual method. As an introduction to this topic we used the Ford-Fulkerson pseudo-polynomial-time algorithm for the maximum flow problem, which is in the section about pseudo-polynomial-time algorithms.

In addition to extending some parts of the book, numerous small improvements and corrections were performed. I am indebted to all those who sent me their comments and suggestions. Especially, I would like to thank Dirk Bongartz, Hans-Joachim Böckenhauer, David Buttgereit, Thomas Deselaers, Bernd Hentschel, Frank Kehren, Thorsten Uthke, Jan van Leeuven, Sebastian Seibert, Koichi Wada, Manuel Wahle, Dieter Weckauf, and Frank Wessel who carefully read and commented on large parts of this book. Special thanks go to Dirk Bongartz and Hans-Joachim Böckenhauer for fruitful discussions on new parts of the book and for their valuable suggestions. The expertise of our LATEX experts Markus Mohr and Manuel Wahle was very helpful and is much appreciated. The excellent cooperation with Ingeborg Mayer and Alfred Hofmann from Springer-Verlag is gratefully acknowledged.

Last but not least I would like to express my deepest thanks to Peter Widmayer for encouraging me to make the work on this book a never-ending story.

Aachen, October 2002   *Juraj Hromkovič*

Preface

Algorithmic design, especially for hard problems, is more essential for success in solving them than any standard improvement of current computer technologies. Because of this, the design of algorithms for solving hard problems is the core of current algorithmic research from the theoretical point of view as well as from the practical point of view. There are many general textbooks on algorithmics, and several specialized books devoted to particular approaches such as local search, randomization, approximation algorithms, or heuristics. But there is no textbook that focuses on the design of algorithms for hard computing tasks, and that systematically explains, combines, and compares the main possibilities for attacking hard algorithmic problems. As this topic is fundamental for computer science, this book tries to close this gap.

Another motivation, and probably the main reason for writing this book, is connected to education. The considered area has developed very dynamically in recent years and the research on this topic discovered several profound results, new concepts, and new methods. Some of the achieved contributions are so fundamental that one can speak about paradigms which should be included in the education of every computer science student. Unfortunately, this is very far from reality. This is because these paradigms are not sufficiently known in the computer science community, and so they are insufficiently communicated to students and practitioners. The main reason for this unpleasant situation is that simple explanations and transparent presentations of the new contributions of algorithmics and complexity theory, especially in the area of randomized and approximation algorithms, are missing on the level of textbooks for introductory courses. This is the typical situation when principal contributions, whose seeping into the folklore of the particular scientific discipline is only a question of time, are still not recognized as paradigms in the broad community, and even considered to be too hard and too special for basic courses by non-specialists in this area. Our aim is to try to speed up this transformation of paradigmatic research results into educational folklore.

This book should provide a "cheap ticket" to algorithmics for hard problems. Cheap does not mean that the matter presented in this introductory material is not precisely explained in detail and in its context, but that it is presented as transparently as possible, and formalized by using mathematics that is as simple as possible for this purpose. Thus, the main goal of this book can be formulated as the following optimization problem.

Input: A computer science student or a practitioner
Constraints:
- To teach the input the main ideas, concepts, and algorithm design techniques (such as pseudo-polynomial-time algorithms, parameterized complexity, local search, branch-and-bound, relaxation to linear programming, randomized algorithms, approximation algorithms, simulated annealing, genetic algorithms, etc.) for solving hard problems in a transparent and well-understandable way.
- To explain the topic on the level of clear, informal ideas as well as on the precise formal level, and to be self-contained with respect to all mathematics used.
- To discuss the possibilities to combine different methods in order to attack specific hard problems as well as a possible speedup by parallelization.
- To explain methods for theoretical and experimental comparisons of different approaches to solving particular problems.

Costs: The expected time that an input needs to learn the topic of the book (particularly, the level of abstractions of mathematics used and the hardness of mathematical proofs).
Objective: Minimization.

I hope that this book provides a feasible solution to this hard optimization problem. To judge the quality (approximation ratio) of the solution provided in this book is left to the reader.

I would like to express my deepest thanks to Hans-Joachim Böckenhauer, Erich Valkema, and Koichi Wada for carefully reading the whole manuscript and for their numerous comments and suggestions. I am indebted to Ivana Černá, Vladimír Černý, Alexander Ferrein, Ralf Klasing, Dana Pardubská, Hartmut Schmeck, Georg Schnitger, Karol Tauber, Ingo Wegener, and Peter Widmayer for interesting discussions or their comments on earlier drafts of this book. Special thanks go to Hans Wössner and the team of Springer-Verlag for their excellent assistance during the whole process of the manuscript preparation. The expertise and helpfulness of our LaTeX expert Alexander Ferrein was very useful and is much appreciated.

Last but not least I would like to thank Tanja for her patience with me during the work on this book.

Aachen, March 2001   *Juraj Hromkovič*

Contents

1

Introduction

*"The advanced reader who skips parts
that appear too elementary
may miss more than
the reader who skips parts
that appear too complex."*

G. PÓLYA

Motivation and Aims

This textbook provides a "cheap ticket" to the design of algorithms for hard computing problems, i.e., for problems for which no low-degree polynomial-time algorithms[1] are known. It focuses on a *systematic* presentation of the fundamental concepts and algorithm design techniques such as pseudo-polynomial-time algorithms, parameterized complexity, branch-and-bound, local search, lowering the worst case complexity of exponential algorithms, dual approximation algorithms, stability of approximation, randomization (foiling an adversary, abundance of witnesses, fingerprinting, random sampling, random rounding), derandomization, simulated annealing, tabu search, genetic algorithms, etc. The presentation of these concepts and techniques starts with some fundamental informal ideas that are later consecutively specified in detail. The algorithms used to illustrate the application of these methods are chosen with respect to their simplicity and transparency rather than with respect to their quality (complexity and reliability). The methods for the design of algorithms are not only presented in a systematic way, they are also combined, compared, and parallelized in order to produce a practical algorithm for the given application. An outlook on possible future technologies such as DNA computing and quantum computing is provided, too.

---

[1] We call attention to the fact that we do not restrict our interpretation of hardness to NP-hardness in this book. Problems like primality testing, that are not known to be NP-hard (but that are also not known to be polynomial-time solvable), are in the center of our interest, too.

The main motivation to write this textbook is related to education. The area of algorithmics for hard problems has developed very dynamically in recent years, and the research on this topic discovered several profound results, new concepts, and new methods. Some of the achieved contributions are so fundamental that one can speak about paradigms which should be broadcasted to every computer scientist. The main aim of this textbook is to try to speed up the process of the transformation of these fundamental research results into educational folklore.

To the Students and Practitioners

Welcome. The textbook is primarily addressed to you and its style follows this purpose. This book contains a fascinating topic that shows the importance of theoretical knowledge for solving practical problems. Several of you consider theory to be boring and irrelevant for computer scientists and too hard in comparison with courses in applied areas of computer science and engineering. Still worse, some of you may view the courses in theory only as a troublesome threshold that must be overcome in order to obtain some qualification certificates. This book tries to show that the opposite of this judgment is true, i.e., that the theory involves exciting ideas that have a direct, transparent connection with practice, and that they are understandable. The realization of this task here is not very difficult because algorithmics, in contrast to several other theoretical areas, has a simple, direct relation to applications.[2] Moreover, we show in this book that several practical problems can be solved only due to some nontrivial theoretical results, and so that success in many applications strongly depends on the theoretical know-how of the algorithm designers.[3] The most fascinating effect occurs when one can jump from millions of years of computer work necessary to execute any naive algorithm to the matter of a few seconds due to an involved theoretical concept. Another crucial fact is that the use of profound theoretical results does not necessarily need to be connected with a difficult study of some complex, abstract, and mysterious mathematics. Mathematics is used here as a formal language and as a tool, but not as a mysterious end in itself. We show here that a very simple formal language on a low level of mathematical abstraction is often sufficient to clearly formulate the main ideas and to prove useful assertions. The *simplicity* is the main educational characteristics of this textbook. All ideas, concepts, algorithms, and proofs are first presented in an informal

---

[2] That is, one does not need to go any long and complicated way in order to show the relevance and usefulness of theoretical results in some applications.

[3] A nice representative example is primality testing for which no deterministic polynomial-time algorithm is known. This problem is crucial for public-key cryptography where one needs to generate large primes. Only due to nontrivial results and concepts of number theory, probability theory, and algorithmics a practical algorithm for this task was developed.

way and then carefully specified in detail. Progressively difficult topics are explained in a step-by-step manner.

At the end of each section we placed an informal summary that once again calls attention to the main ideas and results of the section. After reading a particular section you should confront your knowledge about it with its summary in order to check whether you may have missed some important idea. If new terminology is introduced in a section, the new terms are listed at the end of this section, too.

We hope that this textbook provides you with an enjoyable introduction to the modern theory of algorithmics, and that you will learn paradigms that, in contrast to specific technical knowledge (though useful now, but becomes outdated in a few years), may remain useful for several decades or even become the core of future progress.

To the Teachers

This book is based on the courses *Algorithms* and *Approximation Algorithms and Randomized Algorithms* held regularly at the Christian-Albrechts-University of Kiel and the Technological University (RWTH) Aachen. The course *Algorithms* is a basic course on algorithmics that starts with classical algorithm design techniques but in its kernel mainly focuses on solving *hard* problems. Note that *hard* does not mean only NP-hard here; rather, any problem that does not admit any low-degree polynomial-time algorithm is considered to be hard. Section 2.3 and Chapter 3 about deterministic approaches for designing efficient algorithms and Chapter 6 about heuristics are completely contained in this course. Moreover, the fundamental ideas and concepts of approximation algorithms (Chapter 4) and randomized algorithms (Chapter 5) along with the simplest algorithms illustrating these concepts are included in this fundamental course, too. The advanced course *Approximation Algorithms and Randomized Algorithms* is based on Chapters 4 and 5. The hardest parts of this course are included in Section 4.4 about inapproximability and Section 5.4 about derandomization, and so one can refrain from teaching some subparts of these sections, if any. On the other hand, one can add some more involved analyses of the behavior of randomized heuristics such as simulated annealing and genetic algorithms. Section 7 is devoted to combining, comparing, and parallelizing algorithms designed by different methods and it should be to some extent included in both courses. The fundamentals of DNA computing and of quantum computing (Section 7.6) can be presented in the form of an outlook on possible future technologies at the end of the advanced course.

We preferred the direct placement of the exercises into the text rather than creating sections consisting of exercises only. The exercises presented here are strongly connected with the parts in which they are placed, and they help to understand or deepen the matters presented directly above them. If the solution to an exercise requires some nontrivial new ideas that are not involved in this book, then the exercise is marked with $^{(*)}$. Obviously, this mark $^{(*)}$ is relative with respect to the knowledge of the reader. Note that for an involved training of the presented topics additional exercises have to be formulated.

The book is self-contained and so all mathematics needed for the design and analysis of the presented algorithms can be found in Section 2.2. Reading of additional literature is necessary only if one wants the deepen her/his knowledge in some of the presented or related topics. We call attention to the fact that this book is an introduction to algorithmics for hard problems and so reading of related literature is highly recommended for everybody who wants to become an expert in this field. The pointers to corresponding literature sources in the bibliographical remarks at the end of every chapter should be helpful for this purpose.

There are three main educational features of this material. First, we develop a *systematic* presentation of the topic. To achieve this we did not only revise some known concepts, but we gave names to concepts that were not recognized as concepts in this area so far. Secondly, each particular part of this book starts with an informal, intuitive explanation, continues with a detailed formal specification and examples, and finishes again with an informal discussion (i.e., one returns back to the starting point and confronts the initial aims with their realization). The third feature is *simplicity*. Since the use of mathematics is unavoidable for this topic we tried to use formalisms as simple as possible and on a low level of abstraction. Also in choosing the algorithms for illustrating the particular concepts and design techniques we preferred to show a transparent application of the considered method rather than to present the best (but technical) algorithm for a particular problem. All technical proofs start with a simple proof idea, which is then carefully realized into the smallest details.

In the Preface we formulated the main goal of this book as an optimization problem of learning algorithmics for hard problems in minimal time. Taking a teacher instead of a student as the input, one can reformulate this problem as minimizing time for preparing a course on this topic. We hope that this textbook will not only save you time when preparing such a course, but that it will encourage you to give such courses or use some part of this book in other courses, even if algorithmics is not the main topic of your interest. Most algorithms presented here are jewels of algorithm theory and several of them can be successfully used to illustrate paradigmatic ideas and concepts in other areas of computer science, too.

Organization

This book consists of three parts. The first part is presented in Chapter 2 and it contains elementary fundamentals of mathematics and algorithmics as usually taught in undergraduate courses. Chapters 3, 4, 5, and 6 are devoted to the proper subject of this textbook, i.e., to a systematic presentation of methods for the design of efficient algorithms for hard problems. The third part is covered in Chapter 7 and serves you as a guide in the field of applications of the methods presented in the second part. In what follows we give more details about the contents of the particular chapters.

Chapter 2 consists of two main sections, namely *Fundamentals of Mathematics* and *Fundamentals of Algorithmics*. The aim of the first section is to make this book self-contained in the sense that all formal concepts and arguments needed to design and analyze algorithms presented in Chapters 3, 4, 5, and 6 can be explained in detail. The elementary fundamentals of linear algebra, Boolean logic, combinatorics, graph theory, algebra, number theory, and probability theory are presented here. The only part that is a little bit more difficult is devoted to the proofs of some fundamental results of number theory that are crucial for the design of some randomized algorithms in Chapter 5. In fact, we assume that the reader is familiar with the topic of the section on fundamentals of mathematics. We do not recommend reading this section before starting to read the subsequent chapters devoted to the central topic. It is better to skip this section and to look up specific results if one needs them for the understanding of particular algorithms. Section *Fundamentals of Algorithmics* explains the fundamental ideas and concepts of algorithm and complexity theory and fixes the notation used in this book. For an advanced reader it may also be useful to read this part because it presents the fundamentals of the philosophy of this book. ·

Chapters 3, 4, 5, and 6 are devoted to a systematic presentation of fundamental concepts and algorithm design techniques for solving hard problems. Here, we carefully distinguish between the term *concept* and the term *algorithm design technique*. Algorithm design techniques as divide-and-conquer, dynamic programming, branch-and-bound, local search, simulated annealing, etc., have a well specified structure that even provides a framework for possible implementations or parallelization. Concepts such as pseudo-polynomial-time algorithms, parameterized complexity, approximation algorithms, randomized algorithms, etc., formulate ideas and rough frameworks about how to attack hard algorithmic problems. Thus, to realize a concept for attacking hard problems in a particular case one needs to apply an algorithm design technique (or a combination of different techniques).

Chapter 3 *"Deterministic Approaches"* is devoted to deterministic methods for solving hard problems. It consists of six basic sections. Three sections are devoted to the classical algorithm design techniques for attacking hard problems, namely branch-and-bound, local search, and relaxation to linear programming, and three sections are devoted to the concepts of pseudo-polynomial-time algorithms, parameterized complexity and lowering the worst case exponential complexity. All sections are presented in a uniform way. First, the corresponding method (concept or algorithm design technique) is presented and formally specified. Then, the method is illustrated by designing some algorithms for specific hard problems, and the limits of this method are discussed.

The concepts of *pseudo-polynomial-time algorithms* and *parameterized complexity* are based on the idea to partition the set of all input instances of a particular hard problem into a set of *easy* problem instances and a set of *hard* problem instances, and to design an efficient algorithm for the easy input instances. To illustrate the concept of pseudo-polynomial-time algorithms (Section 3.2) we present the well-known dynamic programming algorithm[4] for the knapsack problem. We use the concept of strong NP-hardness for proving the non-existence of pseudo-polynomial-time algorithms for some hard problems. The concept of parameterized complexity (Section 3.3) is presented as a generalization of the concept of pseudo-polynomial-time algorithms here. To illustrate it we present two algorithms for the vertex cover problem. Strong NP-hardness is again used for showing limits to the applicability of this concept.

Branch-and-bound (Section 3.4) is a classical algorithm design technique used for solving optimization problems and we chose the maximum satisfiability problem and the traveling salesperson problem to illustrate its work and properties. The use of branch-and-bound in combination with other concepts and techniques (approximation algorithms, relaxation to linear programming, heuristics, etc.) is discussed.

The concept of *lowering worst case complexity of exponential algorithms* (Section 3.5) is based on designing algorithms with an exponential complexity in $O(c^n)$ for some $c < 2$. Such algorithms may be practical even for large input sizes. This concept became successful especially for satisfiability problems and we present a simple algorithm from this area. An advanced application of this concept is postponed to Chapter 5 about randomized algorithms.

Section 3.6 presents the basic framework of *local search* and of *Kernighan-Lin's variable depth search*. These techniques are used for attacking optimization problems and provide local optima according to the chosen local neighborhoods. The notions of a polynomial-time searchable neighborhood and of an exact neighborhood are introduced in order to study the tradeoffs between solution quality and computational complexity of local search. Some input instances of the traveling salesperson problem that are pathological for local search are presented. Pathological means that the instances have a unique optimal solution and exponentially many second-best local optima with costs of an exponential size in the optimal cost for a neighborhood of an exponential size.

Section 3.7 is devoted to the technique of *relaxation to linear programming* that is used for solving optimization problems. The realization of this technique is presented here in three steps. The first step (reduction) consists of expressing a given instance of an optimization problem as an input instance of integer programming. For its illustration we use the minimum weighted vertex cover problem and the knapsack problem. The second step (relaxation) consists of solving the instance of integer programming as an instance of linear

---

[4] This algorithm is later used to design an approximation algorithm (FPTAS).

programming. A short, transparent presentation of the simplex algorithm is given in order to show one way this step can be performed. The third step uses the computed optimal solution to the instance of linear programming for computing a high-quality feasible solution to the original instance of the given optimization problem. The concept of rounding, LP-duality, and the primal-dual method are considered here for this purpose.

The main goals of Chapter 4 *"Approximation Algorithms"* are the following:

(1) To give the fundamentals of the concept of the approximation of optimization problems.
(2) To present some transparent examples of the design of efficient approximation algorithms.
(3) To show basic methods for proving limits of the applicability of the concept of approximation (i.e., methods for proving lower bounds on polynomial-time inapproximability).

After a short introduction (Section 4.1), Section 4.2 introduces the basic concept of polynomial-time approximability. This concept shows one of the most fascinating effects occurring in algorithmics. One can jump from a huge inevitable amount of physical work to a few seconds work on a PC due to a small change in the requirements – instead of an optimal solution one demands a solution whose cost differs from the cost of an optimal solution by at most $\varepsilon\%$ of the cost of an optimal solution for some $\varepsilon > 0$. Besides introducing basic terms such as relative error, approximation ratio, approximation algorithm, approximation scheme, a classification of optimization problems with respect to the quality of their polynomial-time approximability is given. In many cases, one can use the standard concepts and algorithm design techniques to design approximation algorithms. Section 4.2 specifies two additional specific concepts (dual approximation algorithm and stability of approximation) that were developed for the design of approximation algorithms.

Section 4.3 is devoted to the design of particular approximation algorithms. Section 4.3.2 presents a simple 2-approximation algorithm for the vertex cover problem and then applies the technique of relaxation to linear programming in order to obtain the approximation ratio 2 for the weighted generalization of this problem, too. Further, the greedy technique is applied to design a $\ln(n)$-approximation algorithm for the set cover problem. Section 4.3.3 shows that a simple local search algorithms provides an at most 2 approximation ratio for the maximum cut problem. In Section 4.3.4 we first use a combination of the greedy method and an exhaustive search to design a polynomial-time approximation scheme for the simple knapsack problem. Using the concept of stability of approximation this approximation scheme is extended to work for the general knapsack problem. Finally, using the pseudo-polynomial-time approximation algorithm for the knapsack problem from Section 3.2 we design a fully polynomial-time scheme for the knapsack problem. Section 4.3.5 is devoted to the traveling salesperson problem (TSP), and to the concept of stability of approximation. First, the spanning tree algorithm and the Christofides algorithm for the metric TSP are presented. Since TSP does not admit any polynomial-time approximation algorithm, we use the concept of stability of approximation in order to partition the set of all instances of TSP into an infinite spectrum of classes with respect to their polynomial-time approximability. The approximation ratios of these classes ranges from 1 to infinity. In Section 4.3.6 an application of the concept of dual approximation algorithms is presented. The design of a dual polynomial-time approximation scheme for the bin-packing problem is used to obtain a polynomial-time approximation scheme for the makespan scheduling problem.

Section 4.4 is devoted to methods for proving lower bounds on polynomial-time approximability of specific optimization problems. We partition these methods into three groups. The first group contains the classical reduction to NP-hard decision problems. The second group is based on specific reductions (between optimization problems) that preserve the quality of solutions. We consider the approximation-preserving reduction and the gap-preserving reduction here. The third method is based on a direct application of the famous PCP-Theorem.

The main goals of Chapter 5 *"Randomized Algorithms"* are the following:

(1) To give the fundamentals of the concept of randomized computation and to classify randomized algorithms with respect to their error probability.
(2) To specify the paradigms of the design of randomized algorithm and to show how they can be applied for solving specific problems.
(3) To explain some derandomization methods for converting randomized algorithms into deterministic ones.

After a short introduction (Section 5.1), Section 5.2 presents the fundamentals of randomized computing, and classifies randomized algorithms into Las Vegas algorithms and Monte Carlo algorithms. Las Vegas algorithms never provide any wrong result, and Monte Carlo ones are further classified with respect to the size and the character of their error probability. Moreover, the paradigms *foiling an adversary, abundance of witnesses, fingerprinting, random sampling, relaxation, and random rounding* for the design of randomized algorithms are specified and discussed.

Section 5.3 is devoted to applications of paradigms of randomized computations to the design of concrete randomized algorithms. The technique of random sampling is used to design a simple Las Vegas algorithm for the problem of finding a quadratic residue in $\mathbb{Z}_p$ for a given prime p in Section 5.3.2. Applying the technique of abundance of witnesses we design the well-known one-sided-error Monte Carlo algorithms, (namely the Solovay-Strassen algorithm and the Miller-Rabin algorithm) for primality testing in Section 5.3.3. We also provide detailed selfcontained proofs of their correctness there. In Section5.3.4 it is shown how the fingerprinting technique can be used to efficiently decide the equivalence of two polynomials over $\mathbb{Z}_p$ for a prime p and the equivalence of two one-time-only branching programs. The concept of randomized optimization algorithms is illustrated on the minimum cut problem in Section 5.3.5. The methods of random sampling and relaxation to linear programming with random rounding are used to design a randomized approximation algorithm for the maximum satisfiability problem in Section 5.3.6. Randomization, local search, and the concept of lowering the worst case complexity are combined in order to design a Monte Carlo $O(1.334^n)$-algorithm for the satisfiability problem of formulas in 3-conjunctive normal form in Section 5.3.7

Section 5.4 is devoted to derandomization. The method of the reduction of the probability space and the method of conditional probabilities are explained and illustrated by derandomizing the randomized approximation algorithms of Section 5.3.6 for the maximum satisfiability problem.

Chapter 6 is devoted to *heuristics*. Here, a heuristic is considered to be a robust technique for the design of randomized algorithms for which one is not able to guarantee at once the efficiency and the quality (correctness) of the computed solutions, not even with any bounded probability. We focus on the presentation of simulated annealing and genetic algorithms here. We provide formal descriptions of the schemes of these techniques, and discuss their theoretical convergence as well as experimental adjustment of their free parameters. Randomized tabu search as a possible generalization of simulated annealing is considered, too.

Chapter 7 provides a guide to solving hard problems. Among others, we discuss the possibilities

- to combine different concepts and techniques in the process of algorithm design,
- to compare algorithms designed by different methods for the same problems with respect to their complexity as well as with respect to the solution quality, and
- to speed up designed algorithms by parallelization.

Besides this, Section 7.6 provides an outlook on DNA computing and quantum computing as hypothetical future technologies. To illustrate these computation modes we present DNA algorithms for the Hamiltonian path problem (Adleman's experiment) and for the 3-colorability problem and a quantum algorithm for generating truly random bits. Chapter 7 finishes with a dictionary of basic terms of algorithmics for hard problems.

2

Elementary Fundamentals

*"One whoose knowledge
is not constantly increasing
is not clever at all."*

JEAN PAUL

2.1 Introduction

We assume that the reader has had undergraduate courses in mathematics and algorithmics. Despite this assumption we present all elementary fundamentals needed for the rest of this book in this chapter. The main reasons to do this are the following ones:

(i) to make the book completely self-contained in the sense that all arguments needed to design and analyze the algorithms presented in the subsequent chapters are explained in the book in detail,

(ii) to explain the mathematical considerations that are essential in the process of algorithm design,

(iii) to informally explain the fundamental ideas of complexity and algorithm theory and to present their mathematical formalization, and

(iv) to fix the notation in this book.

We do not recommend reading this whole chapter before starting to read the subsequent chapters devoted to the central topic of this book. For every at least a little bit experienced reader we recommend skipping the part about the mathematics and looking up specific results if one needs them later for a particular algorithm design. On the other hand, it is reasonable to read the part about the elementary fundamentals of algorithms because it is strongly connected with the philosophy of this book and the basic notation is fixed there. More detailed recommendations about how to use this chapter are given below.

This chapter is divided into two main parts, namely "Fundamentals of Mathematics" and "Fundamentals of Algorithmics". The part devoted to elementary mathematics consists of five sections. Section 2.2.1 contains elementary fundamentals of linear algebra. It focuses on systems of linear equations, matrices, vector spaces, and their geometrical interpretations. The concepts and results presented here are used in Section 3.7 only, where the problem of linear programming and the simplex method are considered. The reader can therefore look at this section when some considerations of Section 3.7 are not clear enough. Section 2.2.2 provides elementary fundamentals of combinatorics, counting, and graph theory. Notions such as permutation and combination are defined and some basic series are presented. Furthermore the notations O, Ω, and Θ for the analysis of the asymptotic growth of functions are fixed and a version of the Master Theorem for solving recurrences is presented. Finally, the basic notions (such as graph, directed graph, multigraph, connectivity, Hamiltonian tour, and Eulerian tour) are defined. The content of Section 2.2.2 is the base for many parts of the book because it is used for the analysis of algorithm complexity as well as for the design of algorithms for graph-theoretical problems. Section 2.2.3 is devoted to Boolean functions and their representations in the forms of Boolean formulae and branching programs. The terminology and basic knowledge presented here are useful for investigating satisfiability problems, which belong to the paradigmatic algorithm problems of main interest. This is the reason why we consider these problems in all subsequent chapters about the algorithm design for hard problems. Section 2.2.4 differs a little bit from the previous sections. While the previous sections are mainly devoted to fixing terminology and presenting some elementary knowledge, Section 2.2.4 contains also some nontrivial, important results like the Fundamental Theorem of Arithmetics, Prime Number Theorem, Fermat's Theorem, and Chinese Remainder Theorem. We included the proofs of these results, too, because the design of randomized algorithms for problems of algorithmic number theory requires a thorough understanding of this topic. If one is not familiar with the contents of this section, it is recommended to look at it before reading the corresponding parts of Chapter 5. Section 2.2.5 is devoted to the elementary fundamentals of probability theory. Here only discrete probability distributions are considered. The main point is that we do not only present fundamental notions like probability space, conditional probability, random variable, and expectation, but we also show their relation to the nature and to the analysis of randomized algorithms. Thus, this part is a prerequisite for Chapter 5 on randomized algorithms.

Section 2.3 is devoted to the fundamentals of algorithm and complexity theory. The main ideas and concepts connected with the primary topic of this book are presented here. It is strongly recommended having at least a short look at this part before starting to read the proper parts of algorithm design techniques for hard problems in Chapters 3, 4, 5, and 6. Section 2.3 is organized as follows. Section 2.3.1 gives the basic terminology of formal language theory (such as alphabet, word, language). This is useful for the representation of data and for the formal description of algorithmic problems and basic concepts of complexity theory. In Section 2.3.2 the formal definitions of all algorithmic problems considered in this book are given. There, we concentrate on decision problems and optimization problems. Section 2.3.3 provides a short survey on the basic concepts of complexity theory as a theory for classification of algorithmic problems according to their computational difficulty. The fundamental notions such as complexity measurement, nondeterminism, polynomial-time reduction, verifier, and NP-hardness are introduced and discussed in the framework of the historical development of theoretical computer science. This provides the base for starting the effort to solve hard algorithmic problems in the subsequent chapters. Finally, Section 2.3.4 gives a concise overview of the algorithms design techniques (divide-and-conquer, dynamic programming, backtracking, local search, greedy algorithms) that are usually taught in undergraduate courses on algorithmics. All these techniques are later applied to attack particular hard problems. Especially, it is reasonable to read this section before reading Section 3, where these techniques are developed or combined with other ideas in order to obtain solutions to different problems.

2.2 Fundamentals of Mathematics

2.2.1 Linear Algebra

The aim of this section is to introduce the fundamental notions of linear algebra such as linear equations, matrices, vectors, and vector spaces and to provide elementary knowledge about them. The terminology introduced here is needed for the study of the problem of linear programming and for introducing the simplex method in Section 3.4. Vectors are also often used to represent data (inputs) of many computing problems and matrices are used to represent graphs, directed graphs, and multigraphs in several following sections.

In what follows, we consider the following fundamental sets:

$\mathbb{N} = \{0, 1, 2, \ldots\}$ ... the set of all natural numbers,
$\mathbb{Z} = \{0, -1, 1, -2, 2, \ldots\}$ ... the set of all integers,
$\mathbb{Q} = \{\frac{m}{n} \mid m, n \in \mathbb{Z}, n \neq 0\}$ ... the set of rational numbers,
$\mathbb{R}$ ... the set of real numbers,
$(a, b) = \{x \in \mathbb{R} \mid a < x < b\}$ for all $a, b \in \mathbb{R}$, $a < b$,
$[a, b] = \{x \in \mathbb{R} \mid a \leq x \leq b\}$ for all $a, b \in \mathbb{R}$, $a < b$,
For every set S, the notation $Pot(S)$ or 2^S denotes the power set of S, i.e.,

$$Pot(S) = \{Q \mid Q \subseteq S\}.$$

An equation of the type

$$y = ax,$$

expressing the variable y in terms of the variable x for a fixed constant a, is called a **linear equation**. The notion of linearity is connected to the geometrical interpretation that corresponds to a straight line.

Definition 2.2.1.1. *Let S be a subset of $\mathbb{R}$ such that if $a, b \in S$, then $a+b \in S$ and $a \cdot b \in S$. The equation*

$$y = a_1 x_1 + a_2 x_2 + \cdots + a_n x_n, \tag{2.1}$$

*where $a_1, \ldots, a_n$ are constants from S and $x_1, x_2, \ldots, x_n$ are variables over S is called a **linear equation** over S. We say that the equation (2.1) expresses y in terms of variables $x_1, x_2, \ldots, x_n$. The variables $x_1, x_2, \ldots, x_n$ are also called the **unknowns** of the linear equation (2.1).*

*For a fixed value of y, a **solution** to the linear equation (2.1) is a sequence $s_1, s_2, \ldots, s_n$ of numbers of S such that*

$$y = a_1 s_1 + a_2 s_2 + \cdots + a_n s_n.$$

For instance $1, 1, 1$ (i.e., $x_1 = 1, x_2 = 1, x_3 = 1$) is a solution to the linear equation

$$x_1 + 2x_2 + 3x_3 = 6$$

over $\mathbb{Z}$. Another solution is $-1, -1, 3$.

Definition 2.2.1.2. *Let S be a subset of $\mathbb{R}$ such that if $a, b \in S$ then $a \cdot b \in S$ and $a + b \in S$. Let m and n be positive integers. A **system of m linear equations in n variables (unknowns)** $x_1, \ldots, x_n$ over S (or simply a **linear system over S**) is a set of m linear equations over S, where each of these linear equations is expressed in terms of the same variables $x_1, x_2, \ldots, x_n$. In other words, a system of m linear equations over S is*

$$a_{11}x_1 + a_{12}x_2 + \cdots + a_{1n}x_n = y_1$$
$$a_{21}x_1 + a_{22}x_2 + \cdots + a_{2n}x_n = y_2$$
$$\vdots \tag{2.2}$$
$$a_{m1}x_1 + a_{m2}x_2 + \cdots + a_{mn}x_n = y_m,$$

where a_{ij} are constants from S for $i = 1, \ldots, n$, $j = 1, \ldots, m$, and $x_1, x_2, \ldots, x_n$ are variables (unknowns) over S. For each $i \in \{1, \ldots, m\}$, the linear equation

$$a_{i1}x_1 + a_{i2}x_2 + \cdots + a_{in}x_n = y_i$$

*is called the **ith equation** of the linear system (2.2). The system (2.2) of linear equations is called **homogeneous** if $y_1 = y_2 = \cdots = y_m = 0$.*

*For given values of $y_1, y_2, \ldots, y_m$, a **solution** to the linear system (2.2) is a sequence $s_1, s_2, \ldots, s_n$ of numbers of S such that each equation of the linear system (2.2) is satisfied when $x_1 = s_1, x_2 = s_2, \ldots, x_n = s_n$ (i.e., such that $s_1, s_2, \ldots, s_n$ is a solution for each equation of the linear system (2.2)).*

The system of linear equations over $\mathbb{Z}$

$$x_1 + 2x_2 = 10$$
$$2x_1 - 2x_2 = -4$$
$$3x_1 + 5x_2 = 26$$

is a system of three linear equations in two variables x_1 and x_2. $x_1 = 2$ and $x_2 = 4$ is a solution to this system.[1]

Note that there are systems of linear equations that do not have any solution. An example is the following linear system:

$$x_1 + \ 2x_2 = 10$$
$$x_1 - \ \ x_2 = -2$$
$$6x_1 + 10x_2 = 40$$

In what follows we define vectors and matrices. They provide a very convenient formalism for representing and manipulating linear systems as well as many other objects in mathematics and computer science.

Definition 2.2.1.3. *Let $S \subseteq \mathbb{R}$ be any set satisfying $a + b \in S$ and $a \cdot b \in S$ for all $a, b \in S$. Let n and m be positive integers. An $m \times n$ **matrix** A over S is a rectangular array of $m \cdot n$ elements of S arranged in m horizontal **rows** and n vertical **columns**:*

$$A = [a_{ij}]_{i=1,\ldots,m,j=1,\ldots,n} = \begin{pmatrix} a_{11} & a_{12} & \cdots & a_{1n} \\ a_{21} & a_{22} & \cdots & a_{2n} \\ \vdots & \vdots & \ddots & \vdots \\ a_{m1} & a_{m2} & \cdots & a_{mn} \end{pmatrix}.$$

*For all $i \in \{1, \ldots, m\}$, $j \in \{1, \ldots, n\}$, a_{ij} is called the (i, j)-**entry**[2] of A. The ith **row** of A is*

$$(a_{i1}, a_{i2}, \ldots, a_{in})$$

*for all $i \in \{1, \ldots, m\}$. For each $j \in \{1, \ldots, n\}$, the jth **column** of A is*

$$\begin{pmatrix} a_{1j} \\ a_{2j} \\ a_{3j} \\ \vdots \\ a_{mj} \end{pmatrix}.$$

For any positive integers n and m, a $1 \times n$ matrix is called an n-dimensional row-vector, and a $m \times 1$ matrix is called an m-dimensional column-vector.

---

[1] We assume that the reader is familiar with the method of elimination that efficiently finds a solution of linear systems. We do not present the method here, because it is not interesting to us from the algorithmic point of view.

[2] Observe that a_{ij} lies on the intersection of the ith row and the jth column.

For any positive integer n, an $n \times n$ matrix is called a **square matrix of order n**. *If $A = [a_{ij}]_{i,j=1,...,n}$ is a square matrix, then we say that the elements $a_{11}, a_{22}, \ldots, a_{nn}$ form the* **main diagonal** *of A.*

A is called the **1-diagonal matrix** *(or the* **identity matrix**), *denoted by I_n, if*

(i) $a_{ii} = 1$ for $i = 1, \ldots, n$, and
(ii) $a_{ij} = 0$ for all $i \neq j$, $i, j \in \{1, \ldots, n\}$.

A is called the **0-diagonal matrix** *if*

(i) $a_{ii} = 0$ for all $i = 1, \ldots, n$, and
(ii) $a_{ij} = 1$ for $i \neq j$, $i, j \in \{1, \ldots, n\}$.

An $m \times n$ matrix $B = [b_{ij}]_{i=1,...,m, j=1,...,n}$ is called a **Boolean matrix** *if $b_{ij} \in \{0, 1\}$ for $i = 1, \ldots, m$, $j = 1, \ldots, n$.*

An $m \times n$ matrix $B = [b_{ij}]_{i=1,...,m,j=1,...,n}$ is called a **zero matrix** *if $b_{ij} = 0$ for all $i \in \{1, \ldots, m\}$, $j \in \{1, \ldots, n\}$. The zero matrix of size $m \times n$ is denoted by $0_{m \times n}$.*

The following matrix $B = [b_{ij}]_{i=1,...,3, j=1,...,4}$ is an example of a 3×4 matrix over $\mathbb{Q}$.

$$B = \begin{pmatrix} 1 & \frac{2}{3} & 4 & -6 \\ \frac{1}{2} & 1 & \frac{3}{4} & -8 \\ -3 & \frac{6}{5} & 2 & 0 \end{pmatrix}.$$

$(\frac{1}{2}, 1, \frac{3}{4}, -8)$ is the second row of B. The $(3, 4)$-entry of B is $b_{34} = 0$.

Definition 2.2.1.4. *Let m, n be two positive integers. Two $m \times n$ matrices $A = [a_{ij}]$ and $B = [b_{ij}]$ are said to be* **equal** *if $a_{ij} = b_{ij}$ for all $i \in \{1, \ldots, m\}$, $j \in \{1, \ldots, n\}$.*

The **sum** *of A and B $(A + B)$ is the matrix $C = [c_{ij}]_{i=1,...,m,j=1,...,n}$ defined by*[3]

$$c_{ij} = a_{ij} + b_{ij}$$

for all $i \in \{1, \ldots, m\}$, $j \in \{1, \ldots, n\}$.

Exercise 2.2.1.5. *Let A, B, C be matrices over $\mathbb{R}$ of same size $m \times n$, $m, n \in \mathbb{N} - \{0\}$. Prove that*

(i) $A + B = B + A$,
(ii) $A + (B + C) = (A + B) + C$,
(iii) *there exists a* **negative** *of A (i.e., there exists a $m \times n$ matrix $D = (-A)$ such that $A + D = 0_{m \times n}$).*

□

---

[3] Observe that the sum of the matrices is defined only when A and B have the same number of rows and the same number of columns.

Definition 2.2.1.6. *Let m, p, n be positive integers. Let*

$$A = [a_{ij}]_{i=1,\ldots,m, j=1,\ldots,p} \text{ and } B = [b_{ij}]_{i=1,\ldots,p, j=1,\ldots,n}$$

be two matrices. The **product (multiplication)** *of A and B is the $m \times n$ matrix $C = [c_{ij}]_{i=1,\ldots,m, j=1,\ldots,n}$ defined by*

$$c_{ij} = a_{i1}b_{1j} + a_{i2}b_{2j} + \cdots + a_{ip}b_{pj} = \sum_{k=1}^{p} a_{ik}b_{kj}$$

for $i = 1,\ldots,m$, $j = 1,\ldots,n$.

To illustrate Definition 2.2.1.6 consider the matrices

$$A = \begin{pmatrix} 1 & 0 & -2 \\ 0 & 3 & -1 \end{pmatrix}, \qquad B = \begin{pmatrix} 1 & -3 \\ 5 & 0 \\ 0 & 4 \end{pmatrix}.$$

Then,

$$A \cdot B = \begin{pmatrix} 1 \cdot 1 + 0 \cdot 5 + (-2) \cdot 0 & 1 \cdot (-3) + 0 \cdot 0 + (-2) \cdot 4 \\ 0 \cdot 1 + 3 \cdot 5 + (-1) \cdot 0 & 0 \cdot (-3) + 3 \cdot 0 + (-1) \cdot 4 \end{pmatrix} = \begin{pmatrix} 1 & -11 \\ 15 & -4 \end{pmatrix}.$$

Observe that $B \cdot A$ is not defined because the product of B and A is defined only if the number of columns of B is equal to the number of rows of A.

Exercise 2.2.1.7. Find two squared matrices A and B over $\mathbb{Z}$ such that $A \cdot B \neq B \cdot A$. □

Exercise 2.2.1.8. Let A, B, C be $m \times p$, $p \times q$, $q \times n$ matrices over $\mathbb{R}$, respectively. Prove:

$$A \cdot (B \cdot C) = (A \cdot B) \cdot C.$$ □

Exercise 2.2.1.9. Prove that for every $n \times n$ matrix A,

$$A \cdot I_n = I_n \cdot A = A.$$ □

Definition 2.2.1.10. *Let r be a real number, and let $A = [a_{ij}]$ be an $m \times n$ matrix over $\mathbb{R}$. The* **scalar multiple of A by r**, $r \cdot A$, *is the $m \times n$ matrix $B = [b_{ij}]$, where*

$$b_{ij} = r \cdot a_{ij}$$

for $i = 1,\ldots,m$, $j = 1,\ldots,n$.

Definition 2.2.1.11. *Let $A = [a_{ij}]$ be an $m \times n$ matrix, $m, n \in \mathbb{N} - \{0\}$. The $n \times m$ matrix $B = [b_{ij}]_{i=1,\ldots,n, j=1,\ldots,m}$, where*

$$a_{ij} = b_{ji} \text{ for } i = 1,\ldots,m, j = 1,\ldots,n$$

is called the **transpose of A** *and denoted by A^{T}. If $A = A^{\mathsf{T}}$, then we say that A is* **symmetric.**

Exercise 2.2.1.12. Prove, for every real number r and any matrices A and B over $\mathbb{R}$, that the following equalities hold:

(i) $\left(A^{\mathsf{T}}\right)^{\mathsf{T}} = A$,

(ii) $(A + B)^{\mathsf{T}} = A^{\mathsf{T}} + B^{\mathsf{T}}$,

(iii) $(A \cdot B)^{\mathsf{T}} = B^{\mathsf{T}} \cdot A^{\mathsf{T}}$,

(iv) $(rA)^{\mathsf{T}} = r \cdot A^{\mathsf{T}}$.

$\square$

Now we show how matrices can be used to represent systems of linear equations. Consider the system (2.2) of Definition 2.2.1.2. Define

$$A = \begin{pmatrix} a_{11} & a_{12} & \cdots & a_{1n} \\ a_{21} & a_{22} & \cdots & a_{2n} \\ \vdots & \vdots & \ddots & \vdots \\ a_{m1} & a_{m2} & \cdots & a_{mn} \end{pmatrix}, \; X = \begin{pmatrix} x_1 \\ x_2 \\ \vdots \\ x_n \end{pmatrix}, \; Y = \begin{pmatrix} y_1 \\ y_2 \\ \vdots \\ y_m \end{pmatrix}.$$

Then, the linear system (2.2) can be written in the matrix form

$$A \cdot X = Y.$$

The matrix A is called the **coefficient matrix** of the linear system (2.2).

For instance, for the system of linear equations

$$-x_1 + 2x_2 - 3x_3 = 7$$
$$6x_1 + x_2 + x_3 = 5$$

the coefficient matrix is

$$A = \begin{pmatrix} -1 & 2 & -3 \\ 6 & 1 & 1 \end{pmatrix} \text{ and } X = \begin{pmatrix} x_1 \\ x_2 \\ x_3 \end{pmatrix}, Y = \begin{pmatrix} 7 \\ 5 \end{pmatrix}.$$

Definition 2.2.1.13. *Let A be a squared $n \times n$ matrix, $n \in \mathbb{N} - \{0\}$. A is called* **nonsingular** *(or* **invertible***) if there exists an $n \times n$ matrix B such that*

$$A \cdot B = B \cdot A = I_n.$$

The matrix B is called the **inverse of A** *and denoted[4] by A^{-1}. If there exists no inverse of A, then A is called* **singular** *(or* **noninvertible***).*

One can easily verify that, for

$$A = \begin{pmatrix} 2 & 3 \\ 2 & 2 \end{pmatrix} \text{ and } B = \begin{pmatrix} -1 & \frac{3}{2} \\ 1 & -1 \end{pmatrix},$$

$A \cdot B = B \cdot A = I_2$, and so $A^{-1} = B$ and $B^{-1} = A$. Observe also that $I_n^{-1} = I_n$ for any positive integer n.

---

[4] Note that if, for a matrix A, there exists a matrix B with the property $A \cdot B = B \cdot A = I_n$, then B is the unique inverse of A.

Exercise 2.2.1.14. Prove the following assertion. If $A_1, A_2, \ldots, A_r$ are $n \times n$ nonsingular matrices, then $A_1 \cdot A_2 \cdots A_r$ is nonsingular, and

$$(A_1 \cdot A_2 \cdots A_r)^{-1} = A_r^{-1} \cdot A_{r-1}^{-1} \cdots A_1^{-1}. \qquad \square$$

Let $A \cdot X = Y$ be a system of linear equations where the coefficient matrix A is an $n \times n$ nonsingular matrix. Then one can solve this system by constructing A^{-1} because multiplying the equality

$$A \cdot X = Y$$

by A^{-1} from the left side one obtains

$$A^{-1} \cdot A \cdot X = A^{-1} \cdot Y.$$

Since $A^{-1} \cdot A = I_n$ and $I_n \cdot X = X$ we obtain

$$X = A^{-1} \cdot Y.$$

Now we look at the geometrical interpretation of systems of linear equations.

Definition 2.2.1.15. *For any positive integer n, we define the **n-dimensional (IR-) vector space***

$$\mathbb{R}^n = \left\{ \begin{pmatrix} a_1 \\ a_2 \\ \vdots \\ a_n \end{pmatrix} \;\middle|\; a_i \in \mathbb{R} \text{ for } i = 1, \ldots, n \right\}.$$

*The vector $0_{n \times 1}$ is called the **origin** of $\mathbb{R}^n$.*

There are two possible geometrical interpretations of the elements of $\mathbb{R}^n$. One possibility is to assign to an element

$$X = \begin{pmatrix} a_1 \\ \vdots \\ a_n \end{pmatrix}$$

of $\mathbb{R}^n$ the point with the coordinates $a_1, a_2, \ldots, a_n$ in $\mathbb{R}^n$. Another possibility is to assign a directed line from $(0, 0, \ldots, 0)^\mathsf{T}$ to $(a_1, \ldots, a_n)^\mathsf{T}$. This directed line is called the **vector** $(a_1, \ldots, a_n)^\mathsf{T}$.

Consider $\mathbb{R}^2$. To build a geometrical interpretation of $\mathbb{R}^2$ one starts from the origin $(0, 0)^\mathsf{T}$. One draws two infinite lines which are orthogonal to each other and which intersect at the origin. One of the lines is usually in a horizontal position and is called the **x-axis**. The other infinite line, the **y-axis**, is taken in a vertical position (see Figure 2.1). Then, the positive reals are placed on the x-axis to the right of the origin in increasing order and the negative reals are placed on the x-axis to the left in decreasing order. Similarly, the y-axis above the origin contains the positive real numbers and the y-axis below the origin contains the negative real numbers. For any point X of the plane one can determine the **coordinates** of X as follows:

(i) Take a line l that contains the point X and is orthogonal (perpendicular) to the x-axis (parallel to the y-axis). The real number a_x associated with the intersection of the x-axis and l is the **x-coordinate** of X.

(ii) Take a line h that contains the point X and is orthogonal to the y-axis (parallel to the x-axis). The real number a_y associated with the intersection of the y-axis and h is the **y-coordinate** of X.

We shall denote the point X by $P(a_x, a_y)$, and the corresponding vector by $(a_x, a_y)^\mathsf{T}$.

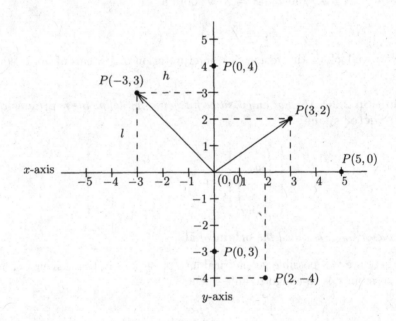

Fig. 2.1.

Definition 2.2.1.16. *Let $P(a_1, a_2)$ and $P(b_1, b_2)$ be two points in $\mathbb{R}^2$. The* **(Euclidean) distance** *between $P(a_1, a_2)$ and $P(b_1, b_2)$ is defined by*

$$\text{distance}\,(P(a_1, a_2), P(b_1, b_2)) = \sqrt{(a_1 - b_1)^2 + (a_2 - b_2)^2}.$$

We observe from Figure 2.2 that the Euclidean distance between two points is exactly the length of the line that connects $P(a_1, a_2)$ and $P(b_1, b_2)$ because of the Pythagorean Theorem.

a_1
a_2

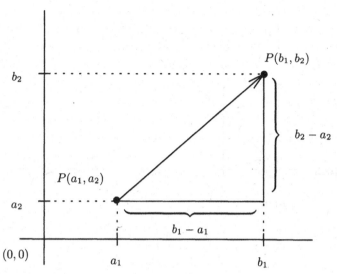

Fig. 2.2.

Exercise 2.2.1.17. Prove that, for every three points $P(a_1, a_2)$, $P(b_1, b_2)$, and $P(c_1, c_2)$,

(i) $distance(P(a_1, a_2), P(a_1, a_2)) = 0$,
(ii) $distance(P(a_1, a_2), P(b_1, b_2)) = distance(P(b_1, b_2), P(a_1, a_2))$, and
(iii) $distance(P(a_1, a_2), P(b_1, b_2)) \leq distance(P(a_1, a_2), P(c_1, c_2))$
$$+ distance(P(c_1, c_2), P(b_1, b_2)).$$

□

We can find a natural interpretation of systems of linear equations of two variables in $\mathbb{R}^2$. One can assign to every linear equation

$$a_1 x_1 + a_2 x_2 = b$$

with $(a_1, a_2) \neq (0, 0)$ the straight line $x_1 = \frac{b - a_2 x_2}{a_1} = \frac{b}{a_1} - \frac{a_2}{a_1} \cdot x_2$ if $a_1 \neq 0$ and the straight line $x_2 = \frac{b}{a_2}$ if $a_1 = 0$. This straight line consists of all points of $\mathbb{R}^2$ that satisfy the given linear equation. So, the set of all solutions to a system of linear equations is exactly the intersection of all straight lines corresponding to the particular equations.

Figure 2.3 contains four lines corresponding to the linear equations $x_1 + 3x_2 = 8$, $x_1 + 3x_2 = 4$, $x_1 + 0 \cdot x_2 = 2$, and $x_1 - x_2 = 0$. The system of linear equations

$$x_1 + 3x_2 = 4$$
$$x_1 - x_2 = 0$$

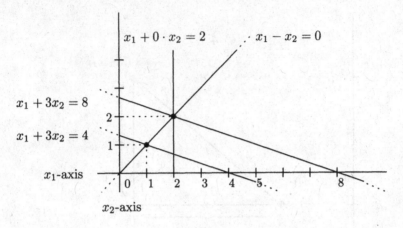

Fig. 2.3.

has the unique solution $P(1,1)$ (i.e., $x_1 = 1$, $x_2 = 1$) that is the intersection point of the lines corresponding to these equations. The system of linear equations

$$x_1 + 3x_2 = 4$$
$$x_1 + 3x_2 = 8$$

does not have any solution because the corresponding lines do not intersect.[5] On the other hand, the system of linear equations

$$x_1 + 3x_2 = 4$$
$$2x_1 + 6x_2 = 8$$

has infinitely many solutions because both linear equalities $x_1 + 3x_2 = 4$ and $2x_1 + 6x_2 = 8$ define the same line, and so every point of this line is a solution. Finally, we observe that the system of linear equations

$$x_1 + \quad 3x_2 = 8$$
$$x_1 + 0 \cdot x_2 = 2$$
$$x_1 - \quad x_2 = 0$$

has exactly one solution $P(2,2)$ (i.e., $x_1 = 2$, $x_2 = 2$), and that the set of linear equations

$$x_1 + \quad 3x_2 = 4$$
$$x_1 + 0 \cdot x_2 = 2$$
$$x_1 - \quad x_2 = 0$$

---

[5] Because they are parallel.

does not have any solution.

We see that any nontrivial linear equation of two variables determines a line that is a one-dimensional part of $\mathbb{R}^2$. In general, any linear equation over n variables determines an $(n-1)$-dimensional subpart[6] of $\mathbb{R}^n$. To understand it geometrically, we present some elementary fundamentals of the theory of vector spaces.

Definition 2.2.1.18. *Let $W \subseteq \mathbb{R}^n$, $n \in \mathbb{N} - \{0\}$. We say that W is a (linear) vector subspace of $\mathbb{R}^n$ if, for all $r_1, r_2 \in \mathbb{R}$ and all $(a_1, a_2, \ldots, a_n)^\mathsf{T}$, $(b_1, b_2, \ldots, b_n)^\mathsf{T} \in W$,*

$$r_1 \cdot (a_1, a_2, \ldots, a_n)^\mathsf{T} + r_2 \cdot (b_1, b_2, \ldots, b_n)^\mathsf{T} \in W.$$

Note that every vector subspace of $\mathbb{R}^n$ contains the origin $0_{n \times 1} = (0, \ldots, 0)^\mathsf{T}$, because one can set $r_1 = r_2 = 0$.

Let $V = \{(a_1, a_2, 0)^\mathsf{T} \mid a_1, a_2 \in \mathbb{R}\}$. We observe that V is a vector subspace of $\mathbb{R}^3$ because

$$r_1 \cdot (b_1, b_2, 0)^\mathsf{T} + r_2 \cdot (d_1, d_2, 0)^\mathsf{T} = (r_1 b_1 + r_2 d_1, r_1 b_2 + r_2 d_2, 0)^\mathsf{T} \in V$$

for all real numbers $r_1, r_2, b_1, b_2, d_1, d_2$.

Definition 2.2.1.19. *Let $A = [a_{ij}]_{i=1,\ldots,m, j=1,\ldots,n}$ be a matrix and let $X = (x_1, x_2, \ldots, x_n)^\mathsf{T}$. For every homogeneous linear system $AX = 0_{1 \times m}$, we define the set of solutions to $A \cdot X = 0_{1 \times m}$ as*

$$\boldsymbol{Sol(A)} = \{Y \in \mathbb{R}^n \mid A \cdot Y = 0_{m \times 1}\}.$$

Analogously, for every A and every $b \in \mathbb{R}^m$, the set of solutions to $A \cdot X = b$ is

$$\boldsymbol{Sol(A, b)} = \{Y \in \mathbb{R}^n \mid A \cdot Y = b\}.$$

Lemma 2.2.1.20. *Let $A \cdot X = 0_{n \times 1}$ be a system of linear equations where A is an $m \times n$ matrix, $m, n \in \mathbb{N} - \{0\}$. The set $Sol(A)$ of all solutions to the linear system $A \cdot X = 0_{n \times 1}$ is a vector subspace of $\mathbb{R}^n$.*

Proof. $X = 0_{n \times 1}$ is a solution of every homogeneous system of linear equations of n variables; thus, $Sol(A)$ is not empty. Let X_1 and X_2 be arbitrary vectors from $Sol(A)$, and let r_1 and r_2 be arbitrary reals. We have to prove that $r_1 X_1 + r_2 X_2 \in Sol(A)$. This can be done by the following simple calculation.

$$A(r_1 X_1 + r_2 X_2) = A r_1 X_1 + A r_2 X_2$$
$$= r_1 A X_1 + r_2 A X_2$$
$$= r_1 \cdot 0_{n \times 1} + r_2 \cdot 0_{n \times 1} = 0_{n \times 1}. \qquad \square$$

---

[6] Later, we shall see it defined by the term "affine subspace" (the term "manifold" is used in some literature, too).

The **trivial vector subspace** of $\mathbb{R}^n$ is $\{0_{n\times1}\}$. Obviously, there is no non-trivial vector subspace W of $\mathbb{R}^n$ with a finite cardinality. This is because for each $X \in W$, $X \neq 0_{n\times1}$, the infinite set

$$\{r \cdot X \mid r \in \mathbb{R}\} \subseteq W.$$

Definition 2.2.1.21. *Let $X, X_1, X_2, \ldots, X_k$ be vectors from $\mathbb{R}^n$, $n, k \in \mathbb{N} - \{0\}$. The vector X is a* **linear combination of the vectors $X_1, X_2, \ldots, X_k$** *if there exist real numbers $c_1, c_2, \ldots, c_k$ such that*

$$X = c_1 X_1 + c_2 X_2 + \cdots + c_k X_k.$$

For instance, $(4, 2, 10, -10)^\mathsf{T}$ is a linear combination of the vectors $(1, 2, 1, -1)^\mathsf{T}$, $(1, 0, 2, -3)^\mathsf{T}$, and $(1, 1, 0, -2)^\mathsf{T}$ because

$$(4, 2, 10, -10)^\mathsf{T} = 2 \cdot (1, 2, 1, -1)^\mathsf{T} + 4 \cdot (1, 0, 2, -3)^\mathsf{T} - 2 \cdot (1, 1, 0, -2)^\mathsf{T}.$$

Definition 2.2.1.22. *Let $S = \{X_1, X_2, \ldots, X_k\} \subseteq \mathbb{R}^n$ be a set of nonzero vectors, and let W be a subset of $\mathbb{R}^n$, $k \in \mathbb{N}, n \in \mathbb{N} - \{0\}$.*

We say that S **spans** *W if every vector from W is a linear combination of vectors from S. The trivial vector subspace $\{0_{n\times1}\} \subseteq \mathbb{R}^n$ is spanned by the empty set S.*

The set S is called **linearly dependent** *if there exist reals $c_1, c_2, \ldots, c_k$ not all zero, such that*

$$c_1 X_1 + c_2 X_2 + \cdots + c_k X_k = 0_{n\times1}.$$

Otherwise, S is called **linearly independent** *(i.e., the equality $c_1 X_1 + c_2 X_2 + \cdots + c_k X_k = 0_{n\times1}$ holds only for $c_1 = c_2 = \cdots = c_k = 0$).*

The set S is called a **basis** *of a vector subspace $U \subseteq \mathbb{R}^n$, if*

(i) S spans U, and
(ii) S is linearly independent.

For any subspace $V \subseteq \mathbb{R}^n$ we say that the **dimension** of V is a $k \in \mathbb{N}$ if there exists a set of vectors $S \subseteq \mathbb{R}^n$ such that

(i) $|S| = k$, and
(ii) S is a basis of V.

We say also that V is a k-dimensional subspace of $\mathbb{R}^n$, or that $\mathbf{dim(V) = k}$.

For instance, $(1, 0, 0, 0)^\mathsf{T}$, $(0, 1, 0, 0)^\mathsf{T}$, $(0, 0, 1, 0)^\mathsf{T}$, $(0, 0, 0, 1)^\mathsf{T}$ is a basis of $\mathbb{R}^4$ because

(i) $(a, b, c, d)^\mathsf{T} = a \cdot (1, 0, 0, 0)^\mathsf{T} + b \cdot (0, 1, 0, 0)^\mathsf{T} + c \cdot (0, 0, 1, 0)^\mathsf{T} + d \cdot (0, 0, 0, 1)^\mathsf{T}$
    for each vector $(a, b, c, d)^\mathsf{T} \in \mathbb{R}^4$, and
(ii) $c_1 \cdot (1, 0, 0, 0)^\mathsf{T} + c_2 \cdot (0, 1, 0, 0)^\mathsf{T} + c_3 \cdot (0, 0, 1, 0)^\mathsf{T} + c_4 \cdot (0, 0, 0, 1)^\mathsf{T} = (0, 0, 0, 0)^\mathsf{T}$
    if and only if $c_1 = c_2 = c_3 = c_4 = 0$.

Let us consider the following linear equation

$$a_1 x_1 + a_2 x_2 + a_3 x_3 + \cdots + a_n x_n = b \qquad (2.3)$$

of n variables (unknowns) $x_1, x_2, \ldots, x_n$. We can also write (2.3) as $A \cdot X = b$, where $A = (a_1, a_2, \ldots, a_n)$ and $X = (x_1, x_2, \ldots, x_n)^\mathsf{T}$.

Lemma 2.2.1.23. *For every linear homogeneous equation* $A \cdot X = 0$, *where* $A = (a_1, \ldots, a_n) \neq (0, \ldots, 0)$, $X = (x_1, \ldots, x_n)^\mathsf{T}$, $Sol(A)$ *is an* $(n-1)$-*dimensional subspace of* $\mathbb{R}^n$.

Proof. Without loss of generality we assume $a_1 \neq 0$. Then

$$x_1 = -\frac{a_2}{a_1} x_2 - \frac{a_3}{a_1} x_3 - \cdots - \frac{a_n}{a_1} x_n$$

is another form of the linear equation (2.3) for $b = 0$. So, $Sol(A) =$

$$\left\{ \left(-\frac{a_2}{a_1} y_2 - \frac{a_3}{a_1} y_3 - \cdots - \frac{a_n}{a_1} y_n, y_2, y_3, \ldots, y_n \right)^\mathsf{T} \,\middle|\, y_2, y_3, \ldots, y_n \in \mathbb{R} \right\}$$

is the set of all solutions of $AX = 0$. We verify this assumption by the following calculation:

$$(a_1, a_2, \ldots, a_n) \cdot \left(-\frac{a_2}{a_1} y_2 - \frac{a_3}{a_1} y_3 - \cdots - \frac{a_n}{a_1} y_n, y_2, y_3, \ldots, y_n \right)^\mathsf{T} =$$
$$(-a_2 y_2 - a_3 y_3 - \cdots - a_n y_n) + a_2 y_2 + a_3 y_3 + \cdots + a_n y_n = 0.$$

Now we claim that

$$S = \left\{ \left(-\frac{a_2}{a_1}, 1, 0, \ldots, 0 \right)^\mathsf{T}, \left(-\frac{a_3}{a_1}, 0, 1, 0, \ldots, 0 \right)^\mathsf{T}, \ldots, \left(-\frac{a_n}{a_1}, 0, \ldots, 0, 1 \right)^\mathsf{T} \right\}$$

is a basis of $Sol(A)$. S spans $Sol(A)$ because

$$\left(-\frac{a_2}{a_1} y_2 - \frac{a_3}{a_1} y_3 - \cdots - \frac{a_n}{a_1} y_n, y_2, y_3, \ldots, y_n \right)^\mathsf{T} =$$
$$y_2 \left(-\frac{a_2}{a_1}, 1, 0, \ldots, 0 \right)^\mathsf{T} + y_3 \left(-\frac{a_3}{a_1}, 0, 1, \ldots, 0 \right)^\mathsf{T} + \cdots + y_n \left(-\frac{a_n}{a_1}, 0, \ldots, 1 \right)^\mathsf{T}$$

for all $y_2, y_3, \ldots, y_n \in \mathbb{R}$.

It remains to be shown that the vectors of S are linearly independent. Assume that

$$c_1 \left(-\frac{a_2}{a_1}, 1, 0, \ldots, 0 \right)^\mathsf{T} + c_2 \left(-\frac{a_3}{a_1}, 0, 1, \ldots, 0 \right)^\mathsf{T} + \cdots$$
$$+ c_{n-1} \left(-\frac{a_n}{a_1}, 0, \ldots, 0, 1 \right)^\mathsf{T} = 0.$$

Then, particularly

$$c_1 \cdot 1 = 0$$
$$c_2 \cdot 1 = 0$$
$$\vdots$$
$$c_{n-1} \cdot 1 = 0.$$

Thus, $c_1 = c_2 = \cdots = c_{n-1} = 0$. $\square$

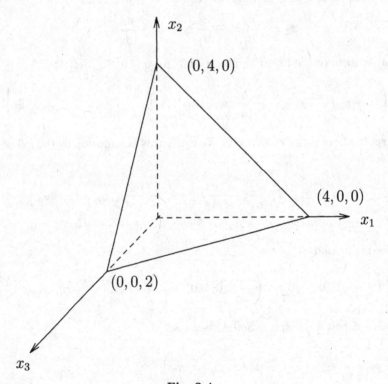

Fig. 2.4.

If one considers the vector space $\mathbb{R}^2$, then the set of solutions of a linear equation $c_1 x_1 + c_2 x_2 = 0$ with $(c_1, c_2) \neq (0, 0)$ is the set of all points of the corresponding straight line $x_1 = \frac{c_2}{c_1} \cdot x_2$. Examples are presented in Figure 2.3. In $\mathbb{R}^3$ the set of all solutions of a linear homogeneous equation is a two-dimensional subspace of $\mathbb{R}^3$ (Fig. 2.4).

As we have already observed, the set $Sol(A)$ of solutions Y to $A \cdot Y = 0_{m \times 1}$ is a subspace of $\mathbb{R}^n$. The question is what is the dimension of $Sol(A)$. Obviously, if A is not a zero matrix, the dimension of $Sol(A)$ is at most $n-1$.

Theorem 2.2.1.24. *Let U be a subset of $\mathbb{R}^n$, $n \in \mathbb{N} - \{0\}$. U is a subspace of $\mathbb{R}^n$ if and only if there exists a matrix A such that $U = Sol(A)$.*

Proof. We have already mentioned that $Sol(A)$ is a subspace of $\mathbb{R}^n$ if A is an $m \times n$ matrix, $m, n \in \mathbb{N} - \{0\}$. It remains to be shown that, for every subspace $U \subseteq \mathbb{R}^m$, there exists such an A that $U = Sol(A)$.

Let $S = \{S_1, S_2, \ldots, S_m\}$ be a basis of U, $S_i = (s_{i1}, \ldots, s_{in})$ for $i = 1, 2, \ldots, m$. Now we construct an $m \times n$ matrix $A = [a_{ij}]$ such that

$$A \cdot S_k^\mathsf{T} = 0_{m \times 1} \tag{2.4}$$

for every $k = 1, 2, \ldots, m$. Setting $S = [s_{ij}]_{i=1,\ldots,m, j=1,\ldots,n}$ we can express (2.4) as

$$A \cdot S^\mathsf{T} = 0_{m \times m}. \tag{2.5}$$

But (2.5) is equivalent to the requirement

$$(a_{l1}, a_{l2}, \ldots, a_{ln}) \cdot S^\mathsf{T} = 0_{1 \times m} \tag{2.6}$$

for every $l \in \{1, \ldots, m\}$. So the lth equation of (2.6) can be seen as a system of linear equations of n unknowns $a_{l1}, a_{l2}, \ldots, a_{ln}$. Solving this linear system one determines the values $a_{l1}, a_{l2}, \ldots, a_{ln}$. Doing it for every l, the matrix A is determined. It remains to be proven that $U = Sol(A)$. Next we prove $U \subseteq Sol(A)$. The opposite direction $Sol(A) \subseteq U$ is left to the reader.

As S is a basis of U, we have, for every $X \in U$,

$$X = c_1 S_1^\mathsf{T} + c_2 S_2^\mathsf{T} + \cdots + c_m S_m^\mathsf{T}$$

for some $c_1, c_2, \ldots, c_m \in R$. Thus

$$
\begin{aligned}
A \cdot X &= A \cdot (c_1 S_1^\mathsf{T} + c_2 S_2^\mathsf{T} + \cdots + c_m S_m^\mathsf{T}) \\
&= c_1 \cdot A \cdot S_1^\mathsf{T} + c_2 \cdot A \cdot S_2^\mathsf{T} + \cdots + c_m \cdot A \cdot S_m^\mathsf{T} \\
&\underset{(2.4)}{=} c_1 \cdot 0_{m \times 1} + c_2 \cdot 0_{m \times 1} + \cdots + c_m \cdot 0_{m \times 1} \\
&= 0_{m \times 1},
\end{aligned}
$$

and so $X \in Sol(A)$. $\qquad\square$

Definition 2.2.1.25. *Let A be an $n \times m$ matrix $[a_{ij}]_{i=1,\ldots,n, j=1,\ldots,m}$. Let $A_i = (a_{i1}, a_{i2}, \ldots, a_{im})^\mathsf{T}$ for $i = 1, \ldots, n$, and let U be the subspace of $\mathbb{R}^n$ that is spanned by $\{A_1, A_2, \ldots, A_n\}$. We define the **rank** of the matrix A as*

$$\mathbf{rank}(A) = dim(U).$$

Obviously, $rank(I_n) = n$ for every positive integer n. The rank of the following matrix

$$M = \begin{pmatrix} 1 & 2 & 0 & 3 \\ 2 & 1 & 1 & 0 \\ 1 & 0 & 0 & 1 \\ 0 & 1 & -1 & 4 \end{pmatrix}$$

is three. This is because

$$(0, 1, -1, 4) = 1 \cdot (1, 2, 0, 3) - 1 \cdot (2, 1, 1, 0) + 1 \cdot (1, 0, 0, 1),$$

and the fact that the set of vectors $\{(1, 2, 0, 3), (2, 1, 1, 0), (1, 0, 0, 1)\}$ is linearly independent.

Exercise 2.2.1.26. Let $A = [a_{ij}]_{i=1,...,n, j=1,...,m}$, $n, m \in \mathbb{N} - \{0\}$. Let $S_i = (a_{i1}, a_{i2}, \ldots, a_{im})^\mathsf{T}$ for $i = 1, \ldots, n$, and let $C_j = (a_{1j}, a_{2j}, \ldots, a_{nj})^\mathsf{T}$ for $j = 1, \ldots, m$. Let U be the subspace of $\mathbb{R}^m$ spanned by $\{S_1, \ldots, S_n\}$ and V be a subspace of $\mathbb{R}^n$ that is spanned by $\{C_1, \ldots, C_m\}$. Prove that

$$dim(U) = dim(V). \qquad \square$$

Exercise 2.2.1.27. Let $B = [b_{ij}]_{i=1,...,m, j=1,...,n}$ be an $m \times n$ matrix, $n, m \in \mathbb{N} - \{0\}$. Prove that

$$dim(B) = n - rank(B). \qquad \square$$

Definition 2.2.1.28. *Let U be a subspace of $\mathbb{R}^n$, $n \in \mathbb{N} - \{0\}$, and let $C \in \mathbb{R}^n$. The vector set*

$$V = \{X \in \mathbb{R}^n \mid X = C + Y, Y \in U\}$$

is called an **affine subspace** *of $\mathbb{R}^n$ translated by C from U.*

Observation 2.2.1.29. Let U be a subspace of $\mathbb{R}^n$, $n \in \mathbb{N} - \{0\}$. For each $C \in \mathbb{R}^n$,

$$dim(\{X \in \mathbb{R}^n \mid X = C + Y, Y \in U\}) = dim(U).$$

The following theorem presents the main relation between systems of linear equations and affine subspaces. We omit its technical proof because it does not contain any idea that would be interesting for an algorithm design in this book.

Theorem 2.2.1.30. *Let n be a positive integer, and let U be a subset of $\mathbb{R}^n$. U is an affine subspace of R^n iff there exist $m \in \mathbb{N}$, $m \leq n$, an $m \times n$ matrix A, and a vector $b \in \mathbb{R}^m$ such that $U = Sol(A, b)$.*

In $\mathbb{R}^3$ the set of all solutions of a linear equation is a two-dimensional subspace of $\mathbb{R}^3$, called also a **plane**. Figure 2.4 contains the plane corresponding to the linear equation $x_1 + x_2 + 2x_3 = 4$. This plane is unambiguously given by the three points $(4, 0, 0)$, $(0, 4, 0)$, $(0, 0, 2)$ in which it crosses the axes of $\mathbb{R}^3$. The line $x_1 + x_2 = 4$ is the intersection of the plane $x_3 = 0$ with the plane

$x_1 + x_2 + 2x_3 = 4$. The line $x_2 + 2x_3 = 4$ is the intersection of the plane $x_1 = 0$ and the plane $x_1 + x_2 + 2x_3 = 4$, and the line $x_1 + 2x_3 = 4$ is the intersection of the plane $x_2 = 0$ and the plane $x_1 + x_2 + 2x_3 = 4$.

A transparent way to work with vector subspaces is to view them as convex sets.

Definition 2.2.1.31. *Let X and Y be two points in $\mathbb{R}^n$. A* **convex combination** *of X and Y is any point*

$$Z = c \cdot X + (1 - c) \cdot Y$$

for any real number c, $0 \le c \le 1$. If $c \notin \{0, 1\}$, then we say that c is a **strict convex combination** *of X and Y.*

Observe that the set

$$\mathbf{Convex}(X, Y) = \{Z \in \mathbb{R}^n \mid Z = c \cdot X + (1 - c) \cdot Y \text{ for a } c \in \mathbb{R}, 0 \le c \le 1\}$$

is exactly the set of points of the straight line that connects X and Y.

Definition 2.2.1.32. *Let n be a positive integer. A set $S \subseteq \mathbb{R}^n$ is* **convex** *if, for all $X, Y \in S$, S contains all convex combinations of X and Y (i.e., $Convex(X, Y) \subseteq S$ for all $X, Y \in S$).*

The trivial example of convex sets in $\mathbb{R}^n$ are $\mathbb{R}^n$, the empty set, and any singleton set. The set of solutions of a linear equation $a_1 x_1 + a_2 x_2 = b$ in $\mathbb{R}^2$ is convex. An important property of convex sets is expressed in the following assertion.

Theorem 2.2.1.33. *The intersection of any number of convex sets is a convex set.*

Proof. Let $\bigcap_{i \in I} S_i$ be the intersection of convex sets S_i for $i \in I$. If $X, Y \in \bigcap_{i \in I} S_i$, then X and Y are in every S_i. Any convex combination of X and Y is then in S_i for every $i \in I$, and therefore in $\bigcap_{i \in I} S_i$.   □

Now we show that the set of solutions of any linear equation

$$a_1 x_1 + a_2 x_2 + \cdots + a_n x_x = b$$

of n unknowns $x_1, \ldots, x_n$ is a convex set in $\mathbb{R}^n$. Let $Y = (y_1, \ldots, y_n)^\mathsf{T}$ and $W = (w_1, \ldots, w_n)^\mathsf{T}$ be arbitrary solutions of this linear equation, i.e.,

$$\sum_{i=1}^{n} a_i y_i = b = \sum_{i=1}^{n} a_i w_i.$$

Let

$$Z = c \cdot Y + (1 - c) \cdot W = (cy_1 + (1 - c)w_1, \ldots, cy_n + (1 - c)w_n)$$

be an arbitrary convex combination of Y and W, $0 \leq c \leq 1$. We prove that Z is a solution to the linear equation

$$\sum_{i=1}^{n} a_i x_i = b,$$

too.

$$(a_1, \ldots, a_n) \cdot Z = \sum_{i=1}^{n} a_i (cy_i + (1-c)w_i)$$

$$= c \cdot \sum_{i=1}^{n} a_i y_i + (1-c) \cdot \sum_{i=1}^{n} a_i w_i$$

$$= c \cdot b + (1-c) \cdot b = b.$$

Because of the above fact and Theorem 2.2.1.33 we obtain that $Sol(A, b)$ is a convex set for every system of linear equations $A \cdot X = b$.

2.2.2 Combinatorics, Counting, and Graph Theory

The aim of this section is to give the definitions of some fundamental objects of combinatorics and graph theory, and to present a few elementary results about them. The terms introduced in this section are useful for the analysis of algorithms as well as for the representation of discrete objects in the subsequent chapters.

More precisely, we first introduce the fundamental categories of combinatorics such as permutation and combination and learn to work with them. Then we define the O, Ω, and Θ notation for the study of the asymptotic behavior of functions, and present some simple summations of some fundamental series, and a simplified version of the Master Theorem for solving some specific recurrences. After that we give the fundamental notions of graph theory such as graphs, multigraphs, directed graphs, planarity, connectivity, matching, cut, etc.

First, we define the basic terms of permutation and combination. The starting point is to have a set of objects that are distinguishable from each other.

Definition 2.2.2.1. *Let n be a positive integer. Let $S = \{a_1, a_2, \ldots, a_n\}$ be a set of n objects (elements). A **permutation of n objects** $a_1, \ldots, a_n$ is an ordered arrangement of the objects of S.*

For instance, if $S = \{a_1, a_2, a_3\}$, then there are the following six different ways to arrange the three objects a_1, a_2, a_3:

$$(a_1, a_2, a_3), (a_1, a_3, a_2), (a_2, a_1, a_3), (a_2, a_3, a_1), (a_3, a_1, a_2), (a_3, a_2, a_1).$$

To denote permutations the simple notation $(i_1, i_2, \ldots, i_n)$ is often used instead of $(a_{i_1}, a_{i_2}, \ldots, a_{i_n})$.

Lemma 2.2.2.2. *For every positive integer n, the number $n!$ of different permutations of n objects is*

$$n! = n \cdot (n-1) \cdot (n-2) \cdot \cdots \cdot 2 \cdot 1 = \prod_{i=1}^{n} i.$$

Proof. The first object in any permutation may be chosen in n different ways (i.e., from n different objects). Once the first object has been chosen, the second object may be chosen in $n-1$ different ways (i.e., from $n-1$ remaining objects), etc. □

By arrangement, we use the notation $0! = 1$.

Definition 2.2.2.3. *Let k and n be non-negative integers, $k \leq n$. A **combination of k objects from n objects** is a selection of k objects without regard to order.*

A combination of four objects from $\{a_1, a_2, a_3, a_4, a_5\}$ is one of the following sets:

$$\{a_1, a_2, a_3, a_4\}, \{a_1, a_2, a_3, a_5\}, \{a_1, a_2, a_4, a_5\}, \{a_1, a_3, a_4, a_5\}, \{a_2, a_3, a_4, a_5\}.$$

Lemma 2.2.2.4. *Let n and k be non-negative integers, $k \leq n$. The number $\binom{n}{k}$ of combinations of k objects from n objects is*

$$\binom{n}{k} = \frac{n \cdot (n-1) \cdot (n-2) \cdot \cdots \cdot (n-k+1)}{k!} = \frac{n!}{k! \cdot (n-k)!}.$$

Proof. Similar to the proof of Lemma 2.2.2.2, we have n possibilities for the choice of the first element, $n-1$ possibilities of the choice of the second element, etc. Hence, there are

$$n \cdot (n-1) \cdot (n-2) \cdot \cdots \cdot (n-k+1)$$

ways of choosing k objects from n objects when order is taken into account. But any order of these k elements provides the same set of k elements. So,

$$\binom{n}{k} = \frac{n \cdot (n-1) \cdot (n-2) \cdot \cdots \cdot (n-k+1)}{k!}.$$

□

Observe that $\binom{n}{0} = \binom{n}{n} = 1$.

Corollary 2.2.2.5. *For all non-negative integers k and n, $k \leq n$,*

$$\binom{n}{k} = \binom{n}{n-k}.$$

Lemma 2.2.2.6. *For all positive integers k and n, $k \le n$,*

$$\binom{n}{k} = \binom{n-1}{k-1} + \binom{n-1}{k}.$$

Proof.

$$\binom{n-1}{k-1} + \binom{n-1}{k} = \frac{(n-1)!}{(k-1)! \cdot (n-k)!} + \frac{(n-1)!}{k! \cdot (n-k-1)!}$$

$$= \frac{k \cdot (n-1)! + (n-k) \cdot (n-1)!}{k! \cdot (n-k)!}$$

$$= \frac{n \cdot (n-1)!}{k! \cdot (n-k)!} = \frac{n!}{k! \cdot (n-k)!} = \binom{n}{k}.$$

$\square$

The values $\binom{n}{k}$ are also known as the **binomial coefficients** from the following theorem.

Theorem 2.2.2.7 (Newton's Theorem). *For every positive integer n,*

$$(1+x)^n = \binom{n}{0} + \binom{n}{1} \cdot x + \binom{n}{2} \cdot x^2 + \cdots + \binom{n}{n-1} \cdot x^{n-1} + \binom{n}{n} \cdot x^n$$

$$= \sum_{i=0}^{n} \binom{n}{i} \cdot x^i.$$

Exercise 2.2.2.8. Prove Newton's Theorem. $\square$

Lemma 2.2.2.9. *For every positive integer n,*

$$\sum_{k=0}^{n} \binom{n}{k} = 2^n.$$

Proof. To prove Lemma 2.2.2.9 it is sufficient to set $x = 1$ in Newton's Theorem. Another argument is that, for each $k \in \{0, 1, \ldots, n\}$, $\binom{n}{k}$ is the number of all k-element subsets of an n-element set. So, $\sum_{k=0}^{n} \binom{n}{k}$ counts the number of all subsets of a set of n elements, and every set of n elements has exactly 2^n different subsets. $\square$

Exercise 2.2.2.10. Prove, for every integer $n \ge 3$,

$$\binom{n}{2} = \prod_{l=3}^{n} \frac{l}{l-2}.$$

$\square$

Next, we fix some fundamental notations concerning elementary functions, and look briefly at the asymptotic behavior of functions.

For any positive real number x, we denote the greatest integer less than or equal to x by $\lfloor x \rfloor$ and call it the **floor of x**. For $x \in \mathbb{R}^{+}$, the **ceiling of x**, denoted by $\lceil x \rceil$, is the least integer greater than or equal to x. We observe that

$$x - 1 < \lfloor x \rfloor \leq x \leq \lceil x \rceil < x + 1$$

for each $x \in \mathbb{R}^{+}$.

Let x be a variable, and let d be a positive integer. A **polynomial in x of degree d** is a function $p(x)$ of the form

$$p(x) = \sum_{i=0}^{d} a_i x^i,$$

where the constants $a_0, a_1, \ldots, a_d$ are the **coefficients** of the polynomial and $a_d \neq 0$.

We use e to denote $\lim_{n \to \infty} \left(1 + \dfrac{1}{n}\right)^n$. Note that e is the base of the natural logarithm function.

Exercise 2.2.2.11. Prove that, for all reals x,

$$e^x = 1 + x + \frac{x^2}{2!} + \frac{x^3}{3!} + \ldots = \sum_{i=0}^{\infty} \frac{x^i}{i!}.$$

$\square$

The assertion of Exercise 2.2.2.11 implies

$$1 + x \leq e^x \leq 1 + x + x^2$$

for every $x \in [-1, 1]$. Note that $e = 2.7182\ldots$, and one can approximate e with an arbitrary precision by using the equation of Exercise 2.2.2.11.

Exercise 2.2.2.12. Prove that for all real x,

$$\lim_{n \to \infty} \left(1 + \frac{x}{n}\right)^n = e^x.$$

$\square$

In this book we use **$\log n$** to denote the binary logarithm $\log_2 n$ and **$\ln n$** $= \log_e n$ to denote the natural logarithm. The equalities of the following exercise provide the elementary rules for working with logarithmic functions.

Exercise 2.2.2.13. Prove, for all positive reals a, b, c and n,

(i) $\log_c(ab) = \log_c(a) + \log_c(b)$,

(ii) $\log_c a^n = n \cdot \log_c a$,

(iii) $a^{\log_b n} = n^{\log_b a}$, and

(iv) $\log_b a = \frac{1}{\log_a b}$. $\qquad\qquad\qquad\qquad\qquad\qquad\qquad\qquad\qquad\qquad$ □

In algorithmics we work with functions from $\mathbb{N}$ to $\mathbb{N}$ in order to measure complexity according to the input size. Here, we are often concerned with how the complexity (running time, for instance) increases with the input size in the limit as the size of the input increases without bound. In this case we say that we are studying the **asymptotic** efficiency of algorithms. This rough characterization of the complexity growth by the order of its growth is usually sufficient to determine the threshold on the input size above which the algorithm is not applicable because of a too huge complexity. In what follows we define the standard asymptotic notation used in algorithmics.

Definition 2.2.2.14. *Let $f : \mathbb{N} \to \mathbb{R}^{\geq 0}$ be a function. We define*

$$O(f(n)) = \{t : \mathbb{N} \to \mathbb{R}^{\geq 0} \mid \exists c, n_0 \in \mathbb{N}, \text{ such that } \forall n \in \mathbb{N}, n \geq n_0 :$$
$$t(n) \leq c \cdot f(n)\}.$$
$$\Omega(f(n)) = \{g : \mathbb{N} \to \mathbb{R}^{\geq 0} \mid \exists d, n_0 \in \mathbb{N}, \text{ such that } \forall n \in \mathbb{N}, n \geq n_0 :$$
$$g(n) \geq \frac{1}{d} \cdot f(n)\}.$$
$$\Theta(f(n)) = O(f(n)) \cap \Omega(f(n))$$
$$= \{h : \mathbb{N} \to \mathbb{R}^{\geq 0} \mid \exists c_1, c_2, n_0 \in \mathbb{N}, \text{ such that } \forall n \in \mathbb{N}, n \geq n_0 :$$
$$\frac{1}{c_1} \cdot f(n) \leq h(n) \leq c_2 \cdot f(n)\}.$$

*If $t(n) \in O(f(n))$ we say that t does **not** grow asymptotically faster than f. If $g(n) \in \Omega(f(n))$ we say that g grows asymptotically at least as fast as f. If $h(n) \in \Theta(f(n))$ we say that h and f are asymptotically equivalent.*

Exercise 2.2.2.15. Let $p(n) = a_0 + a_1 n + a_2 n^2 + \cdots + a_d n^d$ be a polynomial in n for some positive integer d and a positive real number a_d. Prove that

$$p(n) \in \Theta(n^d).$$

$\qquad\qquad\qquad\qquad\qquad\qquad\qquad\qquad\qquad\qquad\qquad\qquad\qquad\qquad\qquad$ □

In the literature one usually uses the notation $t(n) = O(f(n))$, $g(n) = \Omega(f(n))$, and $h(n) = \Theta(f(n))$, respectively, instead of $t(n) \in O(f(n))$, $g(n) \in \Omega(f(n))$, and $h(n) \in \Theta(f(n))$, respectively.

Exercise 2.2.2.16. Which of the following statements are true? Prove your answers.

(i) $2^n \in \Theta(2^{n+a})$ for any positive integer (constant) a.

(ii) $2^{b \cdot n} \in \Theta\left(2^{n}\right)$ for any positive integer (constant) b.

(iii) $\log_{b} n \in \Theta\left(\log_{c} n\right)$ for all $b, c \in \mathbb{R}^{>1}$.

(iv) $(n+1)! \in O(n!)$.

(v) $\log_{2}(n!) \in \Theta(n \cdot \log n)$.

$\square$

In the next part of this section we remind the reader of some elementary series and their sums. For any function $f : \mathbb{N} \to \mathbb{R}$, one can define

$$Sum_{f}(n) = \sum_{i=1}^{n} f(i) = f(1) + f(2) + \cdots + f(n).$$

$Sum_{f}(n)$ is called a **series of** f. In what follows we consider only some fundamental kinds of series.

Definition 2.2.2.17. *Let a, b, and d be some constants. For every function $f : \mathbb{N} \to \mathbb{R}$, defined by $f(n) = a + (n-1) \cdot d$, $Sum_{f}(n)$ is called an* **arithmetic series**. *For every function $h : \mathbb{N} \to \mathbb{R}$ defined by $h(n) = a \cdot b^{n-1}$, $Sum_{h}(n)$ is called a* **geometric series**.

Lemma 2.2.2.18. *Let a and d be some constants. Then*

$$Sum_{a+(n-1) \cdot d}(n) = \sum_{i=1}^{n}(a + (i-1) \cdot d) = an + \frac{d \cdot (n-1) \cdot n}{2}.$$

Proof.

$$\sum_{i=1}^{n}(a + (i-1) \cdot d) = \sum_{i=1}^{n} a + d \cdot \sum_{i=1}^{n}(i-1)$$

$$= a \cdot n + d \cdot \frac{1}{2}\left(2 \cdot \sum_{i=1}^{n}(i-1)\right)$$

$$= a \cdot n + \frac{d}{2}\left(\sum_{i=1}^{n}(i-1) + \sum_{j=n}^{1}(j-1)\right)$$

$$= a \cdot n + \frac{d}{2} \cdot \sum_{i=1}^{n}[((n-1) - (i-1)) + (i-1)]$$

$$= a \cdot n + \frac{d}{2} \cdot \sum_{i=1}^{n}(n-1) = a \cdot n + \frac{d}{2} \cdot (n-1) \cdot n.$$

$\square$

Lemma 2.2.2.19. *Let a and b be some constants, $b \in \mathbb{R}^{+}$, $b \neq 1$. Then*

$$Sum_{a \cdot b^{n-1}}(n) = \sum_{i=1}^{n} a \cdot b^{i-1} = a \cdot \frac{1 - b^{n}}{1 - b}.$$

Proof.

$$Sum_{a \cdot b^{n-1}}(n) = \sum_{i=1}^{n} a \cdot b^{i-1} = a \left(1 + b + b^2 + \cdots + b^{n-1}\right). \tag{2.7}$$

Thus,

$$b \cdot Sum_{a \cdot b^{n-1}}(n) = a \cdot \left(b + b^2 + \cdots + b^n\right). \tag{2.8}$$

Subtracting the equality (2.8) from the equality (2.7) we obtain

$$(1 - b) \cdot Sum_{a \cdot b^{n-1}}(n) = a \left(1 - b^n\right),$$

which directly implies the assertion of Lemma 2.2.2.19.   □

Exercise 2.2.2.20. Prove that for any $b \in (0, 1)$,

$$\lim_{n \to \infty} Sum_{a \cdot b^{n-1}}(n) = \sum_{i=1}^{\infty} a \cdot b^{i-1} = a \cdot \frac{1}{1 - b}.$$

□

Definition 2.2.2.21. *For every positive integer n, the **nth harmonic number** is defined by the series*

$$\textbf{Har}(n) = \sum_{i=1}^{n} \frac{1}{i} = 1 + \frac{1}{2} + \frac{1}{3} + \cdots + \frac{1}{n}.$$

First, we observe that $Har(n)$ tends to infinity with growing n. The simplest way to see it is to partition the term of $Har(n)$ into infinitely many groups of cardinality 2^k as follows:

$$\underbrace{\frac{1}{1}}_{\text{group 1}} + \underbrace{\frac{1}{2} + \frac{1}{3}}_{\text{group 2}} + \underbrace{\frac{1}{4} + \frac{1}{5} + \frac{1}{6} + \frac{1}{7}}_{\text{group 3}}$$

$$\underbrace{+ \frac{1}{8} + \frac{1}{9} + \frac{1}{10} + \frac{1}{11} + \frac{1}{12} + \frac{1}{13} + \frac{1}{14} + \frac{1}{15}}_{\text{group 4}} + \cdots.$$

Both terms in group 2 are between $\frac{1}{4}$ and $\frac{1}{2}$, and so the sum of that group is between $2 \cdot \frac{1}{4} = \frac{1}{2}$ and $2 \cdot \frac{1}{2} = 1$. All four terms in group 3 are between $\frac{1}{8}$ and $\frac{1}{4}$, and so their sum is also between $4 \cdot \frac{1}{8} = \frac{1}{2}$ and $4 \cdot \frac{1}{4} = 1$. In general, for every positive integer k, all 2^{k-1} terms of group k are between 2^{-k} and 2^{-k+1} and hence the sum of the terms of group k is between $\frac{1}{2} = 2^{k-1} \cdot 2^{-k}$ and $1 = 2^{k-1} \cdot 2^{-k+1}$.

This grouping procedure shows us that if n is in group k, then $Har(n) > k/2$ and $Har(n) \leq k$. Thus,

$$\frac{\lfloor \log_2 n \rfloor}{2} + \frac{1}{2} < Har(n) \leq \lfloor \log_2 n \rfloor + 1.$$

Exercise 2.2.2.22. (*) Prove that

$$Har(n) = \ln n + O(1).$$

$\square$

Definition 2.2.2.23. *For any sequence* $a_0, a_1, \ldots, a_n$, $\sum_{k=1}^{n} (a_k - a_{k-1})$ *and* $\sum_{i=0}^{n-1} (a_i - a_{i+1})$ *are called* **telescoping series.**

Obviously,

$$\sum_{k=1}^{n} (a_k - a_{k-1}) = a_n - a_0$$

since each of the terms $a_1, a_2, \ldots, a_{n-1}$ is added exactly once and subtracted out exactly once. Analogously,

$$\sum_{i=1}^{n-1} (a_i - a_{i+1}) = a_1 - a_n.$$

The reason to consider telescoping series is that one can easily simplify a series if one recognizes that the series is telescoping. For instance, consider the series

$$\sum_{k=1}^{n-1} \frac{1}{k(k+1)}.$$

Since $\frac{1}{k \cdot (k+1)} = \frac{1}{k} - \frac{1}{k+1}$ we get

$$\sum_{k=1}^{n-1} \frac{1}{k \cdot (k+1)} = \sum_{k=1}^{n-1} \left(\frac{1}{k} - \frac{1}{k+1} \right) = 1 - \frac{1}{n}.$$

Analyzing the complexity of algorithms one often reduces the analysis to solving a specific recurrence. The typical recurrences are of the form

$$T(n) = a \cdot T\left(\frac{n}{c}\right) + f(n),$$

where a and c are positive integers and f is a function from $\mathbb{N}$ to $\mathbb{R}^+$. In what follows we give a general solution of this recurrence if $f(n) \in \Theta(n)$.

Theorem 2.2.2.24 (Master Theorem). *Let* a, b, *and* c *be positive integers. Let*

$$T(1) = 0,$$
$$T(n) = a \cdot T\left(\frac{n}{c}\right) + b \cdot n.$$

Then,

$$T(n) \in \begin{cases} O(n) & \text{if } a < c \\ O(n \log n) & \text{if } a = c \\ O\left(n^{\log_c a}\right) & \text{if } c < a. \end{cases}$$

Proof. For simplicity, we assume $n = c^k$ for some positive integer k.

$$T(n) = a \cdot T\left(\frac{n}{c}\right) + b \cdot n$$

$$= a \cdot \left[a \cdot T\left(\frac{n}{c^2}\right) + b \cdot \frac{n}{c}\right] + b \cdot n$$

$$= a^2 \cdot T\left(\frac{n}{c^2}\right) + b \cdot \left(\frac{a}{c} \cdot n + n\right)$$

$$= a^k \cdot T(1) + b \cdot n \cdot \sum_{i=0}^{k-1} \left(\frac{a}{c}\right)^i$$

$$= bn \cdot \left(\sum_{i=0}^{(\log_c n)-1} \left(\frac{a}{c}\right)^i\right).$$

Now we distinguish the three cases according to the relation between a and c.

(1) Let $a < c$. Then, following Exercise 2.2.2.20 we obtain

$$\sum_{i=0}^{\log_c n - 1} \left(\frac{a}{c}\right)^i \leq \sum_{i=0}^{\infty} \left(\frac{a}{c}\right)^i = \frac{1}{1 - \frac{a}{c}} \in O(1).$$

Thus, $T(n) \in O(n)$.

(2) Let $a = c$. Obviously,

$$\sum_{i=0}^{\log_c n - 1} \left(\frac{a}{c}\right)^i = \log_c n \in O(\log n).$$

So, $T(n) \in O(n \log n)$.

(3) Let $a > c$. Following Lemma 2.2.2.19 we obtain

$$bn \cdot \sum_{i=0}^{\log_c n - 1} \left(\frac{a}{c}\right)^i = bn \cdot \left(\frac{1 - \left(\frac{a}{c}\right)^{\log_c n}}{1 - \frac{a}{c}}\right)$$

$$= \frac{b}{\frac{a}{c} - 1} \cdot n \cdot \left(\left(\frac{a}{c}\right)^{\log_c n} - 1\right) \in O\left(n \cdot \frac{a^{\log_c n}}{c^{\log_c n}}\right)$$

$$= O\left(a^{\log_c n}\right) = O\left(n^{\log_c a}\right). \qquad \square$$

In what follows we present the fundamental terminology of graph theory.

Definition 2.2.2.25. *An* (**undirected**) **graph** G *is a pair* (V, E), *where*

(i) *V is a finite set called the* **set of vertices** *of G, and*

(ii) *E is a subset of $\{\{u, v\} \mid v, u \in V$ and $v \neq u\}$ called the* **set of edges** *of*
G.

Any element of V is called a **vertex** *of G, and any element of E is called an*
edge *of G. G is called a* **graph** *of $|V|$ vertices.*

An example of a graph is $G' = (V, E)$, where

$$V = \{v_1, v_2, v_3, v_4, v_5\} \text{ and } E = \{\{v_1, v_3\}, \{v_1, v_4\}, \{v_1, v_5\}, \{v_2, v_4\}, \{v_2, v_5\}\}.$$

One usually represents a graph as a picture in the plane. The vertices are points (or small circles) in the plane and an edge $\{u, v\}$ is represented as a curve connecting the points u and v. The graph G' is depicted in Figure 2.5 in two different ways.

A graph G is called **planar** if there exists such a picture representation of G in the plane that the curves representing the edges do not cross in any point of the plane. The graph G' defined above is planar because Figure 2.5b provides its planar representation.

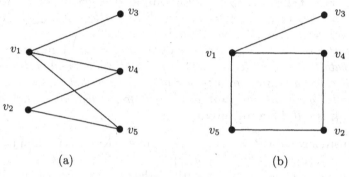

(a)   (b)

Fig. 2.5.

We observe that we have either an edge between two vertices u and v or no edge between u and v. So, another suitable representation of a graph of n vertices $v_1, v_2, \ldots, v_n$ is by a symmetric $n \times n$ Boolean matrix $M_G[a_{ij}]_{i,j=1,\ldots,n}$, called the **adjacency matrix** of G, where $a_{ij} = 1$ iff $\{v_i, v_j\} \in E$. The following matrix represents the graph G' depicted in Figure 2.5.

$$\begin{pmatrix} 0 & 0 & 1 & 1 & 1 \\ 0 & 0 & 0 & 1 & 1 \\ 1 & 0 & 0 & 0 & 0 \\ 1 & 1 & 0 & 0 & 0 \\ 1 & 1 & 0 & 0 & 0 \end{pmatrix}$$

For any edge $\{u, v\}$ of a graph $G = (V, E)$ we say that $\{u, v\}$ is **incident** to the vertices u and v. Two vertices x and y are **adjacent** if the edge $\{x, y\}$ belongs to E. The **degree of a vertex** v of G, $deg_G(v)$, is the number of edges incident to v. The **degree of a graph** $G = (V, E)$ is

$$deg(G) = \max\{deg_G(v) \mid v \in V\}.$$

Exercise 2.2.2.26. Prove that for every graph $G = (V, E)$,

$$\sum_{v \in V} deg_G(v) = 2 \cdot |E|.$$

□

Exercise 2.2.2.27. How many graphs of n vertices $v_1, v_2, \ldots, v_n$ exist?   □

Definition 2.2.2.28. *Let $G = (V, E)$ be a graph. A* **path** *in G is a sequence of vertices $P = v_1, v_2, \ldots, v_m$, $v_i \in V$ for $i = 1, \ldots, m$, such that $\{v_i, v_{i+1}\} \in E$ for $i = 1, \ldots, m - 1$. For $i = 1, \ldots, m$, v_i is a* **vertex of P**, *and for $j = 1, \ldots, m - 1$, $\{v_j, v_{j+1}\}$ is an* **edge of P**. *The* **length of P** *is the number of its edges (i.e., $m - 1$).*

A path $P = v_1, v_2, \ldots, v_m$ is called **simple** *if either all its vertices are distinct (i.e., $|\{v_1, v_2, \ldots, v_m\}| = m$) or all its vertices but v_1 and v_m are distinct (i.e., $|\{v_1, \ldots, v_m\}| = m - 1$ and $v_1 = v_m$).*

A path $P = v_1, v_2, \ldots, v_m$ is called a **cycle** *if $v_1 = v_m$. A* **simple cycle** *is a simple path that is a cycle. A simple cycle that contains all vertices of the graph G is called a* **Hamiltonian tour** *of G. A cycle that contains all edges of G is called an* **Eulerian tour** *of G. If a graph contains a Hamiltonian tour, then it is called* **Hamiltonian**.

Obviously, any path $P = v_1, v_2, \ldots, v_m$ can be viewed as a graph $(\{v_1, \ldots, v_m\}, \{\{v_1, v_2\}, \{v_2, v_3\}, \ldots, \{v_{m-1}, v_m\}\})$.

$P = v_1, v_4, v_2, v_5, v_1, v_4, v_2$ is a path of the graph G' from Figure 2.5. P is not simple. The path v_1, v_4, v_2, v_5, v_1 is a simple cycle. G' does not contain any Eulerian tour nor any Hamiltonian tour.

Exercise 2.2.2.29. Show that if a graph contains a path between two vertices u and v, then it contains a simple path between u and v.   □

For any $n \in \mathbb{N}$, $\boldsymbol{K_n} = (\{v_1, \ldots, v_n\}, \{\{v_i, v_j\} \mid i, j \in \{1, \ldots, n\}, i \neq j\})$ is the **complete** graph of n vertices. Observe that M_{K_n} is the 0-diagonal $n \times n$ Boolean matrix (i.e., the matrix whose diagonal elements are all 0s and non-diagonal elements are all 1s.) A graph $G = (V, E)$ is called **connected** if, for all $x, y \in V$, $x \neq y$, there exists a path between x and y in G. G is called **bipartite** if one can partition V into two sets V_1 and V_2 ($V_1 \cup V_2 = V$, $V_1 \cap V_2 = \emptyset$) such that $E \subseteq \{\{u, v\} \mid u \in V_1, v \in V_2\}$. The graph G' in Figure 2.5 is a connected, bipartite graph. The second property can be viewed especially well in Figure 2.5a with $V_1 = \{v_1, v_2\}$ on the left side and $V_2 = \{v_3, v_4, v_5\}$ on the right side.

Exercise 2.2.2.30. Prove that for every connected graph $G = (V, E)$,

$$|E| \geq |V| - 1.$$

□

Exercise 2.2.2.31. Prove that a connected graph G contains an Eulerian tour if and only if the degree of all vertices of G is even. □

Definition 2.2.2.32. *A* **cut** *of a graph* $G = (V, E)$ *is a triple* (V_1, V_2, E'), *where*

(i) $V_1 \cup V_2 = V$, $V_1 \neq \emptyset$, $V_2 \neq \emptyset$, *and* $V_1 \cap V_2 = \emptyset$, *and*
(ii) $E' = E \cap \{\{u, v\} \mid u \in V_1, v \in V_2\}$.

Obviously, to determine a cut in a given graph $G = (V, E)$ it is sufficient to give (V_1, V_2) or E' only. For instance, $E' = \{\{v_1, v_5\}, \{v_2, v_4\}\}$ determines the cut $(\{v_1, v_3, v_4\}, \{v_2, v_5\}, E')$ of the graph G' in Figure 2.5.

Definition 2.2.2.33. *Let* $G = (V, E)$ *be a graph. A* **matching** *of* G *is any subset* M *of* E *such that if* $\{x, y\}$ *and* $\{u, v\}$ *are two different elements of* M, *then* $\{x, y\} \cap \{u, v\} = \emptyset$. *A matching* M *is called a* **maximal matching** *if, for each edge* $\{r, s\} \in E - M$, $M \cup \{r, s\}$ *is not a matching.*

For instance, $\{\{v_1, v_3\}\}$, $\{\{v_1, v_3\}, \{v_2, v_5\}\}$, and $\{\{v_1, v_5\}, \{v_2, v_4\}\}$ are matchings of G' in Figure 2.5. The last two are maximal matchings in G'.

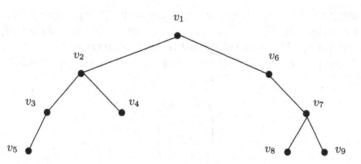

Fig. 2.6.

Definition 2.2.2.34. *A graph* $G = (V, E)$ *is called* **acyclic** *if it does not contain any cycle. An acyclic, connected graph is called a* **tree**. *A* **rooted tree** T *is a tree in which one of the vertices is distinguished from the others. This distinguished vertex is called the* **root** *of the tree.*

Any vertex u *different from the root is called a* **leaf (external vertex)** *of the rooted tree* T *if* $deg_T(u) = 1$. *A vertex* v *of* T *with* $deg_T(v) > 1$ *is called an* **internal vertex.**

The tree $T = (\{v_1, v_2, v_3, v_4, v_5, v_6, v_7, v_8, v_9\}, \{\{v_1, v_2\}, \{v_1, v_6\}, \{v_2, v_3\}, \{v_2, v_4\}, \{v_3, v_5\}, \{v_6, v_7\}, \{v_7, v_8\}, \{v_7, v_9\}\})$ is depicted in Figure 2.6. If v_1 is considered to be the root of T, then v_4, v_5, v_8, and v_9 are the leaves of T. If v_9 is taken as the root of T, then v_4, v_5, and v_8 are the leaves of this rooted tree.

If a tree T is a rooted tree with a root w, then we usually denote the tree T by T_w.

Definition 2.2.2.35. *A* **multigraph** *G is a pair (V, H) where*

(i) V is a finite set called the **set of vertices** *of G, and*
(ii) H is a multiset of elements from $\{\{u, v\} \mid u, v \in V, u \neq v\}$ called the **set of multiedges** *of G.*

An example of a multigraph is $G_1 = (V, H)$, where $V = \{v_1, v_2, v_3, v_4\}$, $H = \{\{v_1, v_2\}, \{v_1, v_2\}, \{v_1, v_2\}, \{v_2, v_3\}, \{v_1, v_3\}, \{v_1, v_3\}, \{v_3, v_4\}\}$. A possible picture representation is given in Figure 2.7.

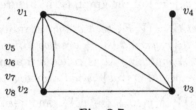

Fig. 2.7.

A multigraph G of n vertices $v_1, v_2, \ldots, v_n$ can be represented by a symmetric $n \times n$ matrix $M_G = [b_{ij}]_{i,j=1,\ldots,n}$ where b_{ij} is the number of edges between v_i and v_j. M_G is called the **adjacency matrix** of G.

The following matrix is the adjacency matrix of the multigraph G_1 in Figure 2.7.

$$\begin{pmatrix} 0 & 3 & 2 & 0 \\ 3 & 0 & 1 & 0 \\ 2 & 1 & 0 & 1 \\ 0 & 0 & 1 & 0 \end{pmatrix}$$

All notions such as degree, path, cycle, connectivity, matching can be defined for multigraphs in the same way as for graphs, and so we omit their formal definitions. For any two graphs or multigraphs $G_1 = (V_1, E_1)$ and $G_2 = (V_1, E_2)$ we say that G_1 is a **subgraph** of G_2 if $V_1 \subseteq V_2$ and $E_1 \subseteq E_2$.

In what follows we define the directed graphs and some fundamental notions connected with them. Informally, a directed graph differs from a graph in having a direction (from one vertex to another vertex) on every edge.

Definition 2.2.2.36. *A* **directed graph** *G is a pair (V, E), where*

(i) V is a finite set called the **set of vertices** *of G, and*
(ii) $E \subseteq (V \times V) - \{(v, v) \mid v \in V\} = \{(u, v) \mid u \neq v, u, v \in V\}$ is the **set of (directed) edges** *of G.*

If $(u, v) \in E$ then we say that (u, v) **leaves** the vertex u and that (u, v) **enters** the vertex v. We also say that (u, v) is **incident** to the vertices u and v.

The directed graph $G_2 = (\{v_1, v_2, v_3, v_4, v_5, v_6\}, \{(v_1, v_2), (v_2, v_1), (v_1, v_3),$ $(v_2, v_4), (v_2, v_5), (v_4, v_2), (v_4, v_5), (v_5, v_3)\})$ is depicted in Figure 2.8. The edge (v_2, v_4) leaves the vertex v_2 and enters the vertex v_4.

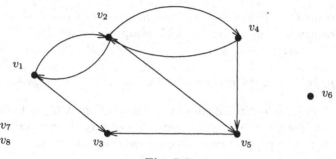

Fig. 2.8.

Again, we can use the notion of an adjacency matrix to represent a directed graph. For any directed graph $G = (V, E)$ of n vertices $v_1, \ldots, v_n$, the **adjacency matrix of G, $M_G = [c_{ij}]_{i,j=1,\ldots,n}$**, is defined by

$$c_{ij} = \begin{cases} 1 \text{ if } (v_i, v_j) \in E \\ 0 \text{ if } (v_i, v_j) \notin E. \end{cases}$$

The following matrix is the adjacency matrix of the directed graph G_2 depicted in Figure 2.8.

$$M_{G_2} = \begin{pmatrix} 0 & 1 & 1 & 0 & 0 & 0 \\ 1 & 0 & 0 & 1 & 1 & 0 \\ 0 & 0 & 0 & 0 & 0 & 0 \\ 0 & 1 & 0 & 0 & 1 & 0 \\ 0 & 0 & 1 & 0 & 0 & 0 \\ 0 & 0 & 0 & 0 & 0 & 0 \end{pmatrix}$$

Exercise 2.2.2.37. How many directed graphs of n vertices $v_1, v_2, \ldots, v_n$ exist? $\qquad \square$

Let $G = (V, E)$ be a directed graph, and let v be a vertex of G. The **indegree of v, $indeg_G(v)$**, is the number of edges of G entering v (i.e., $indeg_G(v) = |E \cap (V \times \{v\})|$). The **outdegree of v, $outdeg_G(v)$**, is the number of edges of G leaving v (i.e., $outdeg_G(v) = |E \cap (\{v\} \times V)|$). For instance, $indeg_{G_2}(v_5) = 2$, $outdeg_{G_2}(v_5) = 1$, and $outdeg_{G_2}(v_2) = 3$. The **degree of v** of G is

$$deg_G(v) = indeg_G(v) + outdeg_G(v).$$

For instance, $deg_{G_2}(v_2) = 5$ and $deg_G(v_6) = 0$. A vertex u with $deg_G(u) = 0$ is called an **isolated vertex** of G. The **degree of a directed graph $G = (V, E)$** is

$$deg(G) = \max\{deg_G(v) \mid v \in V\}.$$

Definition 2.2.2.38. *Let $G = (V, E)$ be a directed graph. A* **(directed) path** *in G is a sequence of vertices $P = v_1, v_2, \ldots, v_m$, $v_i \in V$ for $i = 1, \ldots, m$, such that $(v_i, v_{i+1}) \in E$ for $i = 1, \ldots, m - 1$. We say that P is a path from v_1 to v_m. The* **length of P** *is $m - 1$, i.e., the number of its edges.*

A directed path $P = v_1, \ldots, v_m$ is called **simple** *if either $|\{v_1, \ldots, v_m\}| = m$ or $(v_1 = v_m$, and $|\{v_1, \ldots, v_{m-1}\}| = m - 1)$. A path $P = v_1, \ldots, v_m$ is called a* **cycle** *if $v_1 = v_m$. A cycle $v_1, \ldots, v_m$ is* **simple** *if $|\{v_1, \ldots, v_{m-1}\}| = m - 1$. A directed graph G is called* **acyclic** *if it does not contain any cycle.*

$P = v_1, v_2, v_1, v_3$ is a directed path in the directed graph G_2 in Figure 2.8. v_1, v_2, v_1 is a simple cycle. A directed graph is **strongly connected** if for all vertices $u, v \in V$, $u \neq v$, there are directed paths from u to v and from v to u in G. The graph G_2 is not strongly connected.

Numerous real situations can be represented as graphs. Sometimes we need a more powerful representation formalism called weighted graphs.

Definition 2.2.2.39. *A* **weighted [directed] graph** *G is a triple $(V, E, weight)$, where*

(i) (V, E) is a [directed] graph, and
(ii) weight is a function from E to $\mathbb{Q}^+$.

The **adjacency matrix** *of a weighted graph $G = (V, E, weight)$ is $M_G = [a_{ij}]_{i,j=1,\ldots,|V|}$, where $a_{ij} = weight(\{v_i, v_j\})$ if $\{v_i, v_j\} \in E$ and $a_{ij} = 0$ if $\{v_i, v_j\} \notin E$.*

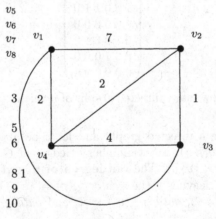

Fig. 2.9.

The weighted graph

$$G = (\{v_1, v_2, v_3, v_4\}, \{\{v_i, v_j\} \mid i, j \in \{1, \ldots, 4\}, i \neq j\}, weight),$$

where $weight(\{v_1, v_2\}) = 7$, $weight(\{v_1, v_3\}) = 1$, $weight(\{v_1, v_4\}) = 2$, $weight(\{v_2, v_3\}) = 1$, $weight(\{v_2, v_4\}) = 2$, and $weight(\{v_3, v_4\}) = 4$, is the graph depicted in Figure 2.9. Its adjacency matrix is the following symmetric matrix:

$$\begin{pmatrix} 0 & 7 & 1 & 2 \\ 7 & 0 & 1 & 2 \\ 1 & 1 & 0 & 4 \\ 2 & 2 & 4 & 0 \end{pmatrix}.$$

Observe that G contains a Hamiltonian tour v_1, v_4, v_3, v_2, v_1.

Keywords introduced in Section 2.2.2

permutation, combination, binomial coefficients, polynomial, asymptotic growth of functions, arithmetic series, geometric series, harmonic numbers, telescoping series, graph, adjacency matrix of a graph, Hamiltonian tour, Eulerian tour, bipartite graph, cut, matching, tree, multigraph, directed graph, weighted graph

2.2.3 Boolean Functions and Formulae

The aim of this section is to give elementary fundamentals of Boolean logic and to present some basic representations of Boolean functions such as Boolean formulae and branching programs. For formulae we consider some special normal forms such as disjunctive normal form (DNF) and conjunctive normal form (CNF).

Boolean logic is a fundamental mathematical system built around the two values "**TRUE**" and "**FALSE**". The values TRUE and FALSE are called **Boolean values**. In what follows we use the value 1 for TRUE and the value 0 for FALSE.

One manipulates Boolean values with logical (Boolean) operations. The fundamental ones are negation, conjunction, disjunction, exclusive or, equivalence, and implication. **Negation** is an unary operation denoted by $\neg$ and defined by $\neg(0) = 1$ and $\neg(1) = 0$. Sometimes we use the notation $\bar{0}$ and $\bar{1}$, respectively, instead of $\neg(0)$ and $\neg(1)$, respectively. **Conjunction** is a binary operation denoted by the symbol $\wedge$, and it corresponds to the logical AND, i.e., the result is 1 if and only if both arguments are 1s. **Disjunction** is a binary operation designated with the symbol $\vee$, and it corresponds to the logical OR, i.e., the result is 1 if and only if at least one of the arguments is 1. Corresponding to their logical meaning, they are defined as follows:

$$\begin{array}{ll} 0 \wedge 0 = 0 & 0 \vee 0 = 0 \\ 0 \wedge 1 = 0 & 0 \vee 1 = 1 \\ 1 \wedge 0 = 0 & 1 \vee 0 = 1 \\ 1 \wedge 1 = 1 & 1 \vee 1 = 1 \end{array}$$

Exclusive or is a binary operation designated by the symbol $\oplus$, and **implication** is designated by the symbol $\Rightarrow$. **Equivalence** is designated by $\Leftrightarrow$. They are defined as follows:

$$
\begin{array}{lll}
0 \oplus 0 = 0 & 0 \Rightarrow 0 = 1 & 0 \Leftrightarrow 0 = 1 \\
0 \oplus 1 = 1 & 0 \Rightarrow 1 = 1 & 0 \Leftrightarrow 1 = 0 \\
1 \oplus 0 = 1 & 1 \Rightarrow 0 = 0 & 1 \Leftrightarrow 0 = 0 \\
1 \oplus 1 = 0 & 1 \Rightarrow 1 = 1 & 1 \Leftrightarrow 1 = 1
\end{array}
$$

So, the result of the exclusive or is 1 if either but not both of its operands are 1. The result of the implication is 0 if and only if the first argument is 1 and the second argument is 0, because the logical TRUE must not imply the logical FALSE.

Exercise 2.2.3.1. Determine which of the above Boolean operations are associative and commutative. $\square$

Exercise 2.2.3.2. Prove that for any Boolean value α and β,

(i) $\alpha \vee \beta = \neg(\neg(\alpha) \wedge \neg(\beta))$,
(ii) $\alpha \wedge \beta = \neg(\neg(\alpha) \vee \neg(\beta))$,
(iii) $\alpha \oplus \beta = \neg(\alpha \Leftrightarrow \beta)$,
(iv) $\alpha \Rightarrow \beta = \neg(\alpha) \vee \beta$, and
(v) $\alpha \Leftrightarrow \beta = (\alpha \Rightarrow \beta) \wedge (\beta \Rightarrow \alpha)$.

$\square$

Definition 2.2.3.3. A **Boolean variable** *is any symbol to which one can associate either of the values 0 or 1.*

Let $X = \{x_1, \ldots, x_n\}$ be a set of Boolean variables for some $n \in \mathbb{N}$. A **Boolean function over** X *is any mapping f from $\{0,1\}^n$ to $\{0,1\}$. One denotes f by $f(x_1, x_2, \ldots, x_n)$ if one wants to call attention to the names of its variables.*

Every argument $\alpha \in \{0,1\}^n$ of f can be also viewed as a mapping $\alpha : X \to \{0,1\}$ that assigns a Boolean value to every variable $x \in X$. Because of this we call α an **input assignment** of f.

The simplest possibility to represent a Boolean function of n variables is to list the values of the function for all 2^n possible arguments (input assignments). Figure 2.10 presents the representation of a Boolean function f of three variables x_1, x_2, and x_3.

Definition 2.2.3.4. *Let $f(x_1, \ldots, x_n)$ be a Boolean function over a set of Boolean variables $X = \{x_1, \ldots, x_n\}$. For every argument $\alpha = (\alpha_1, \alpha_2, \ldots, \alpha_n)$ $\in \{0,1\}^n$ (input assignment α with $\alpha(x_i) = \alpha_i$ for $i = 1, 2, \ldots, n$) such that $f(\alpha) = f(\alpha_1, \alpha_2, \ldots, \alpha_n) = 1$ we say that α satisfies f.*

$$N^1(f) = \{\alpha \in \{0,1\}^n \mid f(\alpha) = 1\}$$

x_1 x_2 x_3	$f(x_1, x_2, x_3)$
0 0 0	1
0 0 1	0
0 1 0	0
0 1 1	1
1 0 0	1
1 0 1	0
1 1 0	1
1 1 1	0

Fig. 2.10.

is the set of all input assignments satisfying f.

$$N^0(f) = \{\beta \in \{0,1\}^n \mid f(\beta) = 0\}$$

is the set of all input assignments that do not satisfy f.

We say that f is **satisfiable** if there exists an input assignment satisfying f (i.e., $|N^1(f)| \geq 1$).

The function $f(x_1, x_2, x_3)$ in Figure 2.10 is satisfiable. The input assignment $\alpha : \{x_1, x_2, x_3\} \to \{0,1\}$ with $\alpha(x_1) = \alpha(x_2) = \alpha(x_3) = 0$ (the argument $(0,0,0)$ of f) satisfies $f(x_1, x_2, x_3)$.

$$N^1(f(x_1, x_2, x_3)) = \{(0,0,0), (0,1,1), (1,0,0), (1,1,0)\}, \text{ and}$$
$$N^0(f(x_1, x_2, x_3)) = \{(0,0,1), (0,1,0), (1,0,1), (1,1,1)\}.$$

Thus, we see that another possibility to represent a Boolean function f is to give $N^1(f)$ or $N^0(f)$.

Definition 2.2.3.5. Let $X = \{x_1, \ldots, x_n\}$ be a set of n Boolean variables, $n \in \mathbb{N} - \{0\}$. Let $f(x_1, \ldots, x_n)$ be a Boolean function over X. We say that $f(x_1, \ldots, x_n)$ **essentially depends on the variable** x_i iff there exist $n-1$ Boolean values

$$\alpha_1, \alpha_2, \ldots, \alpha_{i-1}, \alpha_{i+1}, \ldots, \alpha_n \in \{0,1\}$$

such that

$$f(\alpha_1, \alpha_2, \ldots, \alpha_{i-1}, 0, \alpha_{i+1}, \ldots, \alpha_n) \neq f(\alpha_1, \alpha_2, \ldots, \alpha_{i-1}, 1, \alpha_{i+1}, \ldots, \alpha_n).$$

If $f(x_1, \ldots, x_n)$ does not essentially depend on a variable x_j for some $j \in \{1, \ldots, n\}$, then we say the Boolean variable x_j is **dummy for** f.

For instance, the Boolean function $f(x_1, x_2, x_3)$ from Figure 2.10 essentially depends on x_1 because by fixing $x_2 = 1$ and $x_3 = 0$

$$f(0,1,0) = 0 \neq 1 = f(1,1,0).$$

Exercise 2.2.3.6. Determine whether the Boolean function $f(x_1, x_2, x_3)$ from Figure 2.10 depends essentially on x_2 and x_3.   □

Exercise 2.2.3.7. Prove that every Boolean function f with $|N^1(f)| = 1$ essentially depends on all its input variables.   □

The most usual way to describe Boolean functions is the representation via Boolean formulae.

Definition 2.2.3.8. *Let X be a countable set of Boolean variables and let S be a set of unary and binary Boolean operations. The class of* **Boolean formulae over X and S** *(formulae for short) is defined recursively as follows:*

(i) The Boolean values 0 and 1 are Boolean formulae.
(ii) For every Boolean variable $x \in X$, x is a Boolean formula.
(iii) If F is a Boolean formula and φ is a unary Boolean operation from S, then $\varphi(F)$ is a Boolean formula.
(iv) If F_1 and F_2 are Boolean formulae, and $\triangle \in S$ is a binary operation, then $(F_1 \triangle F_2)$ is a Boolean formula.
(v) Only the expressions constructed by using (i), (ii), (iii), and (iv) are Boolean formulae over X and S.

An example of a Boolean formula over the set of Boolean variables $X = \{x_1, x_2, x_3, \ldots\}$ and $S = \{\vee, \wedge, \Leftrightarrow, \Rightarrow\}$ is

$$F = (((x_1 \vee x_2) \wedge x_7) \Rightarrow (x_2 \Leftrightarrow x_3)).$$

Since the operations $\vee, \wedge, \oplus$, and $\Leftrightarrow$ are commutative and associative, we may sometimes omit the parentheses. Thus, for instance, the following expressions $((x_1 \vee x_2) \vee (x_3 \vee x_4))$, $(((x_1 \vee x_2) \vee x_3) \vee x_4)$, and $x_1 \vee x_2 \vee x_3 \vee x_4$ represent the same Boolean formula. We shall also use $\bigvee_{i=1}^{n} x_i$ [$\bigwedge_{i=1}^{n} x_i, \bigoplus_{i=1}^{n} x_i$] instead of $x_1 \vee x_2 \vee \cdots \vee x_n$ [$x_1 \wedge x_2 \wedge \cdots \wedge x_n, x_1 \oplus x_2 \oplus \cdots \oplus x_n$].

Definition 2.2.3.9. *Let S be a set of unary and binary Boolean operations. Let X be a set of Boolean variables, and let F be a formula over X and S. Let α be an input assignment to X. The* **value of F under the input assignment α, $F(\alpha)$,** *is the Boolean value defined as follows:*

(i) $F(\alpha) = \begin{cases} 0 \text{ if } F = 0 \\ 1 \text{ if } F = 1 \end{cases}$
(ii) $F(\alpha) = \alpha(x)$ if $F = x$ for an $x \in X$,
(iii) $F(\alpha) = \varphi(F_1(\alpha))$ if $F = \varphi(F_1)$ for some unary Boolean operation $\varphi \in S$,
(iv) $F(\alpha) = F_1(\alpha) \triangle F_2(\alpha)$ for some binary $\triangle \in S$ if $F = (F_1 \triangle F_2)$ for some formulae F_1 and F_2.

Let f be a Boolean function over X. If, for each input assignment β from X to $\{0, 1\}$, $f(\beta) = F(\beta)$, then we say **F represents f**. Two formulae F_1 and F_2 are **equivalent**, $F_1 = F_2$, if they represent the same Boolean function (i.e., $F_1(\alpha) = F_2(\alpha)$ for every α).

Obviously, every formula represents exactly one Boolean function. But one Boolean function can be represented by infinitely many formulae. For instance, the formulae $x_1 \vee x_2$, $\neg(\neg(x_1) \wedge \neg(x_2))$, $(x_1 \vee x_2 \vee x_1 \vee x_2) \wedge 1$, and $\neg(x_1 \Rightarrow x_2) \vee \neg(x_2 \Rightarrow x_1) \vee \neg(x_2 \Rightarrow \neg(x_1))$ represent the same Boolean function.

In what follows, we shall use the following notation. Let $\alpha \in \{0,1\}$, and let x be a Boolean variable.

$$x^\alpha = \begin{cases} \neg(x) & \text{if } \alpha = 0 \\ x & \text{if } \alpha = 1. \end{cases}$$

A **clause** over $\{x_1, \ldots, x_n\}$ is a formula $x_{i_1}^{\alpha_1} \vee x_{i_2}^{\alpha_2} \vee \cdots \vee x_{i_m}^{\alpha_m}$ for any $(\alpha_1, \ldots, \alpha_m) \in \{0,1\}^m$, and any $\{i_1, i_2, \ldots, i_m\} \subseteq \{1, 2, \ldots, n\}$. Note that m may be different from n. For instance, $x_1 \vee x_3$, $\overline{x}_1 \vee x_2 \vee x_3 \vee \overline{x}_1 \vee \overline{x}_2$ are clauses over $\{x_1, x_2, x_3\}$.

Exercise 2.2.3.10. Prove the following equivalences between formulae:

(i) $x \vee 0 = x$, $x \wedge 1 = x$, $x \oplus 0 = x$, $x \oplus 1 = \overline{x}$,

(ii) $x \wedge (y \oplus z) = (x \wedge y) \oplus (x \wedge z)$,
    $x \wedge (y \vee z) = (x \wedge y) \vee (x \wedge z)$,
    $x \vee (y \wedge z) = (x \vee y) \wedge (x \vee z)$,

(iii) $x \vee x = x \wedge x = x \vee (x \wedge y) = x \wedge (x \vee y) = x$,

(iv) $x \vee \overline{x} = x \oplus \overline{x} = 1$, and

(v) $x \wedge \overline{x} = x \oplus x = 0$.

$\square$

Definition 2.2.3.11. *Let n be a positive integer, and let $X = \{x_1, \ldots, x_n\}$ be a set of Boolean variables. For every $\alpha = (\alpha_1, \alpha_2, \ldots, \alpha_n) \in \{0,1\}^n$ we define the* **minterm over X** *according to α as the Boolean formula*

$$minterm_\alpha(x_1, \ldots, x_n) = x_1^{\alpha_1} \wedge x_2^{\alpha_2} \wedge \cdots \wedge x_n^{\alpha_n},$$

and the **maxterm over X** *according to α as the clause*

$$maxterm_\alpha(x_1, \ldots, x_n) = x_1^{\neg(\alpha_1)} \vee x_2^{\neg(\alpha_2)} \vee \cdots \vee x_n^{\neg(\alpha_n)}.$$

For instance, $minterm_{(0,1,0)}(x_1, x_2, x_3) = x_1^0 \wedge x_2^1 \wedge x_3^0 = \overline{x}_1 \wedge x_2 \wedge \overline{x}_3$ and $maxterm_{(0,1,0)}(x_1, x_2, x_3) = x_1^1 \vee x_2^0 \vee x_3^1 = x_1 \vee \overline{x}_2 \vee x_3$.

Observation 2.2.3.12. For each $\alpha \in \{0,1\}^n$, $minterm_\alpha(x_1, \ldots, x_n)$ takes the Boolean value 1 iff $x_i = \alpha_i$ for $i = 1, \ldots, n$ (i.e., $N^1(minterm_\alpha(x_1, \ldots, x_n))$ $= \{\alpha\}$), and $maxterm_\alpha(x_1, \ldots, x_n)$ takes the Boolean value 0 iff $x_i = \alpha_i$ for $i = 1, \ldots, n$ (i.e., $N^0(maxterm_\alpha(x_1, \ldots, x_n)) = \{\alpha\}$).

Proof. Obviously, for any $\alpha = (\alpha_1, \alpha_2, \ldots, \alpha_n) \in \{0,1\}^n$,

$$minterm_\alpha(\alpha_1, \ldots, \alpha_n) = \alpha_1^{\alpha_1} \wedge \alpha_2^{\alpha_2} \wedge \cdots \wedge \alpha_n^{\alpha_n} = 1$$

because $\beta^\beta = 1$ for every Boolean value β. Since $\beta^\omega = 0$ for all Boolean values $\beta \neq \omega$,

$$minterm_\alpha(\beta_1, \ldots, \beta_n) = \alpha_1^{\beta_1} \wedge \alpha_2^{\beta_2} \wedge \cdots \wedge \alpha_n^{\beta_n} = 0$$

for all vectors $(\alpha_1, \alpha_2, \ldots, \alpha_n) \neq (\beta_1, \beta_2, \ldots, \beta_n)$.

Similarly, for any $\gamma = (\gamma_1, \gamma_2, \ldots, \gamma_n) \in \{0,1\}^n$,

$$maxterm_\gamma(\gamma_1, \gamma_2, \ldots, \gamma_n) = \gamma_1^{\overline{\gamma_1}} \vee \gamma_2^{\overline{\gamma_2}} \vee \cdots \vee \gamma_n^{\overline{\gamma_n}} = 0$$

because $1^0 = 0^1 = 0$ and so $\beta^{\overline\beta} = 0$ for each $\beta \in \{0,1\}$. Let $(\gamma_1, \ldots, \gamma_n)$, $(\beta_1, \ldots, \beta_n) \in \{0,1\}^n$ and $(\gamma_1, \ldots, \gamma_n) \neq (\beta_1, \ldots, \beta_n)$. Then there exists $i \in \{1, \ldots, n\}$ such that $\beta_i \neq \gamma_i$ (i.e., $\beta_i = \overline{\gamma_i}$). Thus,

$$maxterm_\gamma(\beta_1, \ldots, \beta_n) = \beta_1^{\overline{\gamma_1}} \vee \cdots \vee \beta_n^{\overline{\gamma_n}} = \beta_i^{\overline{\gamma_i}} = \beta_i^{\beta_i} = 1.$$

$\square$

Using the notion of *minterm* and *maxterm* one can assign some unique formula in a special normal form to every Boolean function.

Theorem 2.2.3.13. *Let $f : \{0,1\}^n \to \{0,1\}$, $n \in \mathbb{N} - \{0\}$, be a Boolean function over $X = \{x_1, \ldots, x_n\}$. Then*

$$f(x_1, \ldots, x_n) = \bigvee_{\alpha \in N^1(f)} minterm_\alpha(x_1, \ldots, x_n) = \bigvee_{\alpha \in N^1(f)} (x_1^{\alpha_1} \wedge \cdots \wedge x_n^{\alpha_n}).$$

Proof. Let $\beta = (\beta_1, \ldots, \beta_n) \in \{0,1\}^n$ be a vector such that $f(\beta) = 1$, i.e., $\beta \in N^1(f)$. Then $minterm_\beta(\beta_1, \ldots, \beta_n) = 1$ and so

$$\bigvee_{\alpha \in N^1(f)} minterm_\alpha(x_1, \ldots, x_n) = 1.$$

If $\gamma = (\gamma_1, \ldots, \gamma_n) \in \{0,1\}^n$ is a vector such that $f(\gamma) = 0$, then $\gamma \notin N^1(f)$. Following Observation 2.2.3.12 we have $minterm_\alpha(\gamma) = 0$ for each $\alpha \in N^1(f)$, since $\alpha \neq \gamma$. So,

$$\bigvee_{\alpha \in N^1(f)} minterm_\alpha(\gamma_1, \ldots, \gamma_n) = 0.$$

$\square$

The formula $\bigvee_{\alpha \in N^1(f)} (x_1^{\alpha_1} \wedge \cdots \wedge x_n^{\alpha_n})$ is called the **complete disjunctive normal form**, or **complete DNF**, of f. The complete disjunctive normal form is unique for every Boolean function f because it is unambiguously determined by $N^1(f)$.

The complete DNF of the Boolean function $f(x_1, x_2, x_3)$ from Figure 2.10 is

$$(\overline{x}_1 \wedge \overline{x}_2 \wedge \overline{x}_3) \vee (\overline{x}_1 \wedge x_2 \wedge x_3) \vee (x_1 \wedge \overline{x}_2 \wedge \overline{x}_3) \vee (x_1 \wedge x_2 \wedge \overline{x}_3),$$

because $N^1(f) = \{(0,0,0), (0,1,1), (1,0,0), (1,1,0)\}$.

The next assertion can be proved in an analogous way to how Theorem 2.2.3.13 was proved.

Theorem 2.2.3.14. *Let* $f : \{0,1\}^n \to \{0,1\}$, $n \in \mathbb{N} - \{0\}$ *be a Boolean function over* $X = \{x_1, \ldots, x_n\}$. *Then*

$$f(x_1, \ldots, x_n) = \bigwedge_{\alpha \in N^0(f)} maxterm_\alpha(x_1, \ldots, x_n) = \bigwedge_{\alpha \in N^0(f)} \left(x_1^{\overline{\alpha}_1} \vee \cdots \vee x_n^{\overline{\alpha}_n} \right).$$

Exercise 2.2.3.15. Prove Theorem 2.2.3.14. □

The formula $\bigwedge_{\alpha \in N^0(f)} \left(x_1^{\overline{\alpha}_1} \vee \cdots \vee x_n^{\overline{\alpha}_n} \right)$ is called the **complete conjunctive normal form**, or **complete** CNF, of f. The complete CNF of the Boolean function $f(x_1, x_2, x_3)$ from Figure 2.10 is

$$(x_1 \vee x_2 \vee \overline{x}_3) \wedge (x_1 \vee \overline{x}_2 \vee x_3) \wedge (\overline{x}_1 \vee x_2 \vee \overline{x}_3) \wedge (\overline{x}_1 \vee \overline{x}_2 \vee \overline{x}_3)$$

because $N^0(f) = \{(0,0,1), (0,1,0), (1,0,1), (1,1,1)\}$.

Definition 2.2.3.16. *Let* $X = \{x_1, x_2, x_3, \ldots\}$ *be a set of Boolean variables. Let* $\overline{X} = \{\overline{x} \mid x \in X\} = \{\overline{x}_1, \overline{x}_2, \overline{x}_3, \ldots\}$. *A* **literal** *(over* X*) is any element from* $X \cup \overline{X}$.

Any Boolean formula (over X*) consisting of a conjunction of clauses is called to be in* **conjunctive normal form**, CNF.

For any clause $F = x_{i_1}^{\alpha_1} \vee x_{i_2}^{\alpha_2} \vee \cdots \vee x_{i_n}^{\alpha_n}$, $(\alpha_1, \alpha_2, \ldots, \alpha_n) \in \{0,1\}^n$, *the* **size of** F *is* n, *i.e., the number of its literals. For every positive integer* k, *a formula* Φ *is in* **k-conjunctive normal form**, kCNF, *if* Φ *is in* CNF *and every clause of* Φ *has a size of at most* k.

The formula

$$(x_1 \vee \overline{x}_2) \wedge (\overline{x}_1 \vee x_3 \vee \overline{x}_5) \wedge x_2 \wedge (x_4 \vee x_5)$$

is in 3CNF. Every complete CNF of Boolean functions of n variables is in nCNF.

We observe that Theorem 2.2.3.13 (as well as Theorem 2.2.3.14) implies that every Boolean function can be expressed as a formula over the set of Boolean operations $\{\vee, \wedge, \neg\}$.

Exercise 2.2.3.17. Prove that every Boolean function can be represented by a formula over

(i) $\{\neg, \vee\}$,
(ii) $\{\neg, \wedge\}$, and
(iii) $\{\wedge, \oplus\}$. □

Exercise 2.2.3.18. Find a binary Boolean operation (Boolean function of two variables) φ, such that every Boolean function can be represented by a formula over $\{\varphi\}$. □

Branching programs are currently the standard formalism for the computer representation of Boolean functions. This is because this kind of representation is often more concise than the representations by the value table or formulae, and that some special versions of branching programs can be conveniently handled in several application areas.

Definition 2.2.3.19. *Let $X = \{x_1, \ldots, x_n\}$, $n \in \mathbb{N} - \{0\}$, be a set of Boolean variables. A **branching program** (**BP**) over X is a directed acyclic graph $G = (V, E)$ with the following labeling and properties:*

*(i) There is exactly one vertex with indegree 0 in G, and this vertex is called the **source** (or the **start vertex**) of the branching program.*

(ii) Every vertex of a nonzero outdegree in G is labeled by a Boolean variable from X.

*(iii) There are exactly two vertices with outdegree equal to 0. These vertices are called the **sinks** (or **output vertices**) of the branching program. One of the sinks is labeled by 0 and the other one by 1.*

(iv) Every vertex v of a nonzero outdegree has the outdegree equal to two. One of the two edges leaving v is labeled by 1 and the other one is labeled by 0.

For any input assignment $\alpha : X \to \{0, 1\}$, a branching program A over X computes the Boolean value $A(\alpha)$ in the following way:

(i) A starts the computation on α in its source.

(ii) If A is in a vertex labeled by a Boolean variable $x \in X$, then A moves via the edge labeled by $\alpha(x)$ to the next vertex.

(iii) If A reaches a sink, then $A(\alpha)$ is the label of that sink.

Let $f(x_1, \ldots, x_n) : \{0, 1\}^n \to \{0, 1\}$ be a Boolean function. We say that a BP A **represents** (or **computes**) f if $A(\beta) = f(\beta)$ for every input assignment $\beta : \{x_1, \ldots, x_n\} \to \{0, 1\}$.

For instance, the branching program A in Figure 2.11a follows the computation path

$$x_1 \xrightarrow{0} x_2 \xrightarrow{0} x_1 \xrightarrow{0} x_3 \xrightarrow{0} 1$$

for the input $(0, 0, 0)$ and so $A(0, 0, 0) = 1$. For the inputs $(1, 1, 0)$ and $(1, 0, 0)$ A uses the same computation

$$x_1 \xrightarrow{1} x_3 \xrightarrow{0} 1.$$

One can easily observe, that A computes the Boolean function $f(x_1, x_2, x_3)$ given in Figure 2.10.

Every branching program unambiguously determines the Boolean function that it represents. On the other hand, one can represent a Boolean function by different branching programs. The Boolean function $f(x_1, x_2, x_3)$ in Figure 2.10 is represented by both branching programs from Figure 2.11.

Since the general branching programs are not so easy to handle, one usually uses some restricted normal forms of them. We shall consider only one such

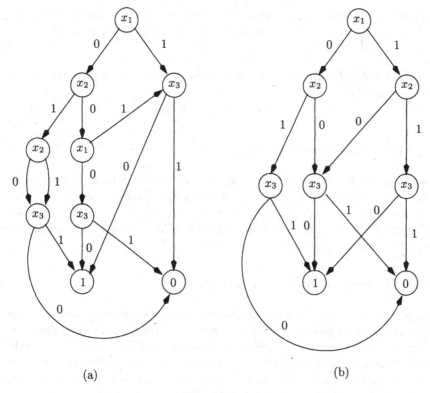

(a)   (b)

Fig. 2.11.

form where the restriction says that every Boolean variable may be asked for its value at most once in any computation of the branching program.

Definition 2.2.3.20. *Let X be a set of Boolean variables, and let A be a branching program over X. We say that A is a **one-time-only branching program** if for every directed path P of A all vertices on P have pairwise different labels.*

We observe that the branching program in Figure 2.11a is not a one-time-only branching program because it asks the Boolean variable x_1 twice on the path

$$x_1 \xrightarrow{\ 0\ } x_2 \xrightarrow{\ 0\ } x_1 \xrightarrow{\ 0\ } x_3.$$

On the other hand, the branching program in Figure 2.11b is a one-time-only one.

Exercise 2.2.3.21. Construct

(a) a branching program with the minimal number of vertices for the Boolean function $f(x_1, x_2, x_3)$ from Figure 2.10.

(b) a minimal (according to the number of vertices) one-time-only branching program representing the function $f(x_1, x_2, x_3)$ from Figure 2.10.

□

Exercise 2.2.3.22. Construct a (one-time-only) branching program for the Boolean function

$$x_1 \oplus x_2 \oplus x_3 \oplus \cdots \oplus x_n.$$

□

Keywords introduced in Section 2.2.3

Boolean values, Boolean operations, negation, conjunction, disjunction, implication, exclusive or, equivalence, Boolean variable, Boolean function, input assignment, satisfiability, Boolean formula, clause, minterm, maxterm, complete DNF, complete CNF, literal, conjunctive normal form (CNF), kCNF, branching program, one-time-only branching program

2.2.4 Algebra and Number Theory

The goal of this section is to recapitulate the definitions of some basic algebraic structures such as groups, semigroups, rings, and fields, and some fundamental results of number theory such as the Fundamental Theorem of Arithmetic, Prime Number Theorem, Fermat's Theorem, and Chinese Remainder Theorem. The proofs of all these results, except for Prime Number Theorem, can be expressed by elementary mathematics and so we present them. The only algorithmic part of this section is devoted to the presentation of Euclid's algorithm for the greatest common divisor. In fact, we are mainly interested in the algebraic structure $\mathbb{Z}_n$ with the set of elements $\{0, 1, 2, \ldots, n-1\}$ and the operations of multiplication and addition modulo n. An important observation is the fact that $\mathbb{Z}_n$ is a field if and only if n is a prime. This and the fundamental theorems mentioned above are very useful for designing efficient randomized algorithms for algorithmic problems from number theory for which no polynomial-time deterministic algorithm is known. These are the most famous and transparent examples certificating the power and the usefulness of randomization in algorithmics. The knowledge of this section is crucial only for some parts of Chapter 5 on randomized algorithms, and so the reader may skip this section for now and look at it immediately before reading the related sections of Chapter 5.

A fundamental mathematical structure, also called algebraic structure or **algebra** for short, is a pair (S, F), where

(i) S is a set of elements, and
(ii) F is a set of functions that map arguments from S to an element of S. More precisely, F is a set of **operations** on S, i.e., for every $f \in F$ there exists an integer m, such that f is a function from S^m into S. If $f : S^m \to S$, then we say that f is an **m-ary operation on S**.

We are not interested in structures with a mapping that for arguments from S produces an element outside S. For our purposes we shall consider only the following fundamental sets of elements (numbers) and some fundamental binary operations corresponding to addition and multiplication:

$\mathbb{N}$	$= \{0, 1, 2, \ldots\}$ the set of all non-negative integers
$\mathbb{Z}$	the set of all integers
$\mathbb{N}^+$	the set of all positive integers
$\mathbb{N}^{\geq k}$	$= \mathbb{N} - \{0, 1, \ldots, k-1\}$ for every positive integer k
$\mathbb{Z}_n$	$= \{0, 1, 2, \ldots, n-1\}$
$\mathbb{Q}$	the set of rational numbers
$\mathbb{Q}^+$	the set of positive rational numbers
$\mathbb{R}$	the set of real numbers
$\mathbb{R}^+$	the set of positive real numbers
$\mathbb{R}^{\geq 0}$	the set of non-negative real numbers

If $F = \{f_1, \ldots, f_k\}$ is a finite set of operations of an algebra (S, F), we simply write $(S, f_1, f_2, \ldots, f_k)$ instead of $(S, \{f_1, f_2, \ldots, f_k\})$ in what follows. For some binary operations f and g we use the notation $\cdot$ and $+$ if f should be interpreted as a "version" of multiplication and if g should be interpreted as a "version" of addition[7] in the algebra considered. Thus, instead of $f(x, y)$ $[g(x, y)]$ we simply write $x \cdot y$ $[x + y]$.

Definition 2.2.4.1. *A **group** is an algebra $(S, *)$, where*

(i) $*$ is a binary operation,

(ii) $*$ is associative, i.e., $\forall x, y, z \in S : (x * y) * z = x * (y * z)$,

*(iii) there exists an $e \in S$ (called the **neutral element according to $*$ in** S), such that $\forall x \in S$:*

$$e * x = x = x * e.$$

(iv) $\forall x \in S$ there exists an element $i(x) \in S$ such that

$$i(x) * x = e = x * i(x),$$

*where $i(x)$ is called the **inverse element of x** according to $*$.*

*A group is said to be **commutative** if $x * y = y * x$ for all $x, y \in S$.*

In what follows, if $*$ is considered to be multiplication, then the neutral element is denoted by 1. If $*$ is considered to be addition, then the neutral element is denoted by 0.

Example 2.2.4.2. $(\mathbb{Z}, +)$ with $+$ as the standard addition of $(\mathbb{Z}, +)$ is a commutative group. The neutral element is 0, and $i(x) = -x$ for every $x \in \mathbb{Z}$. $\qquad \square$

---

[7] Note that this interpretation does not necessarily mean that the considered operations $\cdot$ and $+$ are commutative.

Exercise 2.2.4.3. Prove that we cannot build a group on $\mathbb{N}$ by using the standard addition $+$ on $\mathbb{N}$. □

As already mentioned above we are mainly interested in $\mathbb{Z}_n = \{0, 1, \ldots, n-1\}$. To define a "natural" addition and multiplication on $\mathbb{Z}_n$ one needs the notion of "divisibility" and "remainder of the division". Let a, d be non-negative integers, and let k be a positive integer. If $a = k \cdot d$, then we say that **d divides a** or that a is a **multiple** of d, and write **$d|a$**. By agreement, every positive integer divides 0. Obviously, if a and d are positive integers and $d|a$, then $d \leq a$.

If $d|a$ we also say that d is a **divisor of** a. For every positive integer a the integers 1 and a are so-called **trivial divisors of a**. Nontrivial divisors of a are called **factors of a**. For instance, the factors of 24 are 2, 3, 4, 6, 8, 12.

The definition of the operations on $\mathbb{Z}_n$ is based on the elementary division of a positive integer a by a positive integer b for $b < a$. Using the school algorithm for division one can write

$$a = q \cdot b + r, \tag{2.9}$$

where q is the result of the division procedure and $r < b$ is the remainder of the division. One can easily prove that for every integer a and every positive integer b, there are unique integers q and r such that the equality (2.9) is true. In what follows we use the notation **a div b** for q and **a mod b** for r. For instance, 21 div $5 = 4$ and 21 mod $5 = 1$. For every positive integer n, we define the operations $\oplus_n$ and $\odot_n$ on $\mathbb{Z}_n$ as follows:

For all $x, y \in \mathbb{Z}_n$:

$$x \oplus_n y = (x + y) \bmod n,$$
$$x \odot_n y = (x \cdot y) \bmod n.$$

For instance, $7 \oplus_{13} 10 = (7 + 10) \bmod 13 = 17 \bmod 13 = 4$, $6 \odot_{11} 7 = (6 \cdot 7) \bmod 11 = 42 \bmod 11 = 9$.

Since the remainder of the division by n is smaller than n, $x \oplus_n y$, $x \odot_n y \in \mathbb{Z}_n$ for all $x, y \in \mathbb{Z}_n$. So, $\odot_n$ and $\oplus_n$ are binary operations on $\mathbb{Z}_n$.

Example 2.2.4.4. Consider $\mathbb{Z}_n = \{0, 1, \ldots, n-1\}$ and the operation $\oplus_n$ as defined above:

Then 0 can be considered as the neutral element according to $\oplus_n$. For every $x \in \mathbb{Z}_n$, define $i(x) = (n - x) \bmod n$. We observe that, for every x,

$$x \oplus_n i(x) = (x + (n - x) \bmod n) \bmod n = n \bmod n = 0.$$

Since $\oplus_n$ is associative and commutative, $(\mathbb{Z}_n, \oplus_n)$ with the neutral element 0 is a commutative group. □

The crucial point is that if one has a structure (algebra) like a group $(S, +)$, where there exists an inverse element $i(x)$ according to addition for every $x \in S$, then one also has the **subtraction** in this structure.[8] One can define it as

$$x - y = x + i(y).$$

In what follows we shall also use the notation $-y$ for $i(y)$, where $i(y)$ is the inverse of y according to addition.

A similar situation appears if one considers a group $(S, \cdot)$, where $\cdot$ stands for multiplication. If $i(x)$ is the inverse of x according to $\cdot$, then one also has **division** in the group. Division is defined by

$$x/y = x \cdot i(y),$$

and we often use the notation y^{-1} instead of $i(y)$.

If one considers the set $\mathbb{Z}$ and the standard multiplication operation one can easily observe that it is impossible to build a group. The neutral element according to multiplication is 1 and for no $x \in \mathbb{Z}$, $x \neq 1$, are we able to find an element y such that $x \cdot y = 1$. Thus, we do not have inverse elements according to the multiplication in $\mathbb{Z}$, and we cannot define division in such a structure.

In what follows we also use the abbreviation a^i inductively defined for every $a \in S$ and every $i \in \mathbb{Z}$ in a group $(S, *)$ with the neutral element e as follows:

(i) $a^0 = e$, $a^1 = a$, and $a^{-1} = i(a)$, where a^0 is called the **trivial power of a**;

(ii) $a^{i+1} = a * a^i$ for every positive integer i, and a^i is called a **nontrivial power of a** if $i \geq 1$;

(iii) $a^{-j} = (i(a))^j$ for every positive integer j.

An element $g \in S$ is called a **generator** of a group $(S, *)$ if $S = \{g^i \mid i \in \mathbb{Z}\}$. If a group has a generator, then the group is called **cyclic**. For instance, $(\mathbb{Z}_n, \oplus)$ is a cyclic group for every $n \in \mathbb{N}^+$ because $1^i = i$ for every $i \in \{1, 2, \ldots, n-1\}$, and $1^0 = 0$, i.e., 1 is a generator of $(\mathbb{Z}_n, \oplus_n)$. Observe that 2 is a generator of $(\mathbb{Z}_n, \oplus_n)$ if and only if n is an odd integer.

Definition 2.2.4.5. *A **semigroup** is any algebra $(S, *)$, where $*$ is an associative, binary operation on S.*

*A **monoid** is an algebra $(M, *)$ where*

(i) $*$ is a binary associative operation, and
(ii) there exists an $e \in S$ such that $\forall x \in S : e * x = x = x * e$.

*A **monoid** is commutative if $\forall x, y \in M$, $x * y = y * x$.*

---

[8] Observe that subtraction considered as a function from S^2 to S is not associative. This is the reason why one does not use the notion "operation" for it. The same holds for division.

Example 2.2.4.6. $(\mathbb{N}, +)$ with the neutral element 0, and $(\mathbb{Z}, \cdot)$ with the neutral element 1 are commutative monoids. $(\Sigma^*, \cdot)$, where Σ is an alphabet (a finite set of symbols), Σ^* is the set of all finite sequences over elements of Σ, and $\cdot$ is the concatenation of two sequences is a (noncommutative) monoid. The neutral element of $(\Sigma^*, \cdot)$ is the empty sequence λ. $\square$

Now, we want to consider algebras that contain both addition and multiplication.

Definition 2.2.4.7. *A* **ring** *is an algebra* $(R, +, \cdot)$ *such that*

(i) $(R, +)$ *is a commutative group,*
(ii) $(R, \cdot)$ *is a semigroup, and*
(iii) addition and multiplication are related by the **distributive laws:**

$$x \cdot (y + z) = (x \cdot y) + (x \cdot z)$$
$$(x + y) \cdot z = (x \cdot z) + (y \cdot z)$$

A **ring** $(R, +, \cdot)$ *with neutral element* 0 *according to* + *is called* **zero division free** *if for* $\forall x, y \in R - \{0\}$, $x \cdot y \neq 0$.

Example 2.2.4.8. Each of the sets $\mathbb{Z}$, $\mathbb{Q}$, and $\mathbb{R}$ with standard addition and multiplication is a zero division free ring. $\square$

Because of (i) of Definition 2.2.4.7 we see that subtraction is well defined in every ring. But in general, division is not available for rings. To obtain the possibility to divide in rings one has to add some additional requirements.

Definition 2.2.4.9. *A* **field** *is an algebra* $(R, +, \cdot)$ *satisfying the following conditions:*

(i) $(R, +, \cdot)$ *is a zero division free ring, with the neutral element* 0 *according to* +,
(ii) $a \cdot b = b \cdot a$ *for all* $a, b \in R$,
(iii) there exists an element $1 \in R$ *such that* $1 \cdot a = a \cdot 1 = a$ *for all* $a \in R - \{0\}$,
(iv) for every $b \in R - \{0\}$ *there exists* $i(b) \in R$ *such that* $b \cdot i(b) = 1$.

$\mathbb{Q}$ and $\mathbb{R}$, with respect to addition and multiplication, build fields, but for $\mathbb{Z}$ it is impossible to define division. In what follows we observe that $\mathbb{Z}_n$ can be used to build a field for some ns, but not for every n. The reason for this behavior will be explained later.

Example 2.2.4.10. Consider $\mathbb{Z}_n$ for every positive integer n, with the operations $\oplus_n$ and $\odot_n$. As we already observed, $(\mathbb{Z}_n, \oplus_n)$ is a commutative group. One can easily verify that the distributive laws for $\oplus_n$ and $\odot_n$ hold, too. Obviously, $(\mathbb{Z}_n, \odot_n)$ is a semigroup. Thus, the only properties in question are the zero divisor freeness and the existence of inverse elements according to $\odot_n$.

If $n = 12$, then $3 \odot_{12} 4 = 3 \cdot 4 \bmod 12 = 0$. Since $3 \neq 0$ and $4 \neq 0$ the ring $(\mathbb{Z}_{12}, \oplus_{12}, \odot_{12})$ is not zero division free. Obviously, this is true for every n that can be written as $a \cdot b$, $a, b \in \{2, 3, \ldots, n-1\}$.

Now, let us look for the inverse elements of the elements of $\mathbb{Z}_{12}$ according to $\odot_{12}$. Since $1 \odot_{12} 1 = 1 \cdot 1 \bmod 12 = 1$, 1 can always be considered to be the inverse of 1, i.e., $1^{-1} = 1/1 = 1$ in $\mathbb{Z}_{12}$. Next, consider the element 2. Since $2 \odot_{12} a$ is even for every $a \in \mathbb{Z}_n$, $2 \cdot a \bmod 12$ is even too. Thus, for every $a \in \mathbb{Z}_{12}$, $2 \odot a \neq 1$. The conclusion is that there is no inverse element of 2 according to $\odot_{12}$.

Now we consider $\mathbb{Z}_5$. Realizing all 16 multiplications of nonzero elements, one can see that $(\mathbb{Z}_5, \oplus_5, \odot_5)$ is zero divisor free. In what follows we see the every nonzero element has its inverse according to $\odot_5$.

$1 \odot_5 1 = 1 \cdot 1 \bmod 5 = 1$, i.e., $1^{-1} = 1$

$2 \odot_5 3 = 2 \cdot 3 \bmod 5 = 1$, i.e., $2^{-1} = 3$ and $3^{-1} = 2$ because $2 \odot_5 3 = 3 \odot_5 2$

$4 \odot_5 4 = 4 \cdot 4 \bmod 5 = 16 \bmod 5 = 1$, i.e., $4^{-1} = 4$.

Thus, $(\mathbb{Z}_5, \oplus_5, \odot_5)$ is a field. $\qquad\qquad\qquad\qquad\qquad\qquad\qquad\square$

In what follows, we usually omit the full description of algebras as tuples. For instance, we shall speak shortly about a field $\mathbb{Q}$, $\mathbb{Z}_5$, or $\mathbb{R}$ assuming that the reader automatically considers the standard operations connected with them.

Next, we shall deal with integers, and especially with primes.

Definition 2.2.4.11. *A* **prime** *is an integer $p > 1$, which has no factors (divisors[9] other than itself and one).*

An integer $b > 1$ that is not a prime (i.e., $b = a \cdot c$ for some $a, c > 1$) is called **composite**.

The smallest primes are the numbers $2, 3, 5, 7, 11, 13, 17, 19, 23, \ldots$. The importance of the class of primes is that every positive integer greater that 1 can be expressed as a product of primes (if a number is not itself a prime, it may be successively factorized until all the factors are primes).

Observation 2.2.4.12. For every integer $a \in \mathbb{N}^{\geq 2}$, there exists an integer $k \geq 1$, primes $p_1, p_2, \ldots, p_k$, and $i_1, i_2, \ldots, i_k \in \mathbb{N}^{\geq 1}$ such that

$$a = p_1^{i_1} \cdot p_2^{i_2} \cdot \cdots \cdot p_k^{i_k}.$$

For instance, $720 = 5 \cdot 144 = 5 \cdot 2 \cdot 72 = 5 \cdot 2 \cdot 2 \cdot 36 = 5 \cdot 2 \cdot 2 \cdot 3 \cdot 12 = 5 \cdot 2 \cdot 2 \cdot 3 \cdot 3 \cdot 4 = 5 \cdot 2 \cdot 2 \cdot 3 \cdot 3 \cdot 2 \cdot 2 = 2^4 \cdot 3^2 \cdot 5$.

---

[9] Remember that an integer $a \neq 1$ is said to be a **factor** (or a **nontrivial divisor**) of an integer b if there is some integer $c \neq 1$ such that $b = a \cdot c$ (i.e., $b \bmod a = 0$).

One of the first questions that arises concerning the class of primes is whether there are infinite many different primes or whether the cardinality of the class of all primes is an integer. The following assertion provides the answer to this question.

Theorem 2.2.4.13. *There are infinitely many different primes.*

Proof. We present the old Euclid's proof by the indirect method. Assume there are only finitely many different primes, say $p_1, p_2, \ldots, p_n$. Any other number is composite, and must be divisible by at least one of the primes $p_1, p_2, \ldots, p_n$. Consider the number

$$a = p_1 \cdot p_2 \cdot \cdots \cdot p_n + 1.$$

Since $a > p_i$ for every $i \in \{1, 2, \ldots, n\}$, a must be composite, so, there exists an $i \in \{1, 2, \ldots, n\}$ such that p_i is a factor of a. But $a \bmod p_j = 1$ for every $j \in \{1, 2, \ldots, n\}$, so, our initial assumption that there is only a finite number of primes leads to this contradiction, and hence its contrary must be true.   □

Observe that the proof of Theorem 2.2.4.13 does not provide any method for constructing an arbitrary large sequence of primes. This is because $a = p_1 \cdot p_2 \cdot \cdots \cdot p_n + 1$ does not need to be a prime (for instance, $2 \cdot 3 \cdot 5 \cdot 7 \cdot 11 \cdot 13 + 1 = 30031 = 59 \cdot 509$ is a composite number), and if a is a prime, then it does not need to be the $(n+1)$th prime ($2 \cdot 3 + 1 = 7$ but 7 is the 4th prime number, not the 3rd one). The only right conclusion of the proof of Theorem 2.2.4.13 is that there must exist a prime p greater than p_n and smaller than or equal to a. Therefore, one can search in the interval $(p_n, a]$ for the $(n+1)$th smallest prime.[10]

As already mentioned above the primes are important because each integer can be expressed as a product of primes. The Fundamental Theorem of Arithmetics claims that every integer $a > 1$ can be factorized into primes in only one way.

Theorem 2.2.4.14 (The Fundamental Theorem of Arithmetics). *For every integer $a \in \mathbb{N}^{\geq 2}$, a can be expressed as a product of nontrivial powers of distinct primes[11]*

$$a = p_1^{i_1} \cdot p_2^{i_2} \cdot \cdots \cdot p_k^{i_k}$$

and up to a rearrangement of the factors, this prime factorization is unique.

Proof. The fact that a can be expressed as a product of primes has been formulated in Observation 2.2.4.12. We have to prove that the decomposition of a into a product of primes is unique. We again give an indirect method.

Let m be the smallest integer from $\mathbb{N}^{\geq 2}$ such that

$$m = p_1 \cdot p_2 \cdot \cdots \cdot p_r = q_1 \cdot q_2 \cdot \cdots \cdot q_s, \tag{2.10}$$

---

[10] Note that up till now no simple (efficiently computable) formula yielding an infinite sequence of primes is known.

[11] $p_i \neq p_j$ for $i \neq j$.

where $p_1 \leq p_2 \leq \cdots \leq p_r$ and $q_1 \leq q_2 \leq \cdots \leq q_s$ are primes such that there exists $i \in \{1, \ldots, r\}$ such that $p_i \notin \{q_1, q_2, \ldots, q_s\}$.[12]

First, we observe that p_1 cannot be equal to q_1 (If $p_1 = q_1$, then $m/p_1 = p_2 \cdot \cdots \cdot p_r = q_2 \cdot \cdots \cdot q_s$ is an integer with two different prime factorizations. Since $m/p_1 < m$ this is the contradiction with the minimality of m).

Without loss of generality we assume $p_1 < q_1$. Consider the integer

$$m' = m - (p_1 q_2 q_3 \ldots q_s). \tag{2.11}$$

By substituting for m the two expressions of (2.10) into (2.11) we may write the integer m' in either of the two forms:

$$m' = (p_1 p_2 \ldots p_r) - (p_1 q_2 \ldots q_s) = p_1 (p_2 p_3 \ldots p_r - q_2 q_3 \ldots q_s) \tag{2.12}$$
$$m' = (q_1 q_2 \ldots q_s) - (p_1 q_2 \ldots q_s) = (q_1 - p_1)(q_2 q_3 \ldots q_s). \tag{2.13}$$

Since $p_1 < q_1$, it follows from (2.13) that m' is a positive integer, while (2.11) implies that $m' < m$. Hence, the prime decomposition of m' must be unique.[13] Equality (2.12) implies that p_1 is a factor of m'. So, following (2.12) and the fact that (2.13) is a unique decomposition of m', p_1 is a factor of either $(q_1 - p_1)$ or $(q_2 q_3 \ldots q_s)$. The latter is impossible because $q_s \geq q_{s-1} \geq \cdots \geq q_3 \geq q_2 \geq q_1 > p_1$ and $q_1, \ldots, q_s$ are primes. Hence p_1 must be a factor of $q_1 - p_1$, i.e., there exists an integer a such that

$$q_1 - p_1 = p_1 \cdot a. \tag{2.14}$$

But (2.14) implies $q_1 = p_1(a+1)$ and so p_1 is a factor of q_1. This is impossible because $p_1 < q_1$ and q_1 is a prime.   □

Corollary 2.2.4.15. *If a prime p is a factor of an integer $a \cdot b$, then p must be a factor of either a or b.*

Proof. Assume that p is a factor of neither a nor b. Then the product of the prime decomposition of a and b does not contain p. Since p is a factor of $a \cdot b$, there exists an integer t such that $a \cdot b = p \cdot t$. The product of p and the prime decomposition of t yields a prime decomposition of $a \cdot b$ that contains p. Thus, we have two different prime decompositions of $a \cdot b$ which contradicts the Fundamental Theorem of Arithmetics.   □

A large effort in number theory has been devoted to finding a simple mathematical formula yielding the sequence of all primes. Unfortunately, these attempts have not succeeded up till now and it is questionable whether such a formula exists. On the other hand, another important question:

"How many primes are contained in $\{1, 2, \ldots, n\}$?"

---

[12] So, the case $r = s = 1$ cannot happen because $m = p_1 = q_1$ contradicts $p_1 \notin \{q_1\}$.
[13] Aside from the order of the factors.

has been (at least) approximately answered. The next deep result is one of the most fundamental assertions of number theory and it has numerous applications (see Section 5.2 for some applications in algorithmics). The difficulty of the proof of this assertion is beyond the elementary level of this chapter and we omit it here.

Let $Prim(n)$ denote the number of primes among integers $1, 2, 3, \ldots, n$.

Theorem 2.2.4.16 (Prime Number Theorem).

$$\lim_{n \to \infty} \frac{Prim(n)}{n/\ln n} = 1.$$

$\square$

In other words the Prime Number Theorem says that the density $\frac{Prim(n)}{n}$ of the primes among the first n integers tends to $\frac{1}{\ln n}$ as n increases. The following table shows that $\frac{1}{\ln n}$ is a good "approximation" of $\frac{Prim(n)}{n}$ already for "small" n (for which we are able to compute $Prim(n)$ exactly with a computer).

n	$\frac{Prim(n)}{n}$	$\frac{1}{\ln n}$	$\frac{Prim(n)}{n/\ln n}$
10^3	0.168	0.145	1.159
10^6	0.0885	0.0724	1.084
10^9	0.0508	0.0483	1.053

Note that for $n \geq 100$ one can prove that

$$1 \leq \frac{Prim(n)}{n/\ln n} \leq 1.23,$$

and this is already very useful in the design of randomized algorithms.

In what follows we use the Gauss' congruence notation

$$a \equiv b \pmod{d}$$

for the fact

$$a \bmod d = b \bmod d,$$

and say that **a and b are congruent modulo d**. If a is not congruent to b modulo d, we shall write

$$a \not\equiv b \pmod{d}.$$

Exercise 2.2.4.17. Prove that, for all positive integer $a \geq b$, the following statements are equivalent:

(i) $a \equiv b \pmod{d}$,
(ii) $a = b + nd$ for some integer n,
(iii) d divides $a - b$.

$\square$

Definition 2.2.4.18. *For all integers a and b, $a \neq 0$ or $b \neq 0$, the* **greatest common divisor of a and b,** $gcd(a, b)$, *and the* **lowest common multiple of a and b,** $lcm(a, b)$, *are defined as follows:*

$$gcd(a, b) = \max\{d \mid d \text{ divides both } a \text{ and } b, \text{ i.e., } a \equiv b \equiv 0 (\text{mod } d)\}$$

and

$$lcm(a, b) = \frac{a \cdot b}{gcd(a, b)} = \min\{c \mid a|c \text{ and } b|c\}.$$

By convention, $gcd(0, 0) = lcm(0, 0) = 0$. We say that a and b are **co-primes** *if $gcd(a, b) = 1$, i.e., if a and b have no common factor.*

For instance, the greatest common divisor of 24 and 30 is 6. It is clear that if one has the prime factorizations of two numbers $a = p_1 \cdots \cdot p_r \cdot q_1 \cdots \cdot q_s$ and $b = p_1 \cdots \cdot p_r \cdot h_1 \cdots \cdot h_m$, where $\{q_1, \ldots, q_s\} \cap \{h_1, \ldots, h_m\} = \emptyset$, then $gcd(a, b) = p_1 \cdot p_2 \cdots \cdot p_r$. For instance, $60 = 2 \cdot 2 \cdot 3 \cdot 5$, $24 = 2 \cdot 2 \cdot 2 \cdot 3$, and so $gcd(60, 24) = 2 \cdot 2 \cdot 3 = 12$. Unfortunately, we do not know how to efficiently[14] compute the prime factorization of a given integer, and so this idea is not helpful if one wants to compute $gcd(a, b)$ for two given integers a, b. A very efficient computation of $gcd(a, b)$ is provided by the well-known Euclid's algorithm.[15] To explain it we present some simple assertions. First, we give an important property of common divisors.

Lemma 2.2.4.19. *Let $d > 0$, a, b be some integers.*

(i) If $d|a$ and $d|b$ then

$$d|(ax + by)$$

for any integers x and y.
(ii) For any two positive integers a, b, $a|b$ and $b|a$ imply $a = b$.

Proof. (i) If $d|a$ and $d|b$, then $a = n \cdot d$ and $b = m \cdot d$ for some integers n and m. So,

$$ax + by = n \cdot d \cdot x + m \cdot d \cdot y = d(n \cdot x + m \cdot y).$$

(ii) $a|b$ implies $a \leq b$ and $b|a$ implies $b \leq a$. So, $a = b$. □

The following properties of gcd are obvious.

Observation 2.2.4.20. *For all positive integers a, b, and k,*

(i) $gcd(a, b) = gcd(b, a)$,
(ii) $gcd(a, 0) = a$,
(iii) $gcd(a, ka) = a$ for every integer k,
(iv) if $k|a$ and $k|b$, then $k|gcd(a, b)$.

---

[14] There is no known polynomial-time algorithm for this task.
[15] Euclid's algorithm is one of the oldest algorithms (circa 300 B.C.).

Exercise 2.2.4.21. Prove that the *gcd* operator is associative, i.e., that for all integers a, b, and c,

$$gcd(a, gcd(b, c)) = gcd(gcd(a, b), c).$$

$\square$

Euclid's algorithm is based on the following recursive property of *gcd*.

Theorem 2.2.4.22. *For any non-negative integer a and any positive integer b,*

$$gcd(a, b) = gcd(b, a \bmod b).$$

Proof. We shall prove that $gcd(a, b)$ and $gcd(b, a \bmod b)$ divide each other, and so by the fact (ii) of Lemma 2.2.4.19 they must be equal.

(1) We first prove that $gcd(a, b)$ divides $gcd(b, a \bmod b)$. Following the definition of *gcd*, $gcd(a, b)|a$ and $gcd(a, b)|b$. We know (cf. (2.9)) that we can write a as

$$a = (a \text{ div } b) \cdot b + a \bmod b$$

for some non-negative integer a div b. Thus

$$a \bmod b = a - (a \text{ div } b) \cdot b.$$

This means that $a \bmod b$ is a linear combination of a and b and so every common divisor of a and b must divide $a \bmod b$ due to the fact (i) of Lemma 2.2.4.19. Thus,

$$gcd(a, b)|(a \bmod b).$$

Finally, due to the fact (iv) of Observation 2.2.4.20

$$gcd(a, b)|b \text{ and } gcd(a, b)|(a \bmod b)$$

imply

$$gcd(a, b)|gcd(b, a \bmod b).$$

(2) We shall prove that $gcd(b, a \bmod b)$ divides $gcd(a, b)$. Obviously,

$$gcd(b, a \bmod b)|b \text{ and } gcd(b, a \bmod b)|(a \bmod b).$$

Since

$$a = (a \text{ div } b) \cdot b + a \bmod b,$$

we see that a is a linear combination of b and $a \bmod b$.
Due to the assertion (i) of Lemma 2.2.4.19,

$$gcd(b, a \bmod b)|a.$$

Following the assertion (iv) of Observation 2.2.4.20,

$$gcd(b, a \bmod b)|b \text{ and } gcd(b, a \bmod b)|a$$

together imply

$$gcd(b, a \bmod b)|gcd(a, b).$$

□

Theorem 2.2.4.22 directly implies the correctness of the following recursive algorithm for gcd.

Algorithm 2.2.4.23. EUCLID'S ALGORITHM

$Euclid(a, b)$
Input: two positive integers a, b.
Recursive Step: **if** $b = 0$ **then return** (a)
                    else return $Euclid(b, a \bmod b)$.

Example 2.2.4.24. Consider the computation of EUCLID'S ALGORITHM for $a = 12750$ and $b = 136$.

$$\begin{aligned} Euclid(12750, 136) &= Euclid(136, 102) \\ &= Euclid(102, 34) \\ &= Euclid(34, 0) \\ &= 34. \end{aligned}$$

Thus, $gcd(12750, 136) = 34$. □

Obviously, EUCLID'S ALGORITHM cannot recurse indefinitely, since the second argument strictly decreases in each recursive call. Moreover, the above example shows that three recursive calls are enough to compute gcd of 12750 and 136. Obviously, the computational complexity of EUCLID'S ALGORITHM is proportional to the number of recursive calls it makes. The assertion of the following exercise shows that this algorithm is very efficient.

Exercise 2.2.4.25. Let $a \geq b \geq 0$ be two integers. Prove that the number of recursive calls of $Euclid(a, b)$ is in $O(\log_2 b)$.

There are many situations of both theoretical and practical interest when one asks whether a given integer p is a prime or not. Following the definition of a prime one can decide this by looking at all numbers $2, \ldots, \lfloor\sqrt{p}\rfloor$ and testing whether one of them is a factor of p. If one wants to check that p is a prime in this way, $\Omega(\sqrt{p})$ divisions have to be performed. But the best computer cannot perform $\sqrt{p}$ divisions if p is a number consisting of hundreds of digits, and we need to work with such large numbers in practice. This is one of the reasons why people have searched for other characterizations[16] of primes.

---

[16] Equivalent definitions.

Theorem 2.2.4.26. *For every positive integer $p \geq 2$,*

$$p \text{ is a prime if and only if } \mathbb{Z}_p \text{ is a field.}$$

Proof. In Example 2.2.4.10 we have shown that if p is a composite ($p = a \cdot b$ for $a > 1$, $b > 1$) then $\mathbb{Z}_p$ with $\oplus_n, \odot_n$ cannot build a field since $a \odot_n b = 0$.

We showed already in Example 2.2.4.4 that $(\mathbb{Z}_n, \oplus_n)$ is a commutative group for every positive integer n. So, one can easily see that $(\mathbb{Z}_n, \oplus_n, \odot_n)$ is a ring for all $n \in \mathbb{N}$. Since $\odot_n$ is commutative and $1 \odot_n a = a$ for every $a \in \mathbb{Z}_n$, it is sufficient to prove that the primality of n implies the existence of an inverse element a^{-1} for every $a \in \mathbb{Z}_n - \{0\}$.

Now, let p be a prime. For every $a \in \mathbb{Z}_p - \{0\}$, consider the following $p - 1$ multiples of a:

$$m_0 = 0 \cdot a, \quad m_1 = 1 \cdot a, \quad m_2 = 2 \cdot a, \quad \ldots \quad m_{p-1} = (p-1) \cdot a.$$

First, we prove that no two of these integers can be congruent modulo p. Let there exist two different $r, s \in \{0, 1, \ldots, p-1\}$, such that

$$m_r \equiv m_s (\text{mod } p).$$

Then p is a factor of $m_r - m_s = (r - s) \cdot a$. But this cannot occur because $r - s < p$ (i.e., p is not a factor of $r - s$) and $a < p$.[17] So, $m_0, m_1, m_2, \ldots, m_{p-1}$ are pairwise different in $\mathbb{Z}_p$.

Therefore, the numbers $m_1, m_2, \ldots, m_{p-1}$ must be respectively congruent to the numbers $1, 2, 3, \ldots, p-1$, in some arrangement. So,

$$\{0 \odot_p a, 1 \odot_p a, \ldots, a \odot_p (p-1)\} = \{0, 1, \ldots, p-1\}.$$

Now we are ready because it implies that there exists $b \in \mathbb{Z}_p$ such that $a \odot b = 1$, i.e., b is the inverse element of a in $\mathbb{Z}_p$. Since we have proved it for every $a \in \mathbb{Z}_p - \{0\}$, $\mathbb{Z}_p$ is a field. $\square$

Exercise 2.2.4.27. Prove that if p is a prime, then $(\{1, 2, \ldots, p-1\}, \odot_p)$ is a cyclic group. $\square$

Exercise 2.2.4.28. (*) Define, for every positive integer n,

$$\mathbb{Z}_n^* = \{a \in \mathbb{Z}_n - \{0\} \mid gcd(a, n) = 1\}.$$

Prove that

(i) $(\mathbb{Z}_n^*, \odot_n)$ forms a group.
(ii) the group $(\mathbb{Z}_n^*, \odot_n)$ is cyclic if and only if either $n = 2, 4, p^k$ or $2p^k$, for some non-negative integer k and an odd prime p. $\square$

---

[17] Note that this argument works because of the Fundamental Theorem of Arithmetics.

This nice characterization of primes by Theorem 2.2.4.26 leads directly to the following important result of the number theory.

Theorem 2.2.4.29 (Fermat's Theorem). *For every prime p and every integer a such that $gcd(a, p) = 1$*

$$a^{p-1} \equiv 1 \ (\text{mod } p).$$

Proof. Consider again[18] the numbers

$$m_1 = 1 \cdot a, \quad m_2 = 2 \cdot a, \quad \ldots \quad m_{p-1} = (p-1) \cdot a.$$

We claim by almost the same argument as in the proof above that no two of these integers can be congruent modulo p. Let there exist two different integers $r, s \in \{1, 2, \ldots, p-1\}$, $r > s$, such that

$$m_r \equiv m_s \ (\text{mod } p).$$

Then p is a factor of $m_r - m_s = (r - s) \cdot a$. This cannot occur because $r - s < p$ and p is not a factor of a according to our assumption $gcd(a, p) = 1$. Thus,

$$|\{m_1 \bmod p, m_2 \bmod p, \ldots, m_{p-1} \bmod p\}| = p - 1.$$

Now, we claim that none of the numbers $m_1, m_2, \ldots, m_{p-1}$ is congruent to $0 \bmod p$. Since $\mathbb{Z}_p$ is a field, $m_r = r \cdot a \equiv 0 \ (\text{mod } p)$ forces either $a \equiv 0 \ (\text{mod } p)$ or $r \equiv 0 \ (\text{mod } p)$. But for every $r \in \{1, 2, \ldots, p-1\}$, $m_r = r \cdot a$, $r < p$, and $gcd(a, p) = 1$ (i.e., p is a factor of neither r nor a).

The conclusion is

$$\{m_1 \bmod p, m_2 \bmod p, \ldots, m_{p-1} \bmod p\} = \{1, 2, \ldots, p-1\}. \qquad (2.15)$$

Finally, consider the following number

$$m_1 \cdot m_2 \cdot \cdots \cdot m_{p-1} = 1 \cdot a \cdot 2 \cdot a \cdot \cdots \cdot (p-1) \cdot a = 1 \cdot 2 \cdot \cdots \cdot (p-1) \cdot a^{p-1}. \qquad (2.16)$$

Following (2.15) we get

$$1 \cdot 2 \cdot \cdots \cdot (p-1) \cdot a^{p-1} \equiv 1 \cdot 2 \cdot \cdots \cdot (p-1) \ (\text{mod } p),$$

i.e.,

$$1 \cdot 2 \cdot \cdots \cdot (p-1) \cdot (a^{p-1} - 1) \equiv 0 \ (\text{mod } p).$$

Since $1 \cdot 2 \cdot \cdots \cdot (p-1) \not\equiv 0 \ (\text{mod } p)$ and $\mathbb{Z}_p$ is a field (i.e., $(\mathbb{Z}_p, \oplus_p, \odot_p)$ is zero division free), we obtain

$$a^{p-1} - 1 \equiv 0 \ (\text{mod } p).$$

$\square$

---

[18] As in the proof of Theorem 2.2.4.26

Exercise 2.2.4.30. Check Fermat's Theorem for $p = 5$ and $a = 9$.   □

A nice consequence of Fermat's Theorem is a method for computing the inverse element according to multiplication.

Corollary 2.2.4.31. *Let p be a prime. Then, for every $a \in \mathbb{Z}_p - \{0\}$,*

$$a^{-1} = a^{p-2} \bmod p.$$

Proof. $a \cdot a^{p-2} = a^{p-1} \equiv 1 \pmod{p}$ due to Fermat's Theorem.   □

In what follows we frequently use the notation -1 (the inverse of 1 according to addition) for $p - 1$ in $\mathbb{Z}_p$. The following theorem provides a nice equivalent definition of primality.

Theorem 2.2.4.32. *Let $p > 2$ be an odd integer.*

$$p \text{ is a prime} \Leftrightarrow a^{(p-1)/2} \bmod p \in \{1, -1\} \text{ for all } a \in \mathbb{Z}_p - \{0\}.$$

Proof. (i) Let $p = 2p' + 1$, $p' \geq 1$, be a prime. Following Fermat's Theorem $a^{p-1} \equiv 1 \pmod{p}$ for every $a \in \mathbb{Z}_p - \{0\}$. Since

$$a^{p-1} = a^{2p'} = \left(a^{p'} - 1\right) \cdot \left(a^{p'} + 1\right) + 1,$$

we can write

$$\left(a^{p'} - 1\right) \cdot \left(a^{p'} + 1\right) \equiv 0 \pmod{p}. \tag{2.17}$$

Since $\mathbb{Z}_p$ is a field, (2.17) implies

$$\left(a^{p'} - 1\right) \equiv 0 \pmod{p} \text{ or } \left(a^{p'} + 1\right) \equiv 0 \pmod{p}. \tag{2.18}$$

Inserting $p' = (p-1)/2$ into (2.18) we finally obtain

$$a^{(p-1)/2} \equiv 1 \pmod{p} \text{ or } a^{(p-1)/2} \equiv -1 \pmod{p}.$$

(ii) Let, for all $a \in \mathbb{Z}_p - \{0\}$, $a^{(p-1)/2} \equiv \pm 1 \pmod{p}$. It is sufficient to show that $\mathbb{Z}_p$ is a field. Obviously, $a^{p-1} = a^{(p-1)/2} \cdot a^{(p-1)/2}$.
If $a^{(p-1)/2} \equiv 1 \pmod{p}$, then $a^{p-1} \equiv 1 \pmod{p}$.
If $a^{(p-1)/2} \equiv -1 \equiv (p-1) \pmod{p}$, then

$$a^{p-1} \equiv (p-1)^2 \equiv p^2 - 2p + 1 \equiv 1 \pmod{p}.$$

So, for every $a \in \mathbb{Z}_p - \{0\}$, $a^{p-2} \bmod p$ is the inverse element a^{-1} of a. To prove that $\mathbb{Z}_p$ is a field it remains to show that if $a \cdot b \equiv 0 \pmod{p}$ for some $a, b \in \mathbb{Z}_p$ then $a \equiv 0 \pmod{p}$ or $b \equiv 0 \pmod{p}$. Let $a \cdot b \equiv 0 \pmod{p}$, and let $b \not\equiv 0 \pmod{p}$. Then there exists $b^{-1} \in \mathbb{Z}_p$ such that $b \cdot b^{-1} \equiv 1 \pmod{p}$. Finally,

$$a = a \cdot (b \cdot b^{-1}) = (a \cdot b) \cdot b^{-1} \equiv 0 \cdot b^{-1} \equiv 0 \pmod{p}.$$

In this way we proved that $\mathbb{Z}_p$ is zero division free, and so $\mathbb{Z}_p$ is a field.
    □

In order to use an extension of Theorem 2.2.4.32 for primality testing in Chapter 5, we shall investigate the properties of $\mathbb{Z}_n$ when n is composite. Let, for instance, $n = p \cdot q$ for two primes p and q. We know that $\mathbb{Z}_p$ and $\mathbb{Z}_q$ are fields. Now, consider the direct product $\mathbb{Z}_p \times \mathbb{Z}_q$. The elements of $\mathbb{Z}_p \times \mathbb{Z}_q$ are pairs (a_1, a_2), where $a_1 \in \mathbb{Z}_p$ and $a_2 \in \mathbb{Z}_q$. We define addition in $\mathbb{Z}_p \times \mathbb{Z}_q$ as

$$(a_1, a_2) \oplus_{p,q} (b_1, b_2) = ((a_1 + b_1) \bmod p, (a_2 + b_2) \bmod q),$$

and multiplication as

$$(a_1, a_2) \odot_{p,q} (b_1, b_2) = ((a_1 \cdot b_1) \bmod p, (a_2 \cdot b_2) \bmod q).$$

The idea is to show that $\mathbb{Z}_n$ and $\mathbb{Z}_p \times \mathbb{Z}_q$ are isomorphic, i.e., there exists a bijection $h : \mathbb{Z}_n \to \mathbb{Z}_p \times \mathbb{Z}_q$ such that

$$h(a \oplus_n b) = h(a) \oplus_{p,q} h(b) \text{ and } h(a \odot_n n) = h(a) \odot_{p,q} h(b)$$

for all $a, b \in \mathbb{Z}_n$. If one finds such a h, then one can view $\mathbb{Z}_n$ as $\mathbb{Z}_p \times \mathbb{Z}_q$. We define h simply as follows:

$$\text{For all } a \in \mathbb{Z}_n, \ h(a) = (a \bmod p, a \bmod q).$$

One can simply verify that h is injective.[19] For each $a, b \in \mathbb{Z}_n$, a and b can be written as

$$a = a_1' \cdot p + a_1 = a_2' \cdot q + a_2 \text{ for some } a_1 < p \text{ and } a_2 < q,$$
$$b = b_1' \cdot p + b_1 = b_2' \cdot q + b_2 \text{ for some } b_1 < p \text{ and } b_2 < q,$$

i.e., $h(a) = (a_1, a_2)$ and $h(b) = (b_1, b_2)$. So,

$$h(a \oplus_n b) = h(a + b \bmod n) = ((a + b) \bmod p, (a + b) \bmod q)$$
$$= ((a_1 + b_1) \bmod p, (a_2 + b_2) \bmod q)$$
$$= (a_1, a_2) \oplus_{p,q} (b_1, b_2) = h(a) \oplus_{p,q} h(b).$$

Similarly,

$$h(a \odot_n b) = h(a \cdot b \bmod n) = ((a \cdot b) \bmod p, (a \cdot b) \bmod q)$$
$$= ((a_1' \cdot b_1' \cdot p^2 + (a_1 \cdot b_1' + a_1' \cdot b_1) \cdot p + a_1 \cdot b_1) \bmod p,$$
$$(a_2' \cdot b_2' \cdot q^2 + (a_2' \cdot b_2 + a_2 \cdot b_2') \cdot q + a_2 \cdot b_2) \bmod q)$$
$$= ((a_1 \cdot b_1) \bmod p, (a_2 \cdot b_2) \bmod q)$$
$$= (a_1, a_2) \odot_{p,q} (b_1, b_2) = h(a) \odot_{p,q} h(b).$$

In general, $n = p_1 \cdot p_2 \cdots \cdot p_k$ for primes $p_1, p_2, \ldots, p_k$, and one considers the isomorphism between $\mathbb{Z}_n$ and $\mathbb{Z}_{p_1} \times \mathbb{Z}_{p_2} \times \cdots \times \mathbb{Z}_{p_n}$. This isomorphism is called Chinese remainder and we formulate it in the next theorem.

---

[19] In fact, we do not need to require that p and q are primes; it is sufficient to assume that p and q are coprimes.

Theorem 2.2.4.33 (Chinese Remainder Theorem, first version). *Let* $m = m_1 \cdot m_2 \cdot \; \cdots \; \cdot m_k$, $k \in \mathbb{N}^+$, *where* $m_i \in \mathbb{N}^{\geq 2}$ *are pairwise coprimes (i.e.,* $gcd(m_i, m_j) = 1$ *for* $i \neq j$*). Then, for any sequence of integers* $r_1 \in \mathbb{Z}_{m_1}$, $r_2 \in \mathbb{Z}_{m_2}, \cdots, r_k \in \mathbb{Z}_{m_k}$, *there is an integer* r *such that*

$$r \equiv r_i \; (\text{mod } m_i)$$

for every $i \in \{1, \ldots, k\}$, *and this integer* r *is unique in* $\mathbb{Z}_m$.

Proof. We first show that there exists at least one such r. Since $gcd(m_i, m_j) = 1$ for $i \neq j$, $gcd(\frac{m}{m_l}, m_l) = 1$ for every $l \in \{1, 2, \ldots, k\}$. It follows that there exists a multiplicative inverse n_i for m/m_i in the group $\mathbb{Z}_{m_i}$. Consider, for $i = 1, \ldots, k$,

$$e_i = n_i \frac{m}{m_i} = m_1 \cdot m_2 \cdot \; \cdots \; \cdot m_{i-1} \cdot n_i \cdot m_{i+1} \cdot m_{i+2} \cdot \; \cdots \; \cdot m_k.$$

For every $j \in \{1, \ldots, k\} - \{i\}$, $\frac{m}{m_i} \equiv 0 \; (\text{mod } m_j)$, and so

$$e_i \bmod m_j = 0.$$

Since n_i is the multiplicative inverse for m/m_i in $\mathbb{Z}_{m_i}$,

$$e_i \bmod m_i = n_i \frac{m}{m_i} \bmod m_i = 1.$$

Now, we set

$$r \equiv \left(\sum_{i=1}^{k} r_i \cdot e_i \right) (\text{mod } m)$$

and we see that r has the required properties.

To see that r is uniquely determined modulo m, let us assume that there exist two integers x and y satisfying

$$y \equiv x \equiv r_i \; (\text{mod } m_i)$$

for every $i \in \{1, \ldots, k\}$. Then,

$$x - y \equiv 0 \; (\text{mod } m_i)$$

for every $i \in \{1, \ldots, k\}$, since $m = m_1 \cdot m_2 \cdot \; \cdots \; \cdot m_k$ and $gcd(m_i, m_j) = 1$ for $i \neq j$, $x \equiv y \; (\text{mod } m)$. □

The first version of the Chinese Remainder Theorem can be viewed as a statement about solutions of certain equations. The second version can be regarded as a theorem about the structure of $\mathbb{Z}_n$. Because the proof idea has been already explained for the isomorphism between $\mathbb{Z}_{p \cdot q}$ and $\mathbb{Z}_p \times \mathbb{Z}_q$ above, we leave the proof of the second version of the Chinese Remainder Theorem as an exercise to the reader.

Theorem 2.2.4.34 (Chinese Remainder Theorem, second version).
Let $m = m_1 \cdot m_2 \cdot \cdots \cdot m_k$, $k \in \mathbb{N}^+$, where $m_i \in \mathbb{N}^{\geq 2}$ are pairwise coprimes
(i.e., $gcd(m_i, m_j) = 1$ for $i \neq j$). Then $\mathbb{Z}_m$ is isomorphic to $\mathbb{Z}_{m_1} \times \mathbb{Z}_{m_2} \times \cdots \times \mathbb{Z}_{m_k}$.

Exercise 2.2.4.35. Prove the second version of the Chinese Remainder Theorem. $\qquad\square$

The above stated theorems are fundamental results from number theory
and we will use them in Chapter 5 to design efficient randomized algorithms
for primality testing. For the development of (randomized) algorithms for algorithmic problems arising from number theory, one often needs fundamental
results on (finite) groups, especially on $\mathbb{Z}_n^*$. Hence we present some of the
basic assertions in what follows.

Since $\mathbb{Z}_n^* = \{a \in \mathbb{Z}_n \mid gcd(a, n) = 1\}$, we first look for some basic
properties of greatest common divisors. We start with an equivalent definition
of the greatest common divisor of two numbers.

Theorem 2.2.4.36. Let $a, b \in \mathbb{N} - \{0\}$, and let

$$Com(a, b) = \{ax + by \mid x, y \in \mathbb{Z}\}$$

be the set of linear combinations of a and b. Then

$$gcd(a, b) = \min\{d \in Com(a, b) \mid d \geq 1\},$$

i.e., $gcd(a, b)$ is the smallest positive integer from $Com(a, b)$.

Proof. Let $h = \min\{d \in Com(a, b) \mid d \geq 1\}$, and let $h = ax + by$ for some
$x, y \in \mathbb{Z}$. We prove $h = gcd(a, b)$ by proving the inequalities $h \leq gcd(a, b)$
and $h \geq gcd(a, b)$ separately.

First we show that h divides both a and b and so it divides $gcd(a, b)$.
Following the definition of modulo n, we have

$$a \bmod h = a - \lfloor a/h \rfloor \cdot h = a - \lfloor a/h \rfloor \cdot (ax + by) = a \cdot (1 - \lfloor a/h \rfloor x) + b \cdot (-\lfloor a/h \rfloor y)$$

and so $a \bmod h$ is a linear combination of a and b. Since h is the smallest
positive linear combination of a and b, and $a \bmod h < h$, we obtain

$$a \bmod h = 0, \text{ i.e., } h \text{ divides } a.$$

The same argumentation provides the fact that h divides b. Thus

$$h \leq gcd(a, b).$$

Since $gcd(a, b)$ divides both a and b, $gcd(a, b)$ must divide $au + bv$ for
all $u, v \in \mathbb{Z}$, i.e., $gcd(a, b)$ divides every element in $Com(a, b)$. Since $h \in Com(a, b)$,

$$gcd(a, b) \text{ divides } h, \text{ i.e., } gcd(a, b) \leq h.$$

$\qquad\square$

Theorem 2.2.4.37. *Let $a, n \in \mathbb{N} - \{0\}$, $n \geq 2$, and let $gcd(a, n) = 1$. Then the congruence*

$$ax \equiv 1 \ (\text{mod } n)$$

has a solution $x \in \mathbb{Z}_n$.

Proof. Following Theorem 2.2.4.36, there exists $u, v \in \mathbb{Z}$ such that

$$a \cdot u + n \cdot v = 1 = gcd(a, n).$$

Observe, that for every $k \in \mathbb{Z}$

$$a \cdot u + n \cdot v = a \cdot (u + kn) + n \cdot (v - ka),$$

and so

$$a \cdot (u + kn) + n \cdot (v - ka) = 1.$$

For sure there is a $l \in \mathbb{Z}$ such that $a + ln \in \mathbb{Z}_n$. Set $x = u + ln$. Since

$$a(u + ln) + n(v - la) \equiv a(u + ln) \equiv (1 \ \text{mod} \ n),$$

x is a solution of the congruence $ax \equiv 1 \ (\text{mod } n)$. □

Exercise 2.2.4.38. Let a, n satisfy the assumptions of Theorem 2.2.4.37. Prove that the solution x of the congruence $ax \equiv 1 \ (\text{mod } n)$ is unique in $\mathbb{Z}_n$.

Exercise 2.2.4.39. Let $a, n \in \mathbb{N} - \{0\}$, $n \geq 2$, and let $gcd(a, n) = 1$. Prove, that for every $b \in \mathbb{Z}_n$, the congruence $ax \equiv b \ (\text{mod } n)$ has a unique solution $x \in \mathbb{Z}_n$.

Now, we are ready to prove one of the most useful facts for developing randomized algorithms for primality testing. Later we extend the following assertion for any positive integer n.

Theorem 2.2.4.40. *For every prime n, $(\mathbb{Z}_n^*, \odot \ \text{mod } n)$ is a commutative group.*

Proof. Let a, b be arbitrary elements of $\mathbb{Z}_n^*$. Following the definition of $\mathbb{Z}_n^*$, we have $gcd(a, n) = gcd(b, n) = 1$. Since

$$Com(ab, n) \subseteq Com(a, n) \cap Com(b, n),$$

Theorem 2.2.4.36 implies that the number $1 \in Com(ab, n)$, and so $gcd(ab, n) = 1$. Thus, $ab \in \mathbb{Z}_n^*$, i.e., $\mathbb{Z}_n^*$ is closed under multiplication modulo n. This implies that $(\mathbb{Z}_n^*, \odot \ \text{mod } n)$ is an algebra.

Obviously, 1 is the identity element with respect to $\odot \ \text{mod } n$ and $\odot \ \text{mod } n$ is an associative and commutative operation.

Theorem 2.2.4.37 ensures the existence of an inverse element $x = a^{-1}$ for every $a \in \mathbb{Z}_n^*$ (i.e., for every $a \in \mathbb{Z}_n$ with $gcd(a, n) = 1$). Following Definition 2.2.4.1, the proof is completed. □

Exercise 2.2.4.41. Let $(A, *)$ be a group. Prove the following facts.

(i) For every $a \in A$, $a = (a^{-1})^{-1}$.

(ii) For all $a, b, c \in A$,

$a * b = c * b$ implies $a = c$, and

$b * a = b * c$ implies $a = c$.

(iii) For all $a, b, c \in A$,

$$a \neq b \Leftrightarrow a * c \neq b * c \Leftrightarrow c * a \neq c * b.$$

$\square$

Definition 2.2.4.42. *Let $(A, \circ)$ be a group with the identity element 1. For every $a \in A$, the **order of a**, **order(a)** is the smallest $r \in \mathbb{N} - \{0\}$, such that*

$$a^r = 1,$$

if such an r exists. If $a^i \neq 1$ for all $i \in \mathbb{N} - \{0\}$, then we set order$(a) = \infty$.

We show that Definition 2.2.4.42 is consistent, i.e., that every element of a finite group $(A, \circ)$ has a finite order. Consider the $|A| + 1$ elements

$$a^0, a^1, a^2, \ldots, a^{|A|}$$

from A. There must exist $0 \leq i \leq j \leq |A|$ such that

$$a^i = a^j.$$

This implies

$$1 = a^i \cdot (a^{-1})^i = a^j \cdot (a^{-1})^i = a^{(j-i)}$$

and so order$(a) \leq j - i$, i.e., order$(a) \in \{1, 2, \ldots, |A|\}$.

Definition 2.2.4.43. *Let $(A, \circ)$ be a group. An algebra $(H, \circ)$ is called a **subgroup of A** if $(H, \circ)$ is a group and $H \subseteq A$.*

For instance $(\mathbb{Z}, +)$ is a subgroup of $(\mathbb{Q}, +)$, and $(\{1\}, \odot_{\text{mod } 5})$ is a subgroup of $(\mathbb{Z}_5^*, \odot_{\text{mod } 5})$. But $(\mathbb{Z}_5^*, \odot_{\text{mod } 5})$ is not a subgroup of $(\mathbb{Z}_7^*, \odot_{\text{mod } 7})$ because $\odot_{\text{mod } 5}$ and $\odot_{\text{mod } 7}$ are different operations ($4 \odot_{\text{mod } 5} 4 = 1$ and $4 \odot_{\text{mod } 7} 4 = 2$).

Lemma 2.2.4.44. *Let $(H, \circ)$ be a subgroup of a group $(A, \circ)$. Then the identity elements of both groups are the same.*

Proof. Let e_H be the identity of $(H, \circ)$, and let e_A be the identity of $(A, \circ)$. Since e_H is the identity of $(H, \circ)$,

$$e_H \circ e_H = e_H. \tag{2.19}$$

Since e_A is the identity of $(A, \circ)$ and $e_H \in A$,

$$e_A \circ e_H = e_H. \tag{2.20}$$

Thus, the left sides of the equalities (2.19) and (2.20) are the same, i.e.,

$$e_H \circ e_H = e_A \circ e_H. \tag{2.21}$$

If e_H^{-1} is the inverse of e_H in $(A, \circ)$, then multiplying (2.21) by e_H^{-1} we obtain

$$e_H = e_H \circ e_H \circ e_H^{-1} = e_A \circ e_H \circ e_H^{-1} = e_A.$$

$\square$

Theorem 2.2.4.45. *Let $(A, \circ)$ be a finite group. Every algebra $(H, \circ)$ with $H \subseteq A$ is a subgroup of $(A, \circ)$.*

Proof. Let $H \subseteq A$, and let $(H, \circ)$ be an algebra. To prove that $(H, \circ)$ is a subgroup of $(A, \circ)$, it is sufficient to show that $(H, \circ)$ is a group, i.e., that e_A is the identity of $(H, \circ)$ and that every $b \in H$ has its inverse b^{-1} in H.

Let b be an arbitrary element of H. Since $b \in A$ and A is finite, $order(b) \in \mathbb{N} - \{0\}$. Thus

$$b^{order(b)} = e_A.$$

Since $b^i \in H$ for all positive integers i (remember that H is closed under $\circ$), $e_A \in H$. Since

$$e_A \circ d = d$$

for every $d \in A$ and $H \subseteq A$, e_A is the identity of $(H, \circ)$, too.

Since $b^{order(b)-1} \in H$ for any $b \in H$ and

$$e_A = b^{order(b)} = b \circ b^{order(b)-1},$$

$b^{order(b)-1}$ is the inverse element of b in $(H, \circ)$. $\square$

Theorem 2.2.4.45 is a useful instrument when working with groups, because in order to prove that $(H, \circ)$ is a subgroup of a finite group $(A, \circ)$, it is sufficient to show that $H \subseteq A$ and that H is closed under $\circ$.

Note that the assumption of Theorem 2.2.4.45 that A is finite is essential, because $(\mathbb{N}, +)$ is an algebra, but $(\mathbb{N}, +)$ is not a subgroup of the group $(\mathbb{Z}, +)$.

Exercise 2.2.4.46. Let $(H, \circ)$ and $(G, \circ)$ be two subgroups of a group $(A, \circ)$. Prove that $(H \cap G, \circ)$ is a subgroup of $(A, \circ)$. $\square$

Lemma 2.2.4.47. *Let $(A, \circ)$ be a group with the identity e and let $a \in A$ be an element with a finite order. Then, for $H(a) = \{e, a, a^2, \ldots, a^{order(a)-1}\}$, $(H(a), \circ)$ is the smallest subgroup of $(A, \circ)$ that contains a.*

Proof. First, we prove that $H(a)$ is closed under $\circ$. Let a^i and a^j be two arbitrary elements of $H(a)$. If $i + j < order(a)$, then

$$a^i \circ a^j = a^{i+j} \in H(a).$$

If $i + j > order(a)$, then

$$a^i \circ a^j = a^{i+j} = a^{order(a)} \circ a^{i+j-order(a)}$$
$$= e \circ a^{i+j-order(a)} = a^{i+j-order(a)} \in H(a).$$

Following the definition of $H(a)$, $e \in H(a)$. For every element $a^i \in H(a)$,

$$e = a^{order(a)} = a^i \circ a^{order(a)-i},$$

i.e., $a^{order(a)-i}$ is the inverse element to a^i.

Since every algebra $(G, \circ)$ with $a \in G$ must contain $H(a)$, $(H(a), \circ)$ is the smallest subgroup of $(A, \circ)$ that contains a. $\qquad\square$

Definition 2.2.4.48. *Let $(H, \circ)$ be a subgroup of a group $(A, \circ)$. For every $b \in A$, the set*

$$\boldsymbol{H \circ b} = \{h \circ b \mid h \in H\}$$

*is called a **right coset** of H in $(A, \circ)$, and the set*

$$\boldsymbol{b \circ H} = \{b \circ h \mid h \in H\}$$

*is called a **left coset** of H in $(A, \circ)$. If $H \circ b = b \circ H$, then $H \circ b$ is called a **coset** of H in $(A, \circ)$.*

For instance, $(\{7 \cdot a \mid a \in \mathbb{Z}\}, +)$ is a subgroup of $(\mathbb{Z}, +)$. Let $B_7 = \{7 \cdot a \mid a \in \mathbb{Z}\}$. Then

$$B_7 + i = i + B_7 = \{7 \cdot a + i \mid a \in \mathbb{Z}\} = \{b \in \mathbb{Z} \mid b \bmod 7 = i\}$$

are cosets of B_7 in $(\mathbb{Z}, +)$ for $i = 0, 1, \ldots, 6$. Observe that $\{B_7 + i \mid i = 0, 1, \ldots, 6\}$ is a partition of $\mathbb{Z}$ into 7 disjoint classes.

Observation 2.2.4.49. If $(H, \circ)$ is a subgroup of a commutative group $(A, \circ)$, then all right cosets (left cosets) of H in $(A, \circ)$ are cosets.

An important fact about cosets $H \circ b$ is, that their size is always equal to the size of H.

Theorem 2.2.4.50. *Let $(H, \circ)$ be a subgroup of $(A, \circ)$. Then the following facts hold.*

(i) $H \circ h = H$ for all $h \in H$.
(ii) For all $b, c \in A$,

$$either\ H \circ b = H \circ c\ or\ H \circ b \cap H \circ c = \emptyset.$$

(iii) If H is finite, then

$$|H \circ b| = |H|$$

for all $b \in A$.

Proof. We prove these three claims separately. Let e be the identity of $(A, \circ)$ and $(H, \circ)$.

(i) Let $h \in H$. Since H is closed under $\circ$, we obtain $a \circ h \in H$ for every $a \in H$, i.e.,

$$H \circ h \subseteq H.$$

Since $(H, \circ)$ is a group, $h^{-1} \in H$. Let b be an arbitrary element of H. Then

$$b = b \circ e = b \circ \underbrace{(h^{-1} \circ h)}_{e} = \underbrace{(b \circ h^{-1})}_{\in H} \circ h \in H \circ h, \text{ i.e.,}$$

$$H \subseteq H \circ h.$$

Thus $H \circ h = H$.

(ii) Let $H \circ b \cap H \circ c \neq \emptyset$ for some $b, c \in A$. Then there exists $a_1, a_2 \in H$, such that

$$a_1 \circ b = a_2 \circ c.$$

This implies $c = a_2^{-1} \circ a_1 \circ b$, where $a_2^{-1} \in H$. Then

$$H \circ c = H \circ (a_2^{-1} \circ a_1 \circ b) = H \circ (a_2^{-1} \circ a_1) \circ b. \qquad (2.22)$$

Since $a_2^{-1}, a_1 \in H$, the element $a_2^{-1} \circ a_1$ belongs to H, too. This implies, because of (i), that

$$H \circ (a_2^{-1} \circ a_1) = H. \qquad (2.23)$$

Thus, combining (2.22) and (2.23) we obtain

$$H \circ c = H \circ (a_2^{-1} \circ a_1) \circ b = H \circ b.$$

(iii) Let H be finite, and let $b \in A$. Since $H \circ b = \{h \circ b \mid h \in H\}$, we immediately have

$$|H \circ b| \leq |H|.$$

Let $H = \{h_1, h_2, \ldots, h_k\}$ for some $k \in \mathbb{N}$. We have to show that

$$|\{h_1 \circ b, h_2 \circ b, \ldots, h_k \circ b\}| \geq k, \text{ i.e., that } h_i \circ b \neq h_j \circ b$$

for all $i, j \in \{1, \ldots, k\}$ with $i \neq j$. Since $(A, \circ)$ is a group, $b^{-1} \in A$ and so $h_i \circ b = h_j \circ b$ would imply $h_i \circ b \circ b^{-1} = h_j \circ b \circ b^{-1}$. Thus,

$$h_i = h_i \circ (b \circ b^{-1}) = h_j \circ (b \circ b^{-1}) = h_j$$

would contradict the assumption $h_i \neq h_j$.

$$\square$$

As a consequence of Theorem 2.2.4.50 we obtain that one can partition the set A of every group $(A, \circ)$, that has a proper subgroup $(H, \circ)$, into pairwise disjoint subsets of A, which are the left (right) cosets of H in $(A, \circ)$.

Theorem 2.2.4.51. *Let $(H, \circ)$ be a subgroup of a group $(A, \circ)$. Then $\{H \circ b \mid b \in A\}$ is a partition of A.*

Proof. The claim (ii) of Theorem 2.2.4.50 shows that $H \circ b \cap H \circ c = \emptyset$ or $H \circ b = H \circ c$. So, it remains to show that $A \subseteq \bigcup_{b \in A} H \circ b$. But this is obvious because the identity e of $(A, \circ)$ is also the identity of $(H, \circ)$, and so $b = e \circ b \in H \circ b$ for every $b \in A$.   □

Definition 2.2.4.52. *Let $(H, \circ)$ be a subgroup of a group $(A, \circ)$. We define the* **index of H in $(A, \circ)$** *as*

$$Index_H(A) = |\{H \circ b \mid b \in A\}|,$$

i.e., as the number of different right cosets of H in $(A, \circ)$.

The following theorem of Lagrange is the main reason for our study of group theory. It provides a powerful instrument for proving that there are not too many "bad" elements with some special properties in a group $(A, \circ)$, because it is sufficient to show that all "bad" elements are in a proper subgroup of $(A, \circ)$. The following assertion claims that the size of any proper subgroup of $(A, \circ)$ is at most $|A|/2$.

Theorem 2.2.4.53 (Lagrange's Theorem). *For any subgroup $(H, \circ)$ of a finite group $(A, \circ)$,*

$$|A| = Index_H(A) \cdot |H|,$$

i.e., $|H|$ divides $|A|$.

Proof. Following Theorem 2.2.4.51, A can be divided in $Index_H(A)$ right cosets, which are pairwise disjoint and all of the same size $|H|$.   □

Corollary 2.2.4.54. *Let $(H, \circ)$ be a subalgebra of a finite group $(A, \circ)$. If $H \subset A$, then*

$$|H| \leq |A|/2.$$

Proof. Theorem 2.2.4.45 ensures that $(H, \circ)$ is a subgroup of $(A, \circ)$. Following Lagrange's Theorem

$$|A| = Index_H(A) \cdot |H|.$$

Since $H \subset A$, $1 \leq |H| < |A|$ and so

$$Index_H(A) \geq 2.$$

□

Corollary 2.2.4.55. *Let $(A, \circ)$ be a finite group. Then, for every element $a \in A$, the order of a divides $|A|$.*

Proof. Let a be an arbitrary element of A. Following Lemma 2.2.4.47 $(H(a), \circ)$ with $H(a) = \{e, a, a^2, \ldots, a^{order(a)-1}\} = \{a, a^2, \ldots, a^{order(a)}\}$ is a subgroup of $(A, \circ)$.
Since $|H(a)| = order(a)$, Lagrange's Theorem implies

$$|A| = Index_{H(a)}(A) \cdot order(a).$$

□

In Chapter 5 we will often work with the sets

$$\mathbb{Z}_n^* = \{a \in \mathbb{Z}_n \mid gcd(a, n) = 1\}$$

for $n \in \mathbb{N} - \{0\}$. In Theorem 2.2.4.40 we proved that $(\mathbb{Z}_p^*, \odot_p)$ is a commutative group for every prime p. Note that $\mathbb{Z}_p^* = \mathbb{Z}_p - \{0\} = \{1, 2, \ldots, p-1\}$ for every prime p. We prove now, that $(\mathbb{Z}_n^*, \odot_n)$ is a group for every $n \in \mathbb{N} - \{0\}$.

Theorem 2.2.4.56. *For every* $n \in \mathbb{N} - \{0\}$, $(\mathbb{Z}_n^*, \odot_n)$ *is a commutative group.*

Proof. First of all we have to show that $\mathbb{Z}_n^*$ is closed under $\odot_n$. Let a, b be arbitrary elements of $\mathbb{Z}_n^* = \{a \in \mathbb{Z}_n \mid gcd(a, n) = 1\}$. We have to show that $a \odot_n b \in \mathbb{Z}_n^*$, i.e., that $gcd(a \odot_n b, n) = 1$. Let us assume the contrary, i.e., that $gcd(a \odot_n b, n) = k$ for some $k \geq 2$. Then $n = k \cdot v$ for some $v \in \mathbb{N} - \{0\}$ and $a \cdot b \bmod n = k \cdot d$ for some $d \in \mathbb{N} - \{0\}$. This implies

$$a \cdot b \bmod kv = kd, \text{ i.e., } a \cdot b = kv \cdot s + kd$$

for some $s \in \mathbb{N}$. Since $kvs + kd = k(vs + d)$, k divides $a \cdot b$. Let $k = p \cdot m$, where p is a prime. Obviously, either p divides a or p divides b. But this is the contradiction to the facts $n = p \cdot m \cdot v$, $gcd(a, n) = 1$, and $gcd(b, n) = 1$. Thus, $a \cdot b \bmod p \in \mathbb{Z}_n^*$.

Clearly, 1 is the identity element of $(\mathbb{Z}_n^*, \odot_n)$. Let a be an arbitrary element of $\mathbb{Z}_n^*$. To prove that there exists an inverse a^{-1} with $a \odot_n a^{-1} = 1$, it is sufficient to show that

$$|\{a \odot_n 1, a \odot_n 2, \ldots, a \odot_n (n-1)\}| = n - 1,$$

which implies $1 \in \{a \odot_n 1, \ldots, a \odot_n (n-1)\}$. Let us assume the contrary. Let there exist $i, j, i > j$, such that

$$a \odot_n i = a \odot_n j, \text{ i.e., } a \cdot i \equiv a \cdot j \pmod{n}.$$

This implies

$$a \cdot i = nk_1 + z \text{ and } a \cdot j = n \cdot k_2 + z$$

for some $k_1, k_2, z \in \mathbb{N}$ with $z < n$. Then

$$a \cdot i - a \cdot j = nk_1 - nk_2 = n(k_1 - k_2),$$

and so

$$a \cdot (i - j) = n(k_1 - k_2), \text{ i.e., } n \text{ divides } a \cdot (i - j).$$

Since $gcd(a, n) = 1$, n must divide $(i - j)$. But this is impossible, because $i - j < n$. Thus, we can infer that $(\mathbb{Z}_n^*, \odot_n)$ is a group. Since $\odot_n$ is a commutative operation, $(\mathbb{Z}_n^*, \odot_n)$ is a commutative group.   □

Next we show that $\mathbb{Z}_n^*$ contains all elements of $\mathbb{Z}_n$ that have an inverse element with respect to $\odot_{\bmod\, n}$, i.e., that

$$\begin{aligned} \mathbb{Z}_n^* &= \{a \in \mathbb{Z}_n \mid gcd(a, n) = 1\} \\ &= \{a \in \mathbb{Z}_n \mid \exists a^{-1} \in \mathbb{Z}_n \text{ such that } a \odot_{\bmod\, n} a^{-1} = 1\}. \end{aligned} \qquad (2.24)$$

Theorem 2.2.4.56 implies that $\mathbb{Z}_n^* \subseteq \{a \in \mathbb{Z}_n \mid \exists a^{-1} \in \mathbb{Z}_n\}$ because $(\mathbb{Z}_n^*, \odot_{\bmod\, n})$ is a group. Thus, the following lemma completes the proof of (2.24).

Lemma 2.2.4.57. *Let $a \in \mathbb{Z}_n$. If there exists an $a^{-1} \in \mathbb{Z}_n$ such that $a \odot_n a^{-1} = 1$, then*

$$gcd(a, n) = 1.$$

Proof. Theorem 2.2.4.36 claims that

$$gcd(a, n) = \min\{d \in \mathbb{N} - \{0\} \mid d = ax + by \text{ for } x, y \in \mathbb{Z}\}.$$

Let there exist an element a^{-1} with $a \odot_{\bmod\, n} a^{-1} = 1$. So $a \cdot a^{-1} \equiv 1 \ (\bmod\ n)$, i.e.,

$$a \cdot a^{-1} = k \cdot n + 1 \qquad \text{for a } k \in \mathbb{N}.$$

Choosing $x = a^{-1}$ and $y = -k$ we obtain

$$a \cdot a^{-1} + n \cdot (-k) = k \cdot n + 1 - k \cdot n = 1 \in Com(a, n)$$

and so $gcd(a, n) = 1$. $\qquad\square$

We conclude this section by proving a fundamental result of the group theory. Let $\varphi(n) = |\mathbb{Z}_n^*|$ for any $n \in \mathbb{N}$ be the so-called **Euler's number**.

Theorem 2.2.4.58 (Euler's Theorem). *For all positive integers $n \geq 1$*

$$a^{\varphi(n)} \equiv 1 \ (\bmod\ n)$$

for all $a \in \mathbb{Z}_n^*$.

Proof. Let a be an arbitrary element from $\mathbb{Z}_n^*$. Corollary 2.2.4.55 implies that $order(a)$ in $(\mathbb{Z}_n^*, \odot_n)$ divides $|\mathbb{Z}_n^*| = \varphi(n)$. Since $a^{order(a)} \equiv 1 \ (\bmod\ n)$ we obtain

$$\begin{aligned} a^{\varphi(n)} \bmod n &= (a^{order(a)})^{\varphi(n)/order(a)} \bmod n \\ &= (a^{order(a)} \bmod n)^{\varphi(n)/order(a)} \bmod n \\ &= 1^{\varphi(n)/order(a)} \bmod n = 1 \end{aligned}$$

$\qquad\square$

Exercise 2.2.4.59. Apply Euler's Theorem in order to give an alternative (algebraic) proof of Fermat's Theorem.

Keywords introduced in Section 2.2.4

group, semigroup, ring, field, prime, greatest common divisor, Euclid's algorithm, generator of a group, cyclic group, order of a group element, coset, index of a group

Summary of Section 2.2.4

The main assertions presented in Section 2.2.4 are:

- There are infinitely many primes and the number of primes from $\{2, 3, 4, \ldots, n\}$ is approximately $\frac{n}{\ln n}$ (Prime Number Theorem).
- Every integer greater than 1 can be expressed as a product of nontrivial powers of distinct primes and this prime factorization is unique (Fundamental Theorem of Arithmetics).
- p is prime if and only if $a^{(p-1)/2} \bmod p \in \{1, -1\}$ for every $a \in \{1, 2, \ldots, p-1\}$.
- If $n = p_1 \cdot p_2 \cdot \cdots \cdot p_k$ for primes $p_1, \ldots, p_k$, then $\mathbb{Z}_n$ is isomorphic to $\mathbb{Z}_{p_1} \times \mathbb{Z}_{p_2} \times \cdots \times \mathbb{Z}_{p_k}$ (Chinese Remainder Theorem).
- If $(H, \circ)$ is a subgroup of a group $(A, \circ)$, then $|H|$ divides $|A|$.
- The order of every element a of a group $(A, \circ)$ divides $|A|$.
- For every $\mathbb{Z}_n^*, n \in \mathbb{N}$, and every $a \in \mathbb{Z}_n^*$, $a^{\varphi(n)} \bmod n = 1$ (Euler's Theorem).

2.2.5 Probability Theory

The probability theory has been developed to study experiments with uncertain outcomes. The fundamental notions of probability theory are "sample space" and "elementary event". A **sample space** S is the set of all basic events that may happen in some experiment. Every element of S is called an **elementary event**. Intuitively, in some experiment, the sample space is the set of all results (events) that may be the outcomes of the experiment. For instance, in flipping a coin one can consider the following two outcomes: "head" and "tail". Thus, $\{head, tail\}$ is the sample space and $head$ and $tail$ are the elementary events. If one flips three coins (one after each other or at once), then the sample space is $\{(x, y, z) \mid x, y, z \in \{head, tail\}\}$. The sequences $(head, head, tail)$, $(tail, head, tail)$, and $(head, head, head)$ are examples of elementary events. Intuitively, an elementary event is an event that cannot be expressed as a collection of smaller, more fundamental events, i.e., an elementary event cannot be partitioned into pieces from the point of view of the experiment considered.

Another example of an experiment is a fixed (randomized) algorithm. During the computation on an input x such a randomized algorithm A may have a choice from several possibilities on how to continue. Thus, it may execute different computations depending on the choices. From this point of view, the sample space $S_{A,x}$ is the set of all possible sequences of random choices, or equivalently the set of all computations of A on x (one computation for each sequence of random choices). Thus, every computation of A on x can be considered as an elementary event.

For our purposes it is sufficient to consider that S is countable. So, all following definitions of fundamental notions assume the countability of S. An **event** is any subset of the sample space S. The event S is called the **certain event**, and the event $\emptyset$ is called the **null event**. Two events $S_1, S_2 \subseteq S$ are called **mutually exclusive** if $S_1 \cap S_2 = \emptyset$. For instance, for $S = \{(x, y, z) \mid x, y, z \in \{head, tail\}\}$, $S_1 = \{(head, head, head), (tail, head, tail)\}$ and $S_2 = \{(tail, head, head), (head, tail, head), (head, head, tail)\}$ are two events that are mutually exclusive. Considering $S_{A,x}$, one may be interested in computations of A on x, in which the correct output is computed. Then, one considers the set $Cor(A, x)$ of all computations providing the right output as an event. $Cor(A, x)$ is mutually exclusive to the event consisting of all computations that finish with a wrong output.

In what follows we have the following scenario. For a finite sample space S, we want to assign a probability to every elementary event. This assignment should correspond to the reality, i.e., to the experiment executed. To be fair, we have (among others) the following requirements on this assignment:

(i) Every elementary event has some non-negative probability.
(ii) The sum of the probabilities of all elementary events gives certainty (denoted by 1 in probability theory).
(iii) There is a fair way to compute the probability of any event $S_1 \subseteq S$ (among others, the probability of S_1 plus the probability of the complement $S - S_1$ must be certainty).

This results in the following definition.

Definition 2.2.5.1. *A* **probability distribution** *Prob* *on a sample space S is a mapping from events of S to real numbers (Prob : $2^S \to \mathbb{R}^{\geq 0}$) such that the following* **probability axioms** *are satisfied:*

(1) $Prob(\{x\}) \geq 0$ for every elementary event x,
(2) $Prob(S) = 1$,
(3) $Prob(X \cup Y) = Prob(X) + Prob(Y)$ for any two mutually exclusive events X and Y $(X \cap Y = \emptyset)$.

(If one considers an infinite S, then one requires $Prob\left(\bigcup_{i=1}^{\infty} X_i\right) = \sum_{i=1}^{\infty} Prob(X_i)$ for every countable infinite sequence of mutually exclusive events $X_1, X_2, X_3, \ldots$.)

$Prob(X)$ is called the **probability of the event** X.

If one considers $S = \{head, tail\}$ for a fair coin flipping, then $Prob(\{head\}) = Prob(\{tail\}) = 1/2$. One can simply observe that the probabilities of the elementary events unambiguously determine the probabilities of all events in every sample space. The proof of the following simple observation is left to the reader.

Exercise 2.2.5.2. Prove for all events X, Y of a sample space S and every probability distribution $Prob$ on S that

(i) $Prob(\emptyset) = 0$,

(ii) if $X \subseteq Y$, then $Prob(X) \leq Prob(Y)$,

(iii) $Prob(S - X) = 1 - Prob(X)$,

(iv) $Prob(X \cup Y) = Prob(X) + Prob(Y) - Prob(X \cap Y) \leq Prob(X) + Prob(Y)$.

□

In what follows we always consider that S is finite or countably infinite. In this case we speak about **discrete probability distribution**. If, for every elementary event x of a finite S,

$$Prob(\{x\}) = \frac{1}{|S|},$$

then $Prob$ is called the **uniform probability distribution on S**.

Example 2.2.5.3. Consider the sample space

$$S = \{(x, y, z) \mid x, y, z \in \{head, tail\}\}$$

for the experiment of flipping three coins. If we consider fair coins, it means that $Prob(\{a\}) = \frac{1}{|S|} = \frac{1}{8}$ for every elementary event $a \in S$.

What is the probability of getting at least one head? This event is $Head = \{(head, head, head), (head, head, tail), (head, tail, head), (head, tail, tail), (tail, head, tail), (tail, tail, head), (tail, head, head)\}$. Thus,

$$Prob(Head) = \sum_{a \in Head} Prob(\{a\}) = \frac{7}{8}.$$

A more convenient way to evaluate $Prob(Head)$ is to say that $S - Head = \{(tail, tail, tail)\}$, and so $Prob(S - Head) = \frac{1}{8}$ directly implies

$$Prob(Head) = 1 - Prob(S - Head) = \frac{7}{8}.$$

□

Exercise 2.2.5.4. Let $n \geq k \geq 0$ be two integers. Consider the experiments of flipping n coins and the corresponding sample space $S = \{(x_1, x_2, \ldots, x_n) \mid x_i \in \{head, tail\}$ for $i = 1, \ldots, n\}$. What is the probability of getting exactly k heads if $Prob$ is the uniform probability distribution on S? □

The notions defined above are suitable when looking at an experiment only once, i.e., at the very end. But sometimes we can obtain partial information about the outcome of the experiment by some intermediate observation. For instance, we flip three coins one after each other and we look at the result of the first coin flipping. Knowing this result we ask what the probability is of getting at least two heads in the whole experiment. Or, somebody tells you that the result (x, y, z) contains at least one head and knowing this fact you have to estimate the probability that (x, y, z) contains at least two heads. The tasks of this kind result in the following definition of conditional probabilities.

Definition 2.2.5.5. *Let S be a sample space with a probability distribution Prob. The* **conditional probability** *of an event $X \subseteq S$ given that another event $Y \subseteq S$ occurs (with certainty) is*

$$Prob(X|Y) = \frac{Prob(X \cap Y)}{Prob(Y)},$$

whenever $Prob(Y) \neq 0$. We also say that $Prob(X|Y)$ is **the probability of X given Y**.

Observe that the definition of the conditional probability is natural. $X \cap Y$ consists of elementary events that are in both X and Y. Since we know that Y happens, it is clear that no event from $X - Y$ can happen. Dividing $Prob(X \cap Y)$ by $Prob(Y)$ we normalize the probabilities of all elementary events in Y since

$$\sum_{e \in Y} \frac{Prob(\{e\})}{Prob(Y)} = \frac{1}{Prob(Y)} \cdot \sum_{e \in Y} Prob(\{e\}) = \frac{1}{Prob(Y)} \cdot Prob(Y) = 1.$$

Intuitively it means that we exchange S by Y, because Y happens with certainty. Thus, the conditional probability of X given Y is the ratio of the probability of the event $X \cap Y$ to the probability of Y.

Example 2.2.5.6. Let us again consider the experiment of flipping three coins. Let X be the event that the result contains at least two heads, and let Y be the event that the result contains at least one head. Since $X \cap Y = X$ we obtain

$$Prob(X|Y) \underset{def.}{=} \frac{Prob(X \cap Y)}{Prob(Y)} = \frac{Prob(X)}{Prob(Y)} = \frac{\frac{4}{8}}{\frac{7}{8}} = \frac{4}{7}.$$

$\square$

Definition 2.2.5.7. *Let S be a sample space with a probability distribution Prob. Two events $X, Y \subseteq S$ are* **independent** *if*

$$Prob(X \cap Y) = Prob(X) \cdot Prob(Y).$$

The following observation relates independence with conditional probability, and it provides an equivalent definition of the independence of two events.

Observation 2.2.5.8. Let S be a sample space with a probability distribution Prob. Let $X, Y \subseteq S$, and let $Prob(Y) \neq 0$. Then X and Y are independent if and only if $Prob(X|Y) = Prob(X)$.

Proof. (i) If X and Y are independent, then $Prob(X \cap Y) = Prob(X) \cdot Prob(Y)$. So,

$$Prob(X|Y) \underset{def.}{=} \frac{Prob(X \cap Y)}{Prob(Y)} = \frac{Prob(X) \cdot Prob(Y)}{Prob(Y)} = Prob(X).$$

(ii) Let $Prob(X|Y) = Prob(X)$ and $Prob(Y) \neq 0$. Then

$$Prob(X) = Prob(X|Y) \underset{def.}{=} \frac{Prob(X \cap Y)}{Prob(Y)},$$

which directly implies $Prob(X \cap Y) = Prob(X) \cdot Prob(Y)$.   □

Due to Observation 2.2.5.8 we see that if two events X and Y are independent, then the knowledge that X (Y) occurs with certainty does not change the probability $Prob(Y)$ ($Prob(X)$). This corresponds to our intuitive meaning of the independence of two events X and Y that if one knows that the experiment results in an event X we cannot obtain any partial information about the correspondence between this result and the event Y.

Example 2.2.5.9. Consider our standard experiment of flipping three coins. Let

$X = \{(head, head, head), (head, head, tail), (head, tail, head), (head, tail, tail)\}$

be the event that the result of the first coin flipping is *head*. Obviously, $Prob(X) = \frac{4}{8} = \frac{1}{2}$. Let

$Y = \{(head, tail, head), (head, tail, tail), (tail, tail, head), (tail, tail, tail)\}$

be the event that the result of the second coin flipping is *tail*. Clearly, $Prob(Y) = \frac{1}{2}$. Since $X \cap Y = \{(head, tail, head), (head, tail, tail)\}$,

$$Prob(X \cap Y) = \frac{2}{8} = \frac{1}{4} = \frac{1}{2} \cdot \frac{1}{2} = Prob(X) \cdot Prob(Y).$$

Thus, X and Y are independent and this corresponds to our intuition because the result of the first coin flipping does not have any influence on the result of the second coin flipping.   □

Exercise 2.2.5.10. Determine all pairs of independent events of the experiments from the above example.   □

Exercise 2.2.5.11. Let S be a sample space with a probability distribution $Prob$. Prove that for all events $A_1, A_2, \ldots, A_n \subseteq S$ such that $Prob(A_1) \neq \emptyset, Prob(A_1 \cap A_2) \neq \emptyset, \ldots, Prob(A_1 \cap A_2 \cap \cdots \cap A_{n-1}) \neq \emptyset$,

$$Prob(A_1 \cap A_2 \cap \cdots \cap A_n) = Prob(A_1) \cdot Prob(A_2|A_1)$$
$$\cdot Prob(A_3|A_1 \cap A_2) \cdot \cdots$$
$$\cdot Prob(A_n|A_1 \cap A_2 \cap \cdots \cap A_{n-1}).   \square$$

Theorem 2.2.5.12 (Bayes' Theorem). *Let S be a sample space with a probability distribution Prob. For every two events $X, Y \subseteq S$ with nonzero probability,*

(i) $Prob(X|Y) = \dfrac{Prob(X) \cdot Prob(Y|X)}{Prob(Y)}$,

(ii) $Prob(X|Y) = \dfrac{Prob(X) \cdot Prob(Y|X)}{Prob(X) \cdot Prob(Y|X) + Prob(S - X) \cdot Prob(Y|S - X)}$.

Proof. (i) From the definition of the conditional probability we obtain

$$Prob(X \cap Y) = Prob(Y) \cdot Prob(X|Y) = Prob(X) \cdot Prob(Y|X).$$

The last equality implies directly

$$Prob(X|Y) = \frac{Prob(X) \cdot Prob(Y|X)}{Prob(Y)}.$$

(ii) To get the equality (ii) from (i) we show that $Prob(Y)$ can be expressed as

$$Prob(X) \cdot Prob(Y|X) + Prob(S - X) \cdot Prob(Y|S - X).$$

Since $Y = (Y \cap X) \cup (Y \cap (S - X))$ and $(Y \cap X) \cap (Y \cap (S - X)) = \emptyset$,

$$\begin{aligned}
Prob(Y) &= Prob(Y \cap X) + Prob(Y \cap (S - X)) \\
&= Prob(X) \cdot Prob(Y|X) + Prob(S - X) \cdot Prob(Y|S - X).
\end{aligned}$$

$\square$

Now, we define a notion that is crucial for the analysis of the behavior of randomized algorithms. Let S be a sample space.[20] Any function F from S to $\mathbb{R}$ is called a (**discrete**) **random variable** on S. This means that we associate a real number with every elementary event of S (outcome of the experiment). To see a motivation for this notion one can consider the work of a randomized algorithm A on a fixed input x as the experiment and one run (computation) of A as an elementary event. If F assigns the length (time complexity) of C to every run C of A, then we can analyze the "expected" time complexity of A by the probability distribution induced by F on $\mathbb{R}$. Then one can ask what the probability is that A computes the output in a given time t or vice versa – what is the smallest time t' such that A finishes the work in time t' with the probability of at least $1/2$. Another possibility to define F is that F assigns 1 to a particular run C of A on x if the output produced in this run is correct, and F assigns 0 to C if the output is wrong. Then the "average" (expected) value of F gives us the information about the reliability of the algorithm A. Thus, the free choice of a random variable provides a powerful instrument for the analysis of the behavior of the considered random experiment. Note that an appropriate choice of random variables does not only decide which properties of the experiment will be investigated, but it also may influence the success of this analytic approach as well as the difficulty (efficiency) of the execution of this analysis.

---

[20] Remember that we assume S is either finite or countably infinite.

Definition 2.2.5.13. *Let S be a sample space with a probability distribution Prob, and let F be a random variable on S. For every $x \in \mathbb{R}$, we define the event $F = x$ by*

$$Event(F = x) = \{s \in S \mid F(s) = x\}.$$

The function $f_F : \mathbb{R} \to [0,1]$ defined by

$$\boldsymbol{f_F(x)} = Prob(Event(F = x)),$$

is called the **probability density function** *of the random variable F.*

The **distribution function** *of F is a function $Dis_F : \mathbb{R} \to [0,1]$ defined by*

$$\boldsymbol{Dis_F(x)} = Prob(F \leq x) = \sum_{y \leq x} Prob(Event(F = y)).$$

In what follows, we use the notation $F = x$ instead of $Event(F = x)$, and so the notation $Prob(F = x)$ instead of $Prob(Event(F = x))$.

Observation 2.2.5.14. Let S, $Prob$, and F be as in Definition 2.2.5.13. Then, for every $x \in \mathbb{R}$,

(i) $f_F(x) = Prob(F = x) = \sum_{\{s \in S \mid F(s) = x\}} Prob(\{s\})$,
(ii) $Prob(F = x) \geq 0$, and
(iii) $\sum_{y \in \mathbb{R}} Prob(F = y) = 1$.

$\square$

Example 2.2.5.15. Consider the experiment of rolling three 6-sided dices. The outcome of rolling one dice is one of the numbers 1, 2, 3, 4, 5, and 6 and we consider that the probability distribution is uniform. So, $S = \{(a,b,c) \mid a,b,c \in \{1,2,3,4,5,6\}\}$ and $Prob(\{s\}) = \frac{1}{6^3} = \frac{1}{216}$ for every $s \in S$. Define the random variable F to be the sum of the values on all three dices. For instance, $F((3,1,5)) = 3 + 1 + 5 = 9$. The probability of the event $F = 5$ is

$$Prob(F = 5) = \sum_{\substack{s \in S \\ F(s) = 5}} Prob(\{s\}) = \sum_{\substack{a+b+c=5 \\ a,b,c \in \{1,2,\ldots,6\}}} Prob(\{(a,b,c)\})$$

$$= Prob(\{(1,1,3)\}) + Prob(\{(1,3,1)\})$$
$$+ Prob(\{(3,1,1)\}) + Prob(\{(1,2,2)\})$$
$$+ Prob(\{(2,1,2)\}) + Prob(\{(2,2,1)\})$$

$$= 6 \cdot \frac{1}{216} = \frac{1}{36}.$$

Let G be the random variable defined by $G((a,b,c)) = \max\{a,b,c\}$ for every elementary event $(a,b,c) \in S$.

$$Prob(G = 3) = \sum_{\substack{s \in S \\ G(s) = 3}} Prob(\{s\}) = \sum_{\substack{\max\{a,b,c\}=3 \\ a,b,c \in \{1,2,\ldots,6\}}} Prob(\{(a,b,c)\})$$

$$= \sum_{b,c\in\{1,2\}} Prob(\{(3,b,c)\}) + \sum_{a,c\in\{1,2\}} Prob(\{(a,3,c)\})$$

$$+ \sum_{a,b\in\{1,2\}} Prob(\{(a,b,3)\}) + \sum_{a\in\{1,2\}} Prob(\{(3,3,a)\})$$

$$+ \sum_{b\in\{1,2\}} Prob(\{(3,b,3)\}) + \sum_{a,c\in\{1,2\}} Prob(\{(a,3,3)\})$$

$$+ Prob(\{(3,3,3)\})$$

$$= \frac{4}{216} + \frac{4}{216} + \frac{4}{216} + \frac{2}{216} + \frac{2}{216} + \frac{2}{216} + \frac{1}{216} = \frac{19}{216}.$$

□

Thus, we have seen that one can define several random variables on the same sample space.

Definition 2.2.5.16. *Let S be a sample space with a probability distribution Prob, and let X and Y be two random variables on S. The **joint probability density function** of X and Y is the function $f_{X,Y} : \mathbb{R} \times \mathbb{R} \to [0,1]$ defined by*

$$f_{X,Y}(x,y) = Prob(X = x \text{ and } Y = y) = Prob(Event(X = x) \cap Event(Y = y)).$$

*X and Y are **independent** if, for all $x, y \in \mathbb{R}$,*

$$Prob(X = x \text{ and } Y = y) = Prob(X = x) \cdot Prob(Y = y).$$

□

Note that the above definition of the independence of X and Y is a natural extension of the notion of the independence of two events. Obviously,

$$Prob(X = x) = \sum_{y\in\mathbb{R}} Prob(X = x \text{ and } Y = y),$$

and

$$Prob(Y = y) = \sum_{x\in\mathbb{R}} Prob(X = x \text{ and } Y = y).$$

Applying the notion of conditional probabilities

$$Prob(X = x | Y = y) = \frac{Prob(X = x \text{ and } Y = y)}{Prob(Y = y)}.$$

Thus, if $Event(X = x)$ and $Event(Y = y)$ are independent, then $Prob(X = x | Y = y) = Prob(X = x)$ and we obtain the definition of the independence of X and Y.

The simplest and most useful characterization of the distribution of a random variable is the average of the values it takes on. This average value will be called the expected value in what follows. A good algorithmic motivation for the study of the expected value may be the relation to the analysis of the behavior and the time complexity of randomized algorithms.

Definition 2.2.5.17. *Let S be a sample space with a probability distribution Prob, and let X be a random variable on S. The* **expected value** *(or* **expectation** *of X) is*

$$E[X] = \sum_{x \in \mathbb{R}} x \cdot Prob(X = x)$$

if the sum is finite or converges absolutely.

Exercise 2.2.5.18. Let S be a sample space with a probability distribution *Prob*, and let X be an random variable on S. Prove that

$$E[X] = \sum_{s \in S} X(s) \cdot Prob(\{s\}).$$

$\square$

Example 2.2.5.19. Consider the experiment of rolling one 6-sided dice, and the random variable F defined by $F(a) = a$ for $a \in S = \{1, 2, \ldots, 6\}$.

$$E[F] = \sum_{a \in S} F(a) \cdot Prob(F = a) = \sum_{a \in S} a \cdot \frac{1}{6} = \frac{1}{6} \cdot \sum_{a \in S} a = \frac{1}{6} \cdot \sum_{i=1}^{6} i = \frac{21}{6} = \frac{7}{2}.$$
$\square$

In what follows, a discrete random variable is called an **indicator variable** if it takes only values 0 and 1. An indicator variable X is used to denote the occurrence or nonoccurrence of an event E, where

$$E = \{s \in S \mid X(s) = 1\} \text{ and } S - E = \{s \in S \mid X(s) = 0\}.$$

The above mentioned variable F, with $F(C) = 1$ if the run C of a randomized algorithm computes the right output, is an example of an indicator variable.

If X is a random variable, and g a function from $\mathbb{R}$ to $\mathbb{R}$, then $g(X)$ is a random variable, too. If the expectation of $g(X)$ is defined, then clearly[21]

$$E[g(X)] = \sum_{x \in \mathbb{R}} g(x) \cdot Prob(X = x).$$

Particularly, if $g(x) = r \cdot X$, then

$$E[g(x)] = E[r \cdot X] = r \cdot E[X].$$

Observation 2.2.5.20. Let S be a sample space with a probability distribution *Prob*, and let X and Y be two random variables. Then

$$E[X + Y] = E[X] + E[Y].$$

$\square$

The property of expectations presented in Observation 2.2.5.20 is called **linearity of expectation**.

---

[21] See Exercise 2.2.5.18.

Theorem 2.2.5.21. *Let S be a sample space with a probability distribution Prob. For any two independent random variables X and Y with defined $E[X]$ and $E[Y]$, respectively,*

$$E[X \cdot Y] = E[X] \cdot E[Y].$$

Proof.

$$E[X \cdot Y] = \sum_{z \in \mathbb{R}} z \cdot Prob(X \cdot Y = z)$$

$$= \sum_x \sum_y x \cdot y \cdot Prob(X = x \text{ and } Y = y)$$

$$= \sum_x \sum_y xy \, Prob(X = x) \cdot Prob(Y = y)$$

$$= \left(\sum_x x \, Prob(X = x) \right) \cdot \left(\sum_y y \, Prob(Y = y) \right)$$

$$= E[X] \cdot E[Y]. \qquad \square$$

Definition 2.2.5.22. *Let S be a sample space with a probability distribution Prob. Let $X_1, X_2, \ldots, X_n$, $n \in \mathbb{N}^+$ be random variables on S. We say that $X_1, X_2, \ldots, X_n$ are **mutually independent** if, for all $x_1, x_2, \ldots, x_n \in \mathbb{R}$,*

$$Prob(X_1 = x_1 \text{ and } X_2 = x_2 \text{ and } \ldots \text{ and } X_n = x_n) =$$
$$Prob(X_1 = x_1) \cdot Prob(X_2 = x_2) \cdot \cdots \cdot Prob(X_n = x_n).$$

Exercise 2.2.5.23. Prove the following generalization of Theorem 2.2.5.21. For any n random variables $X_1, X_2, \ldots, X_n$ that are mutually independent,

$$E[X_1 \cdot X_2 \cdot \cdots \cdot X_n] = E[X_1] \cdot E[X_2] \cdot \cdots \cdot E[X_n]. \qquad \square$$

Example 2.2.5.24. Consider the experiment of consecutively rolling three 6-sided dices and the random variable F defined by $F((a, b, c)) = 3a + 2b + c$ for every $(a, b, c) \in S$. We want to compute $E[F]$ without working with the sum $\sum_x x \cdot Prob(F = x)$ consisting of 215 additions. We define three random variables F_1, F_2, and F_3 as follows:

$$F_1(a, b, c) = a, \, F_2(a, b, c) = b, \text{ and } F_3(a, b, c) = c.$$

Obviously, $F = 3F_1 + 2F_2 + F_3$. Since $E[F_i] = 7/2$ for $i = \{1, 2, 3\}$ as computed in the previous example, we obtain

$$E[F] = 3 \cdot E[F_1] + 2 \cdot E[F_2] + E[F_3] = 6 \cdot \frac{7}{2} = 21. \qquad \square$$

The following two examples illustrate the usefulness of the notions random variable and expectation for practical purposes in algorithmics. The first example shows how the investigation of $E[X]$ of a randomized variable X can lead to the design of an efficient algorithm for a given task. The second example shows how to use these notions to analyze the complexity of a randomized algorithm.

Example 2.2.5.25. Let $F = F(x_1, \ldots, x_n)$ be a formula in conjunctive normal form over a set $\{x_1, \ldots, x_n\}$ of n variables. Our aim is to find an assignment to $\{x_1, \ldots, x_n\}$ such that as many as possible clauses are satisfied. Using a simple probabilistic consideration we show that there exists an input assignment that satisfies at least half of the clauses.

Let F consist of m clauses, i.e., $F = F_1 \wedge F_2 \wedge \cdots \wedge F_m$. Suppose the following experiment. We choose the values of $x_1, x_2, \ldots, x_n$ randomly with $Prob(x_i = 1) = Prob(x_i = 0) = 1/2$ for $i = 1, 2, \ldots, n$. Now, we define m random variables $Z_1, \ldots, Z_m$ where, for $i = 1, \ldots, m$, $Z_i(\alpha) = 1$ if F_i is satisfied by the assigned α and $Z_i(\alpha) = 0$ otherwise. For every clause of k distinct literals, the probability that it is not satisfied by a random variable assignment to the set of variables is 2^{-k}, since this event takes place if and only if each literal gets the value 0, and the Boolean values are assigned independently to distinct literals in any clause. This implies that the probability that a clause with k literals is satisfied is at least $1 - 2^{-k} \geq 1/2$ for every $k \geq 1$, i.e.,

$$E[Z_i] \geq 1/2$$

for all $i = 1, \ldots, m$.

Now, we define a random variable Z as $Z = \sum_{i=1}^{m} Z_i$. Obviously, Z counts the number of satisfied clauses. Because of the linearity of expectation

$$E[Z] = E\left[\sum_{i=1}^{m} Z_i\right] = \sum_{i=1}^{m} E[Z_i] \geq \sum_{i=1}^{m} \frac{1}{2} = \frac{m}{2}.$$

Thus, we are sure that there exists an assignment satisfying at least one half of the clauses of F. The following algorithm outputs an assignment whose expected number of satisfied clauses is at least $m/2$.

Algorithm 2.2.5.26 (RANDOM ASSIGNMENT).

Input: A formula $F = F(x_1, \ldots, x_n)$ over n variables $x_1, \ldots, x_n$ in CNF.

Step 1:   Choose uniformly at random n Boolean values $a_1, \ldots, a_n$ and set $x_i = a_i$ for $i = 1, \ldots, n$.

Step 2:   Evaluate each clause of F and set $Z :=$ the number of satisfied clauses.

Output: $(a_1, a_2, \ldots, a_n), Z$.

$\square$

Exercise 2.2.5.27. Let F be a formula in CNF whose every clause consists of at least k distinct variables, $k \geq 2$. Which lower bound on $E[Z]$ can be proved in this case? $\square$

Example 2.2.5.28. Consider the task[22] of sorting a set S of n elements into an increasing order. One of the well-known recursive algorithms for this task is the following RANDOMIZED QUICKSORT RQS(S).

> **Algorithm 2.2.5.29.** RQS(S)
> Input: A set S of numbers.
> Step 1: Choose an element a uniformly at random from S.
>   {every element in S has the probability $\frac{1}{|S|}$ of being chosen}
> Step 2: $S_< := \{b \in S \mid b < a\}$;
> $S_> := \{c \in S \mid c > a\}$;
> Step 3: **output**(RQS($S_<$), a, RQS($S_>$)).
> Output: The sequence of elements of S in increasing order.

The goal of this example is to show that the notions of random variables and expectation can be helpful to estimate the "average" (expected) complexity of this algorithm. As usual for sorting, the complexity is measured in the number of comparisons of pairs of elements of S. Let $|S| = n$ for a positive integer n.

We observe that the complexity of Step 2 is exactly $|S| - 1 = n - 1$. Intuitively, the best random choices of elements from S are choices dividing S into two approximately equal sized sets $S_<$ and $S_>$. In the terminology of recursion this means that the original problems of the size n are reduced to two problems of size $n/2$. So, if $T(n)$ denotes the complexity for this kind of choices, then

$$T(n) \leq 2 \cdot T(n/2) + n - 1.$$

Following Section 2.2.2 we already know that the solution of this recurrence is $T(n) \in O(n \cdot \log n)$. A very bad sequence of random choices is when the smallest element of the given set is always chosen. In this case the number of comparisons is

$$T(n) = \sum_{i=1}^{n-1} i \in \Theta(n^2).$$

Since one can still show that, for the recurrence inequality

$$T(n) \leq T\left(\frac{n}{4}\right) + T\left(\frac{3}{4} \cdot n\right) + n - 1,$$

$T(n) \in O(n \log n)$, RQS will behave well also if the size $|S_<|$ of $S_<$ very roughly approximates $|S_>|$. But this happens when the probability is at least $1/2$ because at least half of the elements are good choices. This is the reason for our hope that algorithm RQS behaves very well in the average. In what follows we carefully analyze the expected complexity of RQS.

---

[22] One of the fundamental computing problems

Let $s_1, s_2, \ldots, s_n$ be the output[23] of the algorithm RQS. Our experiment is the sequence of random choices of RQS. We define the random variable X_{ij} by

$$X_{ij}(C) = \begin{cases} 1 \text{ if } s_i \text{ and } s_j \text{ are compared in the run } C \text{ of RQS} \\ 0 \text{ otherwise} \end{cases}$$

for all $i, j \in \{1, \ldots, n\}$, $i < j$. Obviously, the random variable

$$T = \sum_{i=1}^{n} \sum_{j>i} X_{ij}$$

counts the total number of comparisons. So,

$$E[T] = E\left[\sum_{i=1}^{n} \sum_{j>i} X_{ij}\right] = \sum_{i=1}^{n} \sum_{j>i} E[X_{ij}] \tag{2.25}$$

is the expected complexity of the algorithm RQS.[24] It remains to estimate $E[X_{ij}]$.

Let p_{ij} denote the probability that s_i and s_j are compared in an execution. Since X_{ij} is either 1 or 0,

$$E[X_{ij}] = p_{ij} \cdot 1 + (1 - p_{ij}) \cdot 0 = p_{ij}.$$

Now, consider for every $i, j \in \{1, \ldots, n\}$, $i < j$, the subsequence

$$s_i, s_{i+1}, \ldots, s_{i+j-1}, s_j.$$

If some s_d with $i < d < j$ was randomly chosen by RQS(S) before either s_i or s_j has been randomly chosen, then s_i and s_j were not compared.[25] If s_i or s_j has been randomly chosen to play the splitting role before any of the elements from $\{s_{i+1}, s_{i+2}, \ldots, s_{i+j-1}\}$ have been randomly chosen, then s_i and s_j were compared in the corresponding run[26] of RQS(S). Since each of the elements of $\{s_i, s_{i+1}, \ldots, s_j\}$ is equal likely to be the first in the sequence of random choices,

$$p_{ij} = \frac{2}{j - i + 1}. \tag{2.26}$$

Inserting (2.26) into (2.25) we finally obtain

$$E[T] = \sum_{i=1}^{n} \sum_{j>i} p_{ij}$$

---

[23] That is, $s_1 < s_2 < \cdots < s_n$.

[24] Note that (2.25) holds because of the linearity of expectation.

[25] This is because $s_i \in S_<$ and $s_j \in S_>$ according to d.

[26] In the run corresponding to this sequence of random choices.

$$= \sum_{i=1}^{n} \sum_{j>i} \frac{2}{j-i+1}$$

$$\leq \sum_{i=1}^{n} \sum_{k=1}^{n-i+1} \frac{2}{k}$$

$$\leq 2 \sum_{i=1}^{n} \sum_{k=1}^{n} \frac{1}{k}$$

$$= 2 \sum_{i=1}^{n} Har(n)$$

$$= 2n \cdot Har(n) \approx 2 \cdot n \cdot \ln n + \Theta(n).$$

Thus, as expected, the expected time complexity of RQS is in $O(n \log n)$. $\square$

Keywords introduced in Section 2.2.5

sample space, event, probability distribution, conditional probability, random variable, probability density function, distribution function, expected value (expectation), independence of events, independence of random variables, linearity of expectation

2.3 Fundamentals of Algorithmics

2.3.1 Alphabets, Words, and Languages

All data are represented as strings of symbols. The kind of data representation is often important for the efficiency of algorithm implementations. Here, we present some elementary fundamentals of formal language theory. We do not need to deal too much with details of data representation because we consider algorithms on an abstract design level and do not often work with the details of implementation. The main goal of this section is to give definitions of notions that are sufficient for fixing the representation of some input data and thus to precisely formalize the definitions of some fundamental algorithmic problems. We also need the terms defined here for the abstract considerations of the complexity theory in Section 2.3.3 and for proving lower bounds on polynomial-time inapproximability in Section 4.4.2.

Definition 2.3.1.1. *Any non-empty, finite set is called an* **alphabet**. *Every element of an alphabet Σ is called a* **symbol** *of Σ.*

An alphabet has the same meaning for algorithmics as for natural languages – it is a collection of signs or symbols used in a more or less uniform fashion by a number of people in order to represent information and so to communicate with each other. Thus, alphabets are used for communication between human and machine, between computers, and in algorithmic information processing. A symbol of an alphabet is often considered as a possible content of the computer word. Fixing an alphabet means to fix all possible computer words in this interpretation. Examples of alphabets are

$$\Sigma_{bool} = \{0,1\},$$
$$\Sigma_{lat} = \{a,b,c,\dots,z\},$$
$$\Sigma_{logic} = \{0,1,(,),\wedge,\vee,\neg,x\}.$$

Definition 2.3.1.2. *Let Σ be an alphabet. A* **word** *over Σ is any finite sequence of symbols of Σ. The* **empty word** λ *is the only word consisting of zero symbols. The set of all words over the alphabet Σ is denoted by Σ^*.*

The interpretation of the notion "word over Σ" is a text consisting of symbols of Σ rather than a term representing a notion. So, the contents of a book can be considered as a word over some alphabet including the symbol blank and symbols of Σ_{lat}.

$w = 0,1,0,0,1,0$ is a word over Σ_{bool}. In what follows we usually omit the commas and represent w simply by 010010. So $abcxyzef$ is a word over Σ_{lat}. For $\Sigma = \{a,b\}$, $\Sigma^* = \{\lambda, a, b, aa, ab, ba, bb, aaa, \dots\}$.

Definition 2.3.1.3. *The* **length of a word** w *over an alphabet Σ, denoted by $|w|$, is the number of symbols in w (i.e., the length of w as a sequence). For every word $w \in \Sigma^*$, and every symbol $a \in \Sigma$, $\#_a(w)$ is the number of occurrences of the symbol a in the word w.*

For the word $w = 010010$, $|w| = 6$, $\#_0(w) = 4$, and $\#_1(w) = 2$. We observe that for every alphabet Σ and every word $w \in \Sigma^*$,

$$|w| = \sum_{a \in \Sigma} \#_a(w).$$

Definition 2.3.1.4. *Let Σ be an alphabet. Then, for any $n \in \mathbb{N}$,*

$$\Sigma^n = \{x \in \Sigma^* \mid |x| = n\}.$$

For instance, $\{a,b\}^3 = \{aaa, aab, aba, baa, abb, bab, bba, bbb\}$. We define $\Sigma^+ = \Sigma^* - \{\lambda\}$.

Definition 2.3.1.5. *Given two words v and w over an alphabet Σ, we define the* **concatenation of v and w**, *denoted by vw (or by $v \cdot w$) as the word that consists of the symbols of v in the same order, followed by the symbols of w in the same order.*
For every word $w \in \Sigma^*$, we define

(i) $w^0 = \lambda$, *and*
(ii) $w^{n+1} = w \cdot w^n = ww^n$ *for every positive integer n.*

A **prefix** *of a word* $w \in \Sigma^*$ *is any word* v *such that* $w = vu$ *for some word* u *over* Σ. *A* **suffix** *of a word* $w \in \Sigma^*$ *is any word* u *such that* $w = xu$ *for some word* $x \in \Sigma^*$. *A* **subword** *of a word* w *over* Σ *is any word* $z \in \Sigma^*$ *such that* $w = uzv$ *for some words* $u, v \in \Sigma^*$.

The word $abbcaa$ is the concatenation of the words ab and $bcaa$. The words $abbcaa$, a, ab, bca, and $bbcaa$ are examples of subwords of $abbcaa = ab^2ca^2$. The words a, ab, ab^2, ab^2c, ab^2ca, and ab^2ca^2 are all prefixes of $abbcaa$. caa and a^2 are examples of suffixes of $abbcaa$.

Exercise 2.3.1.6. Prove that, for every alphabet Σ, $(\Sigma^*, \cdot)$, where $\cdot$ is the operation of concatenation, is a monoid. □

In what follows we use words to code data and so to represent input and output data, as well as the contents of the computer memory. Since the complexity of algorithms is measured according to the input length, the first step in the complexity analysis is to fix the alphabet and the data representation over this alphabet. This automatically determines the length of every input. We usually code integers as binary words. For every $u = u_n u_{n-1} \ldots u_2 u_1 \in \Sigma_{bool}^n$, $u_i \in \Sigma_{bool}$ for $i = 1, \ldots, n$,

$$Number(u) = \sum_{i=1}^{n} u_i \cdot 2^{i-1}$$

is the integer coded by u. Thus, for instance, $Number(000) = Number(0) = 0$, and $Number(1101) = 1 \cdot 2^0 + 0 \cdot 2^1 + 1 \cdot 2^2 + 1 \cdot 2^3 = 1 + 0 + 4 + 8 = 13$.

One can use the alphabet $\{0, 1, \#\}$ to code graphs. If $M_G = [a_{ij}]_{i,j=1,\ldots,n}$ is an adjacency matrix of a graph G of n vertices, then the word

$$a_{11}a_{12} \ldots a_{1n} \# a_{21}a_{22} \ldots a_{2n} \# \ldots \# a_{n1}a_{n2} \ldots a_{nn}$$

can be used to code G.

Exercise 2.3.1.7. Design a representation of graphs by words over Σ_{bool}. □

Exercise 2.3.1.8. Design a representation of weighted graphs, where weights are some positive integers, using the alphabet $\{0, 1, \#\}$. □

One can use the alphabet Σ_{logic} to represent formulae over a variable set $X = \{x_1, x_2, x_3, \ldots\}$ and operations $\vee, \wedge$, and $\neg$. Since we have infinitely many variables, we cannot use symbols x_i as symbols of the alphabet. We code a variable x_j by $xbin(j)$, where $bin(j)$ is the shortest word[27] over Σ_{bool} such that $Number(bin(j)) = j$, and x is a symbol of Σ_{logic}. Then, the code of a formula Φ can be obtained by simply exchanging x_i with $xbin(i)$ for every occurrence of x_i in Φ. For instance, the formula

---

[27] Thus, the first (most significant) bit of $bin(j)$ is 1.

$$\Phi = (x_1 \vee \overline{x}_4 \vee x_7) \wedge (x_2 \vee \overline{x}_1) \wedge (x_4 \wedge \overline{x}_8)$$

is represented by the word

$$w_\Phi = (x1 \vee \neg(x100) \vee x111) \wedge (x10 \vee \neg(x1)) \wedge (x100 \wedge \neg(x1000))$$

over $\Sigma_{logic} = \{0, 1, (,), \wedge, \vee, \neg, x\}$.

Definition 2.3.1.9. *Let Σ be an alphabet. Every set $L \subseteq \Sigma^*$ is called a* **language** *over Σ. The* **complement of the language L** *according to Σ is* $L^C = \Sigma^* - L$.

Let Σ_1 and Σ_2 be alphabets, and let $L_1 \subseteq \Sigma_1^*$ and $L_2 \subseteq \Sigma_2^*$ be languages. The **concatenation** *of L_1 and L_2 is*

$$L_1 L_2 = L_1 \circ L_2 = \{uv \in (\Sigma_1 \cup \Sigma_2)^* \mid u \in L_1 \text{ and } v \in L_2\}.$$

$\emptyset$, $\{\lambda\}$, $\{a, b\}$, $\{a, b\}^*$, $\{ab, bba, b^{10}a^{20}\}$, $\{a^n b^{2^n} \mid n \in \mathbb{N}\}$ are examples of languages over $\{a, b\}$. Observe that $L \cdot \emptyset = \emptyset \cdot L = \emptyset$ and $L \cdot \{\lambda\} = \{\lambda\} \cdot L = L$ for every language L. $U = \{1\} \cdot \{0, 1\}^*$ is the language of binary representations of all positive integers.

A language can be used to describe a set of consistent input instances of an algorithmic problem. For instance, the set of all representations of formulae in CNF as a set of words over Σ_{logic} or the set $\{u_1 \# u_2 \# \ldots \# u_m \mid u_i \in \{0, 1\}^m, m \in \mathbb{N}\}$ as the set of representations of all directed graphs over $\{0, 1, \#\}$ are examples of such languages. But words can also be used to code programs and so one can consider the language of codes of all correct programs in a given programming language. Languages can be also used to describe so-called decision problems, but this is the topic of the next section.

The last definition of this section shows how one can define a linear ordering on words over some alphabet Σ, provided one has a linear ordering on the symbols of Σ.

Definition 2.3.1.10. *Let $\Sigma = \{s_1, s_2, \ldots, s_m\}$, $m \geq 1$, be an alphabet, and let $s_1 < s_2 < \cdots < s_m$ be a linear ordering on Σ. We define the* **canonical ordering** *on Σ^* as follows. For all $u, v \in \Sigma^*$,*

$$u < v \text{ if } |u| < |v|$$
$$\text{or } |u| = |v|, u = xs_i u', \text{ and } v = xs_j v'$$
$$\text{for some } x, u', v' \in \Sigma^*, \text{ and } i < j.$$

Keywords introduced in Section 2.3.1

alphabet, symbol, word, empty word, the length of a word, concatenation, prefix, suffix, subword, binary representation of integers, language, the complement of a language, canonical ordering of words

2.3.2 Algorithmic Problems

Thousands of algorithmic problems classified according to different points of view are considered in the literature on algorithmics. In this book we deal with hard problems only. We consider a problem to be hard if there is no known deterministic algorithm (computer program) that solves it efficiently. Efficiently means in a low-degree polynomial time. Our interpretation of hardness here is connected to the current state of our knowledge in algorithmics rather than to the unknown, real difficulty of the problems considered. Thus, a problem is hard if one would need years or thousands of years to solve it by deterministic programs for an input of a realistic size appearing in the current practice. This book provides a handbook of algorithmic methods that attack hard problems. Thousands of problems of great practical relevance are very hard from this point of view. Fortunately, we do not need to define and to consider all of them. There are some crucial, paradigmatic problems such as the traveling salesperson problem, linear (integer) programming, set cover problem, knapsack problem, satisfiability problem, and primality testing that are pattern problems in the sense that solving most of the hard problems can be reduced to solving some of the paradigmatic problems.

The goal of this chapter is to define some of these fundamental pattern problems. The methods for solving them are the topic of the next chapters. Every algorithm (computer program) can be viewed as an execution of a mapping from a subset of Σ_1^* to Σ_2^* for some alphabets Σ_1 and Σ_2. So, every (algorithmic) problem can be considered as a function from Σ_1^* to Σ_2^* or as a relation on $\Sigma_1^* \times \Sigma_2^*$ for some alphabets Σ_1 and Σ_2. We usually do not need to work with this kind of formalism because we consider two classes of problems only – decision problems (to decide for a given input whether it has a prescribed property) and optimization problems (to find the "best" solution from the set of solutions determined by some constraints). In what follows we define the fundamental problems that will be the objects of the algorithmic design in the subsequent chapters. We start with decision problems. If A is an algorithm and x is an input, then $\boldsymbol{A(x)}$ denotes the output of A for the input x.

Definition 2.3.2.1. *A* **decision problem** *is a triple* (L, U, Σ) *where* Σ *is an alphabet and* $L \subseteq U \subseteq \Sigma^*$. *An algorithm A* **solves** (**decides**) *the decision problem* (L, U, Σ) *if, for every* $x \in U$,

(i) $A(x) = 1$ *if* $x \in L$, *and*
(ii) $A(x) = 0$ *if* $x \in U - L$ $(x \notin L)$.

We see that any algorithm A solving a decision problem (L, U, Σ) computes a function from U to $\{0,1\}$. The output "1" is interpreted as the answer "yes" to the question whether a given input belongs to L (whether the input has the property corresponding to the specification of the language L), and the output "0" is equivalent to the answer "no".

An equivalent form of a description of a decision problem is the following form that specifies the input-output behavior.

Problem (L, U, Σ)

Input: An $x \in U$.
Output: "yes" if $x \in L$,
        "no" otherwise.

For many decision problems (L, U, Σ) we assume $U = \Sigma^*$. In that case we shall use the short notation $(\boldsymbol{L}, \boldsymbol{\Sigma})$ instead of (L, Σ^*, Σ).

Next we present the fundamental decision problems that will be studied in Chapter 5.

PRIMALITY TESTING.

Informally, primality testing is to decide, for a given positive integer, whether it is prime or not. Thus, primality testing is a decision problem $(\text{PRIM}, \Sigma_{bool})$, where

$$\text{PRIM} = \{w \in \{0,1\}^* \mid Number(w) \text{ is a prime}\}.$$

Another description of this problem is

Primality testing

Input: An $x \in \Sigma_{bool}^*$
Output: "yes" if $Number(x)$ is a prime,
        "no" otherwise.

One can easily observe that primality testing can also be considered for other integer representations. Using $\Sigma_k = \{0, 1, 2, \ldots, k-1\}$ and the k-ary representation of integers we obtain $(\text{PRIM}_k, \Sigma_k)$, where

$$\text{PRIM}_k = \{x \in \Sigma_k^* \mid x \text{ is the } k\text{-ary representation of a prime}\}.$$

From the point of view of computational hardness it is not essential whether we consider $(\text{PRIM}, \Sigma_{bool})$ or $(\text{PRIM}_k, \Sigma_k)$ for some constant k because one has efficient algorithms for transferring any k-ary representation of an integer to its binary representation, and vice versa. But this does not mean that the representation of integers does not matter for primality testing. If one represents an integer n as

$$\#bin(p_1)\#bin(p_2)\#\ldots\#bin(p_l)$$

over $\{0, 1, \#\}$, where $n = p_1 \cdot p_2 \cdot \cdots \cdot p_l$ and p_is are the nontrivial prime factors of n, then the problem of primality testing becomes easy. This sensibility of hardness of algorithmic problems according to the representation of their inputs is sometimes the reason for taking an exact formal description of the problem that fixes the data representation, too. For primality testing we always consider $(\text{PRIM}, \Sigma_{bool})$ as the formal definition of this decision problem.

EQUIVALENCE PROBLEM FOR POLYNOMIALS.

The problem is to decide, for a given prime p and two polynomials $p_1(x_1, \ldots, x_m)$ and $p_2(x_1, \ldots, x_m)$ over the field $\mathbb{Z}_p$, whether p_1 and p_2 are equivalent, i.e., whether $p_1(x_1, \ldots, x_m) - p_2(x_1, \ldots, x_m)$ is identical 0. The crucial point is that the polynomials are not necessarily given in a normal form such as

$$a_0 + a_1 x_1 + a_2 x_2 + a_{12} x_1 x_2 + a_1^2 x_1^2 + a_2^2 x_2^2 + \cdots$$

but in an arbitrary form such as

$$(x_1 + 3x_2)^2 \cdot (2x_1 + 4x_4) \cdot x_3^2.$$

A normal form may be exponentially long in the length of another representation and so the obvious way to compare two polynomials by transferring them to their normal forms and comparing their coefficients is not efficient.

We omit the formal definition of the representation of polynomials in an arbitrary form over the alphabet $\Sigma_{pol} = \{0, 1, (,), \exp, +, \cdot\}$, because it can be done in a similar way as how one represents formulae over Σ_{logic}. The equivalence problem for polynomials can be defined as follows.

EQ-POL

Input: A prime p, two polynomials p_1 and p_2 over variables from $X = \{x_1, x_2, \ldots\}$.

Output: "yes" if $p_1 \equiv p_2$ in the field $\mathbb{Z}_p$,
         "no" otherwise.

EQUIVALENCE PROBLEM FOR ONE-TIME-ONLY BRANCHING PROGRAMS.

The equivalence problem for one-time-only branching programs, EQ-1BP, is to decide, for two given one-time-only branching programs B_1 and B_2, whether B_1 and B_2 represent the same Boolean function. One can represent a branching program in a similar way as a directed weighted graph[28] and so we omit the formal description of branching program representation.[29]

EQ-1BP

Input: One-time-only branching program B_1 and B_2 over a set of Boolean variables $X = \{x_1, x_2, x_3, \ldots\}$.

Output: "yes"   if B_1 and B_2 are equivalent (represent the same Boolean function),
         "no"   otherwise.

---

[28] Where not only the edges have some labels, but also the vertices are labeled.

[29] Remember that the formal definition of branching programs was given in Section 2.2.3 (Definitions 2.2.3.19, 2.2.3.20, Figure 2.11).

SATISFIABILITY PROBLEM.

The satisfiability problem is to decide, for a given formula in the CNF, whether it is satisfiable or not. Thus, the **satisfiability problem** is the decision problem $(\text{SAT}, \Sigma_{logic})$, where

$$\text{SAT} = \{w \in \Sigma_{logic}^+ \mid w \text{ is a code of a satisfiable formula in CNF}\}.$$

We also consider specific subproblems of SAT where the length of clauses of the formulae in CNF is bounded. For every positive integer $k \geq 2$, we define the **k-satisfiability problem** as the decision problem $(k\text{SAT}, \Sigma_{logic})$, where

$$k\text{SAT} = \{w \in \Sigma_{logic}^+ \mid w \text{ is a code of a satisfiable formula in } k\text{CNF}\}.$$

In what follows we define some decidability problems from graph theory.

CLIQUE PROBLEM.

The clique problem is to decide, for a given graph G and a positive integer k, whether G contains a clique of size k (i.e., whether the complete graph K_k of k vertices is a subgraph of G). Formally, the **clique problem** is the decision problem $(\text{CLIQUE}, \{0, 1, \#\})$, where

$$\text{CLIQUE} = \{x\#w \in \{0, 1, \#\}^* \mid x \in \{0, 1\}^* \text{ and } w \text{ represents a graph}$$
$$\text{that contains a clique of size } Number(x)\}.$$

An equivalent description of the clique problem is the following one.

Clique Problem

Input: A positive integer k and a graph G.
Output: "yes" if G contains a clique of size k,
         "no" otherwise.

VERTEX COVER PROBLEM.

The vertex cover problem is to decide, for a given graph G and a positive integer k, whether G contains a vertex cover of cardinality k. Remember that a vertex cover of $G = (V, E)$ is any set S of vertices of G such that each edge from E is incident to at least one vertex in S.

Formally, the **vertex cover problem (VCP)** is the decision problem $(\text{VCP}, \{0, 1, \#\})$, where

$$\text{VCP} = \{u\#w \in \{0, 1, \#\}^+ \mid u \in \{0, 1\}^+ \text{ and } w \text{ represents a graph that}$$
$$\text{contains a vertex cover of size } Number(u)\}.$$

HAMILTONIAN CYCLE PROBLEM.

The Hamiltonian cycle problem is to determine, for a given graph G, whether G contains a Hamiltonian cycle or not. Remember that a Hamiltonian cycle of G of n vertices is a cycle of length n in G that contains every vertex of G.

Formally, the **Hamiltonian cycle problem (HC)** is the decision problem $(HC, \{0, 1, \#\})$, where

$$HC = \{w \in \{0, 1, \#\}^* \mid w \text{ represents a graph that}$$
$$\text{contains a Hamiltonian cycle}\}.$$

EXISTENCE PROBLEMS IN LINEAR PROGRAMMING.

Here, we consider problems of deciding whether a given system of linear equations has a solution. Following the notation of Section 2.2.1 a system of linear equations is given by the equality

$$A \cdot X = b,$$

where $A = [a_{ij}]_{i=1,\ldots,m,j=1,\ldots,n}$ is an $m \times n$ matrix, $X = (x_1, x_2, \ldots, x_n)^\mathsf{T}$, and $b = (b_1, \ldots, b_m)^\mathsf{T}$ is an m-dimensional column vector. The n elements $x_1, x_2, \ldots, x_n$ of X are called unknowns (variables). In what follows we consider that all elements of A and b are integers. Remember, that

$$Sol(A, b) = \{X \subseteq \mathbb{R}^n \mid A \cdot X = b\}$$

denotes the set of all real-valued solutions of the linear equations system $A \cdot X = b$. In what follows we are interested in deciding whether $Sol(A, b)$ is empty or not (i.e., whether there exist a solution to $A \cdot X = b$) for given A and b. More precisely, we consider several specific decision problems by restricting the set $Sol(A, b)$ to subsets of solutions over $\mathbb{Z}^n$ or $\{0, 1\}^n$ only, or even considering the linear equations over some finite fields instead of $\mathbb{R}$. Let

$$Sol_S(A, b) = \{X \subseteq S^n \mid A \cdot X = b\}$$

for any subset S of $\mathbb{R}$.

First of all observe that the problem of deciding whether $Sol(A, b) = \emptyset$ is one of the fundamental tasks of linear algebra and that it can be solved efficiently. The situation essentially changes if one searches for integer solutions or Boolean solutions. Let $\langle A, b \rangle$ denote a representation of a matrix A and a vector b over the alphabet $\{0, 1, \#\}$, assuming all elements of A and b are integers.

The **problem of the existence of a solution of linear integer programming** is to decide whether $Sol_{\mathbb{Z}}(A, b) = \emptyset$ for given A and b. Formally, this decision problem is $(\text{SOL-IP}, \{0, 1, \#\})$, where

$$\text{SOL-IP} = \{\langle A, b \rangle \in \{0, 1, \#\}^* \mid Sol_{\mathbb{Z}}(A, b) \neq \emptyset\}.$$

The **problem of the existence of a solution of 0/1-linear programming** is to decide whether $Sol_{\{0,1\}}(A,b) = \emptyset$ for given A and b. Formally, this decision problem is $(\text{SOL-0/1-IP}, \{0,1,\#\})$, where

$$\text{SOL-0/1-IP} = \{\langle A,b\rangle \in \{0,1,\#\}^* \mid Sol_{\{0,1\}}(A,b) \neq \emptyset\}.$$

All existence problems mentioned above consider computing over the field $\mathbb{R}$. We are interested in solving the system of linear equations $A \cdot X = b$ over a finite field $\mathbb{Z}_p$ for a prime p. So, all elements of A and b are from $\mathbb{Z}_p = \{0,1,\ldots,p-1\}$, all solutions have to be from $(\mathbb{Z}_p)^n$, and the linear equations are congruences modulo p (i.e., the operation of addition is $\oplus_{\bmod p}$ and the operation of multiplication is $\odot_{\bmod p}$).

The **problem of the existence of a solution of linear programming modulo p** is the decision problem $(\text{SOL-IP}_p, \{0,1,\ldots,p-1,\#\})$ where

$$\begin{aligned}
\text{SOL-IP}_p = \{&\langle A,b\rangle \in \{0,1,\ldots,p-1,\#\}^* \mid \text{if } A \text{ is an } m \times n \text{ matrix} \\
&\text{over } \mathbb{Z}_p, m,n \in \mathbb{N} - \{0\}, \text{ and } b \in \mathbb{Z}_p^m, \text{ then there} \\
&\text{exists } X \in (\mathbb{Z}_p)^n \text{ such that } AX \equiv b \ (\bmod\ p)\}.
\end{aligned}$$

In what follows, we define some fundamental optimization problems. We start with a general framework that describes the formalism for the specification of optimization problems.

Roughly, a problem instance x of an optimization problem specifies a set of constraints. These constraints unambiguously determine the set $\mathcal{M}(x)$ of feasible solutions for the problem instance x. Note that $\mathcal{M}(x)$ may be empty or infinite. The objective, determined by the specification of the problem, is to find a solution from $\mathcal{M}(x)$ that is the "best" one among all solutions in $\mathcal{M}(x)$. Note that there may exist several (even infinitely many) best solutions among the solutions in $\mathcal{M}(x)$.

Definition 2.3.2.2. *An* **optimization problem** *is a 7-tuple* $U = (\Sigma_I, \Sigma_O, L, L_I, \mathcal{M}, cost, goal)$, *where*

(i) Σ_I *is an alphabet, called the* **input alphabet** *of U,*

(ii) Σ_O *is an alphabet, called the* **output alphabet** *of U,*

(iii) $L \subseteq \Sigma_I^*$ *is the* **language of feasible problem instances**,

(iv) $L_I \subseteq L$ *is the* **language of the (actual) problem instances of U,**

(v) $\mathcal{M}$ *is a function from L to $Pot(\Sigma_O^*)$,[30] and, for every $x \in L$, $\mathcal{M}(x)$ is called the* **set of feasible solutions** *for x,*

(vi) *cost is the* **cost function** *that, for every pair (u,x), where $u \in \mathcal{M}(x)$ for some $x \in L$, assigns a positive real number $cost(u,x)$,*

(vii) $goal \in \{minimum, maximum\}$.

---

[30] Remember that $Pot(S)$ is the set of all subsets of the set S, i.e., the power set of S.

For every $x \in L_I$, a feasible solution $y \in \mathcal{M}(x)$ is called **optimal for x and U** *if*

$$cost(y, x) = goal\{cost(z, x) \mid z \in \mathcal{M}(x)\}.$$

For an optimal solution $y \in \mathcal{M}(x)$, we denote $cost(y, x)$ by $\mathbf{Opt_U(x)}$. U is called a **maximization problem** *if goal = maximum, and U is a* **minimization problem** *if goal = minimum. In what follows $\mathbf{Output_U(x)} \subseteq \mathcal{M}(x)$ denotes the set of all optimal solutions for the instance x of U.*

An algorithm A is **consistent** *for U if, for every $x \in L_I$, the output $A(x) \in \mathcal{M}(x)$. We say that an algorithm B* **solves** *the optimization problem U if*

(i) B is consistent for U, and
(ii) for every $x \in L_I$, $B(x)$ is an optimal solution for x and U.

Let us explain the informal meaning of the formal definition of an optimization problem U as a 7-tuple $(\Sigma_I, \Sigma_O, L, L_I, \mathcal{M}, cost, goal)$. Σ_I has the same meaning as the alphabet of decision problems and it is used to code (represent) the inputs. Similarly, Σ_O is the alphabet used to code outputs. On the level of algorithm design used here we usually do not need to specify Σ_I or Σ_O and the coding of inputs and outputs, because these details do not have any essential influence on the hardness of the problems considered. But this formal specification may be useful in the classification of the optimization problems according to their computational difficulty, and especially for proving lower bounds on their polynomial-time approximability.

The language L is the set of codes of all problem instances (inputs) for which U is well defined. L_I is the set of actual problem instances (inputs) and one measures the computational hardness of U according to inputs of L_I. In general, one can simplify the definition of U by removing L and the definition will work as well as Definition 2.3.2.2. The reason to put this additional information into the definition of optimization problems is that the hardness of many optimization problems is very sensible according to the specification of the set of considered problem instances (L_I). Definition 2.3.2.2 enables one to conveniently measure the increase or decrease of the hardness of optimization problems according to the changes of L_I by a fixed L.

Definition 2.3.2.3. *Let $U_1 = (\Sigma_I, \Sigma_O, L, L_{I,1}, \mathcal{M}, cost, goal)$ and $U_2 = (\Sigma_I, \Sigma_O, L, L_{I,2}, \mathcal{M}, cost, goal)$ be two optimization problems. We say that U_1 is a* **subproblem** *of U_2 if $L_{I,1} \subseteq L_{I,2}$.*

The function $\mathcal{M}$ is determined by the constraints given by the problem instances and $\mathcal{M}(x)$ is the set of all objects (solutions) satisfying the constraints given by x. The cost function assigns the cost $cost(\alpha, x)$ to every solution α from $\mathcal{M}(x)$. If the input instance x is fixed, we often use the short notion $\mathbf{cost(\alpha)}$ instead of $cost(\alpha, x)$. If goal = minimum [= maximum], then an optimal solution is any solution from $\mathcal{M}(x)$ with the minimal [maximal] cost.

To make the definitions of specific optimization problems transparent, we often leave out the specification of coding the data over Σ_I and Σ_O. We define the problems simply by specifying

- the set of actual problem instances L_I,
- the constraints given by the input instances, and so $\mathcal{M}(x)$ for every $x \in L_I$,
- the cost function,
- the goal.

TRAVELING SALESPERSON PROBLEM.

Traveling salesperson problem is the problem of finding a Hamiltonian cycle (tour) of the minimal cost in a complete weighted graph. The formal definition follows.

Traveling Salesperson Problem (TSP)

Input: A weighted complete graph (G, c), where $G = (V, E)$ and $c : E \to$ $\mathbb{N}$. Let $V = \{v_1, \ldots, v_n\}$ for some $n \in \mathbb{N} - \{0\}$.

Constraints: For every input instance (G, c), $\mathcal{M}(G, c) = \{v_{i_1}, v_{i_2}, \ldots, v_{i_n},$ $v_{i_1} \mid (i_1, i_2, \ldots, i_n)$ is a permutation of $(1, 2, \ldots, n)\}$, i.e., the set of all Hamiltonian cycles of G.

Costs: For every Hamiltonian cycle $H = v_{i_1} v_{i_2} \ldots v_{i_n} v_{i_1} \in \mathcal{M}(G, c)$, $cost((v_{i_1}, v_{i_2}, \ldots v_{i_n}, v_{i_1}), (G, c)) = \sum_{j=1}^{n} c(\{v_{i_j}, v_{i_{(j \bmod n)+1}}\})$, i.e., the cost of every Hamiltonian cycle H is the sum of the weights of all edges of H.

Goal: *minimum*.

If one wants to specify Σ_I and Σ_O one can take $\{0, 1, \#\}$ for both. The input can be a code of the adjacency matrix of (G, c) and the Hamiltonian paths can be coded as permutations of the set of vertices.

The following adjacency matrix represents the problem instance of TSP depicted in Figure 2.12.

$$\begin{pmatrix} 0 & 1 & 1 & 3 & 8 \\ 1 & 0 & 2 & 1 & 2 \\ 1 & 2 & 0 & 7 & 1 \\ 3 & 1 & 7 & 0 & 1 \\ 8 & 2 & 1 & 1 & 0 \end{pmatrix}$$

Observe that there are $4!/2 = 12$ Hamiltonian tours in K_5. The cost of the Hamiltonian tour $H = v_1, v_2, v_3, v_4, v_5, v_1$ is

$$cost(H) = c(\{v_1, v_2\}) + c(\{v_2, v_3\}) + c(\{v_3, v_4\}) + c(\{v_4, v_5\}) + c(\{v_5, v_1\})$$
$$= 1 + 2 + 7 + 1 + 8 = 19.$$

The unique optimal Hamiltonian tour is

$$H_{Opt} = v_1, v_2, v_4, v_5, v_3, v_1 \text{ with } cost(H_{Opt}) = 5.$$

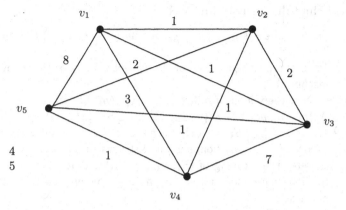

Fig. 2.12.

Now we define two subproblems of TSP.

The **metric traveling salesperson problem**, $\triangle$-**TSP**, is a subproblem of TSP such that every problem instance (G, c) of $\triangle$-TSP satisfies the triangle inequality

$$c(\{u, v\}) \leq c(\{u, w\}) + c(\{w, v\})$$

for all vertices u, w, v of G.

The problem instance depicted in Figure 2.12 does not satisfy the triangle inequality because

$$7 = c(\{v_3, v_4\}) > c(\{v_3, v_5\}) + c(\{v_5, v_4\}) = 1 + 1 = 2.$$

The **geometrical traveling salesperson problem (Euclidean TSP)** is a subproblem of TSP such that, for every problem instance (G, c) of TSP, the vertices of G can be embedded in the two-dimensional Euclidean space in such a way that $c(\{u, v\})$ is the Euclidean distance between the points assigned to the vertices u and v for all u, v of G. A simplified specification of the set of input instances of the geometric TSP is to say that the input is a set of points in the plane and the cost of the connection between any two points is defined by their Euclidean distance.

Since the two-dimensional Euclidean space is a metric space, the Euclidean distance satisfies the triangle inequality and so the geometrical TSP is a subproblem of $\triangle$-TSP.

MAKESPAN SCHEDULING PROBLEM.

The problem of makespan scheduling (MS) is to schedule n jobs with designated processing times on m identical machines in such a way that the whole processing time is minimized. Formally, we define MS as follows.

Makespan Scheduling Problem (MS)

Input: Positive integers $p_1, p_2, \ldots, p_n$ and an integer $m \geq 2$ for some $n \in \mathbb{N} - \{0\}$.

{p_i is the processing time of the ith job on any of the m available machines}.

Constraints: For every input instance $(p_1, \ldots, p_n, m)$ of MS,

$\mathcal{M}(p_1, \ldots, p_n, m) = \{S_1, S_2, \ldots, S_m \mid S_i \subseteq \{1, 2, \ldots, n\}$ for $i = 1, \ldots, m$, $\bigcup_{k=1}^{m} S_k = \{1, 2, \ldots, n\}$, and $S_i \cap S_j = \emptyset$ for $i \neq j\}$.

{$\mathcal{M}(p_1, \ldots, p_n, m)$ contains all partitions of $\{1, 2, \ldots, n\}$ into m subsets. The meaning of $(S_1, S_2, \ldots, S_m)$ is that, for $i = 1, \ldots, m$, the jobs with indices from S_i have to be processed on the ith machine}.

Costs: For each $(S_1, S_2, \ldots, S_m) \in \mathcal{M}(p_1, \ldots, p_n, m)$,

$cost((S_1, \ldots, S_m), (p_1, \ldots, p_n, m)) = \max\left\{\sum_{l \in S_i} p_l \mid i = 1, \ldots, m\right\}$.

Goal: *minimum.*

An example of scheduling seven jobs with the processing times 3, 2, 4, 1, 3, 3, 6, respectively, on 4 machines is given in Figure 4.1 in Section 4.2.1.

COVER PROBLEMS.

Here, we define the minimum vertex cover problem[31] (MIN-VCP), its weighted version, and the set cover problem (SCP). The minimum vertex cover problem is to cover all edges of a given graph G with a minimal number of vertices of G.

Minimum Vertex Cover Problem (MIN-VCP)

Input: A graph $G = (V, E)$.

Constraints: $\mathcal{M}(G) = \{S \subseteq V \mid$ every edge of E is incident to at least one vertex of $S\}$.

Cost: For every $S \in \mathcal{M}(G)$, $cost(S, G) = |S|$.

Goal: *minimum.*

Consider the graph G given in Figure 2.13.

$$\mathcal{M}(G) = \{\{v_1, v_2, v_3, v_4, v_5\}, \{v_1, v_2, v_3, v_4\}, \{v_1, v_2, v_3, v_5\}, \{v_1, v_2, v_4, v_5\}$$
$$\{v_1, v_3, v_4, v_5\}, \{v_2, v_3, v_4, v_5\}, \{v_1, v_3, v_4\}, \{v_2, v_4, v_5\}, \{v_2, v_3, v_5\}\}.$$

The optimal solutions are $\{v_1, v_3, v_4\}$, $\{v_2, v_4, v_5\}$, and $\{v_2, v_3, v_5\}$ and so $Opt_{\mathrm{VCP}}(G) = 3$. Observe that there is no vertex cover of cardinality 2 because

---

[31] Observe that we have two versions of vertex cover problems. One version is the decision problem defined by the language VCP above and the second version MIN-VCP is the minimization problem considered here.

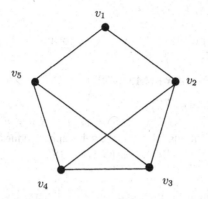

Fig. 2.13.

to cover the edges of the cycle $v_1, v_2, v_3, v_4, v_5, v_1$ one needs at least three vertices.

Set Cover Problem (SCP)

Input: $(X, \mathcal{F})$, where X is a finite set and $\mathcal{F} \subseteq Pot(X)$ such that $X = \bigcup_{S \in \mathcal{F}} S$.

Constraints: For every input $(X, \mathcal{F})$,
$$\mathcal{M}(X, \mathcal{F}) = \{C \subseteq \mathcal{F} \mid X = \bigcup_{S \in C} S\}.$$

Costs: For every $C \in \mathcal{M}(X, \mathcal{F})$, $cost(C, (X, \mathcal{F})) = |C|$.

Goal: *minimum*.

Later we shall observe that MIN-VCP can be viewed as a special subproblem of SCP because, for a given graph $G = (V, E)$, one can assign the set S_v of all edges adjacent to v to every vertex v of G. For the graph in Figure 2.13 it results in the instance $(E, \mathcal{F})$ of SCP where

$\mathcal{F} = \{S_{v_1}, S_{v_2}, S_{v_3}, S_{v_4}, S_{v_5}\}$,
$S_{v_1} = \{\{v_1, v_2\}, \{v_1, v_5\}\}$, $S_{v_2} = \{\{v_1, v_2\}, \{v_2, v_3\}, \{v_2, v_4\}\}$,
$S_{v_3} = \{\{v_3, v_2\}, \{v_3, v_5\}, \{v_3, v_4\}\}$, $S_{v_4} = \{\{v_3, v_4\}, \{v_2, v_4\}, \{v_4, v_5\}\}$, and
$S_{v_5} = \{\{v_1, v_5\}, \{v_3, v_5\}, \{v_4, v_5\}\}$.

The last cover problem that we consider is the weighted generalization of MIN-VCP.

Weighted Minimum Vertex Cover Problem (WEIGHT-VCP)

Input: A weighted graph $G = (V, E, c)$, $c : V \to \mathbb{N} - \{0\}$.

Constraints: For every input instance $G = (V, E, c)$,
$$\mathcal{M}(G) = \{S \subseteq V \mid S \text{ is a vertex cover of } G\}.$$

Cost: For every $S \in \mathcal{M}(G)$, $G = (V, E, c)$,
$$cost(S, (V, E, c)) = \sum_{v \in S} c(v).$$

Goal: *minimum*.

MAXIMUM CLIQUE PROBLEM.

The maximum clique problem (MAX-CL) is to find a clique of the maximal size in a given graph G.

Maximum Clique Problem (MAX-CL)

Input: A graph $G = (V, E)$
Constraints: $\mathcal{M}(G) = \{S \subseteq V \mid \{\{u, v\} \mid u, v \in S, u \neq v\} \subseteq E\}$.
              $\{\mathcal{M}(G)$ contains all complete subgraphs (cliques) of $G\}$
Costs: For every $S \in \mathcal{M}(G)$, $cost(S, G) = |S|$.
Goal: *maximum*.

To present a specific input instance consider the graph G depicted in Figure 2.13.

$$\mathcal{M}(G) = \{\{v_1\}, \{v_2\}, \{v_3\}, \{v_4\}, \{v_5\},$$
$$\{v_1, v_2\}, \{v_1, v_5\}, \{v_2, v_3\}, \{v_2, v_4\}, \{v_3, v_4\}, \{v_3, v_5\}, \{v_4, v_5\},$$
$$\{v_2, v_3, v_4\}, \{v_3, v_4, v_5\}\}.$$

The optimal solutions are $\{v_2, v_3, v_4\}$ and $\{v_3, v_4, v_5\}$ and so $Opt_{\text{MAX-CL}}(G) = 3$.

CUT PROBLEMS.

We introduce the maximum cut problem (MAX-CUT) and the minimum cut problem (MIN-CUT). Remember that a cut of a graph $G = (V, E)$ is any partition of V into (V_1, V_2) such that $V_1 \cup V_2 = V$ and $V_1 \cap V_2 = \emptyset$.

Maximum Cut Problem (MAX-CUT)

Input: A graph $G = (V, E)$.
Constraints:
        $\mathcal{M}(G) = \{(V_1, V_2) \mid V_1 \cup V_2 = V, V_1 \neq \emptyset \neq V_2, \text{and} V_1 \cap V_2 = \emptyset\}$.

Costs: For every cut $(V_1, V_2) \in \mathcal{M}(G)$,

$$cost((V_1, V_2), G) = |E \cap \{\{u, v\} \mid u \in V_1, v \in V_2\}|.$$

Goal: *maximum*.

The **minimum cut problem (MIN-CUT)** can be defined in the same way as MAX-CUT. The only difference is that the goal of MIN-CUT is minimum.

The only optimal solution of **MIN-CUT** for the graph G in Figure 2.13 is $(\{v_1\}, \{v_2, v_3, v_4, v_5\})$, and the optimal solutions of MAX-CUT for the graph G are $(\{v_1, v_2, v_3\}, \{v_4, v_5\})$, $(\{v_1, v_2, v_5\}, \{v_3, v_4\})$, and $(\{v_1, v_4, v_5\}, \{v_2, v_3\})$. So, $Opt_{\text{MIN-CUT}}(G) = 2$ and $Opt_{\text{MAX-CUT}}(G) = 4$.

KNAPSACK PROBLEM.

First, we define the simple knapsack problem(SKP). This optimization task can be described as follows. One has a knapsack whose weight capacity is bounded by a positive integer b (for instance, by b pounds) and n objects of weights $w_1, w_2, \ldots, w_n$, $n \in \mathbb{N} - \{0\}$. The aim is to pack some objects in the knapsack in such a way that the contents of the knapsack are as heavy as possible but not above b.

Simple Knapsack Problem (SKP)

Input: A positive integer b, and positive integers $w_1, w_2, \ldots, w_n$ for some $n \in \mathbb{N} - \{0\}$.

Constraints: $\mathcal{M}(b, w_1, w_2, \ldots, w_n) = \{T \subseteq \{1, \ldots, n\} \mid \sum_{i \in T} w_i \leq b\}$, i.e., a feasible solution for the problem instance $b, w_1, w_2, \ldots, w_n$ is every set of objects whose common weight does not exceed b.

Costs: For each $T \in \mathcal{M}(b, w_1, w_2, \ldots, w_n)$,

$$cost(T, b, w_1, w_2, \ldots, w_n) = \sum_{i \in T} w_i.$$

Goal: *maximum.*

For the problem instance $I = (b, w_1, \ldots, w_5)$, where $b = 29$, $w_1 = 3$, $w_2 = 6$, $w_3 = 8$, $w_4 = 7$, $w_5 = 12$, the only optimal solution is $T = \{1, 2, 3, 5\}$ with $cost(T, I) = 29$. If one considers the problem instance $I' = (b', w_1, \ldots, w_5)$ with $b' = 14$, then the optimal solution is $T' = \{2, 3\}$ with $cost(T', I') = 14$.

The instances of the general knapsack problem contain additionally a cost c_i for every object i. The objective is to maximize the common cost of objects packed into the knapsack[32] by satisfying the constraint b on the weight of the knapsack.

Knapsack Problem (KP)

Input: A positive integer b, and $2n$ positive integers $w_1, w_2, \ldots, w_n$, c_1, $c_2, \ldots, c_n$ for some $n \in \mathbb{N} - \{0\}$.

Constraints:
$$\mathcal{M}(b, w_1, \ldots, w_n, c_1, \ldots, c_n) = \{T \subseteq \{1, \ldots, n\} \mid \sum_{i \in T} w_i \leq b\}.$$

Costs: For each $T \in \mathcal{M}(b, w_1, \ldots, w_n, c_1, \ldots, c_n)$,

$$cost(T, b, w_1, \ldots, w_n, c_1, \ldots, c_n) = \sum_{i \in T} c_i.$$

Goal: *maximum.*

Consider the problem instance I determined by $b = 59$, $w_1 = 12$, $c_1 = 9$, $w_2 = 5$, $c_2 = 4$, $w_3 = 13$, $c_3 = 5$, $w_4 = 18$, $c_4 = 9$, $w_5 = 15$, $c_5 = 9$, $w_6 = 29$, $c_6 = 22$. The optimal solution is $T = \{1, 5, 6\}$. Observe that T satisfies the constraint because $w_1 + w_5 + w_6 = 12 + 15 + 29 = 56 < 59 = b$ and that $Opt_{KP}(I) = c_1 + c_5 + c_6 = 9 + 9 + 22 = 40$.

---

[32] Rather than their weights.

BIN-PACKING PROBLEM.

The bin-packing problem (Bin-P) is similar to the knapsack problem. One has n objects of rational weights $w_1, \ldots, w_n \in [0,1]$. The goal is to distribute them among the knapsacks (bins) of unit size 1 in such a way that a minimal number of knapsacks (bins) is used.

Bin-Packing Problem (BIN-P)

Input: n rational numbers $w_1, w_2, \ldots, w_n \in [0,1]$ for some positive integer n.

Constraints:. $\mathcal{M}(w_1, w_2, \ldots, w_n) = \{S \subseteq \{0,1\}^n \mid$ for every $s \in S$, $s^{\mathsf{T}} \cdot (w_1, w_2, \ldots, w_n) \leq 1$, and $\sum_{s \in S} s = (1,1,\ldots,1)\}$.

{If $S = \{s_1, s_2, \ldots, s_m\}$, then $s_i = (s_{i1}, s_{i2}, \ldots, s_{in})$ determines the set of objects packed in the ith bin. The jth object is packed into the ith bin if and only if $s_{ij} = 1$. The constraint

$$s_i^{\mathsf{T}} \cdot (w_1, \ldots, w_n) \leq 1$$

assures that the ith bin is not overfilled. The constraint

$$\sum_{s \in S} s = (1, 1, \ldots, 1)$$

assures that every object is packed in exactly one bin.}

Cost: For every $S \in \mathcal{M}(w_1, w_2, \ldots, w_n)$,

$$cost(S, (w_1, \ldots, w_n)) = |S|.$$

Goal: *minimum*.

Observe that an alternative way to describe the constraints of BIN-P is to take

$$\mathcal{M}(w_1, \ldots, w_n) = \{(T_1, T_2, \ldots, T_m) \mid m \in \mathbb{N} - \{0\}, T_i \subseteq \{1, 2, \ldots, n\}$$
$$\text{for } i = 1, \ldots, n, T_i \cap T_j = \emptyset \text{ for } i \neq j,$$
$$\bigcup_{i=1}^m T_i = \{1, 2, \ldots, n\}, \text{ and}$$
$$\sum_{k \in T_j} w_k \leq 1 \text{ for } j = 1, \ldots, m\}.$$

MAXIMUM SATISFIABILITY PROBLEM.

The general maximum satisfiability problem (MAX-SAT) is to find an assignment to the variables of a formula Φ such that the number of satisfied clauses is maximized.

Maximum Satisfiability Problem (MAX-SAT)

Input: A formula $\Phi = F_1 \wedge F_2 \wedge \cdots \wedge F_m$ over $X = \{x_1, x_2, \ldots\}$ in CNF
(an equivalent description of this instance of MAX-SAT is to consider
the set of clauses $F_1, F_2, \ldots, F_m$).

Constraints: For every formula Φ over the set $\{x_1, \ldots, x_n\} \subseteq X$, $n \in \mathbb{N} - \{0\}$,
$\mathcal{M}(\Phi) = \{0, 1\}^n$.
{Every assignment of values to $\{x_1, \ldots, x_n\}$ is a feasible solution,
i.e., $\mathcal{M}(\Phi)$ can also be written as $\{\alpha \,|\, \alpha : X \to \{0, 1\}\}$.}

Costs: For every Φ in CNF, and every $\alpha \in \mathcal{M}(\Phi)$,
$cost(\alpha, \Phi)$ is the number of clauses satisfied by α.

Goal: *maximum*.

Observe that, if Φ is a satisfiable formula, an optimal solution is any assignment α that satisfies Φ (i.e., $cost(\alpha, \Phi) = m$ if Φ consists of m clauses).

We consider several subproblems of MAX-SAT here. For every integer $k \geq 2$, we define the **MAX-kSAT** problem as a subproblem of MAX-SAT, where the problem instances are formulae in kCNF[33]. For every integer $k \geq 2$, we define the **MAX-EkSAT** as a subproblem of MAX-kSAT, where the inputs are formulae consisting of clauses of the size k only. Each clause $l_1 \vee \,_2 \vee \cdots \vee l_k$ of such a formula is a Boolean function over exactly k variables, i.e., $l_i \neq l_j$ and $l_i \neq \bar{l}_j$ for all $i, j \in \{1, \ldots, k\}$, $i \neq j$.

LINEAR PROGRAMMING.

First, we define the general version of the linear programming problem and then we consider some special versions of it.

Linear Programming (LP)

Input: A matrix $A = [a_{ij}]_{i=1,\ldots,m, j=1,\ldots,n}$, a vector $b \in \mathbb{R}^m$, and a vector
$c \in \mathbb{R}^n$, $n, m \in \mathbb{N} - \{0\}$.

Constraints: $\mathcal{M}(A, b, c) = \{X \in \mathbb{R}^n \,|\, A \cdot X = b$ and the elements of X
are non-negative reals only$\}$.

Costs: For every $X = (x_1, \ldots, x_n) \in \mathcal{M}(A, b, c)$, $c = (c_1, \ldots, c_n)^{\mathsf{T}}$,
$cost(X, (A, b, c)) = c^{\mathsf{T}} \cdot X = \sum_{i=1}^{n} c_i x_i$.

Goal: *minimum*.

We observe that the set of constraints $A \cdot X = b$ and $x_j \geq 0$ for $i = 1, \ldots, n$ is a system of $m+n$ linear equations of n unknowns. Since every linear equation determines an $(n-1)$-dimensional affine subspace of the vector space $\mathbb{R}^n$, $\mathcal{M}(A, b, c)$ can be considered as the intersection of the $m + n$ affine subspaces determined by the linear equations[34] of $A \cdot X = b$ and $X \in (\mathbb{R}^{\geq 0})^n$.

---

[33] That is, the size of each clause of Φ is at most k.

[34] Note that there exist several other forms of the linear programming problem. For instance, one can exchange the constraint $A \cdot X = b$ with $A \cdot X \geq b$. Or one can take maximization instead of minimization and consider the constraint $A \cdot X \leq b$ instead of $A \cdot X = b$.

In combinatorial optimization we often consider the problem of integer programming. It can be defined by exchanging reals with integers in the problem instances as well as in the feasible solutions.

Integer Linear Programming (IP)

Input: An $m \times n$ matrix $A = [a_{ij}]_{i=1,\ldots,m,j=1,\ldots,n}$, and two vectors $b = (b_1,\ldots,b_m)^{\mathsf{T}}$, $c = (c_1,\ldots,c_n)^{\mathsf{T}}$ for some $n, m \in \mathbb{N} - \{0\}$, a_{ij}, b_i, c_j are integers for $i = 1,\ldots,m$, $j = 1,\ldots,n$.

Constraints: $\mathcal{M}(A, b, c) = \{X = (x_1,\ldots,x_n) \in \mathbb{Z}^n \mid AX = b$ and $x_i \geq 0$ for $i = 1,\ldots,n\}$.

Costs: For every $X = (x_1,\ldots,x_n) \in \mathcal{M}(A, b, c)$, $cost(X, (A, b, c)) = \sum_{i=1}^{n} c_i x_i$.

Goal: *minimum*.

Note that IP is not a subproblem of LP, because we did not restrict the language of inputs only, but also the constraints.

The **0/1-Linear Programming (0/1-LP)** is the optimization problem with the language of input instances of IP and the additional constraints requiring that $X \in \{0, 1\}^n$ (i.e., that $\mathcal{M}(A, b, c) \subseteq \{0, 1\}^n$).

The last problems we consider are maximization problems on systems of linear equations. The objective is to find values for unknowns that satisfy the maximal possible number of linear equations of the given system. Let $k \geq 2$ be prime.

Maximum Linear Equation Problem Mod k (MAX-LINMODk)

Input: A set S of m linear equations over n unknowns, $n, m \in \mathbb{N} - \{0\}$, with coefficients from $\mathbb{Z}_k$.
           (An alternative description of an input is an $m \times n$ matrix over $\mathbb{Z}_k$ and a vector $b \in \mathbb{Z}_k^m$).

Constraints: $\mathcal{M}(S) = \mathbb{Z}_k^m$
           {a feasible solution is any assignment of values from $\{0, 1, \ldots, k-1\}$ to the n unknowns (variables)}.

Costs: For every $X \in \mathcal{M}(S)$,
           $cost(X, S)$ is the number of linear equations of S satisfied by X.

Goal: *maximum*.

Consider the following example of an input instance over $\mathbb{Z}_2$:

$$\begin{aligned} x_1 + x_2 \phantom{{}+ x_3} &= 1 \\ x_1 \phantom{{}+ x_2} + x_3 &= 0 \\ x_2 + x_3 &= 0 \\ x_1 + x_2 + x_3 &= 1. \end{aligned}$$

Observe that this system of linear equations does not have any solution in $\mathbb{Z}_2$. The assignment $x_1 = x_2 = x_3 = 0$ satisfies the second equation and the third equation. The assignment $x_1 = x_2 = x_3 = 1$ satisfies the last three equations and the assignment $x_1 = x_3 = 0$, $x_2 = 1$ satisfies the first two equations and the last equation. The last two assignments are optimal solutions.

For every prime k, and every positive integer m, we define the problem **Max-EmLinMod$k$** as the subproblem of Max-LinMod$k$, where the input instances are sets of linear equations such that every linear equation has at most m nonzero coefficients (contains at most m unknowns).

For instance, the first three linear equations of the system considered above form a problem instance of Max-E2LinMod2.

Keywords introduced in Section 2.3.2

decision problem, primality testing, equivalence problem for polynomials, equivalence problem for one-time-only branching programs, satisfiability problems, clique problem, vertex cover problem, Hamiltonian problem, problems of existence of a solution of systems of linear equations, optimization problem, problem instance, feasible solution, optimal solution, maximization problem, minimization problem, traveling salesperson problem (TSP), metric TSP ($\triangle$-TSP), geometrical TSP, makespan problem (MS), minimum vertex cover problem (Min-VCP), set cover problem (SCP), maximum clique problem (Max-CL), minimum cut problem (Min-Cut), maximum cut problem (Max-Cut), knapsack problem (KP), simple knapsack problem (SKP), bin-packing problem (Bin-P), maximum satisfiability problem (Max-Sat), linear programming (LP), integer programming (IP), 0/1-linear programming (0/1-LP), maximum linear equation problem modulo k (Max-LinMod$k$)

Summary of Section 2.3.2

A decision problem is a problem of deciding whether a given input has a required property. Since the set of all inputs with the required property can be viewed as a language $L \subseteq \Sigma^*$, the decision problem for L is to decide whether a given input $x \in \Sigma^*$ belongs to L or $x \notin L$.

An optimization problem is specified by

- the set of problem instances,
- the constraints determining the set of feasible solutions to every problem instance (input),
- the cost function that assigns a cost to every feasible solution,
- the goal that may be maximization or minimization.

The objective is to find an optimal solution (one of the best feasible solutions according to the cost and the goal) for every input instance.

Currently, there are thousands of hard algorithmic problems considered in the literature on algorithmics and in numerous practical applications. To learn the fundamentals of algorithmics it is sufficient to consider some of them – so-called paradigmatic problems. Paradigmatic problems are some kind of pattern problems in the sense that solving most of the hard problems can be reduced to solving some of the paradigmatic problems. Some of the most fundamental decision problems are satisfiability, primality testing, equivalence problem for polynomials, clique problem, and Hamiltonian problems. Some representatives of the paradigmatic optimization problems are maximum satisfiability, traveling salesperson problem, scheduling problems, set cover problem, maximum clique problem, knapsack problem, bin-packing problem, linear programming, integer programming, and maximum linear equation problems.

2.3.3 Complexity Theory

The aim of this section is to discuss the ways of measuring computational complexity of algorithms (computer programs) and to present the main framework for classifying algorithmic problems according to their computational hardness. The first part of this section is useful for all subsequent chapters. The second part, devoted to the complexity classes and to the concept of NP-completeness, provides (besides the general philosophy of algorithms and complexity) fundamentals for Section 4.4.3 on lower bounds for inapproximability (i.e., for the classification of optimization problems according to their polynomial-time approximability).

This book is devoted to the design of programs on the algorithmic level and so we will not deal with details of algorithm implementations in specific programming languages. To describe algorithms we use either an informal description of its parts such as "choose an edge from the graph G and verify whether the rest of the graph is connected" or a Pascal-like language with instructions such as for, repeat, while, if ... then ... else, etc. Observe that this rough description can be sufficient for the analysis of complexity when the complexity of the implementation of the roughly described parts is well known.

The aim of the complexity analysis is to provide a robust analysis in the sense that the result does not depend on the structural and technological characteristics of concrete sequential computers and their system software. Here, we focus on the time complexity of computations and only sometimes consider the space complexity. We distinguish two basic ways of complexity measurement, namely the **uniform cost** measurement and the **logarithmic cost** measurement.

The approach based on uniform cost measure is the simplest one. The measurement of time complexity consists of determining the overall number of elementary[35] instructions executed in the considered computation, and the measurement of space complexity consists of determining the number of variables used in the computation. The advantage of this measurement is that it is simple. The drawback is that it is not always adequate because it considers

---

[35] Elementary instructions are arithmetic instructions over integers, comparison of two integers, reading, writing, loading integers and symbols, etc.

cost 1 for an arithmetic operation over two integers independently of their size. When the operands are integers whose binary representations consist of several hundreds of bits, none of them can be stored in one computer word (16 or 32 bits). Then the operands must be stored in several computer words (i.e., one needs several space units (variables) to save them) and the execution of the arithmetic operation over these two large integers corresponds to the execution of a special program performing an operation over large integers by several operations over integers of the computer word size. Thus, the uniform cost measurement may be applied in the cases where one can assume that during the whole computation all variables contain values whose size is bounded by a fixed constant (hypothetical computer word length). This is the case for computing over $\mathbb{Z}_p$ for a fixed prime p or working with logical values only (in Boolean algebra).

A serious anomaly of the use of the uniform cost measurement appears in the following example. Let k and $a \geq 2$ be two positive integers of sizes not exceeding the size of a hypothetical computer word. Consider the task of computing the number a^{2^k}. This can be done with the following strategy. Compute

$$a^2 = a \cdot a, \ a^4 = a^2 \cdot a^2, \ a^8 = a^4 \cdot a^4, \ldots, \ a^{2^k} = a^{2^{k-1}} \cdot a^{2^{k-1}}.$$

The uniform cost space complexity is 3 because one additional variable is sufficient to execute the following computation

for $i = 1$ **to** k **do** $a := a * a$.

The uniform cost time complexity is in $O(k)$ because exactly k multiplications are executed. This contrasts with the fact that one needs at least 2^k bits to represent the result a^{2^k} and to write 2^k bits any machine needs $\Omega(2^k)$ operations on its machine words. Since this consideration works for every positive integer k, we have an exponential gap between the uniform cost time complexity and any realistic time complexity, and an unbounded difference between the uniform cost space complexity and any realistic space complexity.

The solution to such a situation, where the values of variables grow unboundedly, is to use the **logarithmic cost** measurement. With respect to this measurement the cost of every elementary operation is the sum of the sizes of the binary representations of the operands.[36] Obviously, this approach to complexity measurement avoids anomalies of the kind mentioned above, and it is generally adopted in the complexity analysis of algorithms. Sometimes one distinguishes the time complexity of distinct operations. Since the best algorithm for multiplying two n-bit integers needs $\Omega(n \cdot \log n)$ binary operations, the complexity of multiplication and division is considered to be $O(n \cdot \log n)$ while addition, subtraction, and assignment have costs linear in the binary size of the arguments.

---

[36] If one wants to be very careful then the binary length of the addresses of the variables in memory (that correspond to the operands) can be added.

Definition 2.3.3.1. *Let Σ_I and Σ_O be alphabets. Let A be an algorithm that realizes a mapping from Σ_I^* to Σ_O^*. For every $x \in \Sigma_I^*$, $\boldsymbol{Time_A(x)}$ denotes the time complexity[37] (according to the logarithmic cost) of the computation of A on the input x, and $\boldsymbol{Space_A(x)}$ denotes the space complexity (according to the logarithmic cost measurement) of the computation of A on x.*

One never considers the time complexity $Time_A$ (the space complexity $Space_A$) as a function from Σ_I^* to $\mathbb{N}$. This is because we usually have exponentially many inputs for every input length and to estimate $Time_A(x)$ for every $x \in \Sigma_I^*$ would often be an unrealistic job. Even if this would succeed, the description of $Time_A$ could be so complex that one would have trouble determining some fundamental characteristics of $Time_A$. The comparison of the complexities of two algorithms for the same algorithmic problem could be a difficult job, too. Thus, the complexity is always considered as a function of the **input size** and one observes the asymptotic growth of this function.

Definition 2.3.3.2. *Let Σ_I and Σ_O be two alphabets. Let A be an algorithm that computes a mapping from Σ_I^* to Σ_O^*. The* **(worst case) time complexity of A** *is a function $Time_A : (\mathbb{N} - \{0\}) \to \mathbb{N}$ defined by*

$$Time_A(n) = \max\{Time_A(x) \mid x \in \Sigma_I^n\}$$

for every positive integer n. The **(worst case) space complexity of A** *is a function $Space_A : (\mathbb{N} - \{0\}) \to \mathbb{N}$ defined by*

$$Space_A(n) = \max\{Space_A(x) \mid x \in \Sigma_I^n\}.$$

$Time_A$ is defined in such a way that we know that every input of size n (i.e., every input from Σ_I^n) is solved by A in time at most $Time_A(n)$ and that there is an input x of size n with $Time_A(x) = Time_A(n)$. This is the reason why one calls this kind of complexity analysis the **worst case analysis**. The drawback of the worst case analysis may occur when one uses it for an algorithm with very different complexities on inputs of the same length.[38] In that case one can consider the average case analysis that consists of determining the average complexity on all instances of size n. There are two problems with this approach. First, to determine the average complexity is usually a much harder problem than to determine $Time_A(n)$, and in many cases we are not able to perform the average complexity analysis. Secondly, the average complexity provides useful information only if the average is taken over a realistic probability distribution over the inputs of any fixed length. Such input distributions may essentially differ from application to application, i.e., an average cost analysis according to a specific input distribution does not

---

[37] Note that one assumes that an algorithm terminates for every input x, and so $Time_A(x)$ is always a non-negative integer.

[38] A famous example is the simplex algorithm for the problem of linear programming.

provide any robust answer according to all possible applications of this algorithm. Even in the case of one specific application, we are often not able to estimate these input distributions for every input size. In this book it will be sufficient to consider the worst case complexity only, because it will provide a good characterization of the behavior of almost all considered algorithms.

An important observation is that we have fixed the input length in Definition 2.3.3.2 in the logarithmic cost manner. There, the length of an input is considered to be the length of its code over Σ_I. Remember that our interpretation of an input alphabet Σ_I is that it is the set of all computer words allowed. Note that sometimes it is sufficient to consider the unit cost of the input size. This means that the size is the number of items (for instance, integers) of the input. This is again acceptable if all the items of the input are of equal size. For instance, for a $n \times n$ matrix over $\mathbb{Z}_p$ for a fixed prime p, one can consider the size n^2 or even n, and measure the complexity according to this parameter. But usually one may not use this approach for problems from algorithmic number theory, where the input consists of one or a few numbers. In such cases we strictly consider the input size as the length of the binary representation of the input.

In this book we shall very rarely perform a very precise analysis of the designed algorithms, because of the following two reasons. First, to make a precise analysis one has to deal with the implementation details that are usually omitted here. Secondly, we consider hard problems only and we will be satisfied with establishing reasonable asymptotic upper bounds on the complexity of designed algorithms.

The main goal of complexity theory is to classify algorithmic problems according to their computational difficulty. Since the time complexity as the number of executed operations seems to be the central measure of algorithm complexity, one prefers to measure the computational difficulty of problems in terms of time complexity. Intuitively, the time complexity of a problem U could be a function $T_U : \mathbb{N} \to \mathbb{N}$ such that $\Theta(T_U(n))$ operations are necessary and sufficient to solve U. But this is still not a consistent definition of the complexity of U because we need to have an algorithmic solution, i.e., an algorithm solving U in the time complexity $O(T_U(n))$. Thus, a natural way to define the time complexity of U seems to be to say that the time complexity of U is the time complexity for the "best" (optimal) algorithm for U. Unfortunately, the following fundamental result of complexity theory shows that this approach to define T_U is not consistent.

Theorem 2.3.3.3. *There is a decision problem* (L, Σ_{bool}) *such that, for every algorithm A deciding L, there exists another algorithm B deciding L, such that*

$$Time_B(n) = \log_2(Time_A(n))$$

for infinitely many positive integers n.

Obviously, Theorem 2.3.3.3 implies that there is no best (optimal) algorithm for L and so it is impossible to define complexity of L as a function from $\mathbb{N}$ to $\mathbb{N}$ in the way proposed above. This is the reason why one does not try to define the complexity of algorithmic problems but rather the lower and upper bounds on the problem complexity.

Definition 2.3.3.4. *Let U be an algorithmic problem, and let f, g be functions from $\mathbb{N}$ to $\mathbb{R}^+$. We say that $O(g(n))$ is an* **upper bound on the time complexity of** U *if there exists an algorithm A solving U with $Time_A(n) \in O(g(n))$.*

We say that $\Omega(f(n))$ is a **lower bound on the time complexity of U** *if every algorithm B solving U has $Time_B(n) \in \Omega(f(n))$.*

An algorithm C is **optimal** *for the problem U if $Time_C(n) \in O(g(n))$ and $\Omega(g(n))$ is a lower bound on the time complexity of U.*

To establish an upper bound on the complexity of a problem U it is sufficient to find an algorithm solving U. Establishing a nontrivial lower bound on the complexity of U is a very hard task because it requires proving that every of the infinitely many known and unknown algorithms solving U must have its time complexity in $\Omega(f(n))$ for some f. This is in fact a nonexistence proof because one has to prove the nonexistence of any algorithm solving U with the time complexity asymptotically smaller than $f(n)$. The best illustration of the hardness of proving lower bounds on problem complexity is the fact that we know thousands of algorithmic problems for which

(i) the time complexity of the best known algorithm is exponential in the input size, and

(ii) no superlinear lower bound such as $\Omega(n \log n)$ is known for any of them.

Thus, we conjecture, for many of these problems, that there does not exist any algorithm solving them in time polynomial in the input size, but we are unable to prove that one really needs more than $O(n)$ time to solve them.

To overcome our disability to prove lower bounds on problem complexity (i.e., to prove that some problems are hard), some concepts providing reasonable arguments for hardness of concrete problems instead of the evidence of their hardness were developed. These concepts are connected with some formal manipulation of algorithms and complexity in terms of Turing machines (TMs) and Turing machine complexity. We assume that the reader is familiar with the Turing machine model. Remember, that following the **Church-Turing thesis**, a Turing machine is a formalization of the intuitive notion of algorithm. This means that a problem U can be solved by an algorithm (computer program in any programming language formalism) if and only if there exists a Turing machine solving U. Using the formalism of TMs it was proved that for every increasing function $f : \mathbb{N} \to \mathbb{R}^+$

(i) there exists a decision problem such that every TM solving it has the time complexity in $\Omega(f(n))$,

(ii) but there is a TM solving it in $O(f(n) \cdot \log f(n))$ time.

This means that there is an infinite hierarchy of the hardness of decision problems. In what follows we shall use the terms algorithm and computer program instead of the term Turing machine whenever possible, and so we omit unnecessary technicalities.

One can say that the main objective of the complexity theory is

to find a formal specification of the class of practically solvable problems

and

to develop methods enabling the classification of algorithmic problems according to their membership in this class.

The first efforts in searching for a reasonable formalization of the intuitive notion of practically solvable problems result in the following definition. Let, for every TM (algorithm) M, $L(M)$ denote the language decided by M.

Definition 2.3.3.5. *We define the complexity class* P *of languages decidable in polynomial-time by*

$$P = \{L = L(M) \mid M \text{ is a TM (an algorithm) with } Time_M(n) \in O(n^c)$$
for some positive integer c *}.*

A language (decision problem) L is called **tractable (practically solvable)** *if $L \in$ P. A language L is called* **intractable** *if $L \notin$ P.*

Definition 2.3.3.5 introduces the class P of decision problems decidable by polynomial-time computations and says that exactly the set P is the specification of the class of tractable (practically solvable) problems. Let us discuss the advantages and the disadvantages of this formal definition of tractability. The two main reasons to connect polynomial-time computations with the intuitive notion of practical solvability are the following:

(1) The definition of the class P is robust in the sense that P is invariant for all reasonable models of computation. The class P remains the same independent of whether it is defined in terms of polynomial-time Turing machines, in terms of polynomial-time computer programs over any programming language, or in terms of polynomial-time algorithms of any reasonable formalization of computation. This is the consequence of another fundamental result of complexity theory saying that all computation models (formalizations of the intuitive notion of algorithm) that are realistic in the complexity measurement are polynomially equivalent. **Polynomially equivalent** means that if there is a polynomial-time algorithm for an algorithmic problem U in one formalism, then there is a polynomial-time algorithm for U in the other formalism, and vice versa. Turing machines and all programming languages used are in this class of polynomially equivalent computing models. Thus, if one designs a polynomial-time algorithm for U in C++, then there is a polynomial-time algorithm for U in any

reasonable computing formalism. On the other hand, if one proves that there is no polynomial-time TM deciding a language L, then one can be sure that there is no polynomial-time computer program deciding L. Note that this kind of robustness is very important and it must be required for any reasonable specification of the class of tractable problems.

(2) While the first reason for choosing P is a theoretical one, the second reason is more connected with intuition about practical solvability and experience in algorithm design. Consider the table in Figure 2.14 that illustrates the growth of complexity functions $10n$, $2n^2$, n^3, 2^n, and $n!$ for input sizes 10, 50, 100, and 300.

n $f(n)$	10	50	100	300
$10n$	100	500	1000	3000
$2n^2$	200	5000	20000	180000
n^3	1000	125000	1000000	27000000
2^n	1024	(16 digits)	(31 digits)	(91 digits)
$n!$	$\approx 3.6 \cdot 10^6$	(65 digits)	(161 digits)	(623 digits)

Fig. 2.14.

Observe that if the values of $f(n)$ are too large we write only the number of digits of the decimal representation of $f(n)$. Assuming that one has a computer that executes $1000000 = 10^6$ operations per second, an algorithm A with $Time_A(n) = n^3$ runs in 27 seconds for $n = 300$. But if $Time_A(n) = 2^n$, then the execution of A for $n = 50$ would take more than 30 years, and for $n = 100$ more than $3 \cdot 10^{16}$ years. If one compares the values of 2^n and $n!$ for a realistic input size between 100 and 300 with the suggested number of seconds since the "Big Bang" that has 21 digits[39], then everybody sees that the execution of algorithms of exponential complexity on realistic inputs is beyond the borders of physical reality. Moreover, observe the following properties of the functions n^3 and 2^n. If M is the time you can wait for the results, then developing a computer that executes twice as many instructions in a time unit as the previous computer, helps you

(i) to increase the size of tractable input instances from $M^{1/3}$ to $\sqrt[3]{2} \cdot M^{1/3}$ for an n^3-algorithm (i.e., one can compute on $\sqrt[3]{2}$ times larger sizes of input instances than before), but

(ii) to increase the size of tractable input instances by 1 bit for a 2^n-algorithm.

Thus, algorithms of exponential complexity cannot be considered practical, and the algorithms of the polynomial-time complexity $O(n^c)$ for small c's can be considered practical. Of course, a running time n^{1000} is unlikely

---

[39] Note that the number of protons in the known universe has 79 digits.

to be of any practical use because $n^{1000} > 2^n$ for all reasonable sizes n of inputs. Nevertheless, experience has proved the reasonability of considering polynomial-time computations to be tractable. In almost all cases, once a polynomial-time algorithm has been found for an algorithmic problem that formerly appeared to be hard, some key insight into the problem has been gained and new polynomial-time algorithms of a low[40] degree of the polynomial have been designed for the problem. There are only a few known exceptions of nontrivial problems where the best polynomial-time algorithm is not of practical utility.

Above we argued that P is a good specification of the class of practically solvable problems. Nevertheless, this whole book is devoted to solving problems that are probably not in P. But this does not destroy the idea of taking polynomial-time as the threshold of practical solvability. Our approaches to solving problems outside P usually change our requirements such as

- using randomized algorithms (providing the right solution with some probability) instead of deterministic ones (providing the right solution with certainty) or
- searching for an approximation of an optimal solution instead of searching for an optimal solution.

Thus, the current view on the specification of the class of tractable problems is more or less connected with randomized polynomial-time (approximation[41]) algorithms.

Having the class P, one would like to have methods of classifying problems according to their membership in P. To prove the membership of a decision problem L to P, it is sufficient to design a polynomial-time algorithm for L. As already mentioned above, we do not have any method that would be able to prove for most of the practical problems of interest that they are not in P, i.e., that they are intractable (hard). To overcome this unpleasant situation the concept of NP-completeness was introduced. This concept provides at least a good reason to believe that a specific problem is hard, when one is unable to prove the evidence of this fact.

To introduce the concept of NP-completeness we have to consider **nondeterministic computation**. Nondeterminism is nothing natural from the computational point of view because we do not know any way it can be efficiently implemented on real computers.[42] For the reader not familiar with nondeterministic Turing machines, one can introduce nondeterminism to every programming language by adding an operation $choice(a, b)$ with the meaning *goto a* or *goto b*. Thus, the computation may branch into two computations.

---

[40] At most 6, but often 3

[41] In the case of optimization problems

[42] In fact we do not believe that there exists an efficient simulation of nondeterministic computations by deterministic ones.

This means that a nondeterministic TM (algorithm) may have a lot of computations on an input x, while any deterministic TM (algorithm) has exactly one computation for every input. One usually represents all computations of a nondeterministic algorithm A on an input x by the so-called **computation tree of A on x**. Such a computation tree is depicted in Figure 2.15 for a nondeterministic algorithm A that accepts SAT (solves the decision problem $(\text{SAT}, \Sigma_{logic})$).

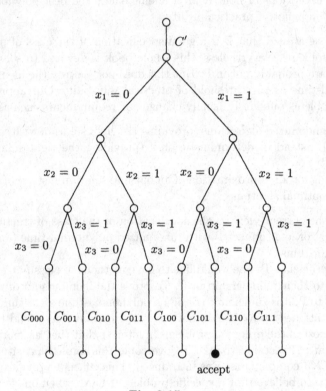

Fig. 2.15.

The tree $T_A(x)$ in Figure 2.15 contains all computations of A on the input x corresponding to the formula

$$\Phi_x = (x_1 \vee x_2) \wedge (x_1 \vee \overline{x}_2 \vee \overline{x}_3) \wedge (\overline{x}_1 \vee x_3) \wedge \overline{x}_2$$

over three variables x_1, x_2, and x_3. A proceeds as follows. First A deterministically (i.e., without computation branching) verifies whether x is a code of a formula Φ_x over Σ_{logic} in the computation part C' (Figure 2.15). If not, A rejects the input. Assume that x codes the above formula Φ_x. The general strategy of A is to nondeterministically guess an assignment that satisfies Φ_x. This is realized in as many branching steps as there are variables in the formula. In our example (Figure 2.15) A first branches into two computations.

The left one corresponds to the guess that $x_1 = 0$ and the right computation corresponds to the guess that $x_1 = 1$. Each of these two computations immediately branches into two computations according to the choice of the Boolean value for x_2. Thus, we obtain 4 computations and each of them immediately branches according to the choice of x_3. Finally, we have 8 computations, each of them corresponding to the choice of one of the 8 assignments for $\{x_1, x_2, x_3\}$. For every assignment α, the corresponding computation C_α deterministically verifies whether α satisfies Φ_x. If α satisfies Φ_x, A accepts x; otherwise, A rejects x. We observe in Figure 2.15 that A accepts x in the computation C_{101} because 101 is the only assignment satisfying Φ_x. All other computations C_α, $\alpha \neq 101$, reject x.

The acceptance and the complexity of nondeterministic algorithms are defined as follows.

Definition 2.3.3.6. *Let M be a nondeterministic TM (algorithm). We say that M **accepts a language** L, $L = L(M)$, if*

(i) for every $x \in L$, there exists at least one computation of M that accepts x, and
(ii) for every $y \notin L$, all computations of M reject y.

*For every input $w \in L$, the **time complexity** $Time_M(w)$ **of** M **on** w is the time complexity of the shortest accepting computation of M on w. The* **time complexity of** M *is the function $Time_M$ from $\mathbb{N}$ to $\mathbb{N}$ defined by*

$$Time_M(n) = \max\{Time_M(x) \mid x \in L(M) \cap \Sigma^n\}.$$

We define the class

$$\mathbf{NP} = \{L(M) \mid M \text{ is a polynomial-time nondeterministic TM}\}$$

as the class of decision problems decided nondeterministically in polynomial time.

We observe that, for a nondeterministic algorithm, it is sufficient that one of the choices is the right way providing the solution (correct answer). For a decision problem (L, Σ), any (deterministic) algorithm B has to decide whether $x \in L$ or $x \notin L$ for every input x. The accepting (rejecting) computation of B on x can be viewed as the proof of the fact $x \in L$ ($x \notin L$). So, the complexity of (deterministic) algorithms for decision problems may be considered as the complexity of producing a proof of the correctness of the output. Following the example of the nondeterministic algorithm for SAT (Figure 2.15), the essential part of the computation from the complexity point of view is the verification whether a guessed assignment satisfies the given formula or not. Thus, the complexity of the nondeterministic algorithm A is in fact the complexity of the verification whether "a given assignment α proves the satisfiability of Φ_x". This leads to the following hypothesis:

The complexity of deterministic computations is the complexity of proving the correctness of the produced output, while the complexity of nondeterministic computation is equivalent to the complexity of deterministic verification of a given proof (certificate) of the fact $x \in L$.

In what follows we show that this hypothesis is true for polynomial-time computations. The fact that deterministic computations can be viewed as proofs of the correctness of the produced outputs is obvious. To prove that the nondeterministic complexity is the complexity of the verification we need the following formal concept.

Definition 2.3.3.7. *Let $L \subseteq \Sigma^*$ be a language. An algorithm A working on inputs from $\Sigma^* \times \{0,1\}^*$ is called a* **verifier for L**, *denoted $L = V(A)$, if*

$$L = \{w \in \Sigma^* \mid A \text{ accepts } (w,c) \text{ for some } c \in \{0,1\}^*\}.$$

If A accepts $(w,c) \in \Sigma^* \times \{0,1\}^*$, we say that c is a **proof (certificate[43])** *of the fact $w \in L$.*

A verifier A for L is called a **polynomial-time verifier** *if there exists a positive integer d such that, for every $w \in L$, $Time_A(w,c) \in O(|w|^d)$ for a proof c of $w \in L$.*

We define the **class of polynomially verifiable languages** *as*

$$\mathbf{VP} = \{V(A) \mid A \text{ is a polynomial-time verifier}\}.$$

We illustrate Definition 2.3.3.7 with the following example. A verifier for SAT is an algorithm that, for each input from $(x,c) \in \Sigma^*_{logic} \times \Sigma^*_{bool}$, interprets x as a representation of a formula Φ_x and c as an assignment of Boolean values to the variables of Φ_x. If this interpretation is possible (i.e., x is a correct code of a formula Φ_x and the length of c is equal to the number of variables in Φ_x), then the verifier checks whether c satisfies Φ_x. Obviously, the verifier accepts (x,c) if and only if c is an assignment satisfying Φ_x. We observe that the verifier is a polynomial-time algorithm because a certificate c for x is always shorter than x and one can efficiently evaluate a formula for a given assignment to its variables.

Exercise 2.3.3.8. Describe a polynomial-time verifier for

  (i) HC,
 (ii) VC, and
(iii) CLIQUE.   □

The following theorem proves our hypothesis in the framework of polynomial-time.

---

[43] Note that a certificate of "$w \in L$" needs not necessarily be a mathematical proof of the fact "$w \in L$". More or less, a certificate should be considered as additional information that essentially simplifies proving the fact "$w \in L$".

Theorem 2.3.3.9.
$$NP = VP.$$

Proof. We prove $NP = VP$ by proving $NP \subseteq VP$ and $VP \subseteq NP$.

(i) We prove $NP \subseteq VP$. Let $L \in NP$, $L \subseteq \Sigma^*$. Then there exists a polynomial-time nondeterministic algorithm (TM) M such that $L = L(M)$. One can construct a polynomial-time verifier A that works as follows:

A: **Input:** $(x, c) \in \Sigma^* \times \Sigma^*_{bool}$.
  (1) A interprets c as a navigator for the simulation of the nondeterministic choices of M. A simulates the work of M (step by step) on w. If M has a choice of two possibilities, then A takes the first one if the next bit of c is 0, and A takes the second possibility if the next bit of c is 1. In this way A simulates exactly one of the computations of M on x.
  (2) If M still has a choice and A used already all bits of c, then A halts and rejects.
  (3) If A succeeds to simulate a complete computation of M on x, then A accepts (x, c) iff M accepts x in this computation.
Obviously, $V(A) = L(M)$ because if M accepts x, then there exists a certificate c that corresponds to a sequence of nondeterministic choices unambiguously determining an accepting computation of A on x. Since A does nothing else than simulating M step by step and, for every $x \in L(M)$, A simulates the shortest accepting computation of M on x, too, V is a polynomial-time verifier for L.

(ii) We prove $VP \subseteq NP$. Let $L \subseteq \Sigma^*$, for an alphabet Σ, be a language from VP. Thus, there exists a polynomial-time verifier A such that $V(A) = L$. One can design a polynomial-time nondeterministic algorithm M that simulates A as follows.
M: **Input:** an $x \in \Sigma^*$.
  (1) M nondeterministically generates a word $c \in \{0, 1\}^*$.
  (2) M simulates step by step the work of A on (x, c).
  (3) M accepts x, if A accepts (x, c), and M rejects x, if A rejects (x, c).
Obviously, $L(M) = V(A)$ and M runs in polynomial-time.   □

Now we have the following situation. We have defined two language classes P and NP. Almost all interesting decision problems appearing in practice are in NP. So, NP is an interesting class from the practical point of view, too. Almost everybody conjectures, if not directly believes, that $P \subset NP$. The two main reasons for this conjecture follow.

(i) *A theoretical reason*
   People do not believe that finding a proof is as easy as verifying the correctness of a given proof. This mathematical intuition supports the hypothesis $P \subset NP = VP$.

(ii) *A practical reason (experience)*
We know more than 3000 problems in NP, many of them have been investigated for 40 years, for which no deterministic polynomial-time algorithm is known. It is not very probable that this is only the consequence of our disability to find efficient algorithms for them. Even if this would be the case, for the current practice the classes P and NP are different because we do not have polynomial-time algorithms for numerous problems from NP.

This provides a new idea for how to "prove" the hardness of some problems, even if we do not have direct mathematical methods for this purpose. Let us try to prove $L \notin P$ for a $L \in NP$ by the additional assumption $P \subset NP$. The idea is to say that a decision problem L is one of the hardest problems in NP if $L \in P$ would immediately imply $P = NP$. Since we do not believe $P = NP$, this is a reasonable argument to believe that $L \notin P$, i.e., that L is hard. Since we want to avoid hard proofs of the nonexistence of efficient algorithms, we define the hardest problems of NP as such problems, so that any hypothetical efficient algorithm for them can be transformed into an efficient algorithm for any other decision problem in NP. The following definition formalizes this idea.

Definition 2.3.3.10. *Let $L_1 \subseteq \Sigma_1^*$ and $L_2 \subseteq \Sigma_2^*$ be two languages. We say that L_1 is* **polynomial-time reducible**[44] *to L_2, $L_1 \leq_p L_2$, if there exists a polynomial-time algorithm A that computes a mapping from Σ_1^* to Σ_2^* such that, for every $x \in \Sigma_1^*$,*

$$x \in L_1 \iff A(x) \in L_2.$$

A is called the **polynomial-time reduction** *from L_1 to L_2.*
A language L is called **NP-hard** *if, for every $U \in NP$, $U \leq_p L$.*
A language L is called **NP-complete** *if*

(i) $L \in NP$, and
(ii) L is NP-hard.

First, observe that $L_1 \leq_p L_2$ means that L_2 is at least as hard as L_1 because if a polynomial-time algorithm M decides L_2 then the "concatenation" of A reducing L_1 to L_2 and M provides a polynomial-time algorithm for L_1. The following claim shows that NP-hardness is exactly the term that we have searched for.

Lemma 2.3.3.11. *If L is NP-hard and $L \in P$, then $P = NP$.*

Proof. Let $L \subseteq \Sigma^*$ be an NP-hard language and let $L \in P$. Then, there is a polynomial-time algorithm M with $L = L(M)$. We prove that for every

---

[44] Note that polynomial-time reducibility of Definition 2.3.3.10 is also called Karp-reducibility or polynomial-time many-to-one reducibility in the literature.

$U \in NP$, $U \subseteq \Sigma_1^*$, there is a polynomial-time algorithm A_U with $L(A_U) = U$, i.e., that $U \in P$. Since $U \leq_p L$, there exists a polynomial-time algorithm B such that $x \in U$ iff $B(x) \in L$. An algorithm A_U with $L(A_U) = U$ can work as follows.

A_U : **Input:** an $x \in \Sigma_1^*$.
Step 1: A_U simulates the work of B on x and computes $B(x)$.
Step 2: A_U simulates the work of M on $B(x) \in \Sigma^*$. A_U accepts x iff M accepts $B(x)$.

Since $x \in U$ iff $B(x) \in L$, $L(A_U) = U$. Since $Time_{A_U}(x) = Time_B(x) + Time_M(B(x))$, B and M work in polynomial time, and $|B(x)|$ is polynomial in $|x|$, we see that A_U is a polynomial-time algorithm.   □

The remaining task is to prove, for a specific language L, that all languages from NP are reducible to L. This so-called master reduction was proved for SAT. We do not present the proof here because we want to omit technical considerations based on the Turing machine formalism.

Theorem 2.3.3.12 (Cook's Theorem). SAT *is* NP-*complete.*[45]

From the practical point of view one is interested in a simple method for establishing that a problem U of interest is NP-hard. One does not need any variation of the master reduction to do it. As claimed in the following observation it is sufficient to take a known NP-hard problem L and to find a polynomial-time reduction from U to L.

Observation 2.3.3.13. Let L_1 and L_2 be two languages. If $L_1 \leq_p L_2$ and L_1 is NP-hard, then L_2 is NP-hard.

Exercise 2.3.3.14. Prove Observation 2.3.3.13. ·
(*Hint:* The proof of Observation 2.3.3.13 is very similar to the proof of Lemma 2.3.3.11.)   □

Historically, by using the claim of Observation 2.3.3.13, SAT has been used to prove NP-completeness for more than 3000 decision problems. An interesting point is that an NP-complete problem can be viewed as a problem that somehow codes any other problem from NP. For instance, the NP-completeness of SAT means that every decision problem (L, Σ) from NP can be expressed in the language of Boolean formulae. This is right because, for each input $x \in \Sigma^*$, we can efficiently construct a Boolean formula Φ_x that is satisfiable iff $x \in L$. Similarly, proving NP-hardness of a problem from graph theory shows that every problem from NP can be expressed in a graph-theoretical language. In what follows we present some examples of reductions between different formal representations (languages).

---

[45] A very detailed, transparent proof of this theorem is given in [Hro03].

Lemma 2.3.3.15. SAT $\leq_p$ CLIQUE.

Proof. Let $\Phi = F_1 \wedge F_2 \wedge \cdots \wedge F_m$ be a formula in CNF, where $F_i = (l_{i1} \vee l_{i2} \vee \cdots \vee l_{ik_i})$, $k_i \in \mathbb{N} - \{0\}$, for $i = 1, 2, \ldots, m$. We construct an input instance (G, k) of the clique problem, such that G contains a k-clique iff Φ is satisfied, as follows.

$k := m;$
$G = (V, E)$, where

$$V := \{[i, j] \mid 1 \leq i \leq m, 1 \leq j \leq k_i\},$$

i.e., we take one vertex for every occurrence of a literal in Φ;

$$E := \{\{[i, j], [r, s]\} \mid \text{ for all } [i, j], [r, s] \in V \text{ such that } i \neq r \text{ and } l_{ij} \neq \bar{l}_{rs}\},$$

i.e., the edges connect vertices corresponding to literals from different clauses only, and additionally if $\{u, v\} \in E$, then the literal corresponding to u is not the negation of the literal corresponding to v.

Observe that the above construction of (G, k) from Φ can be computed efficiently in a straightforward way.

Figure 2.16 shows the graph G corresponding to the formula

$$\Phi = (x_1 \vee x_2) \wedge (x_1 \vee \bar{x}_2 \vee \bar{x}_3) \wedge (\bar{x}_1 \vee x_3) \wedge \bar{x}_2.$$

It remains to be shown that

$$\Phi \text{ is satisfiable } \Longleftrightarrow G \text{ contains a clique of size } k = m. \tag{2.27}$$

The idea of the proof is that two literals (vertices) l_{ij} and l_{rs} are connected by an edge if and only if they are from different clauses and there exists an assignment for which both values of l_{ij} and l_{rs} are 1's. Thus, a clique corresponds to the possibility of finding an input assignment that evaluates all the literals corresponding to the vertices of the clique to 1's.

We prove the equivalence (2.27) by subsequently proving both implications.

(i) Let Φ be satisfiable. Thus, there exists an assignment φ such that $\varphi(\Phi) = 1$. Obviously, $\varphi(F_i) = 1$ for all $i \in \{1, \ldots, m\}$. This implies that, for every $i \in \{1, \ldots, m\}$, there exists a $d_i \in \{1, \ldots, k_i\}$ such that $\varphi(l_{id_i}) = 1$. We claim that the set of vertices $\{[i, d_i] \mid 1 \leq i \leq m\}$ defines an m-clique in G. Obviously, $[1, d_1], [2, d_2], \ldots, [m, d_m]$ are from different clauses. The equality $l_{id_i} = \bar{l}_{jd_j}$ for some $i \neq j$ would imply $\omega(l_{id_i}) \neq \omega(l_{jd_j})$ for every input assignment ω and so $\varphi(l_{id_i}) = \varphi(l_{jd_j})$ would be impossible. Thus, $l_{id_i} \neq \bar{l}_{jd_j}$ for all $i, j \in \{1, \ldots, m\}$, $i \neq j$, and $\{[i, d_i], [j, d_j]\} \in E$ for all $i, j = 1, \ldots, m, i \neq j$.

First clause

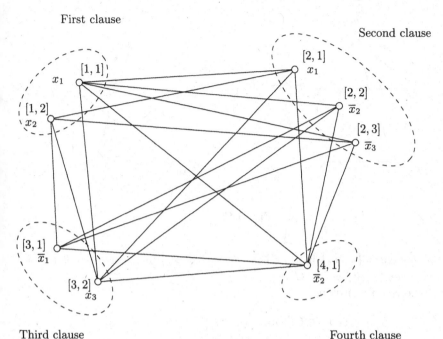

Fig. 2.16.

(ii) Let Q be a clique of G with $k = m$ vertices. Since two vertices are connected in G only if they correspond to two literals from different clauses, there exists $d_1, d_2, \ldots, d_m$, $d_p \in \{1, 2, \ldots, k_p\}$ for $p = 1, \ldots, m$, such that $Q = \{[1, d_1], [2, d_2], \ldots, [m, d_m]\}$. Following the construction there exists an assignment φ such that $\varphi(l_{1d_1}) = \varphi(l_{2d_2}) = \cdots = \varphi(l_{md_m}) = 1$. This directly implies $\varphi(F_1) = \varphi(F_2) = \cdots = \varphi(F_m) = 1$ and so φ satisfies Φ.   □

Lemma 2.3.3.16.

$$\text{CLIQUE} \leq_p \text{VC}.$$

Proof. Let $G = (V, E), k$ be an input of the clique problem. We construct an input $(\overline{G}, m)$ of the vertex cover problem as follows:

$m := |V| - k$,
$\overline{G} = (V, \overline{E})$, where $\overline{E} = \{\{v, u\} \mid v, u \in V, u \neq v, \text{ and } \{u, v\} \notin E\}$.
Obviously, this construction can be executed in linear time.

Figure 2.17 illustrates the construction of the graph $\overline{G}$ from G. The idea of the construction is the following one. If Q is a clique in G, then there is no edge between any pair of vertices from Q in $\overline{G}$. Thus, $V - Q$ must be a vertex cover in $\overline{G}$. So, the clique $\{v_1, v_4, v_5\}$ of G in Figure 2.17 corresponds to the vertex cover $\{v_2, v_3\}$ in $\overline{G}$. The clique $\{v_1, v_2, v_5\}$ of G corresponds to

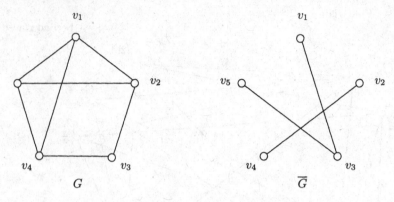

Fig. 2.17.

the vertex cover $\{v_3, v_4\}$ in $\overline{G}$ and the clique $\{v_1, v_2\}$ in G corresponds to the vertex cover $\{v_3, v_4, v_5\}$ in $\overline{G}$.

Obviously, for proving

$$(G, k) \in \text{CLIQUE} \iff (\overline{G}, |V| - k) \in VC$$

it is sufficient to prove

"$S \subseteq V$ is a clique in $G \iff V - S$ is a vertex cover in $\overline{G}$".

We prove this equivalence by proving subsequently the corresponding implications.

(i) Let S be a clique in G. This implies that there is no edge between the vertices of S in $\overline{G}$, i.e., every edge of $\overline{G}$ is incident to at least one vertex in $V - S$. Thus, $V - S$ is a vertex cover in $\overline{G}$.

(ii) Let $C \subseteq V$ be a vertex cover in $\overline{G}$. Following the definition of a vertex cover, every edge of $\overline{G}$ is incident to at least one vertex in C, i.e., there is no edge $\{u, v\} \in \overline{E}$ such that both u and v belong to $V - C$. So, $\{u, v\} \in E$ for all $u, v \in V - C$, $u \neq v$, i.e., $V - C$ is a clique in G.   □

Exercise 2.3.3.17. Prove VC $\leq_p$ CLIQUE.   □

Exercise 2.3.3.18. Prove 3SAT $\leq_p$ VC.   □

The next reduction transforms the language of Boolean formulae into the language of linear equations.

Lemma 2.3.3.19. 3SAT $\leq_p$ SOL-0/1-LP.

Proof. Let $\Phi = F_1 \wedge F_2 \wedge \cdots \wedge F_m$ be a formula in 3CNF over the set of variables $X = \{x_1, \ldots, x_n\}$. Let $F_i = l_{i1} \vee l_{i2} \vee l_{i3}$ for $i = 1, \ldots, m$. First, we construct a system of linear inequalities over X as follows. For every F_i, $i = 1, \ldots, m$, we take the linear inequality LI_i

$$z_{i1} + z_{i2} + z_{i3} \geq 1,$$

where $z_{ir} = x_k$ if $l_{ir} = x_k$ for some $k \in \{1, \ldots, n\}$, and $z_{ir} = (1 - x_q)$ if $l_{ir} = \overline{x}_q$ for some $q \in \{1, \ldots, n\}$. Obviously, $\varphi(F_i) = 1$ for an assignment $\varphi : X \to \{0, 1\}$ if and only if φ is a solution for the linear inequality LI_i. Thus, every φ satisfying Φ is a solution of the system of linear inequalities $LI_1, LI_2, \ldots, LI_m$, and vice versa.

To get a system of linear equations we take $2m$ new Boolean variables (unknowns) $y_1, \ldots, y_m$, $w_1, \ldots, w_m$, and transform every LI_i into the linear equation

$$z_{i1} + z_{i2} + z_{i3} - y_i - w_i = 1.$$

Clearly, the constructed system of linear equations has a solution if and only if the system of linear inequalities $LI_1, LI_2, \ldots, LI_m$ has a solution.   □

Exercise 2.3.3.20. [*] Prove the following reducibilities.

(i) SAT $\leq_p$ SOL-0/1-LP,
(ii) 3SAT $\leq_p$ SOL-IP$_k$ for every prime k,
(iii) SOL-0/1-LP $\leq_p$ SAT,
(iv) SAT $\leq_p$ SOL-IP,
(v) CLIQUE $\leq_p$ SOL-0/1-LP.   □

Above, a successful machinery for proving the hardness of decision problems under the assumption P $\subset$ NP has been introduced. We would like to have a method for proving a similar kind of hardness for optimization problems, too. To develop it, we first introduce the classes PO and NPO of optimization problems that are counterparts of the classes P and NP for decision problems.

Definition 2.3.3.21. **NPO** *is the class of optimization problems, where* $U = (\Sigma_I, \Sigma_O, L, L_I, \mathcal{M}, cost, goal) \in$ NPO *if the following conditions hold:*

(i) $L_I \in$ P,
(ii) there exists a polynomial p_U *such that*
    a) for every $x \in L_I$, *and every* $y \in \mathcal{M}(x)$, $|y| \leq p_U(|x|)$, *and*
    b) there exists a polynomial-time algorithm that, for every $y \in \Sigma_O^*$ *and*
        every $x \in L_I$ *such that* $|y| \leq p_U(|x|)$, *decides whether* $y \in \mathcal{M}(x)$, *and*
(iii) the function cost is computable in polynomial time.

Informally, we see that an optimization problem U is in NPO if

(i) one can efficiently verify whether a string is an instance of U,
(ii) the size of the solutions is polynomial in the size of the problem instances and one can verify in polynomial time whether a string y is a solution to any given input instance x, and
(iii) the cost of any solution can be efficiently determined.

Following the condition (ii), we see the relation to NP that can be seen as the class of languages accepted by polynomial-time verifiers. The conditions (i) and (iii) are natural because we are interested in problems whose kernel is the optimization and not the tractability of deciding whether a given input is a consistent problem instance or the tractability of the cost function evaluation.

We observe that the MAX-SAT problem is in NPO because

(i) one can check in polynomial time whether a word $x \in \Sigma_{logic}^*$ represents a Boolean formula Φ_x in CNF,

(ii) for every x, any assignment $\alpha \in \{0,1\}^*$ to the variables of Φ_x has the property $|\alpha| < |x|$ and one can verify whether $|\alpha|$ is equal to the number of variables of Φ_x even in linear time, and

(iii) for any given assignment α to the variables of Φ_x, one can count the number of satisfied clauses of Φ_x in linear time.

Exercise 2.3.3.22. Prove that the following optimization problems are in NPO:

 (i) MAX-CUT,
 (ii) MAX-CL, and
(iii) MIN-VCP.   □

The following definition corresponds to the natural idea of what tractable optimization problems are.

Definition 2.3.3.23. PO *is the class of optimization problems* $U = (\Sigma_I, \Sigma_O, L, L_I, \mathcal{M}, cost, goal)$ *such that*

(i) $U \in$ NPO, and

(ii) there is a polynomial-time algorithm that, for every $x \in L_I$, computes an optimal solution for x.

In what follows we present a simple method for introducing NP-hardness of optimization problems in the sense that if an NP-hard optimization problem would be in PO, than P would be equal to NP.

Definition 2.3.3.24. *Let* $U = (\Sigma_I, \Sigma_O, L, L_I, \mathcal{M}, cost, goal)$ *be an optimization problem from* NPO. *We define the* **threshold language of** U *as*

$$Lang_U = \{(x,a) \in L_I \times \Sigma_{bool}^* \mid Opt_U(x) \leq Number(a)\}$$

if $goal = minimum$, *and as*

$$Lang_U = \{(x,a) \in L_I \times \Sigma_{bool}^* \mid Opt_U(x) \geq Number(a)\}$$

if $goal = maximum$.

We say that U **is NP-hard** *if* $Lang_U$ *is NP-hard.*

The following lemma shows that to prove the NP-hardness of $Lang_U$ is really a way of showing that U is hard for polynomial-time computations.

Lemma 2.3.3.25. *If an optimization problem $U \in$ PO, then $Lang_U \in$ P.*

Proof. If $U \in$ PO, then there is a polynomial-time algorithm A that, for every input instance x of U, computes an optimal solution for x and so the value $Opt_U(x)$. Then, A can be used to decide $Lang_U$.   □

Theorem 2.3.3.26. *Let U be an optimization problem. If $Lang_U$ is NP-hard and $P \neq NP$, then $U \notin$ PO.*

Proof. Assume the opposite, i.e., $U \in$ PO. Following Lemma 2.3.3.25 $Lang_U \in$ P. Since $Lang_U$ is NP-hard, $Lang_U \in$ P directly implies $P = NP$, a contradiction.   □

To illustrate the simplicity of this method for proving the NP-hardness of optimization problems, we present the following examples.

Lemma 2.3.3.27. MAX-SAT *is NP-hard.*

Proof. Following Definition 2.3.3.23, we have to show that $Lang_{\text{MAX-SAT}}$ is NP-hard. Since we know that SAT is NP-hard, it is sufficient to prove SAT $\leq_p$ $Lang_{\text{MAX-SAT}}$. This reduction is straightforward. Let x code a formula Φ_x of m clauses. Thus, one takes (x, m) as the input for a polynomial-time algorithm for $Lang_{\text{MAX-SAT}}$. Obviously, $(x, m) \in Lang_{\text{MAX-SAT}}$ iff Φ_x is satisfiable.   □

Lemma 2.3.3.28. Max-CL *is NP-hard.*

Proof. Observe that CLIQUE $= Lang_{\text{MAX-CL}}$. Since we have already proved that CLIQUE is NP-hard, the proof is completed.

Exercise 2.3.3.29. Prove that the following optimization problems are NP-hard.

 (i) MAX-3SAT,
 (ii)[*] MAX-2SAT[46],
(iii) MIN-VCP,
 (iv) SCP,
  (v) SKP,
 (vi) MAX-CUT,
(vii) TSP, and
(viii) MAX-E3LINMOD2.   □

Exercise 2.3.3.30. Prove that $P \neq NP$ implies PO $\neq$ NPO.   □

Keywords introduced in Section 2.3.3

uniform cost measurement, logarithmic cost measurement, worst case complexity, time complexity, space complexity, lower and upper bounds on problem complexity, complexity classes P, NP, PO, and NPO, verifiers, polynomial-time reduction, NP-hardness, NP-completeness, NP-hardness of optimization problems

---

[46] Observe that 2SAT $\in$ P

Summary of Section 2.3.3

Time complexity and space complexity are the fundamental complexity measures. Time complexity is measured as the number of elementary operations executed over computer words (operands of constant size). If one executes operations over unboundedly large operands, then one should consider the logarithmic cost measurement, where the cost of one operation is proportional to the length of the representation of the operands.

The (worst case) time complexity of an algorithm A is a function $Time_A(n)$ of the input size. In $Time_A(n)$ complexity A computes the output to each of the inputs of size n and there is an input of size n on which A runs exactly with $Time_A(n)$ complexity.

The class P is the class of all languages that can be decided in polynomial time. Every decision problem in P is considered to be tractable (practically solvable). The class P is an invariant of the choice of any reasonable model of computation.

The class NP is the class of all languages that are accepted by polynomial-time nondeterministic algorithms, or, equivalently, by (deterministic) polynomial-time verifiers. One conjectures that $P \subset NP$, because if P would be equal to NP, then to find a solution would be as hard as to verify the correctness of a given solution for many mathematical problems. To prove or to disprove $P \subset NP$ is one of the most challenging open problems currently in computer science and mathematics.

The concept of NP-completeness provides a method for proving intractability (hardness) of specific decision problems, assuming $P \neq NP$. In fact, an NP-complete problem encodes any other problem from NP in some way, and, for any decision problem from NP, this encoding can be computed efficiently. NP-complete problems are considered to be hard, because if an NP-complete problem would be in P, then P would be equal to NP. This concept can be extended to optimization problems, too.

2.3.4 Algorithm Design Techniques

Over the years people identified several general techniques (concepts) that often yield efficient algorithms for large classes of problems. This chapter provides a short overview on the most important algorithm design techniques; namely,

> divide-and-conquer,
> dynamic programming,
> backtracking,
> local search, and
> greedy algorithms.

We assume that the reader is familiar with all these techniques and knows many specific algorithms designed by them. Thus, we reduce our effort in this section to a short description of these techniques and their basic properties.

Despite the fact that these techniques are so successful that when getting a new algorithmic problem the most reasonable approach is to look whether one of these techniques alone can provide an efficient solution, none of these techniques alone can solve NP-hard problems. Later, we shall see that these methods combined with some new ideas and approaches can be helpful in designing practical algorithms for hard problems. But this is the topic of subsequent chapters.

In what follows we present the above mentioned methods for the design of efficient algorithms in a uniform way. For every technique, we start with its general description and continue with a simple illustration of its application.

DIVIDE-AND-CONQUER.

Informally, the divide-and-conquer technique is based on breaking the given input instance into several smaller input instances in such a way that from the solutions to the smaller input instances one can easily compute a solution to the original input instance. Since the solutions to the smaller problem instances are computed in the same way, the application of the divide-and-conquer principle is naturally expressed by a recursive procedure. In what follows, we give a more detailed description of this technique. Let U be any algorithmic[47] problem.

Divide-and-Conquer Algorithm for U

Input: An input instance I of U, with a $size(I) = n$, $n \geq 1$.

Step 1: **if** $n = 1$ **then** compute the output to I by any method
    else continue with Step 2.

Step 2: Using I, derive problem instances $I_1, I_2, \ldots, I_k$, $k \in \mathbb{N} - \{0\}$ of U
    such that $size(I_j) = n_j < n$ for $j = 1, 2, \ldots, k$.
    {Usually, Step 2 is executed by partitioning I into $I_1, I_2, \ldots, I_k$.
    $I_1, I_2, \ldots, I_k$ are also called **subinstances** of I.}

Step 3: Compute the output $U(I_1), \ldots, U(I_k)$ to the inputs (subinstances)
    $I_1, \ldots, I_k$, respectively, by recursively using the same procedure.

Step 4: Compute the output to I from the outputs $U(I_1), \ldots, U(I_k)$.

In the standard case the complexity analysis results in solving a recurrence. Following our general schema, the time complexity of a divide-and-conquer algorithm A may be computed as follows:

$$Time_A(1) \leq b \text{ if Step 1 can be done in time complexity}$$
$$b \text{ for every input of } U \text{ of size } 1,$$

$$Time_A(n) = \sum_{i=1}^{k} Time_A(n_i) + g(n) + f(n)$$

---

[47] Decision problem, optimization problem, or any other problem

where $g(n)$ is the time complexity of partitioning of I of size n into subinstances $I_1, \ldots, I_k$ (Step 2), and $f(n)$ is the time complexity of computing $U(I)$ from $U(I_1), \ldots, U(I_k)$ (Step 4).

In almost all cases Step 2 partitions I into k input subinstances of the same size $\lceil \frac{n}{m} \rceil$. Thus, the typical recurrence has the form

$$Time_A(n) = k \cdot Time_A\left(\left\lceil \frac{n}{m} \right\rceil\right) + h(n)$$

for a nondecreasing function h and some constants k and m. How to solve such recurrences was discussed in Section 2.2.

Some famous examples of the divide-and-conquer technique are binary search, Mergesort, and Quicksort for sorting, and Strassen's matrix multiplication algorithm. There are numerous applications of this technique, and it is one of the most widely applicable algorithm design technique. Here, we illustrate it on the problem of long integer multiplication.

Example (The problem of multiplying large integers) Let

$$a = a_n a_{n-1} \ldots a_1 \text{ and } b = b_n b_{n-1} \ldots b_1$$

be binary representations of two integers $Number(a)$ and $Number(b)$, $n = 2^k$ for some positive integer k. The aim is to compute the binary representation of $Number(a) \cdot Number(b)$. Recall that the elementary school algorithm involves computing n partial products of $a_n a_{n-1} \ldots a_1$ by b_i for $i = 1, \ldots, n$, and so its complexity is in $O(n^2)$.

A naive divide-and-conquer approach can work as follows. One breaks each of a and b into two integers of $n/2$ bits each:

$$A = Number(a) = \underbrace{Number(a_n \ldots a_{n/2+1})}_{A_1} \cdot 2^{n/2} + \underbrace{Number(a_{n/2} \ldots a_1)}_{A_2}$$

$$B = Number(b) = \underbrace{Number(b_n \ldots b_{n/2+1})}_{B_1} \cdot 2^{n/2} + \underbrace{Number(b_{n/2} \ldots b_1)}_{B_2}.$$

The product of $Number(a)$ and $Number(b)$ can be written as

$$A \cdot B = A_1 \cdot B_1 \cdot 2^n + (A_1 \cdot B_2 + B_1 \cdot A_2) \cdot 2^{n/2} + A_2 \cdot B_2. \tag{2.28}$$

Designing a divide-and-conquer algorithm based on the equality (2.28) we see that the multiplication of two n-bit integers was reduced to

- four multiplications of $\left(\frac{n}{2}\right)$-bit integers ($A_1 \cdot B_1, A_1 \cdot B_2, B_1 \cdot A_2, A_2 \cdot B_2$),
- three additions of integers with at most $2n$ bits, and
- two shifts (multiplications by 2^n and $2^{n/2}$).

Since these additions and shifts can be done in cn steps for some suitable constant c, the complexity of this algorithm is given by the following recurrence:

$$Time(1) = 1$$
$$Time(n) = 4 \cdot Time\left(\frac{n}{2}\right) + cn. \tag{2.29}$$

Following the Master Theorem, the solution of (2.29) is $Time(n) = O(n^2)$. This is no improvement of the classical school method from the asymptotic point of view. To get an improvement one needs to decrease the number of subproblems, i.e., the number of multiplications of $(n/2)$-bit integers. This can be done with the following formula

$$A \cdot B = A_1 B_1 \cdot 2^n + [A_1 B_1 + A_2 B_2 + (A_1 - A_2) \cdot (B_2 - B_1)] \cdot 2^{n/2} + A_2 B_2 \tag{2.30}$$

because

$$(A_1 - A_2) \cdot (B_2 - B_1) + A_1 B_1 + A_2 B_2$$
$$= A_1 B_2 - A_1 B_1 - A_2 B_2 + A_2 B_1 + A_1 B_1 + A_2 B_2$$
$$= A_1 B_2 + A_2 B_1.$$

Although (2.30) looks more complicated than (2.28), it requires only

- three multiplications of $\left(\frac{n}{2}\right)$-bit integers $(A_1 \cdot B_1, (A_1 - A_2) \cdot (B_2 - B_1)$, and $A_2 \cdot B_2)$,
- four additions, and two subtractions of integers of at most $2n$ bits, and
- two shifts (multiplications by 2^n and $2^{n/2}$).

Thus, the divide-and-conquer algorithm C based on (2.30) has the time complexity given by the recurrence

$$Time_C(1) = 1$$
$$Time_C(n) = 3 \cdot Time_C\left(\frac{n}{2}\right) + dn \tag{2.31}$$

for a suitable constant d. According to the Master Theorem (Section 2.2.2) the solution of (2.31) is $Time_C(n) \in O\left(n^{\log_2 3}\right)$, where $\log_2 3 \approx 1.59$. So, C is asymptotically faster than the school method.[48]   □

DYNAMIC PROGRAMMING.

The similarity between divide-and-conquer and dynamic programming is that both these approaches solve problems by combining the solutions to problem subinstances. The difference is that divide-and-conquer does it recursively by dividing a problem instance into subinstances and calling itself on these problem subinstances,[49] while dynamic programming works in a bottom-up

---

[48] Note that the school method is superior to C for integers with fewer than 500 bits because the constant d is too large.

[49] Thus, divide-and-conquer is a top-down method.

fashion as follows. It starts by computing solutions to the smallest (simplest) subinstances, and continues to larger and larger subinstances until the original problem instance has been solved. During its work, any algorithm based on dynamic programming stores all solutions to problem subinstances in a table. Thus, a dynamic-programming algorithm solves every problem subinstance just once because the solutions are saved and reused every time the subproblem is encountered. This is the main advantage of dynamic programming over divide-and-conquer. The divide-and-conquer method can result in solving exponentially many problem subinstances despite the fact that there is only a polynomial number of different problem subinstances. This means that it may happen that a divide-and-conquer algorithm solves some subinstance several times.[50]

Perhaps the most transparent example for the difference between dynamic programming and divide-and-conquer can be seen when computing the nth Fibonacci number $F(n)$. Recall that

$$F(1) = F(2) = 1 \text{ and } F(n) = F(n-1) + F(n-2) \text{ for } n \geq 3.$$

The dynamic-programming algorithm A subsequently computes

"$F(1), \ F(2), \ F(3) = F(1) + F(2), \ \ldots, \ F(n) = F(n-1) + F(n-2)$".

Obviously, $Time_A$ is linear in the value of n. A divide-and-conquer algorithm DCF on the input n will recursively call DCF$(n-1)$ and DCF$(n-2)$. DCF$(n-1)$ will again recursively call DCF$(n-2)$ and DCF$(n-3)$, etc. The tree in Figure 2.18 depicts a part of the recursive calls of DCF. One easily observes that the number of subproblems calls is exponential in n and so $Time_{\text{DCF}}$ is exponential in n.

Exercise 2.3.4.1. Estimate, for every $i \in \{1, 2, \ldots, n-1\}$, how many times the subproblem $F(n-i)$ is solved by DCF.   □

Exercise 2.3.4.2. Consider the problem of computing $\binom{n}{k}$ with the formula

$$\binom{n}{k} = \binom{n-1}{k} + \binom{n-1}{k-1}. \tag{2.32}$$

What is the difference between the time complexities of divide-and-conquer and dynamic programming when both methods use the formula (2.32) to compute the result? (Note that the complexity has to be measured in both input parameters n and k.)   □

Some of the well-known applications of dynamic programming are the Floyd algorithm for the shortest path problem, the algorithm for the minimum

---

[50] Note that it sometimes happens that the natural way of dividing an input instance leads to overlapping subinstances and so the number of different subinstances may even be exponential.

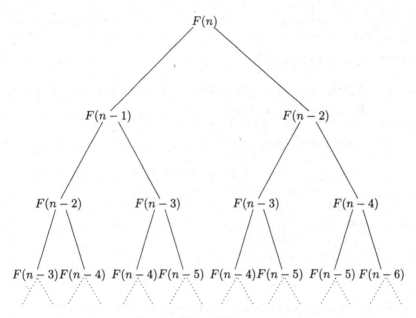

Fig. 2.18.

triangulation problem, the optimal merge pattern algorithm, and the pseudo-polynomial algorithm for the knapsack problem (presented in Chapter 3). We illustrate the method of dynamic programming with the Floyd algorithm.

Example The all-pairs shortest path problem is to find the costs of the shortest path between any pair of vertices of a given weighted graph (G, c), where $G = (V, E)$ and $c : E \to \mathbb{N} - \{0\}$. Let $V = \{v_1, v_2, \ldots, v_n\}$.

The idea is to consecutively compute the values

$$Cost_k(i, j) = \text{the cost of the shortest path between } v_i \text{ and } v_j$$
$$\text{whose internal vertices are from } \{v_1, v_2, \ldots, v_k\}$$

for $k = 0, 1, \ldots, n$. At the beginning one sets

$$Cost_0(i, j) = \begin{cases} c(\{v_i, v_j\}) & \text{if } \{v_i, v_j\} \in E, \\ \infty & \text{if } \{v_i, v_j\} \notin E \text{ and } i \neq j \\ 0 & \text{if } i = j \end{cases}$$

To compute $Cost_k(i, j)$ from $Cost_{k-1}(r, s)$'s for $r, s \in \{1, \ldots, n\}$ one can use the formula

$$Cost_k(i, j) = \min\{Cost_{k-1}(i, j), Cost_{k-1}(i, k) + Cost_{k-1}(k, j)\}$$

for all $i, j \in \{1, 2, \ldots, n\}$. This is because the shortest path between v_i and v_j going via the vertices from $\{v_1, \ldots, v_k\}$ only, either does not go via the vertex v_k or if it goes via v_k then it visits v_k exactly once.[51]

---

[51] The shortest path does not contain cycles because all costs on edges are positive.

The following algorithm is the straightforward implementation of this strategy.

FLOYD'S ALGORITHM

Input: A graph $G = (V, E)$, $V = \{v_1, \ldots, v_n\}$, $n \in \mathbb{N} - \{0\}$, and a cost function $c : E \to \mathbb{N} - \{0\}$.

Step 1:   **for** $i = 1$ **to** n **do**
$\qquad$ **do begin** $Cost[i, i] := 0$;
$\qquad\qquad$ **for** $j := 1$ **to** n **do**
$\qquad\qquad\qquad$ **if** $\{v_i, v_j\} \in E$ **then** $Cost[i, j] := c(\{v_i, v_j\})$
$\qquad\qquad\qquad$ **else if** $i \neq j$ **then** $Cost[i, j] := \infty$
$\qquad$ **end**

Step 2:   **for** $k := 1$ **to** n **do**
$\qquad$ **for** $i := 1$ **to** n **do**
$\qquad\qquad$ **for** $j := 1$ **to** n **do**
$\qquad\qquad\qquad$ $Cost[i, j] := \min\{Cost[i, j], Cost[i, k] + Cost[k, j]\}$.

Obviously, the complexity of the FLOYD'S ALGORITHM is in $O(n^3)$.   □

BACKTRACKING.

Backtracking is a method for solving optimization problems by a possibly exhaustive search of the set of all feasible solutions or for determining an optimal strategy in a finite game by a search in the set of all configurations of the game. Here, we are interested only in the application of backtracking for optimization problems.

In order to be able to apply the backtrack method for an optimization problem one needs to introduce some structure in the set of all feasible solutions. If the specification of every feasible solution can be viewed as an n-tuple $(p_1, p_2, \ldots, p_n)$, where every p_i can be chosen from a finite set P_i, then the following way brings a structure into the set of all feasible solutions.

Let $\mathcal{M}(x)$ be the set of all feasible solutions to the input instance x of an optimization problem.

We define $T_{\mathcal{M}(x)}$ as a labeled rooted tree with the following properties:

(i) Every vertex v of $T_{\mathcal{M}(x)}$ is labeled by a set $S_v \subseteq \mathcal{M}(x)$.
(ii) The root of $T_{\mathcal{M}(x)}$ is labeled by $\mathcal{M}(x)$.
(iii) If $v_1, \ldots, v_m$ are all sons of a father v in $T_{\mathcal{M}(x)}$, then $S_v = \bigcup_{i=1}^m S_{v_i}$ and $S_{v_i} \cap S_{v_j} = \emptyset$ for $i \neq j$.
$\qquad$ {The sets corresponding to the sons define a partition of the set of their father.}
(iv) For every leaf u of $T_{\mathcal{M}(x)}$, $|S_u| \leq 1$.
$\qquad$ {The leaves correspond to the feasible solutions of $\mathcal{M}(x)$.}

If every feasible solution can be specified as described above, one can start to build $T_{\mathcal{M}(x)}$ by setting $p_1 = a$. Then the left son[52] of the root corresponds to the set of feasible solutions with $p_1 = a$ and the right son of the root corresponds to the set of feasible solutions with $p_1 \neq a$. Continuing with this strategy the tree $T_{\mathcal{M}(x)}$ can be constructed in the straightforward way.[53] Having the tree $T_{\mathcal{M}(x)}$ the backtrack method is nothing else than a search (the depth-first-search or the breadth-first-search) in $T_{\mathcal{M}(x)}$. In fact, one does not implement backtracking as a two-phase algorithm, where $T_{\mathcal{M}(x)}$ is created in the first phase, and the second phase is the depth-first-search in $T_{\mathcal{M}(x)}$. This approach would require a too large memory. The tree $T_{\mathcal{M}(x)}$ is considered only hypothetically and one starts the depth-first-search in it directly. Thus, it is sufficient to save the path from the root to the actual vertex only.

In the following example we illustrate the backtrack method on the TSP problem.

Example 2.3.4.3 (Backtracking for TSP). Let $x = (G, c)$, $G = (V, E)$, $V = \{v_1, v_2, \ldots, v_n\}$, $E = \{e_{ij} \mid i, j \in \{1, \ldots, n\}, i \neq j\}$ be an input instance of TSP. Any feasible solution (a Hamiltonian tour) to (G, c) can be unambiguously specified by an $(n-1)$-tuple $(\{v_1, v_{i_1}\}, \{v_{i_1}, v_{i_2}\}, \ldots, \{v_{i_{n-2}}, v_{i_{n-1}}\}) \in E^{n-1}$, where $\{1, i_1, i_2, \ldots, i_{n-1}\} = \{1, 2, \ldots, n\}$ (i.e., $v_1, v_{i_1}, v_{i_2}, \ldots, v_{i_{n-1}}, v_1$ is the Hamiltonian cycle that consists of edges $e_{1i_1}, e_{i_1 i_2}, \ldots, e_{i_{n-2} i_{n-1}}, e_{i_{n-1} 1}$. Let, in what follows, $S_x(h_1, \ldots, h_r, \overline{k}_1, \ldots, \overline{k}_s)$ denote the subset of feasible solutions containing the edges $h_1, \ldots, h_r$ and not containing any of the edges $k_1, \ldots, k_s$, $r, s \in \mathbb{N}$. $T_{\mathcal{M}(x)}$ can be created by dividing $\mathcal{M}(x)$ into $S_x\{e_{12}\} \subseteq \mathcal{M}(x)$ that consists of all feasible solutions that contain the edge e_{12} and $S_x(\overline{e}_{12})$ consisting of all Hamiltonian tours that do not contain the edge e_{12}, etc. Next, we construct $T_{\mathcal{M}(x)}$ for the input instance $x = (G, c)$. $T_{\mathcal{M}(x)}$ is depicted in Figure 2.20. We set $S_x(\emptyset) = \mathcal{M}(x)$.

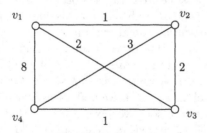

$$v_1 \qquad 1 \qquad v_2$$

Fig. 2.19.

Let us show that all leaves of $T_{\mathcal{M}(x)}$ are one-element subsets of $\mathcal{M}(x)$. $S_x(e_{12}, e_{23}) = \{(e_{12}, e_{23}, e_{34}, e_{41})\}$ because the only possibility of closing a

---

[52] $T_{\mathcal{M}(x)}$ does not necessarily need to be a binary tree, i.e., one may use another strategy to create it.

[53] A specific construction is done in the following example.

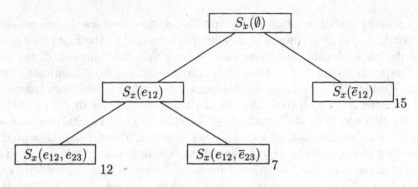

Fig. 2.20.

Hamiltonian tour starting with v_1, v_2, v_3 is to continue with v_4 and then finally with v_1. The cost of this tour is 12. Every Hamiltonian tour that contains e_{12} and does not contain e_{23} must contain e_{24}. So, $S_x(e_{12}, \overline{e}_{23}) = \{e_{12}, e_{24}, e_{43}, e_{31}\}$, and the corresponding Hamiltonian tour v_1, v_2, v_4, v_3, v_1 has cost 7. $S_x(\overline{e}_{12}) = \{(e_{13}, e_{23}, e_{24}, e_{14})\} = S_x\{e_{13}, e_{14}\}$ because if e_{12} is not in a Hamiltonian tour, then e_{13} and e_{14} must be in this tour.[54] Thus, v_4, v_1, v_3 (or v_3, v_1, v_4) must be part of the Hamiltonian tour. Because only the vertex v_2 is missing, one has to take edge e_{24} and e_{23} and the resulting Hamiltonian tour is $v_4, v_1, v_3, v_2, v_4 = v_1, v_3, v_2, v_4, v_1$. Its cost is 15. $\qquad\square$

We shall present a more complex example in Section 3.4, where the use of the backtrack method for solving hard problems will be discussed.

One could ask why we do not simply use the brute force approach by enumerating all $|P_1| \cdot |P_2| \cdot \dots \cdot |P_{n-1}|$ $[(n-1)^{n-1}$ many in the case of TSP] tuples $(p_1, \dots, p_{n-1})$ and to look at the costs of them that specify a Hamiltonian tour. There are at least two possible advantages of backtracking over the brute force method. First of all we generate only feasible solutions by backtracking and so their number may be essentially smaller than $|P_1| \cdot \dots \cdot |P_{n-1}|$. For instance, we have at most $(n-1)!/2$ Hamiltonian tours in a graph of n vertices but the brute force method would generate $(n-1)^{(n-1)}$ tuples candidating for a feasible solution. The main advantage is that we do not necessarily need to realize a complete search in $T_{\mathcal{M}(x)}$. If we find a feasible solution α with a cost m, and if we can calculate in some vertex v of $T_{\mathcal{M}(x)}$ that all solutions corresponding to the subtree T_v rooted by v cannot be better than α (their costs are larger than m if one considers a minimization problem), then we can omit the search in T_v. If one is lucky, the search in many subtrees may be omitted and the backtrack method becomes more efficient. But we stop the discussion on this topic here, because making backtracking more efficient in the applica-

---

[54] Note that, for every vertex v of G, there are two edges adjacent to v in any Hamiltonian tour of G.

tions for hard optimization problems is the main topic of Section 3.4 in the next chapter.

LOCAL SEARCH.

Local search is an algorithm design technique for optimization problems. In contrast to backtracking we do not try to execute an exhaustive search of the set of feasible solutions, but rather a restricted search in the set of feasible solutions. If one wants to design a local search algorithm it is necessary to start by defining a structure on the set of all feasible solutions. The usual way to do it is to say that two feasible solutions α and β are neighbors if α can be obtained from β and vice versa by some local (small) change of their specification. Such local changes are usually called **local transformations**. For instance, two Hamiltonian tours H_1 and H_2 are neighbors if one can obtain H_2 from H_1 by exchanging two edges of H_1 for another two edges. Obviously, any neighborhood on $\mathcal{M}(x)$ is a relation on $\mathcal{M}(x)$. One can view $\mathcal{M}(x)$ as a graph $G(\mathcal{M}(x))$ whose vertices are feasible solutions, and two solutions α and β are connected by the edge $\{\alpha, \beta\}$ if and only if α and β are neighbors. The local search is nothing else but a search in $G(\mathcal{M}(x))$ where one moves via an edge $\{\alpha, \beta\}$ from α to β only if $cost(\beta) > cost(\alpha)$ for maximization problems and $cost(\beta) < cost(\alpha)$ for minimization problems. Thus, local search can be viewed as an iterative improvement that halts with a feasible solution that cannot be improved by the movement to any of its neighbors. In this sense the produced outputs of local search are local optima of $\mathcal{M}(x)$.

If the neighborhood is defined, one can briefly describe local search algorithms by the following scheme.

Input: An input instance x.

Step 1:   Start with a (randomly chosen) feasible solution α from $\mathcal{M}(x)$.

Step 2:   Replace α with a neighbor of α whose cost is an improvement in the comparison with the cost of α.

Step 3:   Repeat Step 2 until a solution α is reached such that no neighbor of α can be viewed as a better solution than α.

Output: α.

The local search technique is usually very efficient and it is one of the most popular methods for attacking hard optimization problems. Besides this, some successful heuristics like simulated annealing are based on this technique. Because of this we postpone deeper discussion and analysis of the local search technique to Chapters 3 and 6, which focus on the topic of this book. We finish the presentation of this method here with a simple application example.

Example 2.3.4.4. A local search algorithm for the minimum spanning tree problem

The minimum spanning tree problem is to find, for a given weighted graph $G = (V, E, c)$, a tree $T = (V, E')$, $E' \subseteq E$, with minimal cost. Let

$T_1 = (V, E_1)$ and $T_2 = (V, E_2)$ be two spanning trees of G. We say that T_1 and T_2 are neighbors if T_1 can be obtained from T_2 by exchanging one edge (i.e., $|E_1 - E_2| = |E_2 - E_1| = 1$). Observe that in looking for neighbors one cannot exchange arbitrary edges because this can destroy the connectivity. The simplest way to do a consistent transformation is to add a new edge e to $T_1 = (V, E_1)$. After this we obtain the graph

$$(V, E_1 \cup \{e\})$$

that contains exactly one cycle since T_1 is a spanning tree. If one removes an edge h of the cycle with $c(h) > c(e)$, then one obtains a better spanning tree

$$(V, (E_1 \cup \{e\}) - \{h\})$$

of G which is a neighbor of T_1.

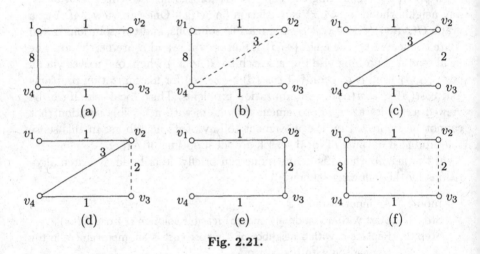

Fig. 2.21.

Consider the graph $G = (V, E, c)$ depicted in Figure 2.19. We start the local search with the tree depicted in Figure 2.21a. Its cost is 10. Let us add the edge $\{v_2, v_4\}$ first. Then, we get the cycle v_1, v_2, v_4, v_1 (Figure 2.21b). Removing the most expensive edge $\{v_1, v_4\}$ of this cycle we get the spanning tree in Figure 2.21c with cost 5. Now, one adds the edge $\{v_2, v_3\}$ (Figure 2.21d). Removing $\{v_2, v_4\}$ from the cycle v_2, v_3, v_4, v_2 we obtain an optimal spanning tree (Figure 2.21e) whose cost is 4. The solution can be also obtained in one iterative step if one would add the edge $\{v_2, v_3\}$ in the first iterative step and then remove the edge $\{v_1, v_4\}$ (Figure 2.21f).   □

Exercise 2.3.4.5. Prove that the local search algorithm for the minimum spanning tree always computes an optimal solution in time $O\left(|E|^2\right)$.   □

GREEDY ALGORITHMS.

The greedy method is perhaps the most straightforward algorithm design technique for optimization problems. A similarity to backtracking and to local algorithms is in that one needs a specification of a feasible solution by a tuple $(p_1, p_2, \ldots, p_n)$, $p_i \in P_i$ for $i = 1, \ldots, n$, and that any greedy algorithm can be viewed as a sequence of local steps. But the greedy algorithms do not move from one feasible solution to another feasible solution. They start with an empty specification and fix one local parameter of the specification (for instance, p_2) forever. In the second step a local algorithm fixes a second parameter (for instance, p_1) of the specification and so on until a complete specification of a feasible solution is reached. The name greedy comes from the way in which the decisions about local specifications are done. A greedy algorithm chooses the parameter that seems to be most promising from all possibilities to make the next local specification. It never reconsiders its decision, whatever situation may arise later. For instance, a greedy algorithm for TSP starts by deciding that the cheapest edge must be in the solution. This is locally the best choice if one has to specify only one edge of a feasible solution. In the next steps it always adds the cheapest new edge that can, together with the already fixed edges, form a Hamiltonian tour to the specification.

Another point about greedy algorithms is that they realize exactly one path from the root to a leaf in the tree $T_{\mathcal{M}(x)}$ created by backtracking. In fact an empty specification means that one considers the set of all feasible solutions $\mathcal{M}(x)$ and $\mathcal{M}(x)$ is the label of the root of the tree $T_{\mathcal{M}(x)}$. Specifying the first parameter p_1 corresponds to restricting $\mathcal{M}$ to a set $S(p_1) = \{\alpha \in \mathcal{M}(x) \mid$ the first parameter of the specification of α is $p_1\}$. Continuing this procedure we obtain the sequence of sets of feasible solutions

$$\mathcal{M}(x) \supseteq S(p_1) \supseteq S(p_1, p_2) \supseteq S(p_1, p_2, p_3) \supseteq \cdots \supseteq S(p_1, p_2, \ldots, p_n),$$

where $|S(p_1, p_2, \ldots, p_n)| = 1$.

Following the above consideration we see that greedy algorithms are usually very efficient, especially in the comparison with backtracking. Another advantage of the greedy method is that it is easy to invent and easy to implement. The drawback of the greedy method is that too many optimization problems are too complex to be solved by such a naive strategy. On the other hand even this simple approach can be helpful in attacking hard problems. Examples documenting this fact will be presented in Chapter 4. The following two examples show that the greedy method can efficiently solve the minimum spanning tree problem and that it can be very weak for TSP.

Example 2.3.4.6 (Greedy for the minimum spanning tree problem). The greedy algorithm for the minimum spanning tree problem can be simply described as follows.

GREEDY-MST

> Input: A weighted connected graph $G = (V, E, c)$, $c : E \to \mathbb{N} - \{0\}$.
>
> Step 1:   Sort the edges according to their costs. Let $e_1, e_2, \ldots, e_m$ be the sequence of all edges of E such that $c(e_1) \leq c(e_2) \leq \cdots \leq c(e_m)$.
>
> Step 2:   Set $E' := \{e_1, e_2\}$; $I := 3$;
>
> Step 3:   **while** $|E'| < |V| - 1$ **do**
> > **begin** add e_I to E' if $(V, E' \cup \{e_I\})$ does not contain any cycle;
> > $I := I + 1$
> > **end**
>
> Output: (V, E').

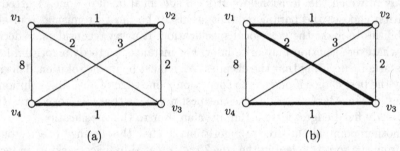

Fig. 2.22.

Since $|E'| = |V| - 1$ and (V, E') does not contain any cycle, it is obvious that (V, E') is a spanning tree of G. To prove that (V, E') is an optimal spanning tree is left to the reader.

We illustrate the work of GREEDY-MST on the input instance G depicted in Figure 2.19. Let $\{v_1, v_2\}, \{v_3, v_4\}, \{v_1, v_3\}, \{v_2, v_3\}, \{v_2, v_4\}, \{v_1, v_4\}$ be the sequence of the edges after sorting in Step 1. Then Figure 2.22 depicts the steps of the specification of the optimal solution $(V, \{\{v_1, v_2\}, \{v_3, v_4\}, \{v_1, v_3\}\})$. $\square$

Exercise 2.3.4.7. Prove that GREEDY-MST always computes an optimal solution. $\square$

Example 2.3.4.8 (Greedy for TSP**).** The greedy algorithm for TSP can be described as follows.

GREEDY-TSP

> Input: A weighted complete graph $G = (V, E, c)$ with $c : E \to \mathbb{N} - \{0\}$, $|V| = n$ for some positive integer n.
>
> Step 1:   Sort the costs of the edges. Let $e_1, e_2, \ldots, e_{\binom{n}{2}}$ be the sequence of all edges of G such that $c(e_1) \leq c(e_2) \leq \cdots \leq c(e_{\binom{n}{2}})$.
>
> Step 2: $E' = \{e_1, e_2\}$, $I := 3$;

Step 3:   **while** $|E'| < n$ **do**
        begin add $\{e_I\}$ to E' if $(V, E' \cup \{e_I\})$ does not contain any
        vertex of degree greater than 2 and any cycle of length
        shorter than n;
        $I := I + 1$;
        end
Output: (V, E').

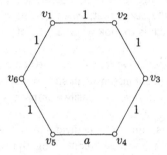

Fig. 2.23.

Since $|E'| = n$ and no vertex has degree greater than 2, (V, E') is a Hamiltonian tour. Considering the graph partially depicted in Figure 2.23 we see that GREEDY-TSP may produce solutions whose costs are arbitrarily far from the optimal cost.

Let the edges $\{v_i, v_{i+1}\}$ for $i = 1, 2, \ldots, 5$ and $\{v_1, v_6\}$ have the costs as depicted in Figure 2.23 and let all missing edges have the cost 2. Assume that a is a very large number. Obviously, GREEDY-TSP takes first all edges of cost 1. Then the Hamiltonian tour $v_1, v_2, v_3, v_4, v_5, v_6, v_1$ is unambiguously determined and its cost is $a + 5$. Figure 2.24 presents an optimal solution $v_1, v_4, v_3, v_2, v_5, v_6, v_1$ whose cost is 8.

Since one can choose an arbitrarily large a, the difference between $a + 5$ and 8 can be arbitrarily large.   □

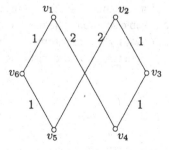

Fig. 2.24.

Keywords introduced in Section 2.3.4

divide-and-conquer, dynamic programming, backtracking, local search, greedy algorithms

Summary of Section 2.3.4

Divide-and-conquer, dynamic programming, backtracking, local search, and greedy algorithms are fundamental algorithm design techniques. These techniques are robust and paradigmatic in the sense that, when getting a new algorithm problem, the most reasonable approach is to look whether one of these techniques alone can provide an efficient solution.

Divide-and-conquer is a recursive technique based on breaking the given problem instance into several problem subinstances in such a way that from the solution to the smaller problem instances one can easily compute a solution to the original problem instance.

Dynamic programming is similar to the divide-and-conquer method in the sense that both techniques solve problems by combining the solutions to subproblems. The difference is that divide-and-conquer does it recursively by dividing problem instances into subinstances and calling itself on these subinstances, while dynamic programming works in a bottom-up fashion by starting with computing solutions to smallest subinstances and continuing to larger and larger subinstances until the original problem instance is solved. The main advantage of dynamic programming is that it solves every subinstance exactly once, while divide-and-conquer may compute a solution to the same subinstance many times.

Backtracking is a technique for solving optimization problems by a possibly exhaustive search of the set of all feasible solutions, in such a systematic way that one never looks twice at the same feasible solution.

Local search is an algorithm design technique for optimization problems. The idea is to define a neighborhood in the set of all feasible solutions $M(x)$ and then to search in $M(x)$ going from a feasible solution to a neighboring feasible solution if the cost of the neighboring solution is better than the cost of the original solution. A local search algorithm stops with a feasible solution that is a local optimum according to the defined neighborhood.

Greedy method is based on a sequence of steps, where in every step the algorithm specifies one parameter of a feasible solution. The name greedy comes from the way it chooses of the parameters. Always, the most promising choice from all possibilities is taken to specify the next parameter, and no decision is reconsidered later.

3

Deterministic Approaches

"As gravity is the essential quality of matter, freedom is the basic quality of the spirit. Human freedom is, first of all, the freedom to perform creative work."

GEORG WILHELM FRIEDRICH HEGEL

3.1 Introduction

In Section 2.3.3 we learned that one does not have any chance to use algorithms of exponential complexity on large input instances because the execution of many, for instance 2^{100}, elementary operations lies beyond the physical reality. Assuming $P \neq NP$, there is no possibility to design polynomial-time (deterministic) algorithms for solving NP-hard problems. The question is what can be done if one wants to attack NP-hard problems with deterministic algorithms in practice. In this chapter we consider the following three approaches:

- *The first approach*
  We try to design algorithms for solving hard problems and we accept their (worst case) exponential complexity if they are efficient and fast enough for most of the problem instances appearing in the specific applications considered. Observe that the success of this approach is possible because we have defined complexity as the worst case complexity and so there may exist many input instances of the given hard problem that can be solved fast by a suitable algorithm. One can roughly view this approach as an effort to partition the set of all instances of a hard problem into two subsets, where one subset contains the easy instances and the other one contains the hard instances.[1] If the problem instances appearing in a specific application

---

[1] In fact, it is not realistic to search for a clear border between easy problem instances and hard problem instances of an algorithmic problem. Realistically we search for large subclasses of easy problem instances only, and we look whether typical problem instances in the considered application belong to these easy subclasses.

are typically (or even always) in the subset of easy problem instances, we have been successful in solving the particular practical task.

- *The second approach*
  We design exponential algorithms for hard problems, and we even accept if their average[2] time complexity is exponential. The crucial idea is that one tries to design algorithms with "slowly increasing" worst case exponential time complexity. For instance, one can design an algorithm A with $Time_A(n) = (1.2)^n$ or $Time_A(n) = 2^{\sqrt{n}}$ instead of a straightforward algorithm of time complexity 2^n. Clearly, this approach moves the border of tractability because one can successfully use $2^{\sqrt{n}}$ and $(1.2)^n$-algorithms for problem instances of sizes for which an 2^n-algorithm does not have any chance of finding a solution in a reasonable time.

- *The third approach*
  We remove the requirement that our algorithm has to solve the given problem. In the case of optimization problems one can be satisfied with a feasible solution that does not need to be optimal if it has some other reasonable properties (for instance, its cost is above some threshold or not too far away from the optimal cost). The typical representative algorithms of this approach are approximation algorithms that provide solutions with costs "close" to the cost of the optimal solutions. Because of the huge success of approximation algorithms for solving hard optimization problems we devote all of Chapter 4 to them and so we do not discuss them here.[3] Sometimes one may essentially relax about requirements and be satisfied only with practical information about the solution. An example is to compute a bound on the cost of an optimal solution instead of searching for an optimal solution. Such information may be very helpful in trying to solve specific problem instances using other approaches and so algorithms based on this approach are often used as a precomputation for other algorithms.

Note that one can combine several different approaches in order to design an algorithm for a given hard problem. The ideas presented in the above approaches lead to the development of several concepts and methods for the design of algorithms for hard problems. In this chapter we present the following ones:

(i) Pseudo-polynomial-time algorithms,
(ii) Parameterized complexity,
(iii) Branch-and-Bound,
(iv) Lowering the worst case exponential complexity,
(v) Local search,
(vi) Relaxation to linear programming.

---

[2] The average is considered over all inputs.
[3] Despite the fact that they belong to the deterministic approaches for solving hard problems.

All these concepts and design methods are of fundamental importance in the same sense as the algorithm design techniques reviewed in Section 2.3.4. If one has a hard problem in a specific application it is reasonable to look first whether some of the above concepts and methods can provide reasonable solutions.

One should be careful to distinguish between concepts and algorithm design techniques here. Branch-and-bound, local search, and relaxation to linear programming are robust algorithm design techniques such as divide-and-conquer, dynamic programming, etc. Pseudo-polynomial-time algorithms, parameterized complexity, and lowering the worst case exponential complexity, on the other hand, are concepts[4] providing ideas and frameworks for attacking hard algorithmic problems. To use these concepts one usually needs to apply some of the algorithm design techniques (alone or in combination) presented in Section 2.3.4.

This chapter is organized as follows. Section 3.2 is devoted to pseudo-polynomial-time algorithms, whose concept is based on the first approach. Here we consider algorithmic problems whose input instances consist of a collection of integers. A pseudo-polynomial-time algorithm is a (potentially exponential-time) algorithm whose time complexity is polynomial in the number of input integers n and in the values of the input integers. Since the value of an integer is exponential in its binary representation, pseudo-polynomial-time algorithms are generally exponential in input size. But pseudo-polynomial-time algorithms work in polynomial time on problem instances, where the values of the input integers are polynomial in the number of the input integers. Thus, such problem instances may be considered easy and one can solve the problem for them.

Section 3.3 is devoted to the concept of parameterized complexity, which is a more general application of the idea of the first approach than the case of pseudo-polynomial-time algorithms. Here, one tries to partition the set of all problem instances into possibly infinitely many subsets according to the value of some input parameter (characteristic) and to design an algorithm that is polynomial in the input length but not in the value of this parameter. For instance, an algorithm can have the time complexity $2^k \cdot n$, where n is the input size and k is the value of the parameter. Obviously, for small k's this algorithm is very efficient, but for $k = n$ it is exponential. Such an algorithm may be very successful if one can reasonably bound k for the problem instances appearing in the specific application.

The branch-and-bound method for optimization problems is presented in Section 3.4. It may be viewed as a combination of the first approach and the third approach in that one tries to make backtracking more efficient due to some additional information about the cost of an optimal solution. This additional information can be obtained by some precomputation based on the third approach.

---

[4] That is, no design techniques

The concept of lowering the worst case exponential complexity presented in Section 3.5 is the pure application of the second approach. Thus, one does not relax any requirement in the formulation of the hard problem. The effort to increase the tractable size of problem instances is based on the design of algorithms with an exponential complexity that does not take too large values for not too large input sizes.

Section 3.6 is devoted to local search algorithms for solving optimization problems. Local algorithms always produce local optima and so they represent a special application of the third approach. In this section we deepen our knowledge about the local search method, whose elementary fundamentals were presented in Section 2.3.4.

Section 3.7 is devoted to the method of relaxation to linear programming. The basic ideas are that many optimization problems can be efficiently reduced to 0/1-linear programming or integer programming, and that linear programming can be solved efficiently while 0/1- and integer programming are NP-hard. But the difference between integer programming and linear programming is only in the domains and so one can relax the problem of solving an instance of integer programming to solving it as an instance of linear programming. Obviously, the resulting solution needs not be a feasible solution of the original problem; rather, it provides some information about an optimal solution to the original problem. For instance, it provides a bound on the cost of optimal solutions. The information obtained by the relaxation method can be used to solve the original problem by another approach.

Sections 3.2, 3.3, 3.4, 3.5, 3.6, and 3.7 are presented in a uniform way. In the first subsection the basic concept of the corresponding algorithm design method is explained and formalized, if necessary. The second subsection illustrates the concepts by designing some algorithms for specific hard problems. In the last subsection, if there is one, the limits of the applicability of the corresponding concepts are discussed.

This chapter finishes with bibliographical remarks in Section 3.8. The main goal of this section is not only to give an overview of the history of the development of the concepts presented, but mainly to provide information about materials deepening the knowledge presented here. Once again we call attention to the fact that in this introductory material we present simple examples transparently illustrating the basic concepts and ideas rather than the best known technical algorithms based on these concepts. Thus, reading additional material is a necessary condition to become an expert in the application of the presented concepts.

3.2 Pseudo-Polynomial-Time Algorithms

3.2.1 Basic Concept

In this section we consider algorithmic problems whose inputs can be viewed as a collection of integers. Such problems are sometimes called **integer-valued**

problems. In what follows we fix the coding of inputs to words over $\{0, 1, \#\}$, where $x = x_1 \# x_2 \# \cdots \# x_n$, $x_i \in \{0, 1\}^*$ for $i = 1, 2, \ldots, n$, is interpreted as a vector of n integers

$$Int(x) = (Number(x_1), Number(x_2), \ldots, Number(x_n)).$$

Obviously, problems such as TSP, the knapsack problem, integer programming, and the vertex-cover problem, can be viewed as integer-valued problems.[5] The size of any input $x \in \{0, 1, \#\}^*$ is considered to be the length $|x|$ of x as a word.[6] Obviously, if $Int(x)$ is a vector of n integers, then $n \leq |x|$. Here, we are still interested in the following characteristic of the input size.

For every $x \in \{0, 1, \#\}^*$, $x = x_1 \# \cdots \# x_n$, $x_i \in \{0, 1\}^*$ for $i = 1, \ldots, n$, we define

$$\boldsymbol{Max\text{-}Int(x)} = \max\{Number(x_i) \mid i = 1, 2, \ldots, n\}.$$

The main idea of the concept of pseudo-polynomial-time algorithms is to design algorithms that are efficient for input instances x with a not too large $Max\text{-}Int(x)$ with respect to $|x|$.

Definition 3.2.1.1. *Let U be an integer-valued problem, and let A be an algorithm that solves U. We say that A is a **pseudo-polynomial-time algorithm for U** if there exists a polynomial p of two variables such that*

$$Time_A(x) = O(p(|x|, Max\text{-}Int(x)))$$

for every instance x of U.

We immediately observe that, for input instances x with $Max\text{-}Int(x) \leq h(|x|)$ for a polynomial h, $Time_A(x)$ of a pseudo-polynomial-time algorithm A is bounded by a polynomial. This can be formally expressed as follows.

Definition 3.2.1.2. *Let U be an integer-valued problem, and let h be a nondecreasing function from $\mathbb{N}$ to $\mathbb{N}$. The **h-value-bounded subproblem of U, Value(h)-U**, is the problem obtained from U by restricting the set of all input instances of U to the set of input instances x with $Max\text{-}Int(x) \leq h(|x|)$.*

Theorem 3.2.1.3. *Let U be an integer-valued problem, and let A be a pseudo-polynomial-time algorithm for U. Then, for every polynomial h, there exists a polynomial-time algorithm for Value(h)-U (i.e., if U is a decision problem then Value(h)-$U \in P$, and if U is an optimization problem then Value(h)-$U \in PO$).*

Proof. Since A is a pseudo-polynomial-time algorithm for U, there exists a polynomial p of two variables such that

---

[5] In fact, every problem whose input instances are (weighted) graphs is an integer-valued problem.

[6] We prefer the precise measurement of the input size here.

$$Time_A(x) \leq O(p(|x|, Max\text{-}Int(x)))$$

for every input instance x of U. Since $Max\text{-}Int(x) \in O(|x|^c)$ for every input instance x of $Value(h)\text{-}U$ with $h(n) \in O(n^c)$, for some positive integer constant c, A is a polynomial-time algorithm for $Value(h)\text{-}U$. $\square$

The concept of pseudo-polynomial-time algorithms follows the first approach that proposes to attack hard problems by searching for large subclasses of easy problem instances. Observe that pseudo-polynomial-time algorithms can be practical in many applications. For instance, for optimization problems whose input instance are weighted graphs, it often happens that the weights are chosen from a fixed interval of values, i.e., the values are independent of the input size. In such cases pseudo-polynomial-time algorithms may be very fast.

In the following Section 3.2.2 we use the method of dynamic programming to design a pseudo-polynomial-time algorithm for the knapsack problem. Section 3.2.4 is devoted to a discussion about the limits of applicability of the concept of pseudo-polynomial-time algorithms. There, we present a simple method that enables us to prove the nonexistence of pseudo-polynomial-time algorithms for some hard problems, unless P = NP.

3.2.2 Dynamic Programming and Knapsack Problem

Remember that the instances of the knapsack problem (KP) are sequences of $2n + 1$ integers $(w_1, w_2, \ldots, w_n, c_1, c_2, \ldots, c_n, b)$, $n \in \mathbb{N} - \{0\}$, where b is the weight capacity of the knapsack, w_i is the weight of the ith object and c_i is the cost of the ith object for $i = 1, 2, \ldots, n$. The objective is to maximize the common cost of objects (the **profit**) packed into the knapsack under the constraint that the common weight of the objects in the knapsack is not above b. Any solution to an input instance $I = (w_1, \ldots, w_n, c_1, \ldots, c_n, b)$ of KP can be represented as a set $T \subseteq \{1, 2, \ldots, n\}$ of indices such that $\sum_{i \in T} w_i \leq b$ and we shall use this representation in what follows. Note that $cost(T, I) = \sum_{i \in T} c_i$. Our aim is to solve KP by the method of dynamic programming. To do this we do not need to consider all 2^n problem subinstances[7] of a problem instance $I = (w_1, \ldots, w_n, c_1, \ldots, c_n, b)$, but only the problem subinstances $I_i = (w_1, w_2, \ldots, w_i, c_1, c_2, \ldots, c_i, b)$ for $i = 1, 2, \ldots, n$. More precisely, the idea is to compute for every I_i, $i = 1, 2, \ldots, n$, and every integer $k \in \{0, 1, 2, \ldots, \sum_{j=1}^{n} c_j\}$, a triple (if it exists)

$$(k, W_{i,k}, T_{i,k}) \in \left\{0, 1, 2, \ldots, \sum_{j=1}^{i} c_j\right\} \times \{0, 1, 2, \ldots, b\} \times Pot(\{1, \ldots, i\}),$$

---

[7] Note that the main art of applying dynamic programming is to find a small subset of input subinstances whose solutions are sufficient to efficiently compute a solution to the original problem instance.

where $W_{i,k} \leq b$ is the minimal weight with which one can achieve exactly the profit k for the input instance I_i, and $T_{i,k} \subseteq \{1, 2, \ldots, i\}$ is a set of indices that provides the profit k under the weight $W_{i,k}$, i.e.,

$$\sum_{j \in T_{i,k}} c_j = k \quad \text{and} \quad \sum_{j \in T_{i,k}} w_j = W_{i,k}.$$

Note that there may be several sets of indices satisfying the above conditions. In such a case it does not matter which one we choose. On the other hand, it may happen that the profit k is not achievable for I_i. In such a case we do not produce any triple for k. In what follows $TRIPLE_i$ denotes the set of all triples produced for I_i. Observe that $|TRIPLE_i| \leq \sum_{j=1}^{i} c_j + 1$ and that $|TRIPLE_i|$ is exactly the number of achievable profits of I_i.

Example 3.2.2.1. Consider the problem instance

$$I = (w_1, \ldots, w_5, c_1, \ldots, c_5, b),$$

where $w_1 = 23$, $c_1 = 33$, $w_2 = 15$, $c_2 = 23$, $w_3 = 15$, $c_3 = 11$, $w_4 = 33$, $c_4 = 35$, $w_5 = 32$, $c_5 = 11$, and $b = 65$. Thus, $I_1 = (w_1 = 23, c_1 = 33, b = 65)$. The only achievable profits are 0 and 33 and so

$$TRIPLE_1 = \{(0, 0, \emptyset), (33, 23, \{1\})\}.$$

$I_2 = (w_1 = 23, w_2 = 15, c_1 = 33, c_2 = 23, b = 65)$. The achievable profits are 0, 23, 33, and 56, and so

$$TRIPLE_2 = \{(0, 0, \emptyset), (23, 15, \{2\}), (33, 23, \{1\}), (56, 38, \{1, 2\})\}.$$

$I_3 = (23, 15, 15, 33, 23, 11, 65)$. The achievable profits are 0, 11, 23, 33, 34, 44, 56, and 67, and

$$
\begin{aligned}
TRIPLE_3 = \{ & (0, 0, \emptyset), (11, 15, \{3\}), (23, 15, \{2\}), (33, 23, \{1\}), \\
& (34, 30, \{2, 3\}), (44, 38, \{1, 3\}), (56, 38, \{1, 2\}), \\
& (67, 53, \{1, 2, 3\}) \}.
\end{aligned}
$$

For $I_4 = (23, 15, 15, 33, 33, 23, 11, 35, 65)$,

$$
\begin{aligned}
TRIPLE_4 = \{ & (0, 0, \emptyset), (11, 15, \{3\}), (23, 15, \{2\}), (33, 23, \{1\}), \\
& (34, 30, \{2, 3\}), (35, 33, \{4\}), (44, 38, \{1, 3\}), (46, 48, \{3, 4\}), \\
& (56, 38, \{1, 2\}), (58, 48, \{2, 4\}), (67, 53, \{1, 2, 3\}), \\
& (68, 56, \{1, 4\}), (69, 63, \{2, 3, 4\}) \}.
\end{aligned}
$$

Finally, for $I = I_5$,

$$
\begin{aligned}
TRIPLE_5 = \{ & (0, 0, \emptyset), (11, 15, \{3\}), (22, 47, \{3, 5\}), (23, 15, \{2\}), \\
& (33, 23, \{1\}), (34, 30, \{2, 3\}), (35, 33, \{4\}), (44, 38, \{1, 3\}), \\
& (45, 62, \{2, 3, 5\}), (46, 48, \{3, 4\}), (56, 38, \{1, 2\}), (58, 48, \{2, 4\}), \\
& (67, 53, \{1, 2, 3\}) \}, (68, 56, \{1, 4\}), (69, 63, \{2, 3, 4\}) \}.
\end{aligned}
$$

Clearly, $\{2,3,4\}$ is an optimal solution because $(69,63,\{2,3,4\})$ is the triple with the maximal profit 69 in $TRIPLE_5$ and $TRIPLE_5$ contains a triple for every profit achievable under the weight constraint b. $\qquad\square$

Computing the set $TRIPLE_n$ for the original input instance $I = I_n$ provides an optimal solution to I. The optimal cost $Opt_{KP}(I)$ is the maximal achievable profit appearing in $TRIPLE_n$ and the corresponding $T_{n,Opt_{KP(I)}}$ is an optimal solution.

The main point is that one can compute $TRIPLE_{i+1}$ from $TRIPLE_i$ in time $O(|TRIPLE_i|)$. First, one computes

$$SET_{i+1} := TRIPLE_i \cup \{(k + c_{i+1}, W_{i,k} + w_{i+1}, T_{i,k} \cup \{i+1\}) \,|$$
$$(k, W_{i,k}, T_{i,k}) \in TRIPLE_i \text{ and } W_{i,k} + w_{i+1} \leq b\}$$

by taking the original set and adding the $(i+1)$th object to the knapsack of every triple if possible. In this way one can get several different triples with the same profit. We put into $TRIPLE_{i+1}$ exactly one triple from SET_{i+1} for every achievable profit k by choosing a triple that achieves the profit k with minimal weight. It does not matter which triple is chosen from SET_{i+1} if several triples have the same profit k and the same weight.

Thus, we can describe our algorithm for the knapsack problem as follows:

Algorithm 3.2.2.2 ((DPKP)).

Input: $I = (w_1, w_2, \ldots, w_n, c_1, c_2, \ldots, c_n, b) \in (\mathbb{N}-\{0\})^{2n+1}$, n a positive integer.

Step 1: $TRIPLE(1) := \{(0,0,\emptyset)\} \cup \{(c_1, w_1, \{1\}) \,|\, \text{if } w_1 \leq b\}$.

Step 2:   **for** $i = 1$ **to** $n-1$ **do**
  begin $SET(i+1) := TRIPLE(i)$;
    for every $(k, w, T) \in TRIPLE(i)$ **do**
      if $w + w_{i+1} \leq b$ **then**
        $SET(i+1) := SET(i+1) \cup \{(k+c_{i+1}, w+w_{i+1}, T \cup \{i+1\})\}$;
      Set $TRIPLE(i+1)$ as a subset of $SET(i+1)$ containing exactly one triple (m, w', T') for every achievable profit m in $SET(i+1)$ by choosing a triple with the minimal weight for the given m
  end

Step 3:   Compute $c := \max\{k \in \{1, \ldots, \sum_{i=1}^{n} c_i\} \,|\, (k, w, T) \in TRIPLE(n)$ for some w and $T\}$.

Output: The index set T such that $(c, w, T) \in TRIPLE(n)$.

Example 3.2.2.1 provides a rough illustration of the work of the algorithm DPKP. Obviously, DPKP solves the knapsack problem.

Theorem 3.2.2.3. *For every input instance I of* KP,

$$Time_{DPKP}(I) \in O\left(|I|^2 \cdot Max\text{-}Int(I)\right),$$

i.e., DPKP *is a pseudo-polynomial-time algorithm for* KP.

Proof. The complexity of Step 1 is in $O(1)$. For $I = (w_1, w_2, \ldots, w_n, c_1, \ldots, c_n, b)$, we have to compute $n - 1$ sets $TRIPLE(i)$ in Step 2. The computation of $TRIPLE(i+1)$ from $TRIPLE(i)$ can be realized in $O(|TRIPLE(i+1)|)$ time. Since $|TRIPLE(i)| \leq \sum_{i=1}^{n} c_i \leq n \cdot Max\text{-}Int(I)$ for every $i \in \{1, 2, \ldots, n\}$, the complexity of Step 2 is in $O(n^2 \cdot Max\text{-}Int(I))$. The complexity of Step 3 is in $O(n \cdot Max\text{-}Int(I))$ because one has only to find the maximum of the set of $|TRIPLE(n)|$ elements.

Since $n \leq |I|$, the time complexity of DPKP on I is in $O(|I|^2 \cdot Max\text{-}Int(I))$.

$\square$

We see that DPKP is an efficient algorithm for KP if the values of the input integers are not too large relative to the number of integers of the input instances. If the values are taken from some fixed interval (independent of the size of input instances), DPKP is a quadratic algorithm for KP. This also happens in several applications. If some of the values of the input instances are very large, one can try to make them smaller by dividing all by the same large integer. Because some rounding is necessary in such a procedure we can lose the guarantee of computing an optimal solution by applying this idea. But, as we shall see later in Chapter 4 on approximation algorithms, one obtains the guarantee of computing a reasonably good approximation of the optimal solution in this way.

3.2.3 Maximum Flow Problem and Ford-Fulkerson Method

In this section we consider the maximum flow problem which is an optimization problem that can be solved in polynomial time. The reason to present a pseudo-polynomial-time algorithm for a problem that is not really hard (at least not NP-hard) lies in the solution method. The Ford-Fulkerson method used here to solve this problem provides a base for a powerful method for attacking NP-hard optimization problems. The investigation of a generalization of the Ford-Fulkerson method will be continued in Section 3.7 which is devoted to the relaxation of hard problems to linear programming and to the LP-duality.

In the maximum flow problem we have a **network** modeled by a directed graph (Fig. 3.1) with two special vertices - the **source** s and the **sink** t. The aim is to send as much material as possible via the network from the source to the sink. The constraints are a bounded capacity of every directed edge of the network (a maximum amount of material that can flow through the edge connection in one time unit) and the zero capacity of all vertices, but the source and the sink. The zero capacity of a vertex means, that the vertex does not have any store and so it cannot collect any material. The consequence is that everything that flows into such a vertex by its ingoing edges must immediately leave this vertex through its outgoing edges. The maximum flow problem is a fundamental problem because it is a model of several different optimization tasks that appear in real life. For instance, it captures the problem of submitting information (messages) from one person (computer) to another in a communication network, the problem of delivering the electric current from a power station to a user, the problem of delivering different kinds of liquid, or the problem of product transportation from a factory to a shop.

In the subsequent definitions we give a formal specification of the maximum flow problem.

Definition 3.2.3.1. *A* **network** *is a 5-tuple* $H = (G, c, A, s, t)$, *where*

(i) $G = (V, E)$ *is a directed graph,*
(ii) c *is a* **capacity function** *from* E *to* A, $c(e)$ *is called the* **capacity of** e *for every* $e \in E$,
(iii) A *is a subset of* $\mathbb{R}^+$,
(iv) $s \in V$ *is the* **source** *of the network, and*
(v) $t \in V$ *is the* **sink** *of the network.*

For every $v \in V$, *the set*

$$In_H(v) = \{(u, v) \mid (u, v) \in E\}$$

is the set of the **ingoing edges** *of* v *and the set*

$$Out_H(v) = \{(v, u) \mid (v, u) \in E\}$$

is the set of the **outgoing edges** *of* v.

A **flow function** *of* H *is any function* $f : E \to A$ *such that*

(1) $0 \le f(e) \le c(e)$ *for all* $e \in E$
    {*the flow* $f(e)$ *of every edge* e *is nonnegative and bounded by the capacity of* e}, *and*
(2) for all $v \in V - \{s, t\}$

$$\sum_{e \in In_H(v)} f(e) - \sum_{h \in Out_H(v)} f(h) = 0$$

    {*for every vertex* v *different from the sink and from the source, the incoming flow to* v *is equal to the outcoming flow of* v}.

The **flow** F_f *of* H *with respect to a flow function* f *is defined as*

$$F_f = \sum_{h \in Out_H(s)} f(h) - \sum_{e \in In_H(s)} f(e),$$

i.e., the flow of H *is measured as the amount of material that leaves the source.*
□

Fig. 3.2 shows the network from Fig. 3.1 with the flow function f described by the labelling $f(e)/c(e)$ for every edge. The corresponding flow F_f is $6 + 10 - 2 = 14$ and this is not optimal.

Exercise 3.2.3.2. Prove, that for every network H and every flow function f of H

$$F_f = \sum_{e \in In_H(t)} f(e) - \sum_{h \in Out_H(t)} f(h),$$

i.e., that F_f can be measured as the amount of material that reaches the sink and remains in it.

The **maximum flow problem** is, for a given network $H = ((V, E), c, A, s, t)$, to find a flow function $f : E \to A$ such that the flow F_f is maximal.

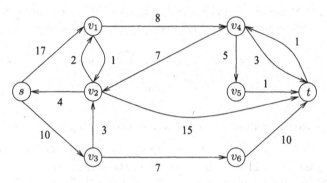

Fig. 3.1.

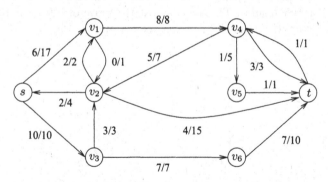

Fig. 3.2.

The first helpful observation we need is that one can measure the flow of a network H not only at the source (or the sink), but at the edges of any cut of H that separates the vertices t and s.

Definition 3.2.3.3. Let $H = ((V, E), c, A, s, t)$ be a network. Let $S \subseteq V$ be a set of vertices such that $s \in S$ and $t \notin S$, and denote $\overline{S} = V - S$.

Let
$$E(S, \overline{S}) = \{(x, y) \mid x \in S \text{ and } y \in \overline{S}\} \cap E$$

and
$$E(\overline{S}, S) = \{(u, v) \mid u \in \overline{S} \text{ and } v \in S\} \cap E.$$

The cut $H(S)$ of H with respect to S is

$$H(S) = E(\overline{S}, S) \cup E(S, \overline{S}).$$

$\square$

Lemma 3.2.3.4. Let $H = ((V, E), c, A, s, t)$, $A \subseteq \mathbb{R}^+$, be a network and let f be a flow function of H. Then, for every $S \subseteq V - \{t\}$ with $s \in S$,

$$F_f = \sum_{e \in E(S, \overline{S})} f(e) - \sum_{e \in E(\overline{S}, S)} f(e). \tag{3.1}$$

Proof. We prove this lemma by induction according to $|S|$.

(i) Let $|S| = 1$, i.e., $S = \{s\}$. Then $Out_H(s) = E(S, \overline{S})$ and $In_H(s) = E(\overline{S}, S)$, and so (3.1) is nothing else than the definition of F_f.

(ii) Let (3.1) be true for all S with $|S| \le k \le |V| - 2$. We prove (3.1) for all S' with $|S'| = k + 1$, $s \in S'$, $t \notin S'$. Obviously, we can write $S' = S \cup \{v\}$, where $|S| = k$ and $v \in \overline{S} - \{t\}$. Following the definition of the flow function (Property (2)), we see that the incoming flow to v is equal to the outcoming flow from v. This means that the move of v from $\overline{S}$ to S does not change anything on the flow between S and $\overline{S}$ (Fig. 3.3). To present

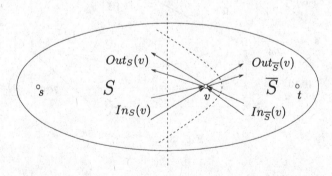

Fig. 3.3.

the formalization of this transparent idea we first express the flow leaving $S \cup \{v\}$ for $\overline{S}$ and the flow leaving $\overline{S} - \{v\}$ for $S \cup \{v\}$ in terms of the flow between S and $\overline{S}$. Let $In_S(v) = In_H(v) \cap E(S, \overline{S})$, $Out_S(v) = Out_H(v) \cap E(\overline{S}, S)$, $In_{\overline{S}}(v) = In_H(v) - In_S(v)$, $Out_{\overline{S}} = Out_H(v) - Out_S(v)$. Then,

$$\sum_{e\in E(S\cup\{v\},\overline{S}-\{v\})} f(e) = \sum_{e\in E(S,\overline{S})} f(e) - \sum_{e\in In_S(v)} f(e) + \sum_{e\in Out_{\overline{S}}(v)} f(e) \quad (3.2)$$

and

$$\sum_{e\in E(\overline{S}-\{v\},S\cup\{v\})} f(e) = \sum_{e\in E(\overline{S},S)} f(e) - \sum_{e\in Out_S(v)} f(e) + \sum_{e\in In_{\overline{S}}(v)} f(e). \quad (3.3)$$

Since $Out_H(v) = Out_S(v) \cup Out_{\overline{S}}(v)$ and $In_H(v) = In_S(v) \cup In_{\overline{S}}(v)$, we have

$$\sum_{e\in Out_{\overline{S}}(v)} f(e) + \sum_{e\in Out_S(v)} f(e) = \sum_{e\in Out_H(v)} f(e) \quad (3.4)$$

and

$$\sum_{e\in In_S(v)} f(e) + \sum_{e\in In_{\overline{S}}(v)} f(e) = \sum_{e\in In_H(v)} f(e). \quad (3.5)$$

Since $v \in V - \{s,t\}$,

$$\sum_{e\in Out_H(v)} f(e) - \sum_{e\in In_H(v)} f(e) = 0. \quad (3.6)$$

Inserting (3.4) and (3.5) into (3.2)−(3.3) we obtain

$$\sum_{e\in E(S\cup\{v\},\overline{S}-\{v\})} f(e) - \sum_{e\in E(\overline{S}-\{v\},S\cup\{v\})} f(e) =$$

$$\sum_{e\in E(S,\overline{S})} f(e) - \sum_{e\in E(\overline{S},S)} f(e) + \sum_{e\in Out_H(v)} f(e) - \sum_{e\in In_H(v)} f(e) \underset{(3.6)}{=}$$

$$\sum_{e\in E(S,\overline{S})} f(e) - \sum_{e\in E(\overline{S},S)} f(e) \underset{ind.}{=} F_f.$$

$\square$

Now we define the minimum network cut problem[8] which is strongly related to the maximum flow problem and will be helpful in the process of designing a pseudo-polynomial-time algorithm for both problems.

Definition 3.2.3.5. *Let* $H = ((V,E), c, A, s, t)$ *be a network and let* $H(S)$ *be a cut of* H *with respect to an* $S \subseteq V$. *The* **capacity of the cut** $H(S)$ *is*

$$c(S) = \sum_{e\in E(S,\overline{S})} c(e).$$

---

[8] Later in Section 3.7 we will call this problem a dual problem to the maximum flow problem.

Thus, the capacity of a cut is the sum of the capacities of all edges going from S to $\overline{S}$. Considering the network in Fig. 3.1, if S contains the 4 leftmost vertices, then $c(S) = 8 + 15 + 7 = 30$. For $S = \{s\}$, $c(S) = 10 + 17 = 27$.

The **minimum network cut problem** is to find, for a given network H, a cut of H with the minimal capacity.

Exercise 3.2.3.6. Find a maximal flow and a minimal cut of the network in Fig. 3.1.

The following results show that both optimization problems for networks are strongly related to each other.

Lemma 3.2.3.7. *Let* $H = ((V, E), c, A, s, t)$, $A \subseteq \mathbb{R}^+$, *be a network. For every flow function f of H and every cut $H(S)$ of H,*

$$F_f \leq c(S).$$

Proof. Following Lemma 3.2.3.4 and the relation $f(e) \leq c(e)$ for every $e \in E$, we directly obtain for every cut $H(S)$

$$F_f = \sum_{e \in E(S,\overline{S})} f(e) - \sum_{e \in E(\overline{S},S)} f(e) \leq \sum_{e \in E(S,\overline{S})} f(e) \leq \sum_{e \in E(S,\overline{S})} c(e) = c(S).$$

□

Lemma 3.2.3.8. *Let* $H = ((V, E), c, A, s, t)$, $A \subseteq \mathbb{R}^+$, *be a network. Let f be a flow function of H and let $H(S)$ be a cut of H. If*

$$F_f = c(S),$$

then

(i) F_f is a maximal flow of H, and
(ii) $H(S)$ is a minimal cut of H.

Proof. This assertion is a direct consequence of Lemma 3.2.3.7. □

To solve the maximum flow problem we present the method of Ford and Fulkerson. The idea is very simple. One starts with an initial flow (for instance, $f(e) = 0$ for all $e \in E$) and tries to improve it step by step. Searching for an improvement of an optimal flow the Ford-Fulkerson algorithm finds a minimal cut of H, and so this algorithm recognizes that the current flow is already an optimal one. The kernel of this algorithm is the way of searching for an improvement possibility. It is based on so called augmenting paths.

A **pseudo-path** of a network H (Fig. 3.4) is a sequence

$$v_0, e_0, v_1, e_1, \ldots, e_k, v_{k+1} \text{ with } v_0, v_1, \ldots, v_{k+1} \in V, \ s = v_0, t = v_{k+1}$$

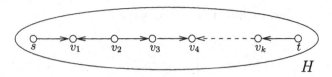

Fig. 3.4.

and $e_0, e_1, \ldots, e_k \in E$, such that it starts in s, finishes in t and does not contain any vertex twice ($|\{v_0, v_1, \ldots, v_{k+1}\}| = k + 2$), and either $e_i = (v_i, v_{i+1})$ or $e_i = (v_{i+1}, v_i)$.

The main difference between a pseudo-path and a directed path is that the edges of a pseudo-path do not need to have the same direction. For instance, s, v_1, v_2, v_3, v_6, t and s, v_2, v_1, v_4, v_5, t determine pseudo-paths of the network in Fig. 3.1 which are no paths.

An **augmenting path** with respect to a network H and a flow function f is a pseudo-path $v_0, e_0, v_1, e_1, \ldots, v_k, e_k, v_{k+1}$ such that

(i) for every edge $e_i = (v_i, v_{i+1})$ directed from s to t, $f(v_i, v_{i+1}) < c(v_i, v_{i+1})$
(ii) for every edge $e_j = (v_{j+1}, v_j)$ directed from t to s, $f(v_{j+1}, v_j) > 0$.

If $e_i = (v_i, v_{i+1})$ is an edge directed from s to t, then the **residual capacity of e_i** is
$$res(e_i) = c(v_i, v_{i+1}) - f(v_i, v_{i+1}).$$

If $e_j = (v_{j+1}, v_j)$ is an edge directed from t to s, then the **residual capacity of e_j** is
$$res(e_j) = f(v_{j+1}, v_j).$$

The **residual capacity of the augmenting path** $P = v_0, e_0, v_1, e_1, \ldots, v_k, e_k, v_{k+1}$ is
$$res(P) = \min\{res(e_i) \mid i = 0, \ldots, k\}$$

If one finds an augmenting path P of H with a flow function f, then there is a possibility to increase F_f by the value $res(P)$.

Lemma 3.2.3.9. *Let $H = ((V, E), c, \mathbb{R}^+, s, t)$ be a network and let f be a flow function of H. Let P be an augmenting path with respect to H and f. Then the function $f' : E \to \mathbb{R}^+$ defined by*

$$f'(e) = f(e) \qquad \text{if } e \text{ is not in } P,$$
$$f'(e) = f(e) + res(P) \text{ if } e \text{ is in } P \text{ and } e \text{ is directed from } s \text{ to } t, \text{ and}$$
$$f'(e) = f(e) - res(P) \text{ if } e \text{ is in } P \text{ and } e \text{ is directed from } t \text{ to } s.$$

is a flow function with $F_{f'} = F_f + res(P)$.

Proof. Following the definition of f' and of $res(e)$ for every edge of P, it is obvious that
$$0 \leq f'(e) \leq c(e)$$

for every $e \in E$. Now, we have to check whether

$$\sum_{e \in In_H(v)} f'(e) - \sum_{e \in Out_H(v)} f'(e) = 0$$

for every $v \in V$. If v is not in P, then this is obviously true. If v is in P, then we consider the four possibilities as depicted in Fig. 3.5.

In the case (a) the augmenting path is $P = s, \ldots, e, v, h, \ldots, t$, where both e and h are directed from s to t. Since $f'(e) = f(e) + res(P)$, the flow incoming to v increases by $res(P)$. Since $f'(h) = f(h) + res(P)$, the flow leaving v increases by $res(P)$, too. The flows of other edges incident to v did not change and so the incoming flow of v and the outcoming flow of v are balanced again.

In the case (b) both the ingoing and the outgoing edge of v are directed from t to s, thus the incoming flow to v and the outcoming flow from v are decreased by the same value $res(P)$ and so they remain balanced.

In the case (c) the edge e is directed from s to t and the edge h is directed from t to s. Thus, the outcoming flow does not change. But the incoming flow does not change as well because $e, h \in In_H(v)$ and the flow of e increases by the same value $res(P)$ as the flow of h decreases.

In the case (d) we have $e, h \in Out_H(v)$. Similarly as in the case (c) both the incoming and the outcoming flow of v do not change.

Thus, f' is a flow function of H. The known facts are that

$$F_{f'} = \sum_{e \in Out_H(s)} f'(e) - \sum_{h \in In_H(s)} f'(h)$$

and that the first edge g of P has to be from $Out_H(s) \cup In_H(s)$. If $g \in Out_H(s)$, then $f'(g) = f(g) + res(P)$ and so $F_{f'} = F_f + res(P)$. If $g \in In_H(s)$, then $f'(g) = f(g) - res(P)$ and again $F_{f'} = F_f + res(P)$.   □

The only remaining question is whether one can efficiently find an augmenting path if such a path exists. The following algorithm shows that this is possible.

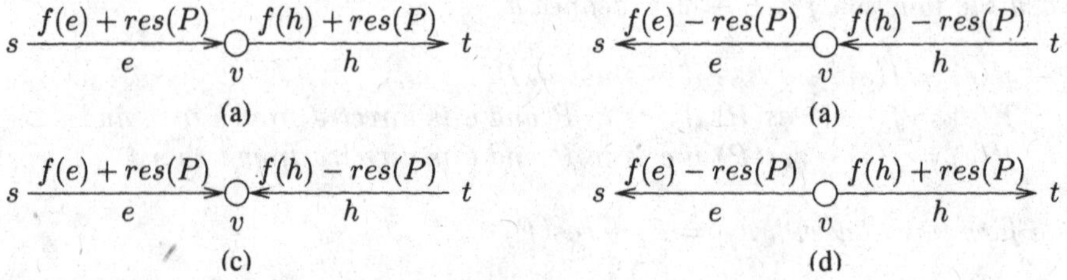

$$s \xrightarrow[e]{f(e)+res(P)} v \xrightarrow[h]{f(h)+res(P)} t \qquad s \xleftarrow[e]{f(e)-res(P)} v \xleftarrow[h]{f(h)-res(P)} t$$

(a)   (a)

$$s \xrightarrow[e]{f(e)+res(P)} v \xleftarrow[h]{f(h)-res(P)} t \qquad s \xleftarrow[e]{f(e)-res(P)} v \xrightarrow[h]{f(h)+res(P)} t$$

(c)   (d)

Fig. 3.5.

Algorithm 3.2.3.10 (The Ford-Fulkerson Algorithm).

Input: $(V, E), c, s, t$ of a network $H = ((V, E), c, \mathbb{Q}^+, s, t)$.

Step 1:   Determine an initial flow function f of H (for instance, $f(e) = 0$ for all $e \in E$); $HALT := 0$

Step 2: $S := \{s\}$; $\overline{S} := V - S$;

Step 3:   **while** $t \notin S$ and $HALT = 0$ **do**

         begin find an edge $e = (u, v) \in E(S, \overline{S}) \cup E(\overline{S}, S)$ such that

                 $res(e) > 0$

                 $-c(e) - f(e) > 0$ if $e \in E(S, \overline{S})$ and $f(e) > 0$ if $e \in E(\overline{S}, S)$";

                 if such an edge does not exist **then** $HALT := 1$

                 else if $e \in E(S, \overline{S})$ **then** $S := S \cup \{v\}$

                                   else $S := S \cup \{u\}$;

            $\overline{S} := V - S$

       end

Step 4:   **if** $HALT = 1$ **then return** (f, S)

         else begin find an augmenting path P from s to t, which consists of vertices of S only; –this is possible because both s and t are in S";

                       compute $res(P)$;

                       determine f' from f as described in Lemma 3.2.3.9

       end;

       goto Step 2

Fig. 3.6 illustrates the work of the Ford-Fulkerson algorithm that starts with the flow function $f(e) = 0$ for every $e \in E$ (Fig. 3.6(a)). The first augmenting path P_1 computed by the algorithm is determined by the sequence of edges $(s, c), (c, d), (d, t)$. Since $res(P_1) = 4$, we obtain $F_{f_1} = 4$ for the flow function f_1 depicted in Fig. 3.6(b). The next augmenting path is P_2 (Fig. 3.6(b)) defined by the sequence of edges $(s, a), (a, b), (b, c), (c, d), (d, t)$. Since $res(P_2) = 3$ we obtain the flow function f_2 with $F_{f_2} = 7$ in Fig. 3.6(c). Now, one can find the augmenting path P_3 (Fig. 3.6(c)) determined by $(s, a), (a, b), (b, t)$. Since $res(P_3) = 7$ we obtain the the flow function f_3 (3.6(d)) with $F_{f_3} = 14$. The flow function f_3 is an optimal solution because reaching $S = \{s, a, b\}$ there is no possibility to extend S and so $H(S)$ is the minimal cut of H.

Theorem 3.2.3.11. *The Ford-Fulkerson algorithm solves the maximum flow problem and the minimum network cut problem and it is a pseudo-polynomial-time algorithm for the input instances with capacity functions from E to $\mathbb{N}$.*

Proof. The Ford-Fulkerson algorithm halts with a flow function f and a set S such that:

(i) for all $e \in E(S, \overline{S})$, $f(e) = c(e)$, and

(ii) for all $h \in E(\overline{S}, S)$, $f(h) = 0$.

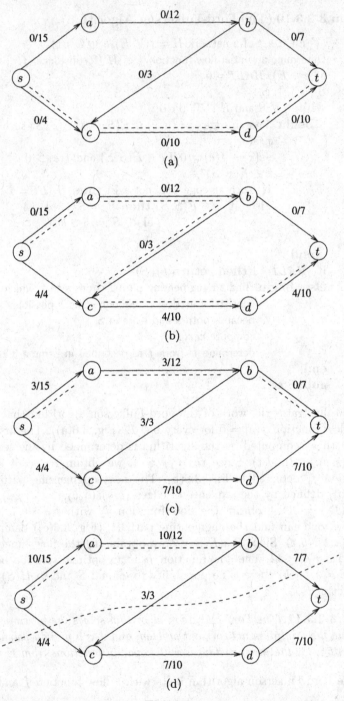

Fig. 3.6.

This implies

$$F_f = \sum_{e \in H(S,\overline{S})} f(e) - \sum_{h \in E(\overline{S},S)} f(h) = \sum_{e \in H(S,\overline{S})} c(e) = c(S)$$

Following Lemma 3.2.3.8, f is a maximal flow function and S induces a minimal cut of H.

Now, we analyze the complexity of the Ford-Fulkerson algorithm. Step 1 can be executed in time $O(|E|)$ and Step 2 in time $O(1)$. One execution of Step 3 costs at most $O(|E|)$ because we look at most once at any edge. Step 4 can be performed in $O(|V|)$ time if one searches for an augmenting path by starting from t and trying to reach s. The number of iterations of the algorithm is at most $F_f = c(S)$. But

$$c(S) \leq \sum_{e \in E} c(e) \leq |E| \cdot \max\{c(e) \mid e \in E\} = |E| \cdot Max\text{-}Int(H).$$

Thus, the algorithm runs in $O(|E|^2 \cdot Max\text{-}Int(H))$ time.   □

A direct consequence of Theorem 3.2.3.11 is the following assertion, which is a stacking point for the important concept of linear-programming duality (Section 3.7.4).

Theorem 3.2.3.12 (Max-Flow Min-Cut Theorem). *For every instance* $I = (G, c, \mathbb{R}^+, s, t)$ *of the maximium flow problem (*MAX-FP*) and of the minimum network cut problem (*MIN-NCP*),*

$$Opt_{\text{MAX-FP}}(I) = Opt_{\text{MIN-NCP}}(I).$$

3.2.4 Limits of Applicability

In this section we are asking for the classification of integer-valued problems according to the existence of pseudo-polynomial-time algorithms for them. We show that by applying the concept of NP-hardness one can easily derive a technique for proving the nonexistence of pseudo-polynomial-time algorithms for some integer-valued problems, if P $\neq$ NP.

Definition 3.2.4.1. *An integer-valued problem* U *is called* **strongly NP-hard** *if there exists a polynomial p such that the problem Value(p)-U is NP-hard.*

The following assertion shows that the strong NP-hardness is exactly the term we are searching for.

Theorem 3.2.4.2. *Let* P $\neq$ NP, *and let* U *be a strongly NP-hard integer-valued problem. Then there does not exist any pseudo-polynomial-time algorithm solving* U.

Proof. Since U is strongly NP-hard, there exists a polynomial p such that $Value(p)$-U is NP-hard. Following Theorem 3.2.1.3, the existence of a pseudo-polynomial-time algorithm for U implies the existence of a polynomial-time algorithm for $Value(h)$-U for every polynomial h (and so for $Value(p)$-U, too). But the existence of a polynomial-time algorithm for the NP-hard problem $Value(p)$-U would immediately imply P $=$ NP.   □

Thus, to prove the nonexistence of any pseudo-polynomial-time algorithm for an integer-valued problem U, it is sufficient to show that $Value(h)$-U is NP-hard for a polynomial h. We illustrate this approach by showing that TSP is strongly NP-hard. Clearly, TSP can be considered an integer-valued problem because any input instance can be viewed as the collection of integer values corresponding to the costs of the edges of a complete graph.

Lemma 3.2.4.3. TSP *is strongly NP-hard.*

Proof. Since HC is NP-hard, it is sufficient to prove HC $\leq_p Lang_{Value(p)\text{-TSP}}$ for the polynomial $p(n) = n$.

Let G be an input of the Hamiltonian problem. The task is to decide whether G contains a Hamiltonian tour. Let $G = (V, E)$, $|V| = n$ for a positive integer n.

We construct a weighted complete graph (K_n, c), where $K_n = (V, E_{com})$, $E_{com} = \{\{u, v\} \mid u, v \in V, u \neq v\}$, and $c : E_{com} \to \{1, 2\}$ is defined by

$$c(e) = 1 \text{ if } e \in E, \text{ and}$$
$$c(e) = 2 \text{ if } e \notin E.$$

We observe that G contains a Hamiltonian tour iff $Opt_{\text{TSP}}(K_n, c) = n$, i.e., iff $((K_n, c), n) \in Lang_{Value(p)\text{-TSP}}$. Thus, solving TSP for the input instance (K_n, c) one decides the Hamiltonian cycle problem for the input G.   □

In the proof of Lemma 3.2.4.3, we showed that TSP is NP-hard even if one restricts the costs of the edges to two values 1 and 2. Since such input instances satisfy the triangle inequality, we obtain that $\triangle$-TSP is strongly NP-hard, too.

We observe that the weighted vertex cover problem (WEIGHT-VCP) is strongly NP-hard because the unweighted version MIN-VCP is NP-hard. In general, every weighted version of an optimization graph problem is strongly NP-hard if the original "unweighted" version is NP-hard.

Keywords introduced in Section 3.2

integer-valued problem, pseudo-polynomial-time algorithm, p-value-bounded sub-problem, strongly NP-hard problem, Ford-Fulkerson Algorithm

Summary of Section 3.2

An integer-valued problem is any problem whose input can be viewed as a collection of integers. A pseudo-polynomial-time algorithm for an integer-valued problem is an algorithm whose running time is polynomial in the input size and in the values of the input integers. So, a pseudo-polynomial-time algorithm runs in polynomial time for the input instances with input values that are polynomial in the size of the whole input.

The dynamic programming method can be used to design a pseudo-polynomial-time algorithm for the knapsack problem. The idea is to subsequently compute a minimal weight solution for every achievable profit in a bottom-up manner. Another nice example of a pseudo-polynomial-time algorithm is the Ford-Fulkerson algorithm for the maximum flow in networks which is based on the concept of duality.

An integer-valued problem is strongly NP-hard if it is also NP-hard for the input instances with small integer values. Thus, strongly NP-hard integer-valued problems are problems that do not admit pseudo-polynomial-time algorithms. TSP is a representative of strongly NP-hard problems, because it is NP-hard even for input instances with the values 1 and 2 only.

3.3 Parameterized Complexity

3.3.1 Basic Concept

The concept of parameterized complexity is similar to the concept of pseudo-polynomial-time algorithms in the sense that both these concepts are based on the first approach of Section 3.1. One tries to analyze a given hard problem more precisely than by taking care of the worst case complexity only. In the concept of parameterized complexity the effort is focused on the search for a parameter that partitions the set of all input instances into possibly infinite many subsets. The idea is to design an algorithm that is polynomial in the input length but possibly not in the value of the chosen parameter. For instance, an algorithm can have the time complexity $2^{k^2} \cdot n^2$, where n is the input size and k is the value of a parameter of the given input. Thus, for small k this algorithm can be considered to be efficient, but for $k = \sqrt{n}$ it is an $n^2 \cdot 2^n$-exponential algorithm. What is important with this concept is that our efforts result in a partition of the set of all input instances into a spectrum of subclasses according to the hardness of particular input instances. In this way one obtains a new insight into the problem by specifying which input instances make the problem hard and for which input instances one can solve the problem efficiently. This moves the thinking about tractability from the classification of problems according to their hardness (as considered in Section 2.3) to the classification of input instances of a particular problem according to their computational difficulty. This is often the best way to attack hard problems in concrete applications.

In what follows we formalize the above described concept.

Definition 3.3.1.1. *Let U be a computing problem, and let L be the language of all instances of U. A* **parameterization of U** *is any function* **Par**: $L \to \mathbb{N}$ *such that*

(i) Par is polynomial-time computable, and
(ii) for infinitely many $k \in \mathbb{N}$, the **k-fixed-parameter set**

$$Set_U(k) = \{x \in L \mid Par(x) = k\}$$

is an infinite set.

We say that A is a **Par-parameterized polynomial-time algorithm for U** *if*

(i) A solves U, and
(ii) there exists a polynomial p and a function $f : \mathbb{N} \to \mathbb{N}$ such that, for every $x \in L$,

$$Time_A(x) \leq f(Par(x)) \cdot p(|x|).$$

If there exists a Par-parameterized polynomial-time algorithm for U, then we say that U is **fixed-parameter-tractable according to Par**.

First of all, we observe that condition (ii) of the definition of a parameterization of U is not necessary but useful for the concept of parameterized complexity. We only consider it to remove dummy parameterizations such as $Par(x)$ is equal to the size $|x|$ of x. In that case if one takes $f(n) = 2^n$, every 2^n-exponential algorithm is Par-parameterized polynomial-time algorithm for U. This is surely not our aim. Forcing the sets $Set_U(k)$ to be infinite means that the parameter given by Par does not depend on the size of input instances in the sense that input instances of an arbitrary large size may have the parameter k. For instance, if the input instances are graphs, the parameter may be the degree of the graph, or its cutwidth (bandwidth). We surely have infinitely many graphs with a fixed degree $k \geq 2$. Obviously, for the choice of Par, it is not sufficient to follow only the requirements (i) and (ii) formulated in Definition 3.3.1.1. A good choice of Par should capture the inherent difficulty of particular input instances, and so looking for Par one has to ask what makes the considered problem difficult.

If one thinks that a chosen parameterization Par corresponds to her/his intuition about the hardness of input instances, the next step is to design an algorithm of a complexity $f(Par(x)) \cdot p(|x|)$. The purpose is to make p a small-degree polynomial and f as a superpolynomial function that grows as slowly as possible. Obviously, p primarily determines the efficiency of the algorithm for fixed parameters, and f decides for which parameters one can make the problem tractable. Thus, one can see the concept of parameterized complexity as the search for a set $\bigcup_{i=1}^{m} Set_U(i)$ such that the subproblem of U obtained by exchanging the set L of all input instances for $\bigcup_{i=1}^{m} Set_U(i)$ becomes tractable. The investigation shows that the hardness of many problems is extremely sensitive with respect to the choice of (the restriction on) the set of input instances, and so studies using this approach may be very successful.

In the following section we illustrate the concept of parameterized complexity with the design of simple parameterized polynomial-time algorithms for some NP-hard problems. Section 3.3.3 contains a short discussion about the applicability of this concept.

3.3.2 Applicability of Parameterized Complexity

First, we observe that the concept of pseudo-polynomial-time algorithms can be viewed as a special case of the concept of parameterized complexity.[9] For any input instance $x = x_1 \# x_2 \# \cdots \# x_n$, $x_i \in \{0,1\}^*$ for $i = 1, \ldots, n$, of an integer-valued problem U, one can define

$$Val(x) = \max\{|x_i| \,|\, i = 1, \ldots, n\}. \tag{3.7}$$

Obviously, $Max\text{-}Int(x) \leq 2^{Val(x)}$ and Val is a parameterization of U. If A is a pseudo-polynomial-time algorithm for U, then there is a polynomial p of two variables, such that

$$Time_A(x) \in O\left(p(|x|, Max\text{-}Int(x))\right) = O\left(p\left(|x|, 2^{Val(x)}\right)\right)$$

for every input instance x of U. So, for sure, $Time_A(x)$ can be bounded by $2^{d \cdot Val(x)} \cdot |x|^c$ for suitable constants c and d. But this implies that A is a Val-parameterized polynomial-time algorithm for U. Thus, we have proved the following result.

Theorem 3.3.2.1. *Let U be an integer-valued problem and let Val be the parameterization of U as defined in (3.7). Then, every pseudo-polynomial-time algorithm for U is a Val-parameterized polynomial-time algorithm for U.*

Note that the opposite (every Val-parameterized polynomial-time algorithm is a pseudo-polynomial-time algorithm) does not hold in general, because a Val-parameterized algorithm with complexity $2^{2^{Val(x)}} \cdot p(|x|)$ for a polynomial p is not a pseudo-polynomial-time algorithm.

Algorithm 3.2.2.2 is a pseudo-polynomial-time algorithm for the knapsack problem. We showed in Theorem 3.2.2.3 that its time complexity is in $O\left(n^2 \cdot Max\text{-}Int(x)\right)$ for every input x. Thus, its complexity is in $O\left(2^k \cdot n^2\right)$ for every input from $Set_{KP}(k) = \{x \in \{0,1,\#\}^* \,|\, Val(x) = k\}$, i.e., Algorithm

---

[9] Note that the original concept of parameterization did not cover the concept of pseudo-polynomial-time algorithms. We have generalized the concept of parameterized complexity here in order to enable exploring the power of the approach of classifying the hardness of input instances. A more detailed discussion about this generalization is given in Section 3.3.3.

3.2.2.2 is an efficient *Val*-parameterized polynomial time algorithm for the knapsack problem.

Now we show a simple example of a parameterized polynomial-time algorithm, where the parameter is simply one item of the input. Consider the vertex cover problem where, for an input (G, k), one has to decide whether G possesses a vertex cover of size at most k. We define $Par(G, k) = k$ for all inputs (G, k). Obviously, *Par* is a parameterization of the vertex cover problem. To design a *Par*-parameterized polynomial-time algorithm for VC we use the following two observations.

Observation 3.3.2.2. For every graph $G = (V, E)$ that possesses a vertex cover $S \subseteq V$ of cardinality at most k, S must contain all vertices of V with a degree greater than k.

Proof. Let u be a vertex adjacent to $a > k$ edges. To cover these a edges without taking u in the cover, one must take all $a > k$ neighbors of u to the cover.   □

Observation 3.3.2.3. Let G have a vertex cover of size at most m and let the degree of G be bounded by k. Then G has at most $m \cdot (k + 1)$ vertices.

The following algorithm is based on the idea of taking (into the vertex cover) all vertices that must be in the vertex cover because of their high degree and then to execute an exhaustive search for covers on the rest of the graph.

Algorithm 3.3.2.4.   Input: (G, k), where $G = (V, E)$ is a graph and k is a positive integer.

Step 1: Let H contain all vertices of G with degree greater than k.
  if $|H| > k$, **then output**("reject") {Observation 3.3.2.2};
  if $|H| \leq k$, **then** $m := k - |H|$ and G' is the subgraph of G obtained
    by removing all vertices of H with their incident edges.

Step 2: **if** G' has more than $m(k + 1)$ vertices $[|V - H| > m(k + 1)]$ **then output**("reject") {Observation 3.3.2.3}.

Step 3: Apply an exhaustive search (by backtracking) for a vertex cover of size at most m in G'.
  if G' has a vertex cover of size at most m, **then output**("accept"),
  else output("reject").

Theorem 3.3.2.5. *Algorithm 3.3.2.4 is a Par-parameterized polynomial-time algorithm for VC.*

Proof. First, observe that Algorithm 3.3.2.4 solves the vertex cover problem. It rejects the input in Step 1 if $|H| > k$, and this is correct because of Observation 3.3.2.2. If $|H| \leq k$, then it takes all vertices of H into the vertex cover because they must be in any vertex cover of size at most k (again because of

Observation 3.3.2.2). So, the question whether G possesses a vertex cover of size k was reduced to the equivalent question whether the rest of the graph G' has a vertex cover of size at most $m = k - |H|$. Following Observation 3.3.2.3 and the fact that the degree of G' is at most k, Algorithm 3.3.2.4 rejects the input in Step 2 if G' has too many vertices (more than $m(k+1)$) to have a vertex cover of size m. Finally in Step 3, Algorithm 3.3.2.4 establishes by an exhaustive search whether G' has a vertex cover of size m or not.

Now we analyze the time complexity of this algorithm. Step 1 can be implemented in time $O(n)$, and Step 2 takes $O(1)$ time. An exhaustive search for a vertex cover of size m in a graph of at most $m(k+1) \le k \cdot (k+1)$ vertices can be performed in time

$$O\left((k \cdot m \cdot (k+1)) \cdot \binom{m \cdot (k+1)}{m}\right) \subseteq O\left(k^3 \cdot \binom{k \cdot (k+1)}{k}\right) \subseteq O\left(k^{2k}\right)$$

because there are at most $\binom{m \cdot (k+1)}{m}$ different subsets of cardinality m of the set of the vertices of G', and G' has at most $k \cdot m \cdot (k+1)$ edges. Thus, the time complexity of Algorithm 3.3.2.4 is in $O\left(n + k^{2k}\right)$ which is included in $O\left(k^{2k} \cdot n\right)$. □

Since k^{2k} can be large already for small parameters k, we present another Par-parameterized polynomial-time algorithm for VC. It is based on the following simple fact.

Observation 3.3.2.6. Let G be a graph. For every edge $e = \{u, v\}$, any vertex cover of G contains at least one of the vertices u and v.

We consider the following divide-and-conquer strategy. Let (G, k) be an input instance of the vertex cover problem. Take an arbitrary edge $\{v_1, v_2\}$ of G. Let G_i be the subgraph of G obtained by removing v_i with all incident edges from G for $i = 1, 2$. Observe that

$$(G, k) \in \text{VC iff } [(G_1, k - 1) \in \text{VC or } (G_2, k - 1) \in \text{VC}].$$

Obviously, $(G_i, k - 1)$ can be constructed from G in time $O(|V|)$. Since, for every graph H, $(H, 1)$ is a trivial problem that can be decided in $O(|V|)$ time and the recursive reduction of (G, k) to subinstances of (G, k) can result in solving at most 2^k subinstances of (G, k), the complexity of this divide-and-conquer algorithm is in $O\left(2^k \cdot n\right)$. Thus, this algorithm is a Par-parameterized polynomial-time algorithm for VC, and it is surely practical for small values of k.

Exercise 3.3.2.7. Combine Algorithm 3.3.2.4 and the above divide-and-conquer algorithm to design a faster algorithm for VC than the presented ones. □

Exercise 3.3.2.8. Let, for every Boolean function Φ in CNF, $Var(\Phi)$ be the number of variables occurring in Φ. Prove that MAX-SAT is fixed-parameter-tractable according to Var. □

Exercise 3.3.2.9. Let $((X, \mathcal{F}), k)$, $\mathcal{F} \subseteq Pot(X)$, be an instance of the decision problem $Lang_{SC}$.[10] Let, for every $x \in X$, $num_{\mathcal{F}}(x)$ be the number of sets in $\mathcal{F}$ that contain x. Define

$$Pat((X, \mathcal{F}), k) = \max\{k, \max\{num_{\mathcal{F}}(x) \mid x \in X\}\}$$

that is a parameterization of $Lang_{SC}$. Find a Pat-parameterized polynomial-time algorithm for $Lang_{SC}$.   □

3.3.3 Discussion

Here, we briefly discuss the applicability of the concept of parameterized complexity. First of all, observe that the fixed-parameter tractability according to a parameterization does not necessarily need to imply the practical solvability (tractability) of the problem. For instance, this approach does not work when the complexity is similar to $2^{2^{2^k}} \cdot n^c$ for some constant c. Such a parameterized polynomial-time algorithm is far from being practical even for small k's, and one has to search for a better one. Another possibility, when this approach fails in an application, is if one chooses a parameter that usually remains large relative to the input size for most of the problem instances to be handled. Such an example is presented in Exercise 3.3.2.8, where the parameter $Var(\Phi)$ is the number of Boolean variables in Φ. Usually, $Var(\Phi)$ is related to the number of literals in Φ. Thus, MAX-SAT remains a very hard optimization problem despite the fact that it is fixed-parameter tractable according to Var. The art of the choice of the parameterization is in finding such a parameterization F that

(i) one can design a practical F-parameterized polynomial-time algorithm, and

(ii) most of the problem instances occurring in the considered application have this parameter reasonably small.

If one is interested in proving negative results one can do it in a way similar to that for pseudo-polynomial-time algorithms. If one proves that the subproblem of U given by the set of input instances $Set_U(k)$ is NP-hard for a fixed constant parameter k, then it is clear that U is not fixed-parameter tractable according to the considered parameterization.

To see an example, consider the TSP problem and the parameterization $Par(G, c) = Max\text{-}Int(G, c)$. The fact that TSP is strongly NP-hard, claimed in Lemma 3.2.4.3, implies that TSP remains NP-hard even for small $Max\text{-}Int(G, c)$. In fact, we have proved that TSP is NP-hard for problem instances with $c : E \to \{1, 2\}$. Thus, there is no Par-parameterized polynomial-time algorithm for TSP.

---

[10] Remember that $Lang_{SC}$ is the threshold language of the set cover problem. $((X, \mathcal{F}), k)$ belongs to $Lang_{SC}$ if there exists a set cover of $(X, \mathcal{F})$ of cardinality at most k.

Another example is the set cover problem and the parameterization *Card* defined by $Card(X, \mathcal{F}) = \max\{num_{\mathcal{F}}(x) \mid x \in X\}$ for every input instance[11] $(X, \mathcal{F})$ of SC. Since SC restricted to the set of input instances $Set_{SC}(2) = \{(X, \mathcal{F}) \mid Card(X, \mathcal{F}) = 2\}$ is exactly the minimum vertex cover problem and MIN-VC is NP-hard, there is no *Card*-parameterized polynomial-time algorithm for SC.

Keywords introduced in Section 3.3

parameterization of a problem, parameterized polynomial-time algorithms, fixed-parameter tractability

Summary of Section 3.3

The concept of parameterized complexity is a generalization of the concept of pseudo-polynomial algorithms. One tries to partition the set of all input instances of a hard problem according to a parameter determined by the input instances in such a way that it is possible to design an algorithm that works in polynomial time according to the input size but not according to the parameter. Then, such an algorithm may provide an efficient solution to input instances whose parameters are reasonably small. This can contribute to the search for the border between the easy input instances and the hard input instances of the given problem.

The main difficulty in applying this approach is to find a parameter that realistically captures the hardness of the problem instances and simultaneously enables the design of an algorithm that is efficient for instances with restricted parameter sizes.

3.4 Branch-and-Bound

3.4.1 Basic Concept

Branch-and-bound is a method for the design of algorithms for optimization problems. We use it if we unconditionally want to find an optimal solution whatever the amount of work (if executable) should be. Branch-and-bound is based on backtracking, which is an exhaustive searching technique in the space of all feasible solutions. The main problem is that the cardinality of the sets of feasible solutions are typically as large as 2^n, $n!$, or even n^n for inputs of size n. The rough idea of the branch-and-bound technique is to speed up backtracking by omitting the search in some parts of the space of feasible solutions, because one is already able to recognize that these parts do not contain any optimal solution in the moment when the exhaustive search would start to search in these parts (to generate the solutions of these parts).

---

[11] See Exercise 3.3.2.9 for the definition of $num_{\mathcal{F}}(x)$.

Remember that backtracking (as described in Section 2.3.4) can be viewed as the depth-first-search or another search strategy in the labeled rooted tree $T_{\mathcal{M}(x)}$, whose leaves are labeled by feasible solutions from $\mathcal{M}(x)$ and every internal vertex of $T_{\mathcal{M}(x)}$ is labeled by the set $S_v \subseteq \mathcal{M}(x)$ that contains all feasible solutions that are labels of the leaves of the subtree T_v rooted by v. Branch-and-bound is nothing else than cutting T_v from $T_{\mathcal{M}(x)}$ if the algorithm is able to determine at the moment when v is visited (generated) that T_v does not contain any optimal solutions. The efficiency of this approach depends on the amount and the sizes of the subtrees of $T_{\mathcal{M}(x)}$ that may be cut during the execution of the algorithm.

The simplest version of the branch-and-bound technique has already been mentioned in Section 2.3.4. Being in a vertex v, one compares the cost of the best solution found up till now with bounds on the minimal or maximal costs of feasible solutions in S_v of T_v. This comparison can be done because the specification of S_v usually enables one to efficiently estimate a range of the costs of feasible solutions in S_v. If the cost of the preliminary best solution is definitely better than any cost in the estimated range, one cuts T_v (i.e., one omits the search in T_v). The efficiency of this naive branch-and-bound approach may significantly vary depending on the input instance as well as on the way in which the tree $T_{\mathcal{M}(x)}$ has been created.[12] This simple approach alone is usually not very successful in solving hard optimization problems in practice.

The standard version of branch-and-bound is based on some precomputation of a bound on the cost of an optimal solution. More precisely, one precomputes a lower bound on the optimal cost for maximization problems and an upper bound on the optimal cost for minimization problems. The standard techniques for computing such bounds are[13]

(i) approximation algorithms (presented in Chapter 4),
(ii) relaxation by reduction to linear programming (presented in Section 3.7),

(iii) random sampling (presented in Chapter 5),
(iv) local search (presented in Section 3.6), and
(v) heuristic methods such as simulated annealing and genetic algorithms (presented in Chapter 6).

The precomputed bound is then used to cut all subtrees of $T_{\mathcal{M}(x)}$ whose solutions are not good enough to reach the cost of this bound. If one succeeded in computing a good bound, then the time complexity of the branch-and-bound procedure may substantially decrease and this approach becomes practical for the problem instance considered.

---

[12] That is, which specification has been used to branch in $\mathcal{M}(x)$.

[13] Note that this is not an exhaustive list and that one can use a combination of several such techniques in order to get a reasonable bound on the cost of an optimal solution.

The rest of this section is organized as follows. The next section shows two examples of applications of the branch-and-bound method for MAX-SAT and TSP. Section 3.4.3 is devoted to a discussion about the advantages and drawbacks of the branch-and-bound method.

3.4.2 Applications for MAX-SAT and TSP

In this section we illustrate the work of the branch-and-bound method on some problem instances of MAX-SAT and TSP. First, we consider the straightforward version of the branch-and-bound technique without any precomputing for MAX-SAT. So, one starts backtracking without any bound on the optimal cost, and after finding a feasible solution one can cut a search subtree if it does not contain any better feasible solutions than the best solution found up till now. Backtracking is very simple for MAX-SAT because every assignment of Boolean values to the variables of the input formula Φ is a feasible solution to Φ. Thus, in every inner vertex of the search tree one branches according to two possibilities,

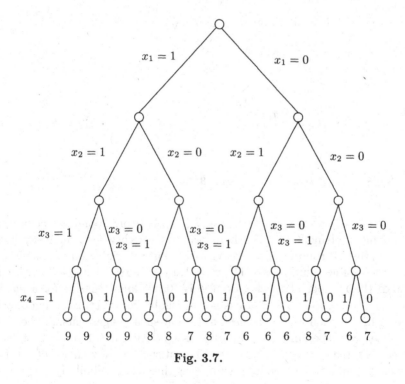

Fig. 3.7.

$x_i = 1$ and $x_i = 0$, for some variable x_i. Figure 3.7 shows the complete search tree $T_{\mathcal{M}(\Phi)}$ for the formula

$$\Phi(x_1, x_2, x_3, x_4) = (x_1 \vee \overline{x}_2) \wedge (x_1 \vee x_3 \vee \overline{x}_4) \wedge (\overline{x}_1 \vee x_2)$$

$$\wedge(x_1 \vee \overline{x}_3 \vee x_4) \wedge (x_2 \vee x_3 \vee \overline{x}_4) \wedge (x_1 \vee \overline{x}_3 \vee \overline{x}_4)$$
$$\wedge x_3 \wedge (x_1 \vee x_4) \wedge (\overline{x}_1 \vee \overline{x}_3) \wedge x_1.$$

Observe that Φ consists of 10 clauses and that it is not satisfiable. Any of the assignments 1111, 1110, 1101, 1100 satisfies 9 clauses, and so all these assignments are optimal solutions for Φ.

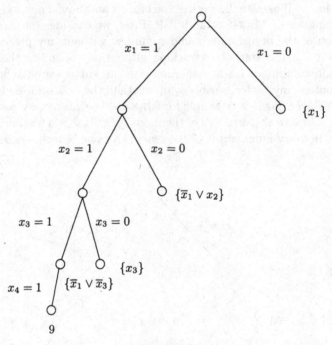

Fig. 3.8.

Figure 3.8 shows the part of the tree $T_{\mathcal{M}(\Phi)}$ that corresponds to the branch-and-bound implementation by the depth-first-search where, for every inner vertex v, the left son of v is always visited before the right son of v is visited. The first feasible solution is 1111 ($x_1 = 1, x_2 = 1, x_3 = 1, x_4 = 1$) with cost 9. Since the number of all clauses of Φ is 10, we know that if, for a partial assignment α, at least one clause is not satisfiable for every extension of α to an assignment, then we do not need to look at such assignments. So, being in the vertex corresponding to the partial assignment $x_1 = 1$, $x_2 = 1$, $x_3 = 1$, we see that the clause $\overline{x}_1 \vee \overline{x}_3$ cannot be satisfied by any choice of the free variable x_4, and we cut the right subtree of this vertex. Similarly:

- for the partial assignment $x_1 = 1$, $x_2 = 1$, $x_3 = 0$, the clause x_3 cannot be satisfied,
- for the partial assignment $x_1 = 1$, $x_2 = 0$, the clause $\overline{x}_1 \vee x_2$ cannot be satisfied,

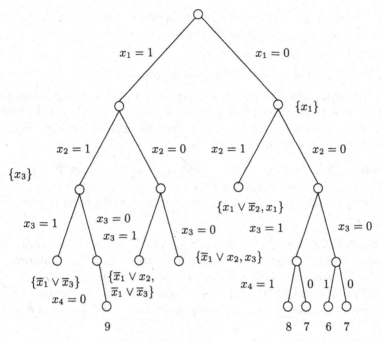

Fig. 3.9.

- for the partial assignment $x_1 = 0$, the clause x_1 cannot be satisfied.

Thus, the branch-and-bound performed by the described depth-first-search has reduced the backtrack method to the visitation (generation) of 8 vertices from the 31 vertices of $T_{\mathcal{M}(\Phi)}$.

In Figure 3.9 we see that we visit (generate) 18 vertices of $T_{\mathcal{M}(\Phi)}$ by the depth-first-search when one first visits the right son (follows the choice $x = 0$ at first) for every inner vertex of $T_{\mathcal{M}(\Phi)}$. This search first finds the feasible solution 0000 with cost 7, then it improves to 0011 with cost 8, and finally it finishes with 1100 with optimal cost 9. For instance, the subtree of the inner vertex corresponding to the partial assignment $x_1 = 0$, $x_2 = 1$ is not visited because

(i) the clauses $x_1 \vee \overline{x}_2$ and x_1 cannot be satisfied by any extension of this partial assignment, and

(ii) the best feasible solutions found up till now satisfies 8 from 10 clauses of Φ.

Comparing Figures 3.8 and 3.9 we see that the efficiency of the branch-and-bound depends on the kind of the search in $T_{\mathcal{M}(\Phi)}$. One can easily observe that using the breadth-first-search one obtains a search subtree of $T_{\mathcal{M}(\Phi)}$ that differs from the subtrees of Figures 3.8 and 3.9. Using another order of variables in building the backtrack tree for Φ leads again to a different complexity of the branch-and-bound method. So, the main observation is that *the efficiency of the branch-and-bound may essentially depend on*

- *the search strategy in the tree* $T_{\mathcal{M}(x)}$,
- *the kind of building of* $T_{\mathcal{M}(x)}$ *by backtracking.*

Moreover, distinct problem instances may require different search and backtrack strategies to be solved efficiently. So, any implementation of the branch-and-bound method for a hard problem may be suitable for some input instances, but not that good for other ones.

Exercise 3.4.2.1. Perform branch-and-bound of $T_{\mathcal{M}(x)}$ in Figure 3.7 by breadth-first-search and compare its time complexity (the number of generated vertices) with the depth-first-search strategies depicted in Figures 3.8 and 3.9. $\qquad\square$

Exercise 3.4.2.2. Take the ordering x_4, x_3, x_1, x_2 of the input variables of the formula $\Phi(x_1, x_2, x_3, x_4)$ and build the backtrack tree $T_{\mathcal{M}(\Phi)}$ according to this variable ordering. Use this $T_{\mathcal{M}(\Phi)}$ as the base for the branch-and-bound method. Considering different search strategies, compare the number of visited vertices with the branch-and-bound implementations presented in Figures 3.8 and 3.9. $\qquad\square$

Now we illustrate the work of the branch-and-bound method with a precomputation on TSP. To decide whether to cut a subtree of $T_{\mathcal{M}(x)}$ in a vertex v we use a straightforward strategy. Consider backtracking for TSP as presented in Section 2.3.4. We calculate the lower bound on the costs of feasible solutions in T_v by the sum of the costs of all edges taken into S_v plus the number of missing edges to build a Hamiltonian tour times the minimal cost over all edges.

Figure 3.11 shows the application of this branch-and-bound strategy for the instance I of TSP depicted in Figure 3.10, starting with a precomputed upper bound 10 on the optimal cost. As introduced in Section 2.3.4 $S_I(h_1, \ldots, h_r, \overline{e}_1, \ldots, \overline{e}_s)$ denote the subset of $\mathcal{M}(I)$ that contains all Hamiltonian tours that contain all edges $h_1, \ldots, h_r$ and do not contain any of the edges $e_1, \ldots, e_s$. In generating the corresponding subtree $T_{\mathcal{M}(I)}$ we frequently use the following two observations.

Observation 3.4.2.3. If a set S of feasible solutions for the problem instance I from Figure 3.10 does not contain any solution containing two given edges incident to the same vertex v, then every solution in S must contain the other two edges incident to v.

Proof. The degree of K_5 is 4 and every Hamiltonian tour contains two edges incident to v for every vertex v of K_5. $\qquad\square$

Observation 3.4.2.4. For all three different edges, e, h, l, determining a path of length 3 in K_5, $|S(e, h, l)| \leq 1$ (i.e., every Hamiltonian tour can be unambiguously determined by any subpath of length 3).

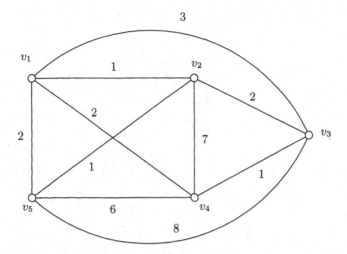

Fig. 3.10.

The first branch in Figure 3.11 is done according to the membership of e_{12} in the solution. $S(e_{12})$ is branched according to the membership of e_{23} in the solution, and $S(e_{12}, e_{23})$ is branched according to the membership of e_{34} in the solution. $|S(e_{12}, e_{23}, e_{34})| = 1$ because $v_1, v_2, v_3, v_4, v_5, v_1$ is the only Hamiltonian tour with the subpath v_1, v_2, v_3, v_4. So, our first feasible solution has cost 12. The next feasible solution $v_1, v_2, v_3, v_5, v_4, v_1$ is the only element of $S(e_{12}, e_{23}, \overline{e}_{34})$ and its cost is 19. The first use of our upper bound 10 is presented when reaching the set $S(e_{12}, e_{24}, \overline{e}_{23})$. $c(e_{12}) + c(e_{24}) = 1 + 7 = 8$ and so any Hamiltonian tour consisting of these two edges must have cost at least 11. Since $S(e_{12}, \overline{e}_{23}, \overline{e}_{24}) = S(e_{12}, e_{25})$ the next branch is done according to e_{35}. Since $c(e_{35}) = 8$, no Hamiltonian tour containing e_{35} can have a cost smaller than 12, and we do not need to work with $S(e_{12}, \overline{e}_{23}, \overline{e}_{24}, e_{35})$ in what follows. The set $S(e_{12}, \overline{e}_{23}, \overline{e}_{24}, \overline{e}_{35}) = S(e_{12}, e_{25}, e_{45})$ contains the only solution $v_1, v_2, v_5, v_4, v_3, v_1$ of the cost 12. Let us now have a short look at the leaves of the right subtree $T_{S(\overline{e}_{12})}$ of $T_{\mathcal{M}(I)}$. We can again bound the cost of the solutions in $S(\overline{e}_{12}, e_{13}, e_{35})$ because $c(e_{13}) + c(e_{35}) = 3 + 8 = 11$ and so any solution in $S(\overline{e}_{12}, e_{13}, e_{35})$ has a cost of at least 14. $S(\overline{e}_{12}, e_{13}, \overline{e}_{35}, e_{23}, e_{24}) = S(e_{13}, e_{23}, e_{24}) = \{(v_1, v_3, v_5, v_4, v_1)\}$ and its cost is 20. $S(\overline{e}_{12}, e_{13}, \overline{e}_{35}, e_{23}, \overline{e}_{24}) = S(e_{13}, e_{23}, e_{25}) = \{(v_1, v_3, v_2, v_5, v_4, v_1)\}$ with cost 14. Following Observation 3.4.2.3 $S(\overline{e}_{12}, e_{13}, \overline{e}_{35}, \overline{e}_{23}) = S(e_{13}, e_{34}, \overline{e}_{12}) = S(e_{13}, e_{34}, e_{15}) = \{(v_1, v_3, v_4, v_2, v_5, v_1)\}$ and its cost is 14. Applying Observation 3.4.2.3 we obtain $S(\overline{e}_{12}, \overline{e}_{13}) = S(e_{14}, e_{15})$, and we branch this set according to e_{24}. The only solution of $S(\overline{e}_{12}, \overline{e}_{13}, e_{24}) = S(e_{14}, e_{15}, e_{24})$ is $v_1, v_4, v_2, v_3, v_5, v_1$ and its cost is 21. On the other hand $S(\overline{e}_{12}, \overline{e}_{13}, \overline{e}_{24}) = S(e_{14}, e_{15}, \overline{e}_{24}) = S(e_{14}, e_{15}, e_{34}) = \{(v_1, v_4, v_3, v_2, v_5, v_1)\}$ and this is the optimal solution whose cost is 8.

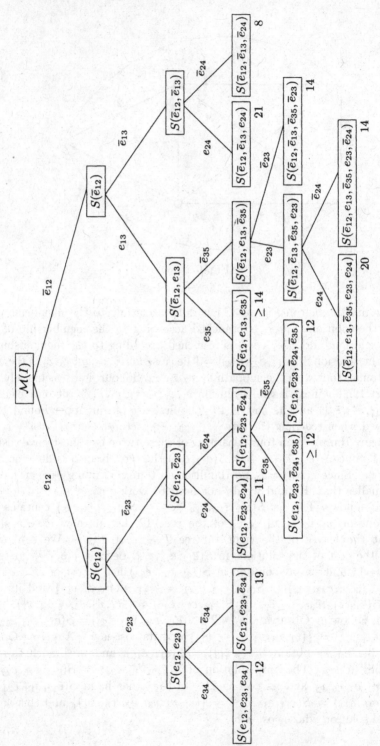

Fig. 3.11.

We observe that the optimal solution $v_1, v_4, v_3, v_2, v_5, v_1$ corresponds to the last leaf visited in the depth-first-search of $T_{\mathcal{M}(I)}$ and so without having the precomputed upper bound 10 one would generate the whole tree $T_{\mathcal{M}(I)}$ by backtracking. Obviously, another ordering of edges taken for the branch procedure or another search strategy in $T_{\mathcal{M}(I)}$ could result in different numbers of visited vertices (different time complexities).

3.4.3 Discussion

Assuming $P \neq NP$, it is clear that the branch-and-bound method with any clever polynomial-time algorithmic precomputation of bounds on the cost of the optimal solutions cannot provide any polynomial-time branch-and-bound algorithm for a given NP-hard problem. The only profit that one can expect from this method is that it works within reasonable time for several of the problem instances appearing in the concrete applications.

Also, if one can precompute very good bounds by some approximation algorithm, there may exist input instances that have exponentially many feasible solutions with their costs between the optimal cost and the precomputed bound. Obviously, this may lead to a high exponential time complexity of the branch-and-bound method.

Since there is no general strategy for searching in $T_{\mathcal{M}(x)}$ that would be the best strategy for every input instance of a hard problem, one has to concentrate on the following two parts of the branch-and-bound method in order to increase its efficiency:

(i) to search for a good algorithm for getting bounds that are as close as possible to $Opt_U(I)$ for any input instance I of the optimization problem U, and

(ii) to find a clever, efficient strategy to compute the bounds on the costs of the solutions in the set of solutions assigned to the vertices of $T_{\mathcal{M}(I)}$.

One can be willing to invest more time in searching for a good bound on $Opt_U(I)$ (because the derived bound could be essential for the complexity of the run of the most complex part (the search in $T_{\mathcal{M}(I)}$)), whereas the procedure of the estimation of a bound on a given set of solutions must be very efficient because it is used in every generated vertex, and the number of generated vertices is usually large.

The last question we want to discuss here is what to do if branch-and-bound runs on some problem instances longer than one can wait for the result. One possibility is simply to break the search and to output the best feasible solution found up till now. Another possibility is to try to prevent runs that are too long from the beginning by changing the hard requirement of finding an optimal solution to the requirement of finding a solution whose cost does not differ more than $t\%$ from the optimal cost.[14] In such cases, one can cut

---

[14] An algorithm producing outputs with this property is called a $(1 + t/100)$-approximation algorithm and such algorithms will be studied in Chapter 4.

any subtree T_v from $T_{\mathcal{M}(I)}$ if the bound on S_v says that no solution in S_v is $t\%$ better than the best feasible solution found up till now.

Summary of Section 3.4

The branch-and-bound method is an algorithm design technique for solving optimization problems. It makes backtracking more efficient by omitting the generation of some subtrees of the backtrack tree, because it is possible to recognize that these parts do not contain any optimal solution or they do not contain any feasible solution that is better than the best solution found up till now. Usually, a branch-and-bound algorithm consists of the following two parts:

(i) The computation of a bound on the optimal cost.
(ii) The use of the bound computed in (i) to cut some subtrees of the backtrack tree.

The precomputed part (i) is usually done by another algorithm design method (approximation algorithms, relaxation to linear programming, local search, etc.), and the quality of the precomputed bound can have an essential influence on the efficiency of the branch-and-bound method.

3.5 Lowering Worst Case Complexity of Exponential Algorithms

3.5.1 Basic Concept

This concept is similar to the idea of branch-and-bound in the sense that one designs an algorithm that solves a given problem and is even prepared to accept an exponential complexity. But in contrast to branch-and-bound, where the complexity substantially varies from input to input, one requires here that the worst case complexity of the designed algorithm is substantially smaller than the complexity of any naive approach. For instance, if one considers an optimization problem with a set of feasible solutions of cardinality 2^n for problem instances of size n, then c^n worst case complexity is an essential improvement for every $c < 2$. Thus, in this approach one accepts algorithms of exponential complexity if they are practical for input sizes for which straightforward exponential algorithms are no more practical.

Figure 3.12 documents the potential usefulness of this approach. If one considers the complexity of 10^{16} as the border between current tractability and nontractability, then we see that an algorithm of the complexity $(1.2)^n$ can be performed in a few seconds for $n = 100$ and that an algorithm of the complexity $n^2 \cdot 2^{\sqrt{n}}$ is even tractable[15] for $n = 300$. On the other hand the application of an algorithm of the complexity 2^n for inputs of size 50 is

---

[15] Even for $n = 300$, $n^2 \cdot 2^{\sqrt{n}}$ is a number consisting of 11 digits only.

already on the border of tractability. Thus, improving the complexity in the exponential range may essentially contribute to the practical solvability of hard problems.

Complexity	$n = 10$	$n = 50$	$n = 100$	$n = 300$
2^n	1024	(16 digits)	(31 digits)	(91 digits)
$2^{n/2}$	32	$\sim 33 \cdot 10^6$	(16 digits)	(46 digits)
$(1.2)^n$	7	9100	$\sim 29 \cdot 10^6$	(24 digits)
$10 \cdot 2^{\sqrt{n}}$	89	1350	10240	$\sim 1.64 \cdot 10^6$
$n^2 \cdot 2^{\sqrt{n}}$	894	~ 336000	$\sim 10.24 \cdot 10^6$	$\sim 14.8 \cdot 10^9$

Fig. 3.12.

This approach can be used for any kind of algorithmic problem. Usually, to be successful with it one needs to analyze the specific problem considered, and use the obtained knowledge in combination with a fundamental algorithm design technique. Currently, there exists no concept that would provide a general strategy for designing practical exponential algorithms for a large class of algorithmic problems, and there is no robust theory explaining which hard problems could possess algorithms of complexity $p(n) \cdot c^n$ for a polynomial p and a constant $c < 2$.

In the next section we show a simple application of this approach for the decision problem 3SAT.

3.5.2 Solving 3SAT in Less than 2^n Complexity

We consider the decision problem (3SAT, Σ_{logic}), i.e., to decide whether a given formula F in 3CNF is satisfiable. If F is over n variables, the naive approach looking for the value $F(\alpha)$ for every assignment α to the variables of F leads to the (worst case) complexity[16] $O(|F| \cdot 2^n)$. Obviously, this simple algorithm works for the general SAT problem. In what follows we show that carefully using the divide-and-conquer method instead of the above-mentioned exhaustive search 3SAT can be decided in $O(|F| \cdot 1.84^n)$ time.

Let F be a formula in CNF, and let l be a literal that occurs in F. Then $F(l = 1)$ denotes a formula that is obtained from F by consecutively applying the following rules:

(i) All clauses containing the literal l are removed from F.
(ii) If a clause of F contains the literal $\bar{l}$ and still at least one different literal from $\bar{l}$, then $\bar{l}$ is removed from the clause.

---

[16] Note that evaluating a CNF F for a given assignment to its variables can be done in $O(|F|)$ time, and that there are exactly 2^n different assignments to n variables of F.

(iii) If a clause of F consists of the literal $\bar{l}$ only, then $F(l = 1)$ is the formula 0 (i.e., an unsatisfiable formula).

Analogously, $F(l = 0)$ denotes a formula that is obtained from F in the following way:

(i) All clauses containing the literal $\bar{l}$ are removed from F.
(ii) If a clause consists of at least two different literals and one of them is l, then l is removed from the clause.
(iii) If a clause contains the literal l only, then $F(l = 0)$ is the formula 0.

In general, for literals $l_1, \ldots, l_c, h_1, \ldots, h_d$, the formula

$$F(l_1 = 1, l_2 = 1, \ldots, l_c = 1, h_1 = 0, h_2 = 0, \ldots, h_d = 0)$$

is obtained from F by constructing $F(l_1 = 1)$, then by constructing $F(l_1 = 1)(l_2 = 1)$, etc. Obviously, $l_1 = 1, l_2 = 1, \ldots, l_c = 1, h_1 = 0, \ldots, h_d = 0$ determine a partial assignment to the variables of F. Thus, the question whether $F(l_1 = 1, l_2 = 1, \ldots, l_c = 1, h_1 = 0, \ldots, h_d = 0)$ is satisfiable is equivalent to the question whether there exists an assignment satisfying $l_1 = 1, \ldots, l_c = 1, h_1 = 0, \ldots, h_d = 0$, and F.

Consider the following formula

$$F = (x_1 \vee \bar{x}_2 \vee x_4) \wedge (\bar{x}_2) \wedge (\bar{x}_2 \vee \bar{x}_3 \vee x_5) \wedge (x_1 \vee \bar{x}_5) \wedge (\bar{x}_1 \vee x_2 \vee x_3)$$

of five variables x_1, x_2, x_3, x_4, x_5 in 3CNF. We see that

$$F(x_1 = 1) = (\bar{x}_2) \wedge (\bar{x}_2 \vee \bar{x}_3 \vee x_5) \wedge (x_2 \vee x_3),$$
$$F(\bar{x}_2 = 1) = (x_1 \vee \bar{x}_5) \wedge (\bar{x}_1 \vee x_3), \text{ and}$$
$$F(\bar{x}_3 = 0) = (x_1 \vee \bar{x}_2 \vee x_4) \wedge (\bar{x}_2) \wedge (\bar{x}_2 \vee x_5) \wedge (x_1 \vee \bar{x}_5).$$

Observe that $F(\bar{x}_2 = 0) \equiv 0$ and $F(x_2 = 1) \equiv 0$ because the second clause consists of the literal $\bar{x}_2$ only. The important fact is that, for every literal l of F and $a \in \{0, 1\}$, $F(l = a)$ contains fewer variables than F. In what follows, for all positive integers n and r,

$$\mathbf{3CNF}(n, r) = \{\Phi \mid \Phi \text{ is a formula over at most } n \text{ variables in 3CNF}$$
$$\text{and } \Phi \text{ contains at most } r \text{ clauses}\}.$$

Consider the following divide-and-conquer strategy for 3SAT. Let $F \in$ 3CNF(n, r) for some positive integers n, r, and let $(l_1 \vee l_2 \vee l_3)$ be some clause of F. Observe that

$$F \text{ is satisfiable} \Longleftrightarrow \text{ at least one of the formulae } F(l_1 = 1),$$
$$F(l_1 = 0, l_2 = 1), F(l_1 = 0, l_2 = 0, l_3 = 1) \quad (3.8)$$
$$\text{is satisfiable.}$$

Following the above rules (i), (ii), and (iii), it is obvious that

$$F(l_1 = 1) \in 3\text{CNF}(n - 1, r - 1),$$
$$F(l_1 = 0, l_2 = 1) \in 3\text{CNF}(n - 2, r - 1),$$
$$F(l_1 = 0, l_2 = 0, l_3 = 1) \in 3\text{CNF}(n - 3, r - 1).$$

In this way we have reduced the question whether a formula F from $3\text{CNF}(n, r)$ is satisfiable to the satisfiability problem for three subinstances of F from $3\text{CNF}(n - 1, r - 1), 3\text{CNF}(n - 2, r - 1)$, and $3\text{CNF}(n - 3, r - 1)$. Thus, we obtain the following recursive algorithm for 3SAT.

Algorithm 3.5.2.1 (D&C-3SAT(F)).

Input: A formula F in 3CNF.
Step 1: **if** $F \in 3\text{CNF}(3, k)$ or $F \in 3\text{CNF}(m, 2)$ for some $m, k \in \mathbb{N} - \{0\}$,
   then decide whether $F \in 3\text{SAT}$ or not by testing all assignments to
   the variables of F;
   if $F \in 3\text{SAT}$ **output**(1) **else** **output**(0).
Step 2: Let H be one of the shortest clauses of F.
   if $H = (l)$ **then** **output**(D&C-3SAT($F(l = 1)$));
   if $H = (l_1 \vee l_2)$
   then output(D&C-3SAT($F(l_1 = 1)$)
             $\vee$D&C-3SAT($F(l_1 = 0, l_2 = 1)$));
   if $H = (l_1 \vee l_2 \vee l_3)$
   then output(D&C-3SAT($F(l_1 = 1)$)
             $\vee$D&C-3SAT($F(l_1 = 0, l_2 = 1)$)
             $\vee$D&C-3SAT($F(l_1 = 0, l_2 = 0, l_3 = 1)$)).

Theorem 3.5.2.2. *The algorithm* D&C-3SAT *solves* 3SAT *and*

$$Time_{\text{D\&C-3SAT}}(F) = O(r \cdot 1.84^n)$$

for every $F \in 3CNF(n, r)$.

Proof. The fact that the algorithm D&C-3SAT solves 3SAT is obvious. If F contains a clause consisting of one literal l only, then it is clear that F is satisfiable iff $F(l = 1)$ is satisfiable. If F contains a clause $(l_1 \vee l_2)$, then F is satisfiable iff $F(l_1 = 1)$ or $F(l_1 = 0, l_2 = 1)$ is satisfiable. In the case when F consists of clauses of length 3 only, the equivalence (3.8) confirms the correctness of D&C-3SAT.

Now let us analyze the complexity $T(n, r) = Time_{\text{D\&C-3SAT}}(n, r)$ of the algorithm D&C-3SAT according to

- the number of variables n, and
- the number of clauses r.

Obviously, $|F|/3 \leq r \leq |F|$ and $n \leq |F|$ for every formula F over n variables that is in 3CNF and that consists of r clauses.[17] Because of Step 1,

$$T(n,r) \leq 8 \cdot |F| \leq 24r \text{ for all pairs of integers } n, r, \qquad (3.9)$$
$$\text{where } n \leq 3 \text{ or } r \leq 2.$$

For $n \in \{1, 2\}$ we may assume $T(2, r) \leq 12r$ and $T(1, r) \leq 3r$.

Since $F(l = a)$ can be constructed from F in time $3 \cdot |F| \leq 9r$, the analysis of Step 2 provides, for all $n > 3, r > 2$,

$$T(n,r) \leq 54r + T(n-1, r-1) + T(n-2, r-1) + T(n-3, r-1). \quad (3.10)$$

We prove by induction according to n that

$$T(n,r) \leq 27r \cdot (1.84^n - 1) \qquad (3.11)$$

satisfies the recurrence given in (3.10), and that for small n and r (3.9) holds. First of all, observe that for $n = 3$ and any $r \geq 1$,

$$T(n,r) = T(3, r) = 27r \cdot (1.84^3 - 1) \geq 100r$$

which is surely more than $24r$ and so Algorithm 3.5.2.1 is executed below the complexity $T(n, r)$ for $n = 3$. The cases $n = 1, 2$ can be shown in a similar way. Thus, (3.9) is satisfied by $T(n, r)$.

Now starting with $n = 4$ we prove by induction according to n that $T(n, r)$ given by formula (3.11) satisfies the recurrence (3.10).

(1) For every $r \geq 2$,

$$\begin{aligned}
T(n,r) \ \underset{(3.10)}{\leq} \ & 54 \cdot r + T(3, r-1) + T(2, r-1) + T(1, r-1) \\
\leq \ & 54r + 24(r-1) + 12(r-1) + 3(r-1) \\
\leq \ & 93r \leq 27r \cdot (1.84^4 - 1).
\end{aligned}$$

(2) Let $T(m, r) \leq 27r \cdot (1.84^m - 1)$ satisfy the recurrence (3.10) for all $m < n$. We prove (3.11) for n now. For every positive integer $r \geq 2$,

$$\begin{aligned}
T(n,r) \ \underset{(3.10)}{=} \ & 54 \cdot r + T(n-1, r-1) + T(n-2, r-1) + T(n-3, r-1) \\
\underset{induc.}{\leq} \ & 54r + 27(r-1) \cdot (1.84^{n-1} - 1) \\
& + 27(r-1) \cdot (1.84^{n-2} - 1) + 27(r-1) \cdot (1.84^{n-3} - 1) \\
\leq \ & 54r + 27 \cdot r \cdot (1.84^{n-1} + 1.84^{n-2} + 1.84^{n-3} - 3) \\
= \ & 27r \cdot 1.84^{n-1} \left(1 + \frac{1}{1.84} + \frac{1}{1.84^2}\right) - 3 \cdot 27r + 54r \\
\leq \ & 27r \cdot 1.84^n - 27r = 27r \cdot (1.84^n - 1).
\end{aligned}$$

---

[17] For simplicity, consider $|F|$ as the number of literals in F.

Thus, there is a function $T(n, r) \in O(r \cdot 1.84^n)$ that satisfies the recurrence (3.10).   □

Exercise 3.5.2.3. Extend the idea of Algorithm 3.5.2.1 for any $k \geq 3$ in such a way that the observed algorithm for kSAT will be of complexity $O\left((c_k)^n\right)$, where $c_k < c_{k+1} < 2$ for every $k \geq 3$.   □

Exercise 3.5.2.4. [*] Improve the algorithm D&C-3SAT in order to obtain an $O(r \cdot 1.64^n)$ algorithm for 3SAT.   □

In Section 5.3.7 we shall combine the concept of lowering the worst case complexity of exponential algorithms with randomization in order to design a randomized $O(r \cdot (1.334)^n)$ algorithm for 3SAT.

Summary of Section 3.5

The concept of lowering the worst case complexity of exponential algorithms focuses on the design of algorithms that may even have an average case exponential complexity. But the worst case complexity of these algorithms should be bounded by an exponential function in $O(c^n)$ for some constant $c < 2$. Such algorithms can be practical because they run fast for realistic input sizes of many particular applications (Figure 3.12). The typical idea behind the design of such algorithms is to search for a solution in a space of an exponential size, but this space should be much smaller than the space searched by any naive approach.

This concept became especially successful for the satisfiablilty problems. Instead of looking for all 2^n assignments to the n variables of a given formula Φ, one decides the satisfiability of Φ by looking for at most $O(c^n)$ assignments for some $c < 2$. The algorithm D&C-3SAT is the simplest known example for an application of this concept.

3.6 Local Search

3.6.1 Introduction and Basic Concept

Local search is an algorithm design technique for optimization problems. The main aim of Section 3.6 is to give a more detailed and formal presentation of this technique than the rough presentation of basic ideas of local search in Section 2.3.4. This is important because of the following two reasons:

(i) The fundamental framework of the classical local search method presented here is the base for advanced local search strategies such as simulated annealing (Chapter 6) or tabu search.

(ii) A formal framework for the study of local search algorithms enables complexity considerations that provide borders on the tractability of searching for local optima of optimization problems.

As already mentioned in Section 2.3.4, a local search algorithm realizes a restricted search in the set of all feasible solutions $\mathcal{M}(I)$ to the given problem instance I. To determine what a "restricted search" means, one needs to introduce a structure in the set of feasible solutions $\mathcal{M}(I)$ by defining a neighborhood for every feasible solution of $\mathcal{M}(I)$.

Definition 3.6.1.1. *Let $U = (\Sigma_I, \Sigma_O, L, L_I, \mathcal{M}, cost, goal)$ be an optimization problem. For every $x \in L_I$, a **neighborhood on $\mathcal{M}(x)$** is any mapping $f_x : \mathcal{M}(x) \to Pot(\mathcal{M}(x))$ such that*

(i) $\alpha \in f_x(\alpha)$ for every $\alpha \in \mathcal{M}(x)$,
(ii) if $\beta \in f_x(\alpha)$ for some $\alpha \in \mathcal{M}(x)$, then $\alpha \in f_x(\beta)$, and
(iii) for all $\alpha, \beta \in \mathcal{M}(x)$ there exists a positive integer k and $\gamma_1, \ldots, \gamma_k \in \mathcal{M}(x)$ such that $\gamma_1 \in f_x(\alpha)$, $\gamma_{i+1} \in f_x(\gamma_i)$ for $i = 1, \ldots, k-1$, and $\beta \in f_x(\gamma_k)$.

*If $\alpha \in f_x(\beta)$ for some $\alpha, \beta \in \mathcal{M}(x)$, we say that α and β are **neighbors in $\mathcal{M}(x)$**. The set $f_x(\alpha)$ is called the **neighborhood of the feasible solution α in $\mathcal{M}(x)$**. The (undirected) graph*

$$G_{\mathcal{M}(x), f_x} = (\mathcal{M}(x), \{\{\alpha, \beta\} \mid \alpha \in f_x(\beta), \alpha \neq \beta, \alpha, \beta \in \mathcal{M}(x)\})$$

*is the **neighborhood graph of $\mathcal{M}(x)$ according to the neighborhood f_x**.*

*Let, for every $x \in L_I$, f_x be a neighborhood on $\mathcal{M}(x)$. The function $f : \bigcup_{x \in L_I}(\{x\} \times \mathcal{M}(x)) \to \bigcup_{x \in L_I} Pot(\mathcal{M}(x))$ with the property $f(x, \alpha) = f_x(\alpha)$ for every $x \in L_I$ and every $\alpha \in \mathcal{M}(x)$ is called a **neighborhood for U**.*

Local search in $\mathcal{M}(x)$ is an iterative movement in $\mathcal{M}$ from a feasible solution to a neighboring feasible solution. Thus, we see the importance of the condition (iii) of Definition 3.6.1.1 that assures us that the neighborhood graph $G_{\mathcal{M}(x),f}$ is connected, i.e., every feasible solution $\beta \in \mathcal{M}(x)$ is reachable from any solution $\alpha \in \mathcal{M}(x)$ by iteratively moving from a neighbor to a neighbor. The following exercise shows that the conditions (i), (ii), and (iii) of the neighborhood function even determine a metrics on $\mathcal{M}(x)$, and so the pair $(\mathcal{M}(x), f_x)$ can be considered as a metric space.

Exercise 3.6.1.2. Let $U = (\Sigma_I, \Sigma_O, L, L_I, \mathcal{M}, cost, goal)$ be an optimization problem, and let $Neigh_x$ be a neighborhood on $\mathcal{M}(x)$ for some $x \in L_I$. Define, for all $\alpha, \beta \in \mathcal{M}(x)$, $distance_{Neigh_x}(\alpha, \beta)$ as the length of the shortest path between α and β in $G_{\mathcal{M}(x), Neigh_x}$. Prove that $distance_{Neigh_x}$ is a metrics on $\mathcal{M}(x)$. $\qquad\qquad\square$

Because of condition (ii) of Definition 3.6.1.1 any neighborhood on $\mathcal{M}(x)$ can be viewed as a symmetric relation on $\mathcal{M}(x)$. When one wants to define a neighborhood on $\mathcal{M}(x)$ in a practical application, then one usually does not work in the formalism of functions or relations. The standard way to introduce a neighborhood on $\mathcal{M}(x)$ is to use so-called **local transformations** on $\mathcal{M}(x)$. Informally, a local transformation transforms a feasible solution α

to a feasible solution β by some local changes of the specification of α. For instance, an example of a local search transformation is the exchange of two edges in the local search algorithm for the minimum spanning tree problem in Example 2.3.4.6. Flipping the Boolean value assigned to a variable is a local transformation for the MAX-SAT problem. Having a set T of reasonable transformations one can define the neighborhood according to T by saying that α is a neighbor of β if there exists a transformation $t \in T$ such that t transforms α into β. Typically, the local transformations are used to define a symmetric relation on $\mathcal{M}(x)$ and so one has to take care of the reachability property (iii) of Definition 3.6.1.1 only.

Note that sometimes one uses neighborhoods that do not satisfy all the conditions (i), (ii), and (iii) of Definition 3.6.1.1. If condition (ii) is violated, then one has to define the neighborhood graph as a directed graph.

Exercise 3.6.1.3. Prove that the transformation of flipping the value of at most one variable defines a neighborhood on $\mathcal{M}(\Phi)$ for every input instance Φ of MAX-SAT. $\qquad\qquad\square$

Introducing a neighborhood on $\mathcal{M}(x)$ enables to speak about local optima in $\mathcal{M}(x)$. Observe that this is impossible without determining any structure on the set $\mathcal{M}(x)$ of all feasible solutions to x.

Definition 3.6.1.4. *Let $U = (\Sigma_I, \Sigma_O, L, L_I, \mathcal{M}, cost, goal)$ be an optimization problem, and let, for every $x \in L_I$, the function f_x be neighborhood on $\mathcal{M}(x)$. A feasible solution $\alpha \in \mathcal{M}(x)$ is a **local optimum for the input instance x of U according to f_x**, if*

$$cost(\alpha) = goal\{cost(\beta) \,|\, \beta \in f_x(\alpha)\}.$$

We denote the set of all local optima for x according to the neighborhood f_x by $LocOPT_U(x, f_x)$.

Having a structure on $\mathcal{M}(x)$ determined by a neighborhood $Neigh_x$ for every $x \in L_I$, one can describe a general scheme of local search as follows. Roughly speaking, a local search algorithm starts off with an initial solution and then continually tries to find a better solution by searching neighborhoods. If there is no better solution in the neighborhood, then it stops.

LSS($Neigh$)-Local Search Scheme according to a neighborhood $Neigh$

Input: An input instance x of an optimization problem U.
Step 1:   Find a feasible solution $\alpha \in \mathcal{M}(x)$.
Step 2:   **while** $\alpha \notin LocOPT_U(x, Neigh_x)$ **do**
          begin find a $\beta \in Neigh_x(\alpha)$ such that
              $cost(\beta) < cost(\alpha)$ if U is a minimization problem and
              $cost(\beta) > cost(\alpha)$ if U is a maximization problem; $\alpha := \beta$
          end
Output: **output**(α).

The following theorem is a direct consequence of the way in which LSS works.

Theorem 3.6.1.5. *Any local search algorithm based on* LSS(*Neigh*) *for an optimization problem U outputs, for every instance x of U, a local optimum for x according to the neighborhood Neigh.*

The success of a local search algorithm mainly depends on the choice of neighborhood. If a neighborhood *Neigh* has the property that $Neigh(\alpha)$ has a small cardinality for every $\alpha \in \mathcal{M}(x)$, then one iterative improvement of Step 2 of LSS(*Neigh*) can be executed efficiently but the risk that there are many local optima (potentially with a cost that is very far from $Opt_U(x)$) can substantially grow. On the other hand, large $|Neigh_x(\alpha)|$ can lead to feasible solutions with costs that are closer to $Opt_U(x)$ than small neighborhoods can, but the complexity of the execution of one run of the while cycle in Step 2 can increase too much. Thus, the choice of the neighborhood is always a game with the tradeoff between the time complexity and the quality of the solution. Small neighborhoods are typical for most applications. But a small neighborhood alone does not provide any assurance of the efficiency of the local search algorithm, because the number of runs of the while cycle of the local search scheme can be exponential in the input size. We can only guarantee a pseudo-polynomial-time algorithm for finding a local optimum of optimization problems for which the costs of feasible solutions are integers whose values are bounded by $p(Max\text{-}Int(x))$ for a polynomial p. The argument for this claim is obvious because

(i) if the costs are integers, then each run of the while cycle improves the cost of the best found solution by at least 1, and

(ii) if the costs are in the set $\{1, 2, \ldots, p(Max\text{-}Int(x))\}$ of positive integers, then at most $p(Max\text{-}Int(x))$ repetitions of the while cycle are possible.

Theorem 3.6.1.6. *Let $U = (\Sigma_O, \Sigma_I, L, L_I, \mathcal{M}, cost, goal)$ be an integer-valued optimization problem with the cost function cost from feasible solutions to positive integers. Let there exist a polynomial p such that $cost(\alpha, x) \leq p(Max\text{-}Int(x))$ for every $x \in L_I$ and every $\alpha \in \mathcal{M}(x)$. For every neighborhood Neigh such that $Neigh_x(\alpha)$ can be generated from α and x in polynomial time in $|x|$ for every $x \in L_I$ and every $\alpha \in \mathcal{M}(x)$, LSS(Neigh) provides a pseudo-polynomial-time algorithm that finds a local optimum according to the neighborhood Neigh.*

Besides the choice of the neighborhood the following two free parameters of LSS(*Neigh*) may influence the success of the local search:

(1) In Step 1 of LSS(*Neigh*) an initial feasible solution is computed. This can be done in different ways. For instance, it can be randomly chosen or it can be precomputed by any other algorithmic method. The choice of an initial solution can essentially influence the quality of the resulting local

optimum. This is the reason why one sometimes performs LSS(*Neigh*) several times starting with different initial feasible solutions. There is no general theory providing a concept for the choice of the initial feasible solution, and so one usually generates them randomly. The execution of several runs of LSS(*Neigh*) from randomly chosen initial feasible solutions is called **multistart local search**.

(2) There are several ways to choose the cost-improving feasible solution in Step 2. The basic two ways are the strategy of the **first improvement** and the strategy of the **best improvement**. The first improvement strategy means that the current feasible solution is replaced by the first cost-improving feasible solution found by the neighborhood search. The best improvement strategy means that the current feasible solution is replaced by the best feasible solution in its neighborhood. Obviously, the first improvement strategy can make one single run of the while cycle faster than the best improvement strategy, but the best improvement strategy may decrease the number of executed runs of the while cycle in the comparison to the first improvement strategy. Note that these two strategies may lead to very different results for the same initial feasible solution.

The rest of this section is organized as follows. Section 3.6.2 shows some examples of neighborhoods for distinct optimization problems and introduces the so-called Kernighan-Lin variable-depth search that can be viewed as an advanced form of local search. Section 3.6.3 discusses the borders of applicability of the local search schemes for searching for optima of optimization problems.

3.6.2 Examples of Neighborhoods and Kernighan-Lin's Variable-Depth Search

A local search algorithm is determined by the definition of the neighborhood and by the local search scheme presented in Section 3.6.1. Thus, if one chooses a neighborhood, then the local search algorithm is more or less determined. This is the reason why we do not present examples of algorithms in this section, but examples of suitable neighborhoods for some optimization problems.

The most famous neighborhoods for TSP are the so-called **2-*Exchange*** and **3-*Exchange*.**[18] The simplest way to define them is to describe the corresponding local transformations. A 2-*Exchange* local transformation consists of removing two edges $\{a, b\}$ and $\{c, d\}$ with $|\{a, b, c, d\}| = 4$ from a given Hamiltonian tour α that visits these 4 vertices in the order a, b, c, d, and of adding two edges $\{a, d\}$ and $\{b, c\}$ to α. We observe (Figure 3.13) that the resulting object is again a Hamiltonian tour (cycle).

---

[18] Note that the notation 2-*Opt* and 3-*Opt* instead of 2-*Exchange* and 2-*Exchange* is also frequently used in the literature. We choose not to use this notation because it is similar to our notation $Opt_U(x)$ for the optimal cost.

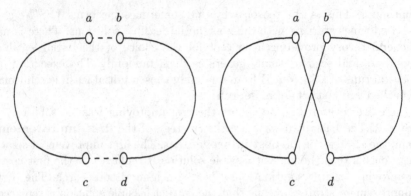

Fig. 3.13. 2-*Exchange* local transformation. The edges $\{a, b\}$ and $\{c, d\}$ are replaced by the edges $\{a, d\}$ and $\{b, c\}$.

For every input instance (K_n, c) of TSP, $n \in \mathbb{N} - \{0\}$, and for every $\alpha \in \mathcal{M}(K_n, c)$, the size of the neighborhood 2-*Exchange*(α) is $\frac{n \cdot (n-3)}{2}$ which is in $\Omega(n^2)$.

Similarly, a 3-*Exchange* local transformation starts by removing 3-edges $\{a, b\}$, $\{c, d\}$, and $\{e, f\}$ such that $|\{a, b, c, d, e, f\}| = 6$ form a Hamiltonian tour α. Then there are several possibilities to add 3 edges in order to get a Hamiltonian tour again. Some of these possibilities are depicted in Figure 3.14.

Observe that 2-*Exchange*$(\alpha) \subseteq$ 3-*Exchange*(α) for every Hamiltonian tour α because putting back one of the removed edges is not forbidden (see the last case of Figure 3.14 where the originally removed edge (a, b) is contained in the resulting tour) and so any 2-*Exchange* local transformation can be performed. The cardinality of 3-*Exchange*(α) is in $\Omega(n^3)$.

One can also consider the k-*Exchange*(α) neighborhood that is based on the exchange of k edges. But k-*Exchange* neighborhoods are rarely used in practical applications for $k > 3$ because the k-*Exchange*(α) neighborhoods are too large. Observe that the number of possibilities to choose k edges is in $\Omega(n^k)$ and additionally the number of possibilities to add new edges in order to get a Hamiltonian tour grows exponentially with k. A report on the experiments with the application of local search schemes with the neighborhoods 2-*Exchange* and 3-*Exchange* is given in Chapter 7.

The most natural neighborhood for the maximum satisfiability problem is based on flipping one bit of the input assignment. For a formula Φ of n variables, this *Flip* neighborhood has cardinality exactly n. One can easily see that the neighborhood graph $G_{\mathcal{M}(\Phi), Flip}$ is the well-known hypercube.

For the cut problems, there is a very simple way to define a neighborhood. For a given $cut(V_1, V_2)$ one simply moves one vertex from V_1 (assuming V_1 contains at least two vertices) to V_2 or from V_2 to V_1. This neighborhood is of

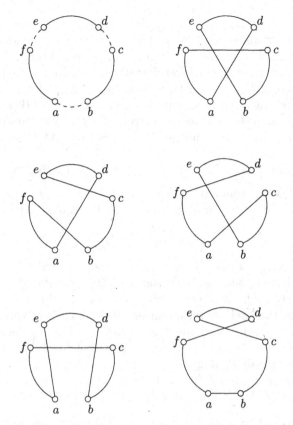

Fig. 3.14. Examples of the 3-*Exchange* local transformation. The edges $\{a, b\}$, $\{c, d\}$, and $\{e, f\}$ are replaced by other three edges. The last figure shows that replacing only two edges is also allowed. In this case the edges $\{e, f\}$ and $\{c, d\}$ are replaced by the edges $\{e, c\}$ and $\{d, f\}$.

linear size and one can extend it to larger neighborhoods by moving several vertices between V_1 and V_2.

A crucial drawback of local search is the fact that LSS(*Neigh*) may get stuck in very poor local optima.[19] It is obvious that large neighborhoods can be expected to yield feasible solutions of higher quality, but the time complexity to verify local optimality becomes too large to be profitable.[20] To overcome the difficulty with this tradeoff between time complexity and solution quality, we introduce the so-called **variable-depth search** algorithm that may realize several local transformations in order to find an improvement

---

[19] That is, in local optima whose costs substantially differ from the cost of optimal solutions.

[20] The effort to get a better solution may be too large in the comparison to the size of its quality improvement.

in a larger distance in $G_{\mathcal{M}(I), Neigh}$ but does not realize the exhaustive search of all feasible solutions in any large distance from the current solution.

Let us explain this concept in detail for an optimization problem U. Remember that working with local transformations, we overall assume that the feasible solutions of U are specified as lists $(p_1, \ldots, p_n)$ of n local specifications, and a local transformation may change one or a few of them. Let $Neigh$ be a neighborhood for U that can be determined by a local transformation. If one defines, for every positive integer k and every $\alpha \in \mathcal{M}(I)$ for an instance I of U,

$$Neigh_I^k(\alpha) = \{\beta \in \mathcal{M}(I) \mid distance_{Neigh_I}(\alpha, \beta) \leq k\}$$

as the set of feasible solutions that can be achieved from α by at most k applications of the local transformation, then there obviously exists an m such that

$$Neigh_I^m(\alpha) = \mathcal{M}(I)$$

for every $\alpha \in \mathcal{M}(I)$. Typically, m is approximately equal to n, where n is the length of the specification. For instance, $Flip_\Phi^n(\alpha) = \mathcal{M}(\Phi)$ for MAX-SAT and this property holds for all other neighborhoods introduced in this section. The variable-depth search algorithm can enable us to efficiently find a solution β from $Neigh_I^n(\alpha)$ as the next iterative improvement of the current feasible solution α without realizing the exhaustive search of $Neigh_I^n(\alpha)$. This is done by the following greedy strategy.

Let, for any minimization [maximization] problem U, and all feasible solutions $\alpha, \beta \in \mathcal{M}(I)$ for an instance I of U,

$$gain(\alpha, \beta) = cost(\alpha) - cost(\beta) \quad [cost(\beta) - cost(\alpha)].$$

Note that $gain(\alpha, \beta)$ may be also negative if β is not an improvement of α. In order to overcome a local optima of $Neigh$ the idea is to apply the local transformation at most n times in such a way that:

(i) if starting with a feasible solution $\alpha = (p_1, p_2, \ldots, p_n)$, then the resulting feasible solution $\gamma = (q_1, q_2, \ldots, q_n)$ has the property $q_i \neq p_i$ for all $i = 1, \ldots, n$,

(ii) if $\alpha_0, \alpha_1, \alpha_2, \ldots, \alpha_m$, where $\alpha = \alpha_0$, $\alpha_m = \gamma$, is the sequence of created feasible solutions (α_{i+1} is obtained from α_i by the application of one local transformation), then

    a) $gain(\alpha_i, \alpha_{i+1}) = \max\{gain(\alpha_i, \delta) \mid \delta \in Neigh(\alpha_i)\}$
       (i.e., we use the greedy strategy to go from α_i to α_{i+1}), and

    b) if the step from α_i to α_{i+1} changes the parameter p_j of the initial feasible solution α to some q_j, then q_j is never changed later (i.e., q_j is a parameter of γ).

After creating the sequence $\alpha_0, \alpha_1, \ldots, \alpha_m$, the algorithm replaces α by such an α_l that

$$gain(\alpha, \alpha_l) = \max\{gain(\alpha, \alpha_i) \mid i = 1, \ldots, m\}$$

if $gain(\alpha, \alpha_l) > 0$. If $gain(\alpha, \alpha_l) \leq 0$, then the algorithm halts with the output α. The main idea of the above-described approach is that a few steps in the wrong direction (when $gain(\alpha, \alpha_1)$, $gain(\alpha_1, \alpha_2), \ldots, gain(\alpha_r, \alpha_{r+1})$ are all negative) may ultimately be redeemed by a large step in the right direction ($gain(\alpha_r, \alpha_{r+1}) \gg |\sum_{i=0}^{r} gain(\alpha_i, \alpha_{i+1})|$, for instance).

KL(*Neigh*) Kernighan-Lin Variable-Depth Search Algorithm with respect to the neighborhood *Neigh*

Input: An input instance I of an optimization problem U.

Step 1:   Generate a feasible solution $\alpha = (p_1, p_2, \ldots, p_n) \in \mathcal{M}(I)$ where $(p_1, p_2, \ldots, p_n)$ is such a parametric representation of α that the local transformation defining *Neigh* can be viewed as an exchange of a few of these parameters.

Step 2:   *IMPROVEMENT* := *TRUE*;
    EXCHANGE := $\{1, 2, \ldots, n\}$; $J := 0$; $\alpha_J := \alpha$;
    while *IMPROVEMENT* = *TRUE* **do begin**
      while *EXCHANGE* $\neq \emptyset$ **do**
        begin $J := J + 1$;
        $\alpha_J :=$ a solution from $Neigh(\alpha_{J-1})$ such that $gain(\alpha_{J-1}, \alpha_J)$
        is the maximum of
        $\{gain(\alpha_{J-1}, \delta) \,|\, \delta \in Neigh(\alpha_{J-1}) - \{\alpha_{J-1}\}$ and δ differs
        from α_{J-1} in the parameters of *EXCHANGE* only$\}$;
        EXCHANGE := *EXCHANGE*$-\{$the parameters in which
        α_J and α_{J-1} differ$\}$
    end;
    Compute $gain(\alpha, \alpha_i)$ for $i = 1, \ldots, J$;
    Compute $l \in \{1, \ldots, J\}$ such that

$$gain(\alpha, \alpha_l) = \max\{gain(\alpha, \alpha_i) \,|\, i \in \{1, 2, \ldots, J\}\};$$

    if $gain(\alpha, \alpha_l) > 0$ **then**
      begin $\alpha := \alpha_l$;
        EXCHANGE := $\{1, 2, \ldots, n\}$
    end
    else *IMPROVEMENT* := *FALSE*
  end

Step 3:   **output**(α).

An important point is that KL(*Neigh*) uses the greedy strategy to find an improvement of the current feasible solution α in $Neigh^l(\alpha)$ for some $l \in \{1, \ldots, n\}$ and so it avoids the running time $|Neigh^l(\alpha)|$ that may be exponential in $|\alpha|$. In fact, the time complexity of the search for one iterative improvement of α in KL(*Neigh*) is at most

$$n \cdot f(|\alpha|) \cdot |Neigh(\alpha)|,$$

where f is the complexity of the execution of one local transformation.

For the problems MIN-CUT and MAX-CUT and the neighborhood consisting of the move of one vertex from V_1 to V_2 or vice versa, the complexity of the local transformation is $O(1)$ and so one iterative improvement of KL costs $O(n)$ only. We have the same situation for KL($Flip$) for MAX-SAT.

Note that in some applications one can consider a modification of the Kernighan-Lin's variable-depth search algorithm. For instance, an additional exchange of an already changed parameter or the return to the original value of a parameter may be allowed under some circumstances. If $Neigh(\alpha)$ is too large (for instance, superlinear as in the case of 2-*Exchange* and 3-*Exchange*) one can omit the exhaustive search of $Neigh(\alpha)$ by using an efficient greedy approach to find a candidate $\beta \in Neigh(\alpha)$ with a reasonable gain.

3.6.3 Tradeoffs Between Solution Quality and Complexity

As we already noted, local search can be efficient for small neighborhoods, but there are optimization problems such that, for every small neighborhood, one can find hard input instances for which local search either provides arbitrarily poor solutions or needs exponential time to compute at least a local optimum.[21] Local search assures a good solution only if local optima according to the given neighborhood have costs close to the optimal cost. As we will see in Section 4 this can lead to the design of good approximation algorithms for some optimization problems.

Assuming $P \neq NP$, there is no polynomial-time local search algorithm for any NP-hard optimization problem, because there is no polynomial-time algorithm for any NP-hard problem at all. Thus, one does not need to develop any special method to prove that an optimization algorithm cannot be solved in polynomial time by local search. But using local search for a hard optimization problem one can at least ask where the difficulty from the local search point of view lies. We already observed that the time complexity of any local search algorithm can be roughly bounded by

$$(\textit{time of the search in a local neighborhood}) \cdot (\textit{the number of improvements}).$$

We pose the following question:

For which NP-hard optimization problems can one find a neighborhood Neigh of polynomial size such that LSS(Neigh) always outputs an optimal solution?

Note that such neighborhoods may exist for hard problems because local search algorithms with small neighborhoods may have an exponential complexity due to a large number of iterative improvements needed in the worst case. To formulate our question more precisely, we need the following definition.

---

[21] Such an example will be presented at the end of this section.

Definition 3.6.3.1. *Let $U = (\Sigma_I, \Sigma_O, L, L_I, \mathcal{M}, cost, goal)$ be an optimization problem, and let f be a neighborhood for U. f is called an **exact neighborhood**, if, for every $x \in L_I$, every local optimum for x according to f_x is an optimal solution to x (i.e., $LocOPT_U(x, f_x)$ is equal to the set of optimal solutions for x).*

*A neighborhood f is called **polynomial-time searchable** if there is a polynomial-time[22] algorithm that, for every $x \in L_I$ and every $\alpha \in \mathcal{M}(x)$, finds one of the best feasible solutions in $f_x(\alpha)$.*

We note that if a neighborhood is polynomial-time searchable, this does not mean that this neighborhood is of polynomial size. There exist neighborhoods of exponential size that are searchable even in linear time.

Thus, for a given optimization problem $U \in$ NPO, our question can be formulated in the terminology introduced in Definition 3.6.3.1 as follows:

Does there exist an exact polynomial-time searchable neighborhood for U?

The positive answer to this question means that the hardness of the problem from the local search point of view is in the number of iterative improvements needed to reach an optimal solution. In many cases, it means that local search may be suitable for U. For instance, if U is an integer-valued optimization problem, then the existence of a polynomial-time searchable exact neighborhood *Neigh* for U can imply that LSS(*Neigh*) is a pseudo-polynomial-time algorithm[23] for U.

The negative answer to this question implies that no polynomial-time searchable neighborhood can assure the success of local search in searching for an optimal solution. So, one can at most try to obtain a local optimum in polynomial time in such a case.

We present two techniques for proving that an optimization problem is hard for local search in the above-mentioned sense. Both are based on suitable polynomial-time reductions.

Definition 3.6.3.2. *Let $U = (\Sigma_I, \Sigma_O, L, L_I, \mathcal{M}, cost, goal)$ be an integer-valued optimization problem. U is called **cost-bounded**, if for every input instance $I \in L_I$, $Int(I) = (i_1, i_2, \ldots, i_n)$, $i_j \in \mathbb{N}$ for $j = 1, \ldots, n$,*

$$cost(\alpha) \leq \sum_{j=1}^{n} i_j$$

for every $\alpha \in \mathcal{M}(I)$.

---

[22] Note that we consider optimization problems from NPO only, and so any algorithm working in polynomial time according to $|\alpha|$ is a polynomial-time algorithm according to $|x|$.

[23] See Theorem 3.6.1.6.

Observe that the condition $cost(\alpha) \leq \sum_{j=1}^{n} i_j$ is natural because it is satisfied for all integer-valued optimization problems considered in this book. The costs of feasible solutions are usually determined by some subsum of the sum $\sum_{j=1}^{n} i_j$.

Theorem 3.6.3.3. *Let $U \in NPO$ be a cost-bounded integer-valued optimization problem such that there is a polynomial-time algorithm that, for every instance x of U, computes a feasible solution for x. If $P \neq NP$ and U is strongly NP-hard, then U does not possess an exact, polynomial-time searchable neighborhood.*

Proof. Assume the opposite that $U = (\Sigma_I, \Sigma_O, L, L_I, \mathcal{M}, cost, goal)$ is strongly NP-hard and has an exact, polynomial-time searchable neighborhood $Neigh$. Then we design a pseudo-polynomial-time algorithm A_U for U which together with the strong NP-hardness of U contradicts $P \neq NP$.

Without loss of generality assume $goal = minimum$. Since there exists an exact polynomial-time searchable neighborhood $Neigh$ for U, there exists a polynomial-time algorithm A that, for every feasible solution $\alpha \in \mathcal{M}(x)$, $x \in L_I$, finds a feasible solution $\beta \in Neigh_x(\alpha)$ with $cost(\alpha, x) > cost(\beta, x)$ or verifies that α is a local optimum with respect to $Neigh_x$. We assume that there exists a polynomial-time algorithm B that, for every $x \in L_I$, outputs an $\alpha_0 \in \mathcal{M}(x)$. Thus, a pseudo-polynomial-time algorithm A_U for U can work as follows.

Algorithm A_U

Input: An $x \in L_I$.
Step 1:   Use B to compute an feasible solution $\alpha_0 \in \mathcal{M}(x)$.
Step 2:   Use A to iteratively improve α_0, until a local optimum with respect to $Neigh_x$ was found.

As mentioned above, both A and B work in polynomial time according to the sizes of their inputs. Since $U \in NPO$, $|\alpha|$ is polynomial in $|x|$ for every $x \in L_I$ and every $\alpha \in \mathcal{M}(x)$ and so A works in polynomial time according to $|x|$, too. Let $Int(x) = \{i_1, \ldots, i_n\}$. Since U is a cost-bounded integer-valued problem, the costs of all feasible solutions of $\mathcal{M}(x)$ lie in $\{1, 2, \ldots, \sum_{j=1}^{n} i_j\} \subseteq \{1, 2, \ldots, n \cdot Max\text{-}Int(x)\}$. Since every iterative improvement improves the cost at least by 1, the number of runs of the algorithm A is bounded by $n \cdot Max\text{-}Int(x) \leq |x| \cdot Max\text{-}Int(x)$. Thus, A_U is a pseudo-polynomial-time algorithm for U. $\square$

Corollary 3.6.3.4. *If $P \neq NP$, then there exists no exact polynomial-time searchable neighborhood for TSP, $\triangle$-TSP, and WEIGHT-VCP.*

Proof. In Section 3.2.4 we proved that the cost-bounded integer-valued problems TSP, $\triangle$-TSP, and WEIGHT-VCP are strongly NP-hard. $\square$

The second method for proving the nonexistence of an exact polynomial-time searchable neighborhood is independent of the presented concept of pseudo-polynomial-time algorithms. The idea is to prove the NP-hardness of deciding the optimality of a given solution α to an input instance x.

Definition 3.6.3.5. *Let $U = (\Sigma_I, \Sigma_O, L, L_I, \mathcal{M}, cost, goal)$ be an optimization problem from NPO. We define the* **suboptimality decision problem** *to U as the decision problem $(SUBOPT_U, \Sigma_I \cup \Sigma_O)$, where*

$$SUBOPT_U = \{(x, \alpha) \in L_I \times \Sigma_O^* \mid \alpha \in \mathcal{M}(x) \text{ and } \alpha \text{ is not optimal}\}.$$

Theorem 3.6.3.6. *Let $U \in$ NPO. If P $\neq$ NP, and $SUBOPT_U$ is NP-hard, then U does not possess any exact, polynomial-time searchable neighborhood.*

Proof. It is sufficient to show that if U possesses an exact, polynomial-time searchable neighborhood *Neigh*, then there exists a polynomial-time algorithm A that decides $(SUBOPT_U, \Sigma_I \cup \Sigma_O)$, i.e., $SUBOPT_U \in$ P.

Let $(x, \alpha) \in L_I \times \mathcal{M}(x)$ be the input of A. A starts by searching in $Neigh_x(\alpha)$ in polynomial time. If A finds a better solution than α, then A accepts (x, α). If A does not find a feasible solution better than α in $Neigh_x(\alpha)$, then α is an optimal solution to x because *Neigh* is an exact neighborhood. In this case A rejects (x, α).   □

We again use TSP to illustrate the applicability of this approach. Our aim is to prove that $SUBOPT_{\text{TSP}}$ is NP-hard. To do it we first need to prove that the following decision problem is NP-hard. The **restricted Hamiltonian cycle problem (RHC)** is to decide, for a given graph $G = (V, E)$ and a Hamiltonian path P in G, whether there exists a Hamiltonian cycle in G. Formally,

$$\mathbf{RHC} = \{(G, P) \mid P \text{ is a Hamiltonian path in } G \text{ that cannot}$$
$$\text{be extended to any Hamiltonian cycle,}$$
$$\text{and } G \text{ contains a Hamiltonian cycle}\}.$$

In comparison to inputs of HC, we see that one has additional information in RHC, namely a Hamiltonian path in G. The question is whether this additional information makes this decision problem easier. The answer is negative. We show it by reducing the NP-complete HC problem to RHC.

Lemma 3.6.3.7.
$$\text{HC} \leq_p \text{RHC}$$

Proof. To find a polynomial-time reduction from HC to RHC we need the special-purpose subgraph called **diamond** that is depicted in Figure 3.15. When we take this graph as a subgraph of a more complicated graph G, then we shall allow the connection of the diamond with the rest of G only via edges incident to the four **corner vertices** N (north), E (east), S (south),

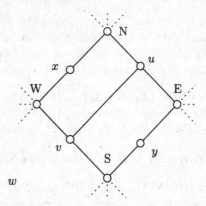

Fig. 3.15.

and W (west). With this assumption we call attention to the fact that if G has a Hamiltonian cycle then the diamond can be traversed only in one of the two ways depicted in Figure 3.16. The path depicted in Figure 3.16a is called the North-South mode of traversing the diamond and the path depicted in Figure 3.16b is called the East-West mode of traversing the diamond.

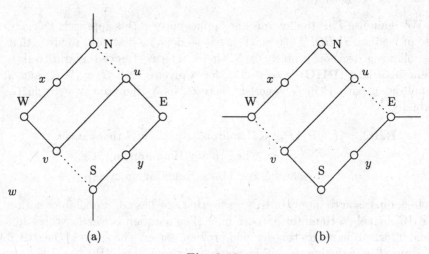

(a)   (b)

Fig. 3.16.

To see this, consider that a Hamiltonian cycle C of G enters the diamond at the vertex N. The only possibility of visiting x is to continue to x immediately and then to W. C cannot leave the diamond at the vertex W because if C would do that, then neither v nor u could be visited later or y cannot be visited. Thus, C must continue to v. From v, C must go to u because in the opposite case u remains unvisited by C. Being in u, the only possibility to visit the rest of the diamond is to continue to E, y, and S and to leave the diamond at S. Thus, we have obtained the North-South mode of traversing the diamond. When C enters the diamond at any of the vertices W, E, or S, the argument is an analogous one.

Now we describe a polynomial-time reduction from HC to RHC. Let $G = (V, E)$ be an instance of HC. We shall construct a graph G' with a Hamiltonian path P in G' in such a way that G has a Hamiltonian cycle if and only if G' has a Hamiltonian cycle.

The idea of the construction of G' is to replace the vertices of G with diamonds and to embed the edges of G into G' by connecting east and west vertices of the corresponding diamonds. The Hamiltonian path P is created by connecting all diamonds using the North-South mode.

More precisely, let $V = \{w_1, w_2, \dots, w_n\}$. We set $G' = (V', E')$, where $V' = \bigcup_{i=1}^{n} \{N_i, W_i, E_i, S_i, x_i, u_i, v_i, y_i\}$ and E' contains the following sets of edges:

(i) all edges of the n diamonds, i.e., $\{\{N_i, x_i\}, \{N_i, u_i\}, \{W_i, x_i\}, \{W_i, v_i\},$ $\{v_i, u_i\}, \{E_i, u_i\}, \{E_i, y_i\}, \{S_i, v_i\}, \{S_i, y_i\}\}$ for $i = 1, 2, \dots, n$,
(ii) $\{\{S_i, N_{i+1}\} \mid i = 1, 2, \dots, n - 1\}$ (Figure 3.17a),
(iii) $\{\{W_i, E_j\}, \{E_i, W_j\} \mid$ for all $i, j \in \{1, \dots, n\}$ such that $\{w_i, w_j\} \in E\}$ (Figure 3.17b)

Following Figure 3.17a we see that the edges of the class (ii) unambiguously determine a Hamiltonian path P that starts in N_1, finishes in S_n, and traverses all diamonds in the North-South mode in the order $N_1, S_1, N_2, S_2, \dots, N_{n-1}, S_{n-1}, N_n, S_n$. Since N_1 and S_n are not connected in G', P cannot be completed to get a Hamiltonian cycle.

It is obvious that G' together with P can be constructed from G in linear time.

It remains to be shown that G' contains a Hamiltonian cycle if and only if G contains a Hamiltonian cycle. Let $H = v_1, v_{i_1}, v_{i_2}, \dots, v_{i_{n-1}}, v_1$ be a Hamiltonian cycle in G. Then

$$W_1, \dots, E_1, W_{i_1}, \dots, E_{i_1}, W_{i_2}, \dots, E_{i_2}, \dots W_{i_{n-1}}, \dots, E_{i_{n-1}}, W_1$$

mimics H by visiting the diamonds in G' in the same order as in H and moving through every diamond in the East-West mode.

Conversely, let H' be a Hamiltonian cycle in G'. First of all, observe that H' cannot enter any diamond in its north vertex or in its south vertex. If this would happen then this diamond must be traversed in the North-South mode. But this directly implies that all diamonds must be traversed in the North-South mode. Since there is no edge incident to N_1 or S_n besides the inner diamond edges, this is impossible. Thus, H' must traverse all diamonds in the East-West mode. Obviously, the order in which the diamonds are visited determines the order of vertices of G that corresponds to a Hamiltonian cycle of G. □

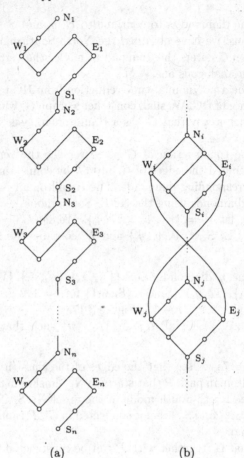

Fig. 3.17.

Now we use the NP-hardness of RHC to show that $SUBOPT_{\text{TSP}}$ is NP-hard.

Lemma 3.6.3.8.

$$\text{RHC} \leq_p SUBOPT_{\text{TSP}}.$$

Proof. Let (G, P), where $G = (V, E)$ is a graph and P is a Hamiltonian path in G, be an instance of RHC. Let G have n vertices $v_1, \ldots, v_n$. We construct an instance (K_n, c) of TSP and a feasible solution $\alpha \in \mathcal{M}(K_n, c)$ as follows:

(i) $K_n = (V, E_{com})$ is the complete graph of n vertices,

(ii) $c(\{v_i, v_j\}) = \begin{cases} 1 \text{ if } \{v_i, v_j\} \in E \\ 2 \text{ if } \{v_i, v_j\} \notin E \end{cases}$, and

(iii) α visits the vertices of K_n in the order given by P.

Now we have to prove that $(G, P) \in$ RHC if and only if α is not an optimal solution for (K_n, c).

Let $(G, P) \in$ RHC. Then the cost of the Hamiltonian cycle α (determined by P) in K_n is exactly $(n - 1) + 2 = n + 1$ because P cannot be extended to a Hamiltonian cycle. Since G contains a Hamiltonian cycle β, the cost of β in K_n is exactly n and so the feasible solution α is not optimal.

Let α be no optimal solution for (K_n, c). Then $cost(\alpha) = n + 1$. ($cost(\alpha) \geq n + 1$ because the minimal possible cost of a Hamiltonian cycle is n and α is not optimal. $cost(\alpha) \leq n + 1$ because α contains $n - 1$ edges of P of cost 1.) Since α is not optimal there exists a Hamiltonian cycle β in K_n with $cost(\beta) = n$. Obviously, β is a Hamiltonian cycle in G, and so $(G, P) \in$ RHC. $\quad\square$

Corollary 3.6.3.9. *SUBOPT*$_{\text{TSP}}$ *is NP-hard, and so, if* $P \neq NP$, *TSP does not possess any exact, polynomial-time searchable neighborhood.*

Proof. This assertion is a direct consequence of Lemma 3.6.3.8 and Theorem 3.6.3.6. $\quad\square$

Now we use the structure of the diamond in order to show that there are pathological instances of TSP for large k-*Exchange* neighborhoods. Here, pathological means that the instances have a unique optimal solution and exponentially many second-best (local) optima whose costs are exponential in the optimal cost. So, starting from a random initial solution there is a large probability that the scheme LSS($n/3$-*Exchange*) outputs a very poor feasible solution.

Theorem 3.6.3.10. *For every positive integer* $k \geq 2$, *and every large number* $M \geq 2^{8k}$, *there exists an input instance* (K_{8k}, c_M) *of* TSP *such that*

(i) *there is exactly one optimal solution to* (K_{8k}, c_M) *with the cost* $Opt_{\text{TSP}}(K_{8k}, c_M) = 8k$,
(ii) *the cost of the second-best feasible solutions (Hamiltonian cycle) is* $M + 5k$,
(iii) *there are* $2^{k-1}(k - 1)!$ *second-best optimal solutions for* (K_{8k}, c_M), *and*
(iv) *every second-best optimal solution differs from the unique optimal solution in exactly* $3k$ *edges (i.e., every second-best optimal solution is a local optimum with respect to the* $(3k - 1)$-*Exchange neighborhood).*

Proof. To construct (K_{8k}, c_M) for every k, we have to determine the function c_M from the edges of the complete graph of $8k$ vertices to the set of non-negative integers. To see the idea behind it we take the following four steps.

(1) First view K_{8k} as a graph consisting of k diamonds $D_i = (V_i, F_i) = (\{N_i, W_i, E_i, S_i, x_i, u_i, v_i, y_i\}, \{\{N_i, u_i\}, \{N_i, x_i\}, \{W_i, x_i\}, \{W_i, v_i\}, \{v_i, u_i\}, \{E_i, y_i\}, \{E_i, u_i\}, \{S_i, v_i\}, \{S_i, y_i\}\})$ for $i = 1, 2, \ldots, k$ as depicted in Figure 3.15. Following the proof of Lemma 3.6.3.7 we know that if a graph G contains a diamond D_i whose vertices x_i, u_i, v_i, y_i have degree two in G (i.e., there are no edges connecting the vertices x, u, v, y of the diamond with the rest of the graph) and G does not contain any edge between the vertices V_i aside from the edges in F_i, then every Hamiltonian cycle in G

traverses D_i either in the North-South mode or in the East-West mode (Figure 3.16). Obviously, we cannot force some edges to be missing in K_{8k}, but we can use very expensive costs for these edges in order to be sure that no optimal solution and no second-best feasible solution could contain any of these edges.

Set the cost of all edges incident to vertices x_i, u_i, v_i, y_i, but the edges of the diamond D_i, to $2 \cdot M$ for $i = 1, 2, \ldots, k$. Set the cost of all edges in $\{\{r, s\} \mid r, s \in V_i, r \neq s\} - F_i$ to $2 \cdot M$ for $i = 1, 2, \ldots k$.

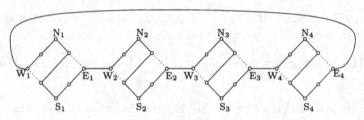

Fig. 3.18.

(2) Connect the $8k$ diamonds D_i by edges $\{E_i, W_{(i \bmod k)+1}\}$. This results in a graph with exactly one Hamiltonian cycle[24]

$$H_{E-W} = W_1, \ldots, E_1, W_2, \ldots, E_2, W_3, \ldots, W_k, \ldots, E_k, W_1$$

that traverses all diamonds in the East-West mode (see Figure 3.18). Assign to every edge of H_{E-W} the cost 1. Thus, $cost(H_{E-W}) = 8k$. In this way we have assigned the weight 1 to all edges of the diamond D_i, apart from the edges $\{W_i, v_i\}$, $\{u_i, E_i\}$. Set $c_M(\{W_i, v_i\}) = c_M(\{u_i, E_i\}) = 0$. Since we assign the value $2M$ to all remaining edges between the vertices of the set

$$W\text{-}E = \{W_1, \ldots, W_k, E_1, \ldots, E_k\},$$

every Hamiltonian cycle different from H_{E-W} and traversing the diamonds in the East-West mode has cost at least $2 \cdot M \geq 2^{8k+1}$.

(3) Let
$$N\text{-}S = \{N_1, N_2, \ldots, N_k, S_1, S_2, \ldots, S_k\}.$$

Assign to all edges between the vertices of $N\text{-}S - \{N_1\}$ the cost 0 (i.e., the vertices of $N\text{-}S - \{N_1\}$ build a clique of edges of cost 0 in K_{8k}). Set $c_M(\{N_1, N_i\}) = c_M(\{N_1, S_j\}) = M$ for all $i \in \{2, 3, \ldots, k\}$, $j \in \{1, 2, \ldots, k\}$. This implies that every Hamiltonian cycle traversing all diamonds in the North-South mode has the cost exactly $M + 5k$. (Observe that in traversing a diamond in North-South mode one goes via 5 edges of cost 1 and via 2 edges of cost 0).

(4) For all edges not considered in (1), (2), and (3), assign the value $2 \cdot M$.

---

[24] If one assumes that the edges of cost $2M$ are forbidden.

Now let us show that the constructed problem instance has the required properties.

(i) Consider the subgraph G' obtained from K_{8k} by removing all edges of cost M and $2M$. G' contains only one Hamiltonian cycle H_{E-W} of cost $8k$. This is because G' consists of diamonds connected via vertices N_i, W_i, E_i, and S_i, only and so every Hamiltonian tour in G' must traverse all diamonds either in the North-South mode or in the East-West mode. H_{E-W} is the only Hamiltonian cycle going in the East-West mode because $c_M(\{E_i, W_j\}) = 2M$ for all $i, j \in \{1, \ldots, k\}$, $j \neq (i \bmod k) + 1$ (see Figure 3.18). There is no Hamiltonian tour traversing diamonds in the North-South mode because $c_M(\{N_1, S_j\}) = c_M(\{N_1, N_j\}) = M$ for all $j \in \{2, \ldots, n\}$.

Thus, H_{E-W} is the unique optimal solution to (K_{8k}, c_M) because any other Hamiltonian cycle must contain at least one edge with cost at least $M \geq 2^{8k} > 8k$ for $k \geq 2$.

(ii) Consider the subgraph G'' obtained from K_{8k} by removing all edges of cost $2M$ (i.e., consisting of edges of cost 0, 1, and M). G'' also has the property that every Hamiltonian cycle traverses all diamonds either in the North-South mode or in the East-West mode. As already claimed above in (i), every Hamiltonian cycle different from H_{E-W} and traversing diamonds in the East-West mode must contain at least one edge of cost $2M$, and so it is not in G''. Consider the North-South mode now. To traverse a diamond in this mode costs exactly 5 (5 edges with cost 1 and 2 edges with cost 0), and so, to traverse all diamonds costs exactly $5k$. The edges $\{N_i, S_j\}$ between the diamonds have cost 0 for $i \neq 1$. Since the Hamiltonian cycle must use an edge connecting N_1 with some S_j or N_j with $j \in \{2, 3, \ldots, k\}$, one edge of cost M must be used. Thus, the cost of any Hamiltonian path of G'' in the North-South mode is at least $M + 5k$, and

$$N_1, S_2, \ldots, N_2, S_3, \ldots, N_3, S_4, \ldots, S_k, \ldots, N_k, S_1, \ldots, N_1$$

is a Hamiltonian cycle of cost $M + 5k$. Since any Hamiltonian cycle of K_{8k} that is not a Hamiltonian cycle of G'' must contain at least one edge of cost $2M > M + 5k$; the cost of the second-best solution for the input instance (K_{8k}, c_M) is $M + 5k$.

(iii) We count the number of Hamiltonian cycles of G'' that traverse all diamonds in the North-South direction and that contain exactly one edge of the cost M (i.e., cycles with the cost $M + 5k$). Since the vertices of $N\text{-}S - \{N_1\}$ build a clique of edges of cost 0 in G'', we have $(k-1)!$ orders in which the diamonds $D_1, D_2, \ldots, D_k$ can be traversed. Additionally, for each of the k diamonds one can choose one of the two directions of the North-South mode (from N_i to S_i or from S_i to N_i). Thus, there are altogether

$$2^{k-1}(k-1)!$$

distinct Hamiltonian cycles of cost $M + 5k$.

(iv) Finally, observe that the optimal solution H_{E-W} and any second-best Hamiltonian cycle have exactly $5k$ edges in common – the inner edges of the diamonds of cost 1. Thus, they differ in exactly $3k$ edges. Since the only better solution than a second-best solution is the optimal solution, H_{E-W}, every second-best solution is a local optimum with respect to any neighborhood *Neigh* where the optimal solution is not in $Neigh(\alpha)$ for any second-best solution α. Since $(3k-1)$-$Exchange(\alpha)$ does not contain H_{E-W} for any second-best solution α, all second-best solutions of (K_{8k}, c_M) are local optima with respect to $(3k-1)$-$Exchange$.   □

Keywords introduced in Section 3.6

neighborhood on sets of feasible solutions, local transformation, local optimum with respect to a neighborhood, local search scheme, *k-Exchange* neighborhood for TSP, Lin-Kernighan's variable-depth neighborhood, exact neighborhood, polynomial-time searchable neighborhood, cost-bounded integer valued problem, suboptimality decision problem, restricted Hamiltonian cycle problem

Summary of Section 3.6

Local search is an algorithm design technique for optimization problems. The first step of this technique is to define a neighborhood on the sets of feasible solutions. Usually, neighborhoods are defined by so-called local transformations in such a way that two feasible solutions α and β are neighbors if one can obtain α from β (and vice versa) by some local change of the specification of β (α). Having a neighborhood, local search starts with an initial feasible solution and then iteratively tries to find a better solution by searching the neighborhood of the current solution. The local algorithm halts with a solution α whose neighborhood does not contain any feasible solution better than α. Thus, α is a local optimum according to the neighborhood considered.

Any local search algorithm finishes with a local optimum. In general, there is no guarantee that the costs of local optima are not very far from the cost of an optimal solution and that local search algorithms work in polynomial time. For TSP there exist input instances with a unique optimal solution and exponentially many very poor second-best solutions that are all local optima according to the $(n/3)$-$Exchange$ neighborhood.

The following tradeoff exists between the time complexity and the quality of local optima. Small neighborhoods usually mean an efficient execution of one improvement but they often increase the number of local optima (i.e., the probability of getting stuck in a very weak local optimum). On the other hand, large neighborhoods can be expected to yield feasible solutions of higher quality, but the time complexity to verify local optimality becomes too large to be profitable. A way to overcome this difficulty may be the Kernighan-Lin's variable-depth search, which is a compromise between the choice of a small neighborhood and the choice of a large neighborhood. Very roughly, one uses a greedy approach to search for an improvement in a large neighborhood instead of an exhaustive search of a small neighborhood.

A neighborhood is called exact if every local optimum with respect to this neighborhood is a total optimum, too. A neighborhood is called polynomial-time searchable, if there exists a polynomial-time algorithm that finds the best solution in the neighborhood of every feasible solution. The existence of an exact polynomial-time neighborhood for an integer-valued optimization problem U usually implies that the local search algorithm according to this neighborhood is a pseudo-polynomial time algorithm for U.

If an optimization problem is strongly NP-hard, then it does not possess any exact, polynomial-time searchable neighborhood. TSP, $\triangle$-TSP, and WEIGHT-VCP are examples of such problems.

TSP has pathological input instances for the local search with the large $n/3$-$Exchange$ neighborhood. These pathological instances have a unique optimal solution and exponentially many second-best local optima whose costs are exponential in the optimal cost.

3.7 Relaxation to Linear Programming

3.7.1 Basic Concept

Linear Programming (LP) is an optimization problem that can be solved in polynomial time, while 0/1-linear programming (0/1-LP) and integer linear programming (IP) are NP-hard. All these optimization problems have the common constraints $A \cdot X = b$ for a matrix A and a vector b and the same objective to minimize $X \cdot c^{\mathsf{T}}$ for a given vector c. The only difference is that one minimizes over reals in LP, while the feasible solutions are over $\{0, 1\}$ for 0/1-LP and over integers for IP. This difference is essential because it determines the hardness of these problems.

Another important point is that many hard problems can be easily reduced to 0/1-LP or to IP; it is even better to say, many hard problems are naturally expressible in the form of linear programming problems. Thus, linear programming problems become the paradigmatic problems of combinatorial optimization and operations research.

Since LP is polynomial-time solvable, and 0/1-LP and IP are NP-hard, a very natural idea is to solve problem instances of 0/1-LP and IP as program instances of the efficiently solvable LP. This approach is called **relaxation** because one relaxes the requirement of finding an optimal solution over $\{0, 1\}$ or over positive integers by searching for an optimal solution over reals. Obviously, the computed optimal solution α for LP does not need to be a feasible solution for 0/1-LP or IP. Thus, one can ask what is the gain of this approach. First of all, $cost(\alpha)$ is a lower bound[25] on the cost of the optimal solutions

---

[25] Remember that we consider minimization problems. For maximization problems $cost(\alpha)$ is an upper bound on the optimal cost.

with respect to 0/1-LP and IP. Thus, $cost(\alpha)$ can be a good approximation of the cost of the optimal solutions of the original problem. This combined with the prima-dual method in Section 3.7.4 can be very helpful as a precomputation for a successful application of the branch-and-bound method as described in Section 3.4. Another possibility is to use α to compute a solution β that is feasible for 0/1-LP or IP. This can be done, for instance, by (randomized) rounding of real values to the values 0 and 1 or to positive integers. For some problems, such a feasible solution β is of high quality in the sense that it is reasonably close to an optimal solution. Some examples of such successful applications are given in Chapters 4 and 5.

Summarizing the considerations above, the algorithm design technique of relaxation to linear programming consists of the following three steps:

(1) *Reduction*
   Express a given instance x of an optimization problem U as an input instance $I(x)$ of 0/1-LP or IP.
(2) *Relaxation*
   Consider $I(x)$ as an instance of LP and compute an optimal solution α to $I(x)$ by an algorithm for linear programming.
(3) *Solving the original problem*
   Use α to either compute an optimal solution for the original problem, or to find a high-quality feasible solution for the original problem.

While the first two steps of the above schema can be performed in polynomial time, there is no guarantee that the third step can be executed efficiently. If the original optimization problem is NP-hard, the task to find an optimal solution in Step 3 is NP-hard, too. Anyway, this approach is very practical because for many problems the bound given by the value $cost(\alpha)$ helps to speed up the branch-and-bound method. This approach may be especially successful if one relaxes the requirement to find an optimal solution to the requirement to compute a reasonably good solution. For several problems one even can give some guarantee on the quality of the output of this scheme. Examples of this kind are presented in the next two chapters.

Section 3.7 is organized as follows. In Section 3.7.2 we illustrate the first step of the above scheme by presenting a few "natural" reductions to 0/1-LP and to IP. Natural means that there is a one-to-one correspondence between the feasible solutions of the original optimization problem and the resulting 0/1-linear programming (integer programming) problem. If we say that IP expresses another optimization problem U, then this even means that the formal representation of solutions in U and IP are exactly the same[26] $(\mathcal{M}_U(x) = \mathcal{M}_{IP}(y)$ for every pair of corresponding input instances x and $y)$ and two corresponding solutions of IP and U specified by the same representation α also have the same cost. In order to give at least an outline of how to

---

[26] For instance, vectors from $\{0, 1\}^n$

solve the linear programming problems, Section 3.7.3 gives a short description
of the simplex method that is in fact a simple local search algorithm.

3.7.2 Expressing Problems as Linear Programming Problems

In Section 2.3.2 we introduced the linear programming problem as to minimize

$$c^{\mathsf{T}} \cdot X = \sum_{i=1}^{n} c_i x_i$$

under the constraints

$$A \cdot X = b, \text{ i.e., } \sum_{i=1}^{n} a_{ji} x_i = b_j \text{ for } j = 1, \ldots, m.$$

$$x_i \geq 0 \text{ for } i = 1, \ldots, n \text{ (i.e., } X \in (\mathbb{R}^{\geq 0})^n)$$

for every input instance $A = [a_{ji}]_{j=1,\ldots,m, i=1,\ldots,n}$, $b = (b_1, \ldots, b_m)^{\mathsf{T}}$, and $c = (c_1, \ldots, c_n)^{\mathsf{T}}$ over reals. In what follows we call this form of LP the **standard
(equality) form**. Remember that the integer linear programming problem
is defined in the same way except that all coefficients are integers and the
solution $X \in \mathbb{Z}^n$.

There are several forms of LP. First of all we show that all these forms can
be expressed in the standard form. The **canonical form** of LP is to minimize

$$c^{\mathsf{T}} \cdot X = \sum_{i=1}^{n} c_i \cdot x_i$$

under the constraints

$$AX \geq b, \text{ i.e., } \sum_{i=1}^{n} a_{ji} x_i \geq b_j \text{ for } j = 1, \ldots, m$$

$$x_i \geq 0 \text{ for } i = 1, \ldots, n$$

for every input instance (A, b, c).

Now we transform the canonical form to the standard (equality) form. This
means we have to replace the inequality constraint $AX \geq b$ by some equality
constraint $B \cdot Y = d$. For every inequality constraint

$$\sum_{i=1}^{n} a_{ji} x_i \geq b_j$$

we add a new variable s_j (called **surplus** variable) and require

$$\sum_{i=1}^{n} a_{ji} x_i - s_j = b_j, \ s_j \geq 0.$$

Thus, if $A = [a_{ji}]_{j=1,\ldots,m,\, i=1,\ldots,n}$, $b = (b_1, \ldots, b_m)^\mathsf{T}$, and $c = (c_1, \ldots, c_n)^\mathsf{T}$ is the problem instance in the canonical form with the set of feasible solutions $\{X \in (\mathbb{R}^{\geq 0})^n \mid AX \geq b\}$, then the corresponding instance is expressed in the standard equality form (B, b, d), where

$$B = \begin{pmatrix} a_{11} & a_{12} & \cdots & a_{1n} & -1 & 0 & \cdots & 0 \\ a_{21} & a_{22} & \cdots & a_{2n} & 0 & -1 & \cdots & 0 \\ \vdots & \vdots & \vdots & \vdots & \vdots & \vdots & \ddots & \vdots \\ a_{m1} & a_{m2} & \cdots & a_{mn} & 0 & 0 & \cdots & -1 \end{pmatrix}$$

$$d = (c_1, \ldots, c_n, 0, \ldots, 0)^\mathsf{T} \in \mathbb{R}^{m+n}.$$

The set of solutions for (B, b, d) is

$$Sol(B, b, d) = \{Y = (x_1, \ldots, x_n, s_1, \ldots, s_m)^\mathsf{T} \in (\mathbb{R}^{\geq 0})^{m+n} \mid B \cdot Y = b\}.$$

We observe that

(i) for every feasible solution $\alpha = (\alpha_1, \ldots, \alpha_n)^\mathsf{T}$ to (A, b, c) in the canonical form of LP, there exists $(\beta_1, \ldots, \beta_m)^\mathsf{T} \in (\mathbb{R}^{\geq 0})^m$ such that

$$(\alpha_1, \ldots, \alpha_n, \beta_1 \ldots, \beta_m)^\mathsf{T} \in Sol(B, b, d),$$

and vice versa,

(ii) for every feasible solution $(\delta_1, \ldots, \delta_n, \gamma_1, \ldots, \gamma_m)^\mathsf{T} \in Sol(B, b, d)$, the vector $(\delta_1, \ldots, \delta_n)^\mathsf{T}$ is a feasible solution to the input instance (A, b, c) in the canonical form of LP.

Since $d = (c_1, \ldots, c_n, 0, \ldots, 0)^\mathsf{T}$, the minimization in $Sol(B, b, d)$ is equivalent to the minimization of $\sum_{i=1}^{n} c_i x_i$ in $\{X \in (\mathbb{R}^{\geq 0})^n \mid AX \geq b\}$. We see that the size of (B, b, d) is linear in the size of (A, b, c) and so the above transformation can be realized efficiently.

Next, we consider the **standard inequality form** of LP. One has to minimize

$$c^\mathsf{T} \cdot X = \sum_{i=1}^{n} c_i x_i$$

under the constraints

$$AX \leq b, \text{ i.e., } \sum_{i=1}^{n} a_{ji} x_i \leq b_j \text{ for } j = 1, \ldots, m, \text{ and}$$

$$x_i \geq 0 \text{ for } i = 1, \ldots, n$$

for a given instance A, b, c. To transform $\sum_{i=1}^{n} a_{ji} x_j \leq b_j$ to an equation we introduce a new variable s_j (called **slack** variable) and set

$$\sum_{i=1}^{n} a_{ji} x_i + s_j = b_j, \text{ and } s_j \geq 0.$$

Analogously to the transformation from the canonical form to the standard form, it is easy to observe that the derived standard equality form is equivalent to the given standard form.

Exercise 3.7.2.1. Consider the **general form** of LP defined as follows. For an input instance

$$A = [a_{ji}]_{j=1,\ldots,m,\, i=1,\ldots,n}, b = (b_1,\ldots,b_m)^\mathsf{T}, c = (c_1,\ldots,c_n)^\mathsf{T},$$

$$M \subseteq \{1,\ldots,m\}, Q \subseteq \{1,\ldots,n\},$$

one has to minimize

$$\sum_{i=1}^{n} c_i x_i$$

under the constraints

$$\sum_{i=1}^{n} a_{ji} x_i = b_j \ \text{ for } j \in M$$

$$\sum_{i=1}^{n} a_{ri} x_i \geq b_r \ \text{ for } r \in \{1,\ldots,m\} - M$$

$$x_i \geq 0 \ \text{ for } i \in Q.$$

Reduce this general form to

(i) the standard form,
(ii) the standard inequality form, and
(iii) the canonical form.

□

The reason to consider all these forms of LP is to simplify the transformation of combinatorial problems to LP. Thus, one can choose the form that naturally expresses the original optimization problem. Since all forms of LP introduced above are minimization problems, one could ask how to express maximization problems as linear programming problems. The answer is very simple. Consider, for an input A, b, c, the task

$$\text{maximize } c^\mathsf{T} \cdot X = \sum_{i=1}^{n} c_i x_i$$

under some linear constraints given by A and b. Then this task is equivalent to the task

$$\text{minimize } [(-1) \cdot c^\mathsf{T}] \cdot X = \sum_{i=1}^{n} (-c_i) \cdot x_i$$

under the same constraints.

In what follows we express some optimization problems as LP problems.

MINIMUM WEIGHTED VERTEX COVER.

Remember that the input instances of WEIGHT-VCP are weighted graphs $G = (V, E, c)$, $c : V \to \mathbb{N} - \{0\}$. The goal is to find a vertex cover S with a minimal cost $\sum_{v \in S} c(v)$. Let $V = \{v_1, \ldots, v_n\}$.

To express this instance of the WEIGHT-VCP problem as an instance of LP (in fact of 0/1-LP), we represent the sets $S \subseteq V$ by its characteristic vectors $X_S = (x_1, \ldots, x_n) \in \{0, 1\}^n$, where

$$x_i = 1 \text{ iff } v_i \in S.$$

The constraint to cover all edges from E can be expressed by

$$x_i + x_j \geq 1 \text{ for every } \{v_i, v_j\} \in E$$

because for every edge $\{v_i, v_j\}$ one of the incident vertices must be in the vertex cover. The goal is to minimize

$$\sum_{i=1}^{n} c(v_i) \cdot x_i.$$

Relaxing $x_i \in \{0, 1\}$ to $x_i \geq 0$ for $i = 1, \ldots, n$ we obtain the canonical form of LP.

KNAPSACK PROBLEM.

Let $w_1, w_2, \ldots, w_n, c_1, c_2, \ldots, c_n$, and b be an instance of the knapsack problem. We consider n Boolean variables $x_1, x_2, \ldots, x_n$ where $x_i = 1$ indicates that the ith object was packed in the knapsack. Thus, the task is to maximize

$$\sum_{i=1}^{n} c_i x_i$$

under the constraints

$$\sum_{i=1}^{n} w_i x_i \leq b, \text{ and}$$

$$x_i \in \{0, 1\} \text{ for } i = 1, \ldots, n.$$

Exchanging the maximization of $\sum_{i=1}^{n} c_i x_i$ for the minimization of

$$\sum_{i=1}^{n} (-c_i) \cdot x_i$$

one obtains a standard input instance of 0/1-LP. Relaxing $x_j \in \{0, 1\}$ to $x_i \geq 0$ for $i = 1, \ldots, n$ results in the standard inequality form of LP.

MAXIMUM MATCHING PROBLEM.

The maximum matching problem is to find a matching of maximal cardinality in a given graph $G = (V, E)$. Remember that a matching in G is a set of edges $H \subseteq E$ with the property that, for all edges $\{u, v\}, \{x, y\} \in H$, $\{u, v\} \neq \{x, y\}$ implies $|\{u, v, x, y\}| = 4$ (i.e., no two edges of a matching share a common vertex).

To express the instances of this problem as instances of 0/1-LP we consider Boolean variables x_e for every $e \in E$, where

$$x_e = 1 \text{ iff } e \in H.$$

Let, for every $v \in V$, $E(v) = \{\{v, u\} \in E \mid u \in V\}$ be the set of all edges incident to v. Now, the task is to maximize

$$\sum_{e \in E} x_e$$

under the $|V|$ constraints

$$\sum_{e \in E(v)} x_e \leq 1 \text{ for every } v \in V,$$

and the following $|E|$ constraints

$$x_e \in \{0, 1\} \text{ for every } e \in E.$$

Relaxing $x_e \in \{0, 1\}$ to $x_e \geq 0$ one obtains an instance of LP.

Exercise 3.7.2.2. Express the instances of the maximum matching problem as instances of LP in the canonical form.   □

Exercise 3.7.2.3. (*) Consider the maximum matching problem for bipartite graphs only. Prove, that every optimal solution of its relaxation to LP is also a feasible (Boolean) solution to the original instance of the maximum matching problem (i.e., that this problem is in P).   □

Exercise 3.7.2.4. Consider the following generalization of the maximum matching problem. Given a weighted graph (G, c), $G = (V, E)$, $c : E \to \mathbb{N}$, find a perfect matching with the minimal cost, where the cost of a matching H is $cost(H) = \sum_{e \in H} c(e)$.

A **perfect** matching is a matching H, where every edge e from E is either in H or shares one common vertex with an edge in H. Express the input instances of this minimization problem as input instances of 0/1-LP.   □

Exercise 3.7.2.5. Express the minimum spanning tree problem as a 0/1-LP problem.   □

Exercise 3.7.2.6. A cycle cover of a graph $G = (V, E)$ is any subgraph $C = (V, E_C)$ of G, where every vertex has the degree exactly two. The minimum cycle cover problem is to find, for every weighted complete graph (G, c), a cycle cover of G with the minimal cost with respect to c. Relax this problem to the standard form of LP.   □

MAKESPAN SCHEDULING PROBLEM.

Let $(p_1, p_2, \ldots, p_n, m)$ be an instance of MS. Remember that $p_i \in \mathbb{N} - \{0\}$ is the processing time of the i-th job on any of the m identical machines for $i = 1, \ldots, n$. The task is to distribute the n jobs to m machines in such a way that the whole processing time (makespan - the time after which every machine processed all its assigned jobs) is minimized.

We consider the Boolean variables $x_{ij} \in \{0, 1\}$ for $i = 1, \ldots, n$, $j = 1, \ldots, m$ with the following meaning:

$x_{ij} = 1$ iff the i-th job was assigned to the j-th machine.

The n linear equalities

$$\sum_{j=1}^{m} x_{ij} = 1 \text{ for all } i \in \{1, \ldots, n\}$$

guarantee that each job is assigned to exactly one machine and so each job will be processed exactly once. Now, we take an integral variable t for the makespan. Thus, the objective is to

minimize t

under the n linear equalities above and the constraints

$$t - \sum_{i=1}^{n} p_i \cdot x_{ij} \leq 0 \text{ for all } j \in \{1, \ldots, m\}$$

that assures that every machine finishes the work on its job in time at most t.

There is also a possibility to express $(p_1, p_2, \ldots, p_n, m)$ as a problem instance of $0/1 - LP$, that looks more natural than the use of the integer variable t. Since the m machines are identical, we may assume without loss of generality that the first machine has always the maximal load. Then we minimize

$$\sum_{i=1}^{n} p_i \cdot x_{i1},$$

which is the makespan of the first machine. The constraints

$$\sum_{j=1}^{m} x_{ij} \geq 1 \text{ for all } i \in \{1, \ldots, n\}$$

guarantee that each job is assigned to at least one machine, and the constraints

$$\sum_{i=1}^{n} p_i \cdot x_{ij} \leq \sum_{i=1}^{n} p_i \cdot x_{i1} \text{ for all } j \in \{2, \ldots, m\}$$

ensure that the makespan of the first machine is at least as large as the makespan of any other machine.

Maximum Satisfiability Problem.

Let $F = F_1 \wedge F_2 \wedge \ldots \wedge F_m$ be a formula in CNF over the set of variables $X = \{x_1, x_2, \ldots, x_n\}$. We use the same boolean variables $x_1, x_2, \ldots, x_n$ for the relaxation and take the additional variables $z_1, z_2, \ldots, z_m$ with the meaning

$$z_i = 1 \text{ iff the clause } F_i \text{ is satisfied.}$$

Let, for each $i \in \{1, \ldots, m\}$, $\text{In}^+(F_i)$ be the set of indices of the variables from X which appear as positive literals in F_i and let $\text{In}^-(F_i)$ be the set of the indices of the negated variables that appear as literals in F_i. For instance, for the clause $C = x_1 \vee \overline{x}_3 \vee x_7 \vee \overline{x}_9$ we have $\text{In}^+(C) = \{1, 7\}$ and $\text{In}^-(C) = \{3, 9\}$. Now, the instance $\text{LP}(F)$ of IP that expresses F is to

$$\text{maximize} \sum_{j=1}^{m} z_j$$

subject to the following $2m + n$ constraints

$$z_j - \sum_{i \in \text{In}^+(F_j)} x_i - \sum_{l \in \text{In}^-(F_j)} (1 - x_l) \leq 0 \text{ for } j = 1, \ldots, m$$
$$x_i \in \{0, 1\} \text{ for } i = 1, \ldots, n$$
$$z_j \in \{0, 1\} \text{ for } j = 1, \ldots, m.$$

The linear inequality

$$z_j \leq \sum_{i \in \text{In}^+(F_j)} x_i + \sum_{l \in \text{In}^-(F_j)} (1 - x_l)$$

assures that z_j may take the value 1 only if at least one of the literals in F takes the value 1. Relaxing $x_i \in \{0, 1\}$ to

$$x_i \geq 0 \text{ and } x_i \leq 1$$

for $i = 1, \ldots, n$ and relaxing $z_j \in \{0, 1\}$ to

$$z_j \geq 0$$

for $j = 1, \ldots, m$ results in an irregular form of LP. Using additional variables one can obtain any of its normal forms.

Set Cover Problem.

Let $(X, \mathcal{F})$ with $X = \{a_1, \ldots a_n\}$ and $\mathcal{F} = \{S_1, S_2, \ldots, S_m\}$, $S_i \subseteq X$ for $i = 1, \ldots, m$, be an instance of the set cover problem (SCP). For $i = 1, \ldots, m$ we consider the Boolean variable x_i with the meaning

$$x_i = 1 \text{ iff } S_i \text{ is picked for the set cover.}$$

Let $\text{Index}(k) = \{d \in \{1, \ldots, m\} \mid a_k \in S_d\}$ for $k = 1, \ldots, n$. Then we have to

$$\text{minimize} \sum_{i=1}^{m} x_i$$

under the following n linear constraints

$$\sum_{j \in \text{Index}(k)} x_j \geq 1 \text{ for } k = 1, \ldots, n.$$

Exercise 3.7.2.7. Consider the set multicover problem as the following minimization problem. For a given instance $(X, \mathcal{F}, r)$ where $(X, \mathcal{F})$ is an instance of SCP and $r \in \mathbb{N} - \{0\}$, a feasible solution to $(X, \mathcal{F}, r)$ is a r-multicover of X that covers each element of X at least r times. In a multicover it is allowed to use the same set $S \in \mathcal{F}$ several times. Find a relaxation of the set multicover problem to LP. □

Exercise 3.7.2.8. Consider the **hitting set problem** (HSP) defined as follows.

Input: $(X, \mathcal{S}, c)$, where X is a finite set, $\mathcal{S} \subseteq Pot(X)$, and c is a function from X to $\mathbb{N} - \{0\}$.

Constraints: For every input $(X, \mathcal{S}, c)$,

$$\mathcal{M}(X, \mathcal{S}, c) = \{Y \subseteq X \mid Y \cap S \neq \emptyset \text{ for every } S \in \mathcal{S}\}.$$

Costs: For every $Y \in \mathcal{M}(X, \mathcal{S}, c)$, $cost(Y, (X, \mathcal{S}, c)) = \sum_{x \in Y} c(x)$.
Goal: minimum.

Show that

(i) the hitting set problem (HSP) can be relaxed to a LP problem, and
(ii)[*] the set cover problem can be polynomial-time reduced to HSP. □

3.7.3 The Simplex Algorithm

There is a large amount of literature on methods for solving the linear programming problem and there are a couple of thick books devoted to the study of different versions of the simplex method. We do not want to go in detail here. The aim of this section is only to roughly present the SIMPLEX ALGORITHM in a way that enables us to see it as a local search algorithm in some geometrical interpretation. The reason for restricting our attention to solving the linear programming problem in the above sense is that this is a topic of operations research, and the main interest from the combinatorial optimization point of view focuses on the possibility to express (combinatorial) optimization problems as problems of linear programming[27] (and so to use algorithms of operations research as a subpart of algorithms for solving the original combinatorial optimization problems).

Consider the standard form of the linear programming problem that is, for given $A = [a_{ji}]_{j=1,\ldots,m, i=1,\ldots,n}$, $b = (b_1, \ldots, b_m)^\mathsf{T}$, and $c = (c_1, \ldots, c_n)^\mathsf{T}$

$$\text{to minimize } \sum_{i=1}^{n} c_i x_i$$

under the constraints

$$\sum_{i=1}^{n} a_{ji} x_i = b_j \text{ for } j = 1, \ldots, m, \text{ and}$$

$$x_i \geq 0 \quad \text{for } i = 1, \ldots, n.$$

Recall Section 2.2.1, where we showed that the set $Sol(A)$ of solutions of a system of homogeneous linear equations $AX = 0$ is a subspace of $\mathbb{R}^n$ and that the set $Sol(A, b)$ of solutions of a system $AX = b$ of linear equations is an affine subspace of $\mathbb{R}^n$. A subspace of $\mathbb{R}^n$ always contains the origin $(0, 0, \ldots, 0)^\mathsf{T}$ and any affine subspace can be viewed as a subspace that is shifted (translated) by a vector from $\mathbb{R}^n$. So, the dimension of $Sol(A, b)$ is that of $Sol(A)$. Remember that the dimension of $Sol(A, b)$ for one linear equation ($m = 1$) is exactly $n - 1$. In general, $\dim(Sol(A, b)) = n - \text{rank}(A)$.

Definition 3.7.3.1. *Let n be a positive integer. An affine subspace of $\mathbb{R}^n$ of dimension $n - 1$ is called a* **hyperplane**. *Alternatively, a hyperplane of $\mathbb{R}^n$ is a set of all $X = (x_1, \ldots, x_n)^\mathsf{T} \in \mathbb{R}^n$ such that*

$$a_1 x_1 + a_2 x_2 + \ldots + a_n x_n = b$$

for some $a_1, a_2, \ldots, a_n, b$, where not all a's are equal to zero. The sets

$$\text{HS}_{\geq}(a_1, \ldots, a_n, b) = \left\{ X = (x_1, \ldots, x_n)^\mathsf{T} \in \mathbb{R}^n \ \middle|\ \sum_{i=1}^{n} a_i x_i \geq b \right\}, \text{ and}$$

$$\text{HS}_{\leq}(a_1, \ldots, a_n, b) = \left\{ X = (x_1, \ldots, x_n)^\mathsf{T} \in \mathbb{R}^n \ \middle|\ \sum_{i=1}^{n} a_i x_i \leq b \right\}$$

are called **halfspaces**.

Obviously, any halfspace of $\mathbb{R}^n$ has the same dimension n as $\mathbb{R}^n$, and it is a convex set. Since any finite intersection of convex sets is a convex set (see Section 2.2.1), the set

---

[27] Recently, even as problems of semidefinite programming or other generalizations of linear programming that can be solved in polynomial time.

$$\{X \in \mathbb{R}^n \mid A \cdot X \leq b\} = \bigcap_{j=1}^{m} \mathrm{HS}_{\leq}(a_{j1}, \ldots, a_{jn}, b_j)$$

$$= \bigcap_{j=1}^{m} \{X = (x_1, \ldots, x_n)^{\mathsf{T}} \in \mathbb{R}^n \mid \sum_{i=1}^{n} a_{ji}x_i \leq b_j\},$$

called $\boldsymbol{Polytope(AX \leq b)}$, is a convex set.

Definition 3.7.3.2. *Let n be a positive integer. The intersection of a finite number of halfspaces of $\mathbb{R}^n$ is a* **(convex) polytope** *of $\mathbb{R}^n$. For given constraints $\sum_{j=1}^{n} a_{ji}x_i \leq b_j$, for $j = 1 \ldots, m$, and $x_i \geq 0$ for $i = 1, \ldots, n$,*

$$\boldsymbol{Polytope(AX \leq b, X \geq 0_{n \times 1})} = \{X \in (\mathbb{R}^{\geq 0})^n \mid A \cdot X \leq b\}.$$

Observe that

$$Polytope(AX \leq b, X \geq 0_{n \times 1}) =$$

$$\bigcap_{j=1}^{m} \mathrm{HS}_{\leq}(a_{j1}, \ldots, a_{jn}, b_j) \cap \left(\bigcap_{j=1}^{n} \{(x_1, \ldots, x_n)^{\mathsf{T}} \in \mathbb{R}^n \,\middle|\, x_j \geq 0\} \right)$$

is the intersection of $m + n$ halfspaces.

Consider the following constraints in $\mathbb{R}^2$.

$$\begin{aligned}
x_1 + x_2 &\leq 8 \\
x_2 &\leq 6 \\
x_1 - x_2 &\leq 4 \\
x_1 &\geq 0 \\
x_2 &\geq 0.
\end{aligned}$$

The polytope depicted in Figure 3.19 is exactly the polytope that corresponds to these constraints. Consider minimizing $x_1 - x_2$ and maximizing $x_1 - x_2$ for these constraints. Geometrically, one can solve this problem as follows. One takes the straight line $x_1 - x_2 = 0$ (see Figure 3.20) and moves it via the polytope in both directions perpendicular to this straight line. In the "upwards" direction (i.e., in the direction of the halfplane $x_1 - x_2 \leq 0$) this straight line leaves the polytope at the point $(0, 6)$ and this $(x_1 = 0, x_2 = 6)$ is the unique optimal solution of this minimization problem. In the "downwards" direction (i.e., in the direction of the halfplane $x_1 - x_2 \geq 0$) this straight line leaves the polytope in the set of points corresponding to the points of the line connecting the points $(4, 0)$ and $(6, 2)$. Thus,

$$\{(x_1, x_2)^{\mathsf{T}} \in \mathbb{R}^2 \mid (4, 0) \leq (x_1, x_2) \leq (6, 2), x_1 - x_2 = 4\}$$

is the infinite set of all optimal solutions of the maximization problem.

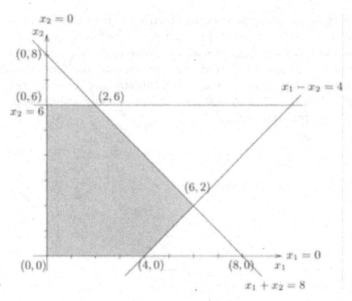

Fig. 3.19.

Consider the same polytope (constraints) and the tasks of minimizing $3x_1 - 2x_2$ and maximizing $3x_1 - 2x_2$. Then, one can take the straight line $3x_1 - 2x_2 =$

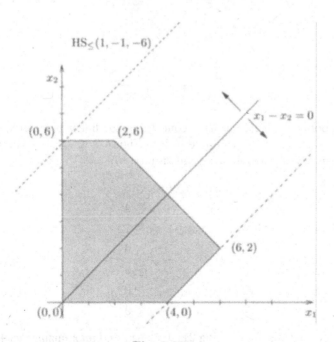

Fig. 3.20.

0 and move it in both perpendicular directions to this straight line as depicted in Figure 3.21. Moving in the direction of the halfplane $3x_1 - 2x_2 \leq 0$, this straight line leaves the polytope at the point $(0,6)$, and so $x_1 = 0$, $x_2 = 6$ is the unique optimal solution of the minimization problem. Moving in the direction of the halfplane $3x_1 - 2x_2 \geq 0$, this straight line leaves the polytope at the point $(6,2)$, and so $x_1 = 6$, $x_2 = 2$ is the unique optimal solution of the maximization problem.

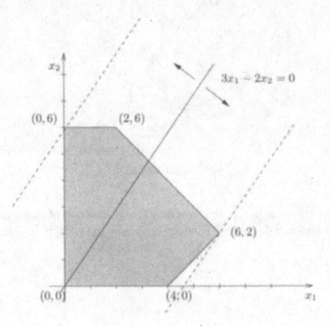

Fig. 3.21.

To support our visual intuition concerning the linear programming problem we consider one more example in $\mathbb{R}^3$. Figure 3.22 involves the polytope that corresponds to the following constraints:

$$x_1 + x_2 + x_3 \leq 8$$
$$x_1 \qquad\qquad \leq 4$$
$$3x_2 + x_3 \leq 12$$
$$x_3 \leq 6$$
$$x_1 \qquad\qquad \geq 0$$
$$x_2 \qquad \geq 0$$
$$x_3 \geq 0$$

Consider the tasks minimizing $2x_1 + x_2 + x_3$ and maximizing $2x_1 + x_2 + x_3$. Take the hyperplane $2x_1 + x_2 + x_3 = 0$. To minimize $2x_1 + x_2 + x_3$ move

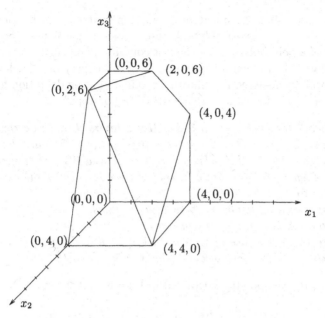

Fig. 3.22.

this hyperplane via the polytope in the direction to the halfspace given by $2x_1 + x_2 + x_3 \leq 0$. Obviously, it leaves the polytope immediately in the origin, and so $(0,0,0)^\mathsf{T}$ is the unique solution of this optimization problem. To maximize $2x_1 + x_2 + x_3$ move the hyperplane $2x_1 + x_2 + x_3 = 0$ in the direction of the halfspace $2x_1 + x_2 + x_3 \geq 0$. It leaves the polytope at the set $S = \{(x_1, x_2, x_3)^\mathsf{T} \in \mathbb{R}^3 \mid (4,4,0) \leq (x_1, x_2, x_3) \leq (4,0,4) \text{ and } x_1 + x_2 + x_3 = 8\}$, which corresponds to the line connecting the points $(4,4,0)^\mathsf{T}$ and $(4,0,4)^\mathsf{T}$ in $\mathbb{R}^3$. Thus, S is the set of optimal solutions of this maximization problem.

Exercise 3.7.3.3. Solve the following optimization problem for the constraints (polytope) depicted in Figure 3.19 by playing the geometrical game explained above:

(i) minimize $-3x_1 + 7x_2$,
(ii) maximize x_2,
(iii) maximize $10x_1 + x_2$, and
(iv) minimize $-x_1 - 2x_2$.   □

Exercise 3.7.3.4. Solve the following optimization problem for the constraints represented by the polytope of Figure 3.22:

(i) minimize $-x_1 + 2x_2 - x_3$,
(ii) maximize x_3,
(iii) maximize $x_1 + x_2 + x_3$, and
(iv) minimize $7x_1 - 13x_2 + 5x_3$.   □

We have seen that the linear programming problem can be viewed as a problem to find, for a given polytope and a given hyperplane, the points on the surface of a polytope that are the last common points of the polytope and the hyperplane when moving the hyperplane via the polytope. Such a general situation in $\mathbb{R}^3$ is depicted in Figure 3.23. The following definition formalizes our informal expression "a hyperplane leaves a polytope".

Definition 3.7.3.5. *Let d and n be positive integers. Let A be a convex polytope of dimension d in $\mathbb{R}^n$. Let H be a hyperplane of $\mathbb{R}^n$ and let HS be a halfspace defined by H. If $A \cap \mathrm{HS} \subseteq H$ (i.e., A and HS just "touch in their exteriors"), then $A \cap \mathrm{HS}$ is called a* **face of A,** *and H is called the* **supporting hyperplane defining $A \cap \mathrm{HS}$.**
We distinguish the following three kinds of faces.
A **facet of A** *is a face of dimension $n - 1$.*
An **edge of A** *is a face of dimension 1 (i.e., a line segment).*
A **vertex of A** *is a face of dimension 0 (i.e., a point).*

Let A denote the polytope depicted in Figure 3.19. Then

$$A \cap \mathrm{HS}_{\geq}(0, 1, 6) \subseteq \{(x_1, x_2)^{\mathsf{T}} \in \mathbb{R}^2 \mid x_2 = 6\}$$

is the edge of A that corresponds to the line segment connecting the points $(0, 6)$ and $(2, 6)$.

$$A \cap \mathrm{HS}_{\geq}(1, 1, 8) \subseteq \{(x_1, x_2)^{\mathsf{T}} \in \mathbb{R}^2 \mid x_1 + x_2 = 8\}$$

is the edge that corresponds to the line segment connecting the points $(2, 6)$ and $(6, 2)$.

$$A \cap \mathrm{HS}_{\leq}(1, -1, -6) \subseteq \{(x_1, x_2)^{\mathsf{T}} \in \mathbb{R}^2 \mid x_1 - x_2 = -6\}$$

is the vertex of P (see Figure 3.20) that is the point $(0, 6)$.

Exercise 3.7.3.6. Determine, for every of the following faces of A, at least one of its supporting hyperplanes:

(i) the vertex $(6, 2)$,
(ii) the edge corresponding to the line segment between the points $(4, 0)$ and $(6, 2)$.

$\square$

Let B denote the polytope depicted in Figure 3.22. Then

$$B \cap \mathrm{HS}_{\geq}(1, 1, 1, 8) \subseteq \{(x_1, x_2, x_3)^{\mathsf{T}} \in \mathbb{R}^3 \mid x_1 + x_2 + x_3 = 8\}$$

is the facet of B that is the convex hull of the points $(0, 2, 6)$, $(2, 0, 6)$, $(4, 0, 4)$, and $(4, 4, 0)$.

$$B \cap \mathrm{HS}_{\geq}(0, 1, 0, 4) \subseteq \{(x_1, x_2, x_3)^{\mathsf{T}} \in \mathbb{R}^3 \mid x_2 = 4\}$$

is the edge of B that corresponds to the line segment connecting the points $(0, 4, 0)$ and $(4, 4, 0)$.

$$B \cap \mathrm{HS}_{\leq}(1, 1, 1, 0) \subseteq \{(x_1, x_2, x_3)^{\mathsf{T}} \in \mathbb{R}^3 \mid x_1 + x_2 + x_3 = 0\}$$

is the vertex of B that is the point $(0, 0, 0)$.

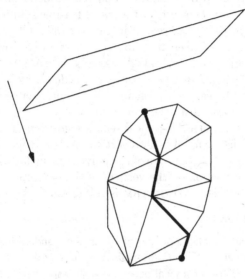

Fig. 3.23.

Exercise 3.7.3.7. Determine, for every of the following faces of B, at least one of its supporting hyperplanes.

(i) the edge that is the line segment that connects the points $(2, 0, 6)$ and $(4, 0, 4)$,

(ii) the edge that is the line segment that connects the points $(0, 0, 0)$ and $(4, 0, 0)$,

(iii) the edge that is the line segment that connects the points $(0, 2, 6)$ and $(4, 4, 0)$, and

(iv) the vertex $(4, 4, 0)$.   □

Assuming the constraints $A \cdot X \leq b$, $X \geq 0_{n \times 1}$, the set of all feasible solutions of this linear programming problem is the set of all points of the polytope $Polytope(AX \leq b, X \geq 0)$. Playing our geometrical game with moving hyperplanes we see that all optimal solutions must lie on the exteriors of $Polytope(AX \leq b, X \geq 0)$. More precisely, the set of the optimal solutions of every linear programming problem is a face of the corresponding polytope.[28] This directly implies the following crucial observation.

---

[28] We omit the formal proof of this fact hoping that the geometrical intuition that was built above is sufficient to accept this assertion.

Observation 3.7.3.8. Let (A, b, c) be an instance of the linear programming problem in the standard inequality form with satisfiable constraints. Then the set of optimal solutions to (A, b, c) contains at least one solution that is a vertex of $Polytope(AX \leq b, X \geq 0)$.

The importance of Observation 3.7.3.8 is in the fact that to solve LP it is sufficient to find an optimal solution and not to determine the set of all optimal solutions. So, we can restrict the set of feasible solutions to the set of vertices of the polytope and to search in this set only. Observe that the set of vertices of a polytope (given by a finite number of constraints) is always finite. Thus, one can have a look at all vertices of the polytope in order to find an optimal solution. But this is not a good idea because the number of vertices of the polytope can be exponential in the number of constraints and so in the size of the input instance.

The SIMPLEX ALGORITHM is a local search algorithm. The neighborhood *Neigh* is simply defined as follows. Two vertices of a polytope are called neighbors if they are connected by an edge of the polytope. Obviously, this neighborhood is a symmetric relation on the vertices of the polytope. The geometrical description of the SIMPLEX ALGORITHM is as follows.

SIMPLEX ALGORITHM

Input: An input instance (A, b, c) of LP in the standard inequality form.
Step 1:   Find a vertex X of $Polytope(AX \leq b, X \geq 0)$.
Step 2:   **while** X is not a local optimum with respect to *Neigh*
          do   replace X by a vertex (neighbor) $Y \in Neigh_{(A,b,c)}(X)$
          if $c^\mathsf{T} \cdot X > c^\mathsf{T} \cdot Y$.
Step 3:   **output**(X).

Thus, being in a vertex X of the polytope, the SIMPLEX ALGORITHM moves from X via an edge (of the polytope) that continously improves the cost of the solutions to another vertex Y. If there are several edges that (starting from X) decrease the costs, the SIMPLEX ALGORITHM can follow the strategy to choose the edge that falls down as steep as possible (i.e., with the maximal cost improvement per unit length). Such a sequence of chosen edges can be seen as depicted in Figure 3.23.

If one considers minimizing $3x_1 - 2x_2$ for the polytope depicted in Figure 3.21 and the SIMPLEX ALGORITHM starts in the vertex $(6, 2)$, then it first moves to the vertex $(2, 6)$, and then finally to the optimal solution $(0, 6)$. If $(4, 0)$ is the initial feasible solution then one moves to $(0, 0)$ first, and then to $(0, 6)$.

Now consider minimizing $-2x_1 - x_2 - x_3$ starting at the vertex $(0, 0, 6)$ of the polytope depicted in Figure 3.22. The best possibility to decrease $cost(0, 0, 6) = -6$ is to move to the vertex $(2, 0, 6)$ with the cost -10. Now the only possibility to decrease the cost is to move to the vertex $(4, 0, 4)$ with cost -12. Since there is no possible improvement, the simplex algorithm outputs $(4, 0, 4)$.

Here we do not deal with the implementation details that show that the above search for a better solution in the neighborhood can be done very efficiently. We only formulate two crucial observations.

Observation 3.7.3.9. The neighborhood defined above for the vertices of a polytope is an exact neighborhood of LP, and so the SIMPLEX ALGORITHM always computes an optimal solution.

The Observation 3.7.3.9 must be intuitively clear from the visual (geometrical) point of-view. If a local optimum is a vertex of a polytope A, and X would not be an optimal solution, then there would exist a vertex Y with $cost(Y) < cost(X)$. Let us connect X and Y by the line segment L. Then, a part of L in the close neighborhood of X cannot lie in the polytope because there is no better solution than X in the neighborhood of X and the solutions on L continously improve if moving from X to Y. Thus, L is not included in the polytope which is a contradiction with the convexity of the polytope.

Observation 3.7.3.10. There exists an input instance of LP such that the simplex algorithm executes exponentially many[29] consecutive improvements until reaching a local optimum.

Thus, the worst case complexity of the SIMPLEX ALGORITHM is exponential. But the SIMPLEX ALGORITHM is the most famous example of a very efficient, practical algorithm of exponential worst case complexity, because the situations when the SIMPLEX ALGORITHM is not fast enough are very rare. Despite the fact that there are polynomial-time algorithms for LP, one often prefers using the SIMPLEX ALGORITHM because of its excellent average case behavior.

3.7.4 Rounding, LP-Duality and Primal-Dual Method

We have introduced the method of relaxation to linear programming in order to obtain some information that could be useful in efficiently attacking the original, unrelaxed optimization problem. There are several possibilities how to use an optimal solution α of the relaxed problem instance to get a reasonable feasible (integral or Boolean) solution β of the original problem instance. One often successful attempt is to round α (possibly in a random way). We call this method **rounding** or **LP-rounding**. The goal is to execute rounding in such a way that one can ensure that the obtained rounded integral solution is a feasible solution of the original input instance and that the cost has not been changed too much in the rounding processes. Now, we are presenting an example of deterministic rounding. Some further applications of this approach are presented in the chapters 4 and 5 about approximation algorithms and randomized algorithms.

---

[29] In the number of constraints

Let us consider a special version SCP(k) of the set cover problem (SCP) for every integer $k \geq 2$. The minimization problem **SCP(k)** is a subproblem of SCP, where, for every input instance $(X, \mathcal{F})$, each element of X is contained in at most k sets of $\mathcal{F}$.

Exercise 3.7.4.1. Prove, that SCP(k) is NP-hard for every $k \geq 2$. $\square$

Let $(X, \mathcal{F})$, $X = \{a_1, \ldots, a_n\}$, $\mathcal{F} = \{S_1, S_2, \ldots S_m\}$ be an instance of SCP(k) for an integer $k \geq 2$. If one defines $Index(a_j) = \{d \in \{1, \ldots, m\} \mid a_j \in S_d\}$ for $j = 1, \ldots, n$, we know that $|Index(a_j)| \leq k$ for every $j = 1, \ldots, n$. The corresponding instance of 0/1-LP is to

$$\text{minimize} \sum_{i=1}^{m} x_i$$

under the constraints

$$\sum_{h \in Index(a_j)} x_h \geq 1 \text{ for } j = 1, \ldots, n,$$

$$x_i \in \{0, 1\} \text{ for } i = 1, \ldots, m,$$

where the meaning of x_i is

$$x_i = 1 \text{ iff } S_i \text{ is picked for the set cover.}$$

Following the schema of the method of relaxation to linear programming we obtain the following algorithm A_k for SCP(k) for every $k \in \mathbb{N} - \{0\}$.

Algorithm 3.7.4.2.   Input: An instance $(X, \mathcal{F})$, $X = \{a_1, \ldots, a_n\}$, $\mathcal{F} = \{S_1, \ldots, S_m\}$ of SCP(k).

Step 1: −Reduction″

Express $(X, \mathcal{F})$ as an instance $I(X, \mathcal{F})$ of 0/1-LP in the way described above.

Step 2: −Relaxation″

Relax $I(X, \mathcal{F})$ to an instance LP($X, \mathcal{F}$) of LP by relaxing $x_i \in \{0, 1\}$ to $0 \leq x_i \leq 1$ for every $i = 1, \ldots, m$.

Solve LP($X, \mathcal{F}$) by an algorithm for linear programming.

Let $\alpha = (\alpha_1, \alpha_2, \ldots, \alpha_m)$ [i.e., $x_i = \alpha_i$] be an optimal solution for LP($X, \mathcal{F}$).

Step 3: −Solving the original problem″

Set $\beta_i = 1$ iff $\alpha_i \geq 1/k$.

Output: $\beta = (\beta_1, \ldots, \beta_m)$.

First of all we observe that $\beta \in \mathcal{M}(X, \mathcal{F})$. Since $|Index(a_j)| \leq k$ for every $j \in \{1, \ldots, m\}$, the constraint

$$\sum_{h \in Index(a_j)} x_h \geq 1$$

ensures that there exists an $r \in Index(a_j)$ with $\alpha_r \geq 1/k$ (i.e., at least one of the $|Index(a_j)| \leq k$ variables in this inequality must take a value which is at least $1/|Index(a_j)|$). Following Step 3 of Algorithm 3.7.4.2, $\beta_r = 1$, and so a_j is covered in the solution $\beta = (\beta_1, \ldots, \beta_m)$. Since this holds for all $j = 1, \ldots, m$, β is a feasible solution for $LP(X, \mathcal{F})$.

Our rounding procedure sets $\beta_i = 1$ if $\alpha_i \geq 1/k$ and $\beta_i = 0$ if $\alpha_i < 1/k$. This implies that $\beta_i \leq k \cdot \alpha_i$ for all $i = 1, \ldots, m$, and hence

$$cost(\beta) \leq k \cdot cost(\alpha).$$

Thus, we have proved the following assertion.

Lemma 3.7.4.3. *Algorithm 3.7.4.2 is consistent for* SCP(k), *i.e., it computes a feasible solution* β *for every instance* $(X, \mathcal{F})$ *of* SCP(k), *and*

$$cost(\beta) \leq k \cdot Opt_{\mathrm{SCP}(k)}(X, \mathcal{F}).$$

One of the main advantages of the method of relaxation to linear programming is that the designed algorithms often work even if one generalizes the problem by assigning weights (costs) to the objects considered.

Exercise 3.7.4.4. Consider the weighted generalization of SCP(k), where each set $S \in \mathcal{F}$ has assigned a weight from $\mathbb{N} - \{0\}$. The cost of a solution (set cover) is the sum of the weights of the sets in this solution.

Prove, that if one applies algorithm 3.7.4.2 (without any change in the rounding strategy) to this generalized problem, then the assertion of Lemma 3.7.4.3 holds, too.   □

Another powerful concept related to the method of relaxation to linear programming is based on the so-called **linear-programming duality** (**LP-duality**). Originally, duality was used to prove that a result of an optimization algorithm is really an optimal solution, and the identification of a dual problem to a given problem typically resulted in the discovery of a polynomial algorithm.

The main idea of the LP-duality concept is to search for a **dual** problem $Dual(U)$ to a given optimization problem U. (U is called the **primal** problem when its dual problem can be identified.) If U is a minimization [maximization] problem, then $Dual(U)$ has to have the following properties:

(i) $Dual(U)$ is a maximization [minimization] problem that can be obtained by converting every instance I of U to an instance $Dual(I)$ of $Dual(U)$. There should exist an efficient algorithm for computing $Dual(I)$ for a given I.

(ii) For every instance I of U, the cost of any feasible solution for U is not smaller [not greater] than the cost of any feasible solution for $Dual(U)$, i.e.,

$$cost(\alpha) \geq cost(\beta) \ [cost(\alpha) \leq cost(\beta)]$$
for all $\alpha \in \mathcal{M}(I)$ and for all $\beta \in \mathcal{M}(Dual(I))$.

(iii) For every instance I of U,

$$Opt_U(I) = Opt_{Dual(U)}(Dual(I)).$$

Following Fig. 3.24 one can see how a dual problem can be useful in searching for a high-quality feasible solution of the primal problem. One starts with a feasible solution α for I and with a feasible solution β for $Dual(I)$ and tries to alternatingly improve both α and β. When $cost(\alpha) = cost(\beta)$, one knows that α is an optimal solution for the primal problem instance and that β is an optimal solution for the dual problem instance. One can also stop when $|cost(\alpha) - cost(\beta)|$ or $|cost(\alpha) - cost(\beta)|/cost(\alpha)$ is small, guaranteeing that the computed solution α is almost as good as an optimal one, i.e., that

$$|cost(\alpha) - Opt_U(I)| \le |cost(\alpha) - cost(\beta)|.$$

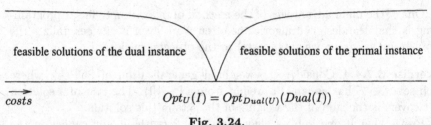

feasible solutions of the dual instance   feasible solutions of the primal instance

$\overrightarrow{costs}$ $Opt_U(I) = Opt_{Dual(U)}(Dual(I))$

Fig. 3.24.

A famous illustration of the concept of LP-duality is the Ford-Fulkerson algorithm for the maximum flow problem presented in Section 3.2. The primal problem is the maximum flow problem and its dual problem is the minimum network cut problem. A network is an input instance for both problems and so the requirement (i) for the dual problems is trivially satisfied. Lemma 3.2.3.7 ensures that the property (ii) of dual problems is satisfied and the **Max-Flow Min-Cut Theorem** (Theorem 3.2.3.12) shows that the property (iii) holds, too. Remember, that we have used the properties (ii) and (iii) to prove that the Ford-Fulkerson algorithm computes an optimal solution for both optimization problems, because this algorithm halts when it finds a flow and a cut whose costs are equal.

The next question of our interest is:

"Which optimization problems do admit dual problems?"

In what follows we show that for any linear programming problem we always have a dual problem, which is a linear programming problem also. This is of importance, because we know that many optimization problems can be expressed as (relaxed) linear programming problems.

First, we explain the idea of constructing $Dual(I)$ for any instance I of LP in the canonical form by the following specific instance I'. Let I' be

$$\text{to minimize } f(x_1, x_2, x_3) = 13x_1 + 2x_2 + 3x_3$$

under the constraints

$$2x_1 - x_2 + 3x_3 \geq 7$$
$$3x_1 + x_2 - x_3 \geq 4$$
$$x_1, x_2, x_3 \geq 0.$$

Our aim is to formulate a task of proving lower bounds on the cost of any feasible solution in $\mathcal{M}(I')$, i.e., lower bounds on $Opt_{LP}(I')$. We use the following notation for the constraints of I':

$$lrow_1 = 2x_1 - x_2 + 3x_3, \ rrow_1 = 7,$$
$$lrow_2 = 3x_1 + x_2 - x_3, \ rrow_2 = 4.$$

Consider $2 \cdot lrow_1 + 3 \cdot lrow_2 = 13x_1 + x_2 + 3x_3$. Since

$$13x_1 + x_2 + 3x_3 \leq 13x_1 + 2x_2 + 3x_3$$

we obtain

$$2lrow_1 + 3lrow_2 \leq f(x_1, x_2, x_3). \tag{3.12}$$

Since the constraints are $lrow_1 \geq rrow_1$ and $lrow_2 \geq rrow_2$, (3.12) implies

$$f(x_1, x_2, x_3) \geq 2lrow_1 + 3lrow_2 \geq 2rrow_1 + 3rrow_2 = 2 \cdot 7 + 3 \cdot 4 = 26.$$

Thus, $Opt_{LP}(I') \geq 26$, because $f(x_1, x_2, x_3) \geq 26$ for all values of x_1, x_2, x_3 that satisfy the constraints.

What do we do? We look for values y_1 and y_2 such that

$$f(x_1, x_2, x_3) \geq y_1 \cdot lrow_1 + y_2 \cdot lrow_2 \tag{3.13}$$

and then we know

$$f(x_1, x_2, x_3) \geq y_1 \cdot rrow_1 + y_2 \cdot rrow_2 = 7y_1 + 4y_2$$

and so we try to maximize

$$7y_1 + 4y_2. \tag{3.14}$$

The constraint (3.13) can be expressed (at the level of coefficients) as

$$2y_1 + 3y_2 \leq 13$$
$$-y_1 + y_2 \leq 2$$
$$3y_1 - y_2 \leq 3$$
$$y_1, y_2, y_3 \geq 0. \tag{3.15}$$

Thus, $Dual(I')$ specified by (3.14) and (3.15) can be efficiently derived from I' and it fulfills $Opt_{LP}(Dual(I')) \leq Opt_{LP}(I')$.

Clearly, this construction of $Dual(I)$ from I can be performed for any instance I of LP in the canonical form

$$\text{minimize } \sum_{i=1}^{n} c_i x_i \tag{3.16}$$

under the constraints

$$\sum_{i=1}^{n} a_{ji} x_i \geq b_j \quad \text{for } j = 1, \ldots, m \tag{3.17}$$
$$x_i \geq 0 \quad \text{for } i = 1, \ldots, n.$$

We define **Dual(I)** as to

$$\text{maximize } \sum_{j=1}^{m} b_j y_j \tag{3.18}$$

under the constraints

$$\sum_{j=1}^{m} a_{ji} y_j \leq c_j \quad \text{for } i = 1, \ldots, n \tag{3.19}$$
$$y_j \geq 0 \quad \text{for } j = 1, \ldots, m.$$

Exercise 3.7.4.5. Construct $Dual(I)$ for the following instance I of LP:

$$\text{minimize } 3x_1 + 7x_2$$

under the constraints

$$\begin{aligned}
x_1 - x_2 &\geq 4 \\
x_1 + x_2 &\geq 8 \\
x_1 + 2x_2 &\geq 6 \\
x_1, x_2 &\geq 0.
\end{aligned}$$

□

We see that the construction of $Dual(I)$ for a given instance I is straightforward and so our requirement (i) is satisfied. Following the idea of the construction, the requirement (ii) is satisfied, too. The following theorem provides a formal proof of this fact.

Theorem 3.7.4.6 (Weak LP-Duality Theorem). *Let I be an instance of LP in the canonical form as given by (3.16) and (3.17). Let $Dual(I)$ be the dual instance of I given by (3.18) and (3.19). Then, for every feasible solution $\alpha = (\alpha_1, \ldots, \alpha_n) \in \mathcal{M}(I)$ and every feasible solution $\beta = (\beta_1, \ldots, \beta_m) \in \mathcal{M}(Dual(I))$,*

$$cost(\alpha) = \sum_{i=1}^{n} c_i \alpha_i \geq \sum_{j=1}^{m} b_j \beta_j = cost(\beta),$$
$$i.e., \ Opt_{\mathrm{LP}}(I) \geq Opt_{\mathrm{LP}}(Dual(I)).$$

Proof. Since $\beta \in \mathcal{M}(Dual(I))$ is a feasible solution for $Dual(I)$,

$$\sum_{j=1}^{m} a_{ji}\beta_j \leq c_i \qquad \text{for } i = 1, \ldots, n. \tag{3.20}$$

Since $\alpha_i \geq 0$ for $i = 1, \ldots, n$, the constraints (3.20) imply

$$cost(\alpha) = \sum_{i=1}^{n} c_i\alpha_i \geq \sum_{i=1}^{n} \left(\sum_{j=1}^{m} a_{ji}\beta_j \right) \alpha_i. \tag{3.21}$$

Similarly, because $\alpha \in \mathcal{M}(I)$,

$$\sum_{i=1}^{n} a_{ji}\alpha_i \geq b_j \qquad \text{for } j = 1, \ldots, m$$

and so[30]

$$cost(\beta) = \sum_{j=1}^{m} b_j\beta_j \leq \sum_{j=1}^{m} \left(\sum_{i=1}^{n} a_{ji}\alpha_i \right) \beta_j. \tag{3.22}$$

Since

$$\sum_{i=1}^{n} \left(\sum_{j=1}^{m} a_{ji}\beta_j \right) \alpha_i = \sum_{j=1}^{m} \left(\sum_{i=1}^{n} a_{ji}\alpha_i \right) \beta_j,$$

the inequalities (3.21) and (3.22) imply

$$cost(\alpha) \geq cost(\beta).$$

$\square$

Corollary 3.7.4.7. *Let I and $Dual(I)$ be instances of* LP *as described in Theorem 3.7.4.6. If $\alpha \in \mathcal{M}(I)$, $\beta \in \mathcal{M}(Dual(I))$, and $cost(\alpha) = cost(\beta)$, then α is an optimal solution for I and β is an optimal solution for $Dual(I)$.*

The following assertion shows, that $Dual(I)$ satisfies the requirement (iii) for being a dual problem of I, too. We omit the proof of this theorem, because it is technical and is not important from the algorithm design point of view.

Theorem 3.7.4.8 (LP-Duality Theorem). *Let I and $Dual(I)$ be instances of* LP *as described in Theorem 3.7.4.6. Then*

$$Opt_{\mathrm{LP}}(I) = Opt_{\mathrm{LP}}(Dual(I)).$$

---

[30] Note that $\beta_j \geq 0$ for $j = 1, \ldots, m$.

Since every instance I of LP can be transformed into the canonical form, our construction of $Dual(I)$ can be applied to any form of LP. But in the case of the **standard maximization form** given by

$$\text{maximize } f(x_1, \ldots, x_n) = c_1 x_1 + c_2 x_2 + \ldots + c_n x_n \qquad (3.23)$$

under the constraints

$$\begin{aligned} \sum_{i=1}^{n} a_{ji} x_i \leq b_j & \quad \text{for } j = 1, \ldots, m, \\ x_i \geq 0 & \quad \text{for } i = 1, \ldots, n, \end{aligned} \qquad (3.24)$$

one can directly construct the dual instance by an analogous strategy as for the canonical form. The dual problem is to prove upper bounds on the objective function $f(x_1, \ldots, x_n)$. If

$$lrow_k = \sum_{i=1}^{n} a_{ki} x_i \quad \text{and} \quad rrow_k = b_k,$$

then one searches for $y_1, y_2, \ldots, y_m$ such that

$$y_1 \cdot lrow_1 + y_2 \cdot lrow_2 + \ldots + y_m \cdot lrow_m \geq f(x_1, \ldots x_n) = \sum_{i=1}^{n} c_i x_i. (3.25)$$

Clearly, if (3.25) holds, then $\sum_{j=1}^{m} y_j \cdot rrow_j = \sum_{j=1}^{m} y_j b_j$ is an upper bound on $f(x_1, \ldots, x_n)$ for all values of $x_1, \ldots, x_n$ satisfying the constraints (3.24).

Thus, we can define, for every instance I of LP in the standard maximization form defined by (3.23) and (3.24), the dual instance $Dual(I)$ to I as to

$$\text{minimize } b_1 y_1 + b_2 y_2 + \ldots + b_m y_m \qquad (3.26)$$

under the constraints

$$\begin{aligned} \sum_{j=1}^{m} a_{ji} y_j \geq c_i & \quad \text{for } i = 1, \ldots, n, \\ y_j \geq 0 & \quad \text{for } j = 1, \ldots, m. \end{aligned} \qquad (3.27)$$

Exercise 3.7.4.9. Prove another version of the Weak LP-Duality Theorem, when I is in the standard maximization form given by (3.23) and (3.24), and $Dual(I)$ is given by (3.26) and (3.27). $\qquad \square$

Exercise 3.7.4.10. Let I be an instance of LP in the canonical (standard maximization) form. Prove, that

$$I = Dual(Dual(I)).$$

$\qquad \square$

Exercise 3.7.4.11. [*] Consider the maximum flow problem. Let $H = ((V, E), c, A, s, t)$ be a network. Let, for every $e \in H$, x_e be a real-valued variable such that $x_e = f(e)$, i.e., the value of x_e is the flow of e. Express the instance H of the maximum flow problem as an instance I of LP over variables x_e, for all $e \in E$. Determine $Dual(I)$ and explain its relation to the minimum network cut problem. $\qquad \square$

To illustrate the usefulness of the concept of LP-duality for combinatorial optimization problems, consider the maximum matching problem (MMP). MMP is the problem of finding a matching of maximal size in a given graph G. Let $G = (V, E)$ be an instance of this problem. Denote, for every $v \in V$,

$$Inc(v) = \{e \in E \mid e = \{v, u\} \text{ for some } u \in V\}$$

and consider, for every $e \in E$, the Boolean variable x_e with the meaning

$$x_e = 1 \Leftrightarrow e \text{ is in the matching.}$$

The relaxed version $I(G)$ of the instance G of the maximum matching problem can be expressed as

$$\text{maximize } \sum_{e \in E} x_e \tag{3.28}$$

under the constraints

$$\sum_{e \in Inc(v)} x_e \leq 1 \quad \text{for every } v \in V, \tag{3.29}$$

$$x_e \geq 0. \tag{3.30}$$

Obviously, (3.28), (3.29), and (3.30) is the description of an instance $I(G)$ of LP in the standard maximization form. Thus, following our construction of $Dual(I(G))$ we have $|V|$ variables y_v for every $v \in V$ and $Dual(I(G))$ is to

$$\text{minimize } \sum_{v \in V} y_v \tag{3.31}$$

under the constraints

$$y_u + y_w \geq 1 \quad \text{for every } \{u, w\} \in E \tag{3.32}$$
$$y_v \geq 0. \tag{3.33}$$

If one requires $y_v \in \{0, 1\}$ for every $v \in V$ instead of (3.33), then these constraints together with (3.31) and (3.32) describe an instance of IP that expresses the instance G of the minimum vertex cover problem[31]. Thus, the relaxed maximum matching problem is dual to the relaxed vertex cover problem.

Exercise 3.7.4.12. Prove, that if G is a bipartite graph, then

(i) any optimal solution to $I(G)$ is a Boolean solution, i.e.,
   $Opt_{LP}(I(G)) = Opt_{MMP}(G)$, and

---

[31] Observe, that the meaning of $y_v \in \{0, 1\}$ is that $y_v = 1$ iff v was taken into the vertex cover.

(ii) any optimal solution to $Dual(I(G))$ is a Boolean solution, i.e.,
$$Opt_{LP}(Dual(I(G))) = Opt_{VCP}(G).$$ □

The Max-Flow Min-Cut Theorem (Theorem 3.2.3.12) and Exercise 3.7.4.12 show that the concept of LP-duality can be useful in designing efficient algorithms for problems in PO, or even in proving that some optimization problems (whose membership in PO was not known) are in PO. What about NP-hard optimization problems? Can the concept of LP-duality help, too? The answer is positive in the sense that LP-duality similarly as rounding can help to design efficient algorithms that compute high-quality feasible solutions. Here, we present the basic idea of the so-called **primal-dual method**, which is based on LP-duality and which has already found many applications.

Primal-dual scheme

Input: An input instance G of an optimization problem U.

Step 1:   Express G as an instance $I(G)$ of IP and relax $I(G)$ to an instance $I_{rel}(G)$ of LP.

Step 2:   Construct $Dual(I_{rel}(G))$ as an instance of LP.

Step 3:   Try to solve at once $Dual(I_{rel}(G))$ as an instance of LP and $I(G)$ as an instance of IP.

Output: A feasible solution α for $I(G)$ and $Opt_{LP}(Dual(I_{rel}(G)))$.

The situation is depicted in Fig. 3.25. On one side we try to find a nice solution for the original (NP-hard) optimization problem U and on the other side we solve the dual program of the relaxed version of U. Then the difference

$$|cost(\alpha) - Opt_{LP}(Dual(I_{rel}(G)))|$$

is an upper bound on

$$|cost(\alpha) - Opt_U(G)|.$$

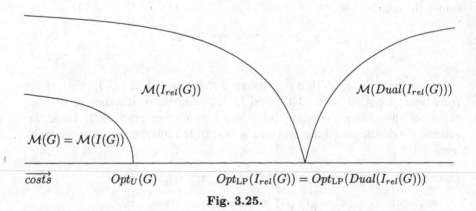

Fig. 3.25.

We have already explained how to execute Step 1 and Step 2 of the primal-dual schema. But the kernel of the primal-dual method is the execution of

Step 3. We know that $Dual(I_{rel}(G))$ as an instance of LP can be solved in polynomial time. The idea is to find a reasonable mechanism, how to use a feasible solution $\beta \in \mathcal{M}(Dual(I_{rel}(G)))$ to determine a "related" feasible solution $\alpha_\beta \in \mathcal{M}(I(G)) = \mathcal{M}(G)$.

If one can do it in such a way that $cost(\alpha_\beta)$ and $cost(\beta)$ are reasonably related, the primal-dual method can guarantee high-quality solutions to the original instances of the hard problem U. The following theorem shows us the relationship between the optimal solutions for $I_{rel}(G)$ and $Dual(I_{rel}(G))$. This relation will serve as the base for determining α_β from β.

Theorem 3.7.4.13 (Complementary slackness conditions). *Let I be an instance of LP in the canonical form as given by (3.16) and (3.17). Let $Dual(I)$ be the dual instance of I given by (3.18) and (3.19). Let $\alpha = \{\alpha_1, \ldots, \alpha_n\}$ be a feasible solution for I and let $\beta = \{\beta_1, \ldots, \beta_m\}$ be a feasible solution for $Dual(I)$.*

Then both α and β are optimal if and only if the following conditions are satisfied:

Primal complementary slackness conditions:

For each $i \in \{1, 2, \ldots, n\}$, either $\alpha_i = 0$ or $\sum_{j=1}^{m} a_{ji}\beta_j = c_i$

Dual complementary slackness conditions:

For each $j \in \{1, 2, \ldots, m\}$, either $\beta_j = 0$ or $\sum_{i=1}^{n} a_{ji}\alpha_i = b_j$.

Proof. Let α be optimal for I and let β be optimal for $Dual(I)$, i.e.,

$$cost(\alpha) = \sum_{i=1}^{n} c_i \alpha_i = \sum_{j=1}^{m} b_j \beta_j = cost(\beta).$$

Following the proof of the Weak LP-Duality Theorem, the inequality (3.21) becomes an equality, i.e.,

$$cost(\alpha) = \sum_{i=1}^{n} c_i \alpha_i = \sum_{i=1}^{n} \left(\sum_{j=1}^{m} a_{ji}\beta_j \right) \alpha_i. \tag{3.34}$$

Because of the constraints (3.20), the fact

$$\sum_{j=1}^{m} a_{ji}\beta_j < c_i$$

and the equality (3.34) force $\alpha_i = 0$.

Similarly, the inequality (3.22) becomes an equality, too, and so the constraints

$$\sum_{i=1}^{n} a_{ji}\alpha_i \geq b_j \qquad \text{for } j = 1, \ldots, m$$

imply the dual complementary slackness conditions.

The opposite direction is straightforward. If

$$\sum_{j=1}^{m} a_{ji}\beta_j = c_i \qquad \text{for all } i \in \{1,\ldots,n\} \text{ with } \alpha_i \neq 0$$

then the equality (3.34) holds. Similarly, the dual complementary conditions imply

$$\bar{cost}(\beta) = \sum_{j=1}^{m} b_j\beta_j = \sum_{j=1}^{m}\left(\sum_{i=1}^{n} a_{ji}\alpha_i\right)\beta_j. \qquad (3.35)$$

The equalities (3.34) and (3.35) imply $cost(\alpha) = cost(\beta)$, and thus following the Weak LP-Duality Theorem, both α and β are optimal. $\square$

The following consequence of Theorem 3.7.4.13 is useful for algorithm design. Let α be optimal for I and let β be optimal for $Dual(I)$. Then, for every $i \in \{1,\ldots,n\}$,

$$\begin{array}{l} \alpha_i > 0 \text{ implies } \sum_{j=1}^{m} a_{ji}\beta_j = c_i, \text{ and} \\ \sum_{j=1}^{m} a_{ji}\beta_j < c_i \text{ implies } \alpha_i = 0, \end{array} \qquad (3.36)$$

and, for every $j \in \{1,2,\ldots,m\}$,

$$\begin{array}{l} \beta_j > 0 \text{ implies } \sum_{i=1}^{n} a_{ji}\alpha_i = b_j, \text{ and} \\ \sum_{i=1}^{n} a_{ji}\alpha_i < b_j \text{ implies } \beta_j = 0. \end{array} \qquad (3.37)$$

Let us illustrate the usefulness of the conditions (3.36) and (3.37) by the design of an algorithm for the minimum vertex cover problem. For a given graph $G = (V,E)$, the corresponding relaxed instance $I_{rel}(G)$ of LP is given by (3.31), (3.32), and (3.33). The corresponding dual instance $Dual(I_{rel}(G))$ is given by (3.28), (3.29), and (3.30). Remember that the variables of $I_{rel}(G)$ are $y_{v_1}, y_{v_2}, \ldots, y_{v_n}$, if $V = \{v_1,\ldots,v_n\}$, with the meaning

$$y_{v_i} = 1 \Leftrightarrow v_i \text{ is in the vertex cover.}$$

The variables of $Dual(I_{rel}(G))$ are $x_{e_1}, x_{e_2}, \ldots, x_{e_m}$, if $E = \{e_1,\ldots,e_m\}$. Their meaning is

$$x_{e_j} = 1 \Leftrightarrow e_j \text{ is in the matching.}$$

We design the following algorithm for MIN-VCP.

Algorithm 3.7.4.14 (Primal-Dual(MIN-VCP)).

Input: A graph $G = (V,E)$, $V = \{v_1,\ldots,v_n\}$, $E = \{e_1,\ldots,e_m\}$ for some $n \in \mathbb{N} - \{0\}$, $m \in \mathbb{N}$.

Step 1:   Relax G to the instance $I_{rel}(G)$ of LP given by (3.31), (3.32), and (3.33).

Step 2:   Construct the dual instance $Dual(I_{rel}(G))$ to the primal instance $I_{rel}(G)$.

Step 3:   Solve $Dual(I_{rel}(G))$.
Let $\beta = (\beta_1, \beta_2, \ldots, \beta_m)$ be an optimal solution for $Dual(I_{rel}(G))$.

Step 4:   **for** $i := 1$ **to** n **do**
   if $\sum_{e_r \in Inc(v_i)} \beta_r < 1$
   (i.e., the i-th constraint of (3.29) is not sharp)
   then $\alpha_i := 0$
   else $\alpha_i := 1$

Output: $\alpha = (\alpha_1, \alpha_2, \ldots, \alpha_n)$.

Lemma 3.7.4.15. *The output α of Primal-Dual*(MIN-VCP) *is a feasible solution for G, and*

$$cost(\alpha) \leq 2 \cdot Opt_{\text{MIN-VCP}}(G).$$

Proof. First, we prove that $\alpha \in \mathcal{M}(G)$. Let us assume that there is an edge $e_j = \{v_r, v_s\}$ that is not covered by α, i.e., that $\alpha_r = \alpha_s = 0$. Following Step 4 of Algorithm 3.7.4.14, this is possible only if

$$\sum_{e_k \in Inc(v_r)} \beta_k < 1 \quad \text{and} \quad \sum_{e_h \in Inc(v_s)} \beta_h < 1, \tag{3.38}$$

where $e_j \in Inc(v_r) \cap Inc(v_s)$. Since β_j occurs only in the r-th constraint and in the s-th constraint of (3.29), (3.38) implies that there is a possibility to increase β_j without breaking the constraints (3.29). But this contradicts the assumption that β is optimal (maximal) for $Dual(I_{rel}(G))$. Thus, α is a feasible solution for the instance G of MIN-VCP.

To show that $cost(\alpha) \leq 2 \cdot Opt_{\text{MIN-VCP}}(G)$ it is sufficient[32] to prove $cost(\alpha) \leq 2 \cdot cost(\beta)$.

Since $\alpha_i = 1$ iff $\sum_{e_r \in Inc(v_i)} \beta_r = 1$, we have

$$\alpha_i = \sum_{e_r \in Inc(v_i)} \beta_r \quad \text{for all } i \text{ with } \alpha_i = 1. \tag{3.39}$$

Thus,

$$cost(\alpha) = \sum_{i=1}^{n} \alpha_i = \sum_{\substack{i \\ \alpha_i = 1}} \alpha_i \underset{(3.39)}{=} \sum_{\substack{i \\ \alpha_i = 1}} \left(\sum_{e_r \in Inc(v_i)} \beta_r \right). \tag{3.40}$$

Since every x_{e_r} appears in at most two constraints, every β_r appears at most twice in

---

[32] because of the Weak LP-Duality Theorem (see also Fig. 3.25)

$$\sum_{i=1}^{n} \left(\sum_{e_r \in Inc(v_i)} \beta_r \right).$$

Thus,

$$\sum_{\substack{i \\ \alpha_i = 1}} \left(\sum_{e_r \in Inc(v_i)} \beta_r \right) \leq 2 \cdot \sum_{j=1}^{m} \beta_j. \tag{3.41}$$

The equality (3.40) and the inequality (3.41) imply

$$cost(\alpha) \leq 2 \cdot \sum_{j=1}^{m} \beta_j = cost(\beta).$$

$\square$

Exercise 3.7.4.16. Use the idea of Algorithm 3.7.4.14 to design an algorithm for SCP(k) for any $k \geq 3$. $\square$

The complexity of Algorithm 3.7.4.14 is related to the complexity of solving $Dual(I_{rel}(G))$, which may be time consuming. In what follows we give a faster $O(n + m)$ algorithm with the same guarantee on the quality of all computed solutions. The idea is to start with a feasible solution $\beta = (0, \ldots, 0)$ for $Dual(I_{rel}(G))$ and with an $\alpha = (0, \ldots, 0) \notin \mathcal{M}(I(G))$. In at most n iteration steps we alternatingly improve β and change α, until α becomes a feasible solution in $\mathcal{M}(I(G)) = \mathcal{M}(G)$. To update α and β in each iteration step we use the forms (3.36) and (3.37) of the slackness conditions.

Algorithm 3.7.4.17 (Primal-Dual(MIN-VCP)-2).

Input: A graph $G = (V, E)$, $V = \{v_1, \ldots, v_n\}$, $E = \{e_1, \ldots, e_m\}$ for some $n \in \mathbb{N} - \{0\}, m \in \mathbb{N}$.

Step 1:   Express G as an instance $I(G)$ of 0/1-LP.

Step 2:   Take $I_{rel}(G)$ as the relaxed LP instance of $I(G)$ and construct $Dual(I_{rel}(G))$.

Step 3:   **for** $j = 1$ **to** m **do** $\beta_i := 0$;
  for $i = 1$ **to** n **do**
  begin $\alpha_i := 0$;
    $gap[i] := 1$ $\{gap[i]$ has the meaning $1 - \sum_{e_d \in Inc(v_i)} \beta_d$
          during the whole computation.$\}$
  end

Step 4:   **while** there is an unsatisfied constraint of (3.32),
        i.e., $\alpha_r + \alpha_s = 0 < 1$ for some $e_j = \{v_r, v_s\} \in E$
      do
        if $gap[r] \leq gap[s]$ **then**

begin $\alpha_r := 1$;

$\quad \beta_j := \beta_j + gap[r]$;

$\quad gap[s] := gap[s] - gap[r]$

$\quad$–because β_j is in $\sum\limits_{e_d \in Inc(v_s)} \beta_d$ and $gap[s] = 1 - \sum\limits_{e_d \in Inc(v_s)} \beta_d"$;

$\quad$ $gap[r] := 0$

$\quad$–because $gap[r]$ was $1 - \sum\limits_{e_c \in Inc(v_r)} \beta_c$, $e_j \in Inc(v_r)$, and

$\quad$ β_j was increased exactly by $gap[r]$, and so the r-th

$\quad$ constraint of (3.29) is satisfied"

end

else begin $\alpha_s := 1$;

$\quad \beta_j := \beta_j + gap[s]$;

$\quad gap[r] := gap[r] - gap[s]$;

$\quad gap[s] := 0$

$\quad$–the s-th constraint of (3.29) is satisfied"

end

Output: $\alpha = (\alpha_1, \ldots, \alpha_n)$

Lemma 3.7.4.18. *The output α of Primal-Dual(MIN-VCP)-2 is a feasible solution for G with*

$$cost(\alpha) \leq 2 \cdot Opt_{\text{MIN-VCP}}(G),$$

and Primal-Dual(MIN-VCP)-2 works in time $O(n+m)$ for any G with n nodes and m edges.

Proof. First, we show that α is a feasible solution for G. Following the condition of the while loop in Step 4, we have that the algorithm halts only if α satisfies all constraints of (3.32). Thus, if Primal-Dual(MIN-VCP)-2 terminates, then α is a feasible solution. But the algorithm must terminate because each run of the while loop increases the number of satisfied constraints of (3.32) by at least one. Since $\alpha \in \{0,1\}^n$, α is a feasible solution for G.

To prove $cost(\alpha) \leq 2 \cdot Opt_{\text{MIN-VCP}}(G)$ it is sufficient to show that $cost(\alpha) \leq 2 \cdot cost(\beta)$. Since $\alpha_i = 1$ implies $\sum\limits_{e_d \in Inc(v_i)} \beta_d = 1$, we have again

$$\alpha_i = \sum_{e_d \in Inc(v_i)} \beta_i \quad \text{for all } i \text{ with } \alpha_i = 1$$

and the argumentation leading to $cost(\alpha) \leq 2 \cdot cost(\beta)$ is exactly the same as in the proof of Lemma 3.7.4.15.

Now, let us analyze the time complexity of Primal-Dual(MIN-VCP)-2. Steps 1,2 and 3 can be executed in $O(n + m)$ time. The while loop runs at most m times (the number of constraints in (3.32)). Using suitable data structures, each run can be performed in $O(1)$ steps and so Step 4 can be executed in $O(m)$ steps. Thus, the time complexity of Primal-Dual(MIN-VCP)-2 is in $O(n + m)$.   □

Exercise 3.7.4.19. Use the concept of Primal-Dual(MIN-VCP)-2 to design an algorithm for SCP(k) for any integer $k \geq 3$.   □

Keywords introduced in Section 3.7

standard form of linear programming, canonical form of LP, surplus variable, standard inequality form of LP, slack variable, general form of LP, hitting set problem, hyperplane, halfspace, polytope, face of a polytope, facet of a polytope, edge and vertex of a polytope, simplex algorithm, rounding, LP-duality, dual problem, primal problem, primal-dual method, primal complementary slackness conditions, dual complementary slackness conditions

Summary of Section 3.7

The problem of linear programming is solvable in polynomial time, while 0/1-linear programming (0/1-LP) and integer linear programming (IP) are NP-hard. This difference in hardness is caused by the restrictions of 0/1-LP and IP on the domains over which the sets of feasible solutions are defined. The method of relaxation to linear programming consists of the following three steps.

1. *Reduction*
   Express instances of a given optimization problem as instances of 0/1-LP or IP.
2. *Relaxation*
   Relax the requirement of finding an optimal solution for 0/1-LP or IP to finding an optimal solution for LP and solve this LP problem instance.
3. *Solving the original problem*
   Use the optimal solution for LP to search for a high-quality solution for the original problem.

   Many optimization problems (for instance, KP, WEIGHT-VCP) can be "naturally expressed" as 0/1-LP (IP) problems. Naturally expressed means that there is a one-to-one correspondence between the feasible solutions of the original optimization problem and the feasible solutions of the resulting 0/1-LP or IP.

   LP can be solved efficiently. The most famous algorithm for LP is the SIMPLEX ALGORITHM. It is a local search algorithm where one restricts the set of feasible solutions to the vertices of the polytope (determined by the constraints) and the neighborhood is defined by considering two vertices to be neighbors if they are connected by an edge of the polytope. Then, the SIMPLEX ALGORITHM can be viewed as an iterative movement from a vertex of the polytope to another vertex of the polytope via the edges of the polytope. Since the polytope is convex, the considered neighborhood is exact and so the SIMPLEX ALGORITHM always computes an optimal solution. The worst case complexity of the SIMPLEX ALGORITHM is exponential, because there exist input instances of LP where a path of iterative improvements from a vertex of the polytope to an optimum has an exponential length. But such extreme situations appear very rarely and so the SIMPLEX ALGORITHM is usually very fast in practical applications.

There are two basic techniques that use optimal solutions of relaxed input instances in order to solve the original problem instance. The method of rounding simply takes an optimal solution of the relaxed problem instance and rounds its values to integers or Boolean values in such a way that the resulting solution is feasible for the original problem and its cost does not differ too much from the cost of the optimal solution for the relaxed problem instance. A more sophisticated method is the primal-dual method based on LP-duality. For each instance of LP (called primal problem instance) there exists a dual problem instance with the following properties:

(i) if the primal instance is a maximization [minimization] problem, then the dual instance is a minimization [maximization] problem, and

(ii) the costs of the optimal solutions for the primal problem and of the optimal solutions for the dual problem are the same.

Under the primal-dual method, an integral solution for the primal problem and a feasible solution for the relaxed dual problem can be constructed iteratively. The difference between the costs of these two iteratively computed solutions is an upper bound on the difference between the cost of the integral solution for the original problem and the optimal cost.

3.8 Bibliographical Remarks

The concept of pseudo-polynomial-time algorithms for integer-valued problems is one of the first approaches to attack NP-hard problems. The ideas presented in Section 3.2 (including the concept of strong NP-hardness) are from Garey and Johnson [GJ78]. The pseudo-polynomial-time dynamic programming algorithm for the knapsack problem was used by Ibarra and Kim [IK75] to design polynomial-time algorithms producing feasible solutions of arbitrarily high quality for KP (such algorithms are called polynomial-time approximation schemes and will be introduced and discussed in Chapter 4). For more examples of solving knapsack problems using dynamic programming, see Hu [Hu82]. The pseudo-polynomial-time algorithm for the maximum flow problem is due to Ford and Fulkerson [FF62].

An excellent and exhaustive overview about the concept of parameterized complexity is given in Downey and Fellows [DF99], who introduced the complexity theory for this concept in [DF92, DF95a, DF95b]. They call a problem **parameterized** if its inputs can be divided into two parts and **fixed-parameter tractable** if it can be solved by an algorithm that is polynomial in the size of the first input part. The second part of the input is called a parameter and there is no restriction on the growth of the complexity with this

parameter. In Section 3.3 we generalized this concept by allowing the parameter to be an arbitrary, efficiently computable characteristic of the input and
not a local part of the input only. In this way the concept of pseudo-polynomial
algorithms becomes a special case of the parameterized complexity concept
because $Val(x)$ (or $Max\text{-}Int(x)$) can be viewed as a parameter (as a parameterization, when following the terminology of this book).

Perhaps the vertex cover problem provides the nicest known history of developing parameterized polynomial-time algorithms for a problem. Remember
that (G, k) is an instance of this decision problem, and k is considered to be the
parameter. First, Fellows and Langston [FL88] designed an $O(f(k) \cdot n^3)$ algorithm, and Johnson [Joh87] designed an $O(f(k) \cdot n^2)$ algorithm for VC. In 1988
Fellows [Fel88] gave the $O(2^k \cdot n)$ algorithm presented as the second algorithm
for VC in the Section 3.3.2. In 1989 Buss [Bu89] described an $O(kn + 2^k \cdot k^{2k+2})$
algorithm. Combining the ideas of [Fel88, Bu89], Balasubramanian, Downey,
Fellows, and Raman [BDF$^+$92] developed an $O(kn + 2^k \cdot k^2)$ algorithm for VC.
In 1993 Papadimitriou and Yannakakis [PY93] presented an algorithm for VC
that runs in polynomial time whenever k is in $O(\log n)$. The same result has
been rediscovered by Bonet, Steel, Warnow, and Yooseph [BSW$^+$98]. Finally,
combining and refining previous techniques, Balasubramanian, Fellows, and
Raman designed an $O(kn + (4/3)^k k^2)$ algorithm for VC.

Backtracking and branch-and-bound belong among the oldest fundamental algorithm design techniques. For some early sources on backtracking one
can look for Walker [Wal60] and Golomb and Baumert [GB65]. The techniques for analyzing the complexity of backtrack programs were first proposed
by Hall and Knuth [HK65] and later explained in greater detail by Knuth
[Knu75]. Some early references on branch-and-bound are Eastman [Eas58],
Little, Murtly, Sweeny, and Karel [LMS$^+$63], Ignall and Schrage [IS65], and
Lawler and Wood [LW66]. Branch-and-bound algorithms for TSP have been
proposed by many researchers. See, for instance, some of the early papers
by Bellmore and Nemhauser [BN68], Garfinkel [Gar73], and Held and Karp
[HK70, HK71]. An involved discussion about the branch-and-bound method
is given in Horowitz and Sahni [HS78].

Monien and Speckenmeyer [MS79, MS85] gave the first improvement of
the naive $O(|F| \cdot 2^n)$ algorithm for the satisfiability problems. They designed
algorithms of the worst case complexity $O(2^{(1-\varepsilon_k)n})$ with $\varepsilon_k > 0$ for kSAT for
every $k \geq 3$. The $O(|F| \cdot 1.84^n)$ algorithms for 3SAT presented in Section 3.5.2
is a simplification of the algorithm in [MS79], where an $O(|F| \cdot 1.62^n)$ algorithm
for 3SAT is presented. This result was also independently achieved by Dantsin
[Dan82]. There are several papers improving these early results. For instance,
Kullman [Kul99] designed an $O(2^{0.589n})$ algorithm for 3SAT, Paturi, Pudlák,
and Zane [PPZ97] proposed an $O(2^{(1-1/k)n})$ randomized algorithm for kSAT,
Schöning [Schö99] developed an $O\left(\frac{2k}{k+1}\right)^n$ randomized algorithm for kSAT,
and finally Paturi, Pudlák, Saks, and Zane [PPSZ98] presented a randomized
algorithm for SAT that works in time $O(2^{0.448n})$ for 3SAT. In all the results

mentioned above n is the number of variables in the given formula. There also exist several papers measuring the complexity in the length of the formula or in the number of clauses (see, for instance, Hirsch [Hir98, Hir00], Niedermeier and Rossmanith [NR99], and Gramm and Niedermeier [GN00]). The concept of lowering the worst case exponential complexity was also used for problems different from the satisfiability problems. Examples can be found in Beigel and Eppstein [BE95], Bach, Condon, Glaser, and Tangway [BCG+98], and Gramm and Niedermeier [GN00].

Similarly to backtracking and branch-and-bound, local search is one of the oldest methods of attacking NP-hard problems. An early description of local search can be found by Dunham, Fridshal, and Fridshal, and North [DFF+61]. An excellent in-depth source for the study of local search is the book *"Local Search in Combinatorial Optimization"*, edited by Aarts and Lenstra [AL97]. To mention at least some chapters of this book, Aarts and Lenstra [AL97a] provide in [AL97] an excellent overview of the basic concepts in this area, Yannakakis [Yan97] presents a complexity theory for local search, and Johnson and McGeoch [JG97] present the experience with local search applications for TSP. Several further chapters are devoted to heuristics that are based on local search. Some of them will be discussed later in Chapter 6.

Here we presented some of the first edge-exchange neighborhoods introduced by Bock [Boc58a], Croes [Cro58], Lin [Lin65], and Reiter and Sherman [RS65]. The concept of variable-depth search algorithms (Section 3.6.2) was first introduced by Kernighan and Lin [KL70, LK73]. The complexity considerations related to the nonexistence of any exact polynomial-time searchable neighborhood for TSP follow the paper of Papadimitriou and Steiglitz [PS77] (mainly the presentation in [PS82]). Papadimitriou and Steiglitz [PS78] also constructed examples of TSP instances with a unique optimal solution and exponentially many second-best solutions that are locally optimal with respect to k-exchange neighborhoods for $k < \frac{3}{8} \cdot n$, while their costs are arbitrarily far from the cost of the optimal solution (see Section 3.6.3).

Another fundamental complexity question related to local search is to estimate, for an optimization problem U and a given neighborhood for U, whether a polynomial number of iterative improvements is always sufficient to reach a local optimum with respect to this neighborhood. One can view this as a relaxation of the requirement of finding an optimal solution to finding at least a local optimum according to a given neighborhood. Lueker [Lue75] constructed instances of TSP with an initial feasible solution such that an exponential number of improvements is necessary for the simple 2-*Exchange* neighborhood. There are also other problems and neighborhoods with this bad behavior (see, for instance, Rödl and Tovey [RT87]). In 1988, Johnson, Papadimitriou, and Yannakakis [JPY88] developed a complexity theory for the local search for local optimization problems by defining the class of polynomial-time local search problems and some kind of completeness in this class. An involved survey and transparent explanation of this theory is given in Yannakakis [Yan97].

Relaxation to linear programming is one of the most successful methods for the design of approximation algorithms. In Section 3.7 we focused on the use of this method in order to get a bound on the cost of optimal solutions. The simplex algorithm was invented in 1947 by Dantzig [Dan49] and we highly recommend his comprehensive text [Dan63]. There are many monographs and textbooks devoted to linear programming. Here, we followed the excellent textbook of Papadimitriou and Steiglitz [PS82]. Examples of input instances of LP that may cause an exponential-time complexity of the simplex method (i.e., the proof that the simplex method is not a polynomial-time algorithm) were presented by Klee and Minty [KM72]. (Note that the dichotomy between polynomial-time computations and exponential-time computations was first considered by John von Neumann [vNe53] and the term polynomial-time computation was first explicitly used by Cobham [Cob64] and Edmonds [Edm65].) Other pathological input instances with exponentially many pivots were given in Jeroslow [Jer73] and Zadeh [Zad73, Zad73a]. The long stated open problem whether LP can be solved in polynomial time was solved in 1979 by Chachian [Cha79] who developed the so-called ellipsoid algorithm. The existence of a polynomial-time algorithm for linear programming has essentially influenced the development in combinatorial optimization (see, for instance, Grötschel, Lovász, and Schrijver [GLS81], Karp and Papadimitriou [KP80], Lovász [Lov79, Lov80], and Nešetřil and Poljak [NP80]).

In Section 3.7.4 we considered some simple, transparent applications of linear programming in combinatorial optimization, where an optimization problem is reduced and relaxed to a fundamental form of LP. We presented the method of rounding, which is based on rounding an optimal solution for the relaxed problem instance in order to obtain a feasible "high-quality" solution for the original problem instance. A more sophisticated approach presented here is the primal-dual method, which is based on LP-duality. The primal-dual method was first described by Dantzig, Ford, and Fulkerson [DFF56] (who were inspired by Kuhn [Kuh55]). The main advantage of this method is that it allows a weighted optimization problem to be reduced to a purely combinatorial unweighted problem. In this way it provides a general methodology for solving optimization problems (for instance, for designing approximation algorithms for NP-hard problems or polynomial-time algorithms for the shortest path problem (Dijkstra [Dij59]) and the maximum flow problem (Ford and Fulkerson [FF62])). A very successful generalization of the method of reduction and relaxation to linear programming is the relaxation to semidefinite programming. The power of this method was discovered in Goemans and Williamson [GW95], who applied it to several NP-hard problems. A very nice survey on this method is given by Hofmeister and Hühne [HH98] in [MPS98].

4

Approximation Algorithms

"Progress is brought about by those who are corageous enough to change all the time everything that is not in order."

BERNARD BOLZANO

4.1 Introduction

Currently, approximation algorithms seem to be the most successful approach for solving hard optimization problems. Immediately after introducing NP-hardness (completeness) as a concept for proving the intractability of computing problems, the following question was posed:

"If an optimization problem does not admit any efficient[1] algorithm computing an optimal solution, is there a possibility to efficiently compute at least an approximation of the optimal solution?"

Already in the middle of the 1970s a positive answer was given for several optimization problems. It is a fascinating effect if one can jump from exponential complexity (a huge inevitable amount of physical work for inputs of a realistic size) to polynomial complexity (a tractable amount of physical work) due to a small change in the requirements – instead of an exact optimal solution one demands a solution whose cost differs from the cost of an optimal solution by at most $\varepsilon\%$ of the cost of an optimal solution for some $\varepsilon > 0$. This effect is very strong, especially, if one considers problems for which these approximation algorithms are of large practical importance. It should not be surprising when one currently considers an optimization problem to be tractable if there exists a polynomial-time approximation algorithm that solves it with a reasonable relative error.

The goals of this chapter are the following:

---

[1] Polynomial time

(i) To give the fundamentals of the concept of the approximation of optimization problems, including the classification of the optimization problems according to their polynomial-time approximability.

(ii) To present some transparent examples of approximation algorithms in order to show successful techniques for the design of efficient approximation algorithms.

(iii) To give some of the basic ideas of methods for proving lower bounds on polynomial-time inapproximability.

To reach these goals we organize this chapter as follows. Section 4.2 provides the definition of fundamental notions like δ-approximation algorithms, polynomial-time approximation scheme, dual approximation algorithms, stability of approximation, and a fundamental classification of hard optimization problems according to their polynomial-time approximability (inapproximability).

Section 4.3 includes several examples of polynomial-time approximation algorithms. We are very far from trying to present a survey on approximation algorithms for fundamental optimization problems. Rather, we prefer to give algorithms that transparently present some fundamental ideas for the design of approximation algorithms. We shall see that the classical algorithm design methods like greedy algorithms, local algorithms, and dynamic programming can be successful in the design of approximation algorithms, too. On the other hand, we also present new methods and concepts that are specific for the design of efficient approximation algorithms. The main ones are the method of dual approximation and the concept of the stability of approximation.

Section 4.4 is devoted to readers who are also interested in the theory and may be skipped by anybody who is interested in the algorithm design only. Here, an introduction to the concepts for proving lower bounds on polynomial-time inapproximability of optimization problems (under the assumption P $\neq$ NP or a similar one) is given. The approaches considered here contribute to the classification of optimization problems according to the hardness of getting an approximate solution. This topic is of practical importance, too, because the information provided could be very helpful for the choice of a suitable algorithmic approach to solve the given problem in a concrete application.

4.2 Fundamentals

4.2.1 Concept of Approximation Algorithms

We start with the fundamental definition of approximation algorithms. Informally and roughly, an approximation algorithm for an optimization problem is an algorithm that provides a feasible solution whose quality does not differ too much from the quality of an optimal solution.

Definition 4.2.1.1. *Let $U = (\Sigma_I, \Sigma_O, L, L_I, \mathcal{M}, cost, goal)$ be an optimization problem, and let A be a consistent algorithm for U. For every $x \in L_I$, the* **relative error** $\varepsilon_A(x)$ *of A on x is defined as*

$$\varepsilon_A(x) = \frac{|cost(A(x)) - Opt_U(x)|}{Opt_U(x)}.$$

For any $n \in \mathbb{N}$, we define **the relative error of A** *as*

$$\varepsilon_A(n) = \max \{\varepsilon_A(x) \,|\, x \in L_I \cap (\Sigma_I)^n\}.$$

For every $x \in L_I$, the **approximation ratio** $R_A(x)$ *of A on x is defined as*

$$R_A(x) = \max \left\{ \frac{cost(A(x))}{Opt_U(x)}, \frac{Opt_U(x)}{cost(A(x))} \right\}.$$

For any $n \in \mathbb{N}$, we define the **approximation ratio of A** *as*

$$R_A(n) = \max \{R_A(x) \,|\, x \in L_I \cap (\Sigma_I)^n\}.$$

*For any positive real $\delta > 1$, we say that A is a δ-**approximation algorithm** for U if $R_A(x) \leq \delta$ for every $x \in L_I$.*

*For every function $f : \mathbb{N} \to \mathbb{R}^+$, we say that A is an $f(n)$-**approximation algorithm** for U if $R_A(n) \leq f(n)$ for every $n \in \mathbb{N}$.*

We observe that

$$R_A(x) = \frac{cost(A(x))}{Opt_U(x)} = 1 + \varepsilon_A(x)$$

if U is a minimization problem, and that

$$R_A(x) = \frac{Opt_U(x)}{cost(A(x))}$$

if U is a maximization problem. Note that unfortunately there are many different terms used to refer to R_A in the literature. The most frequent ones, besides the term approximation ratio used here, are worst case performance, approximation factor, performance bound, performance ratio, and error ratio.

To illustrate Definition 4.2.1.1 we give an example of a 2-approximation algorithm.

Example 4.2.1.2. Consider the problem of makespan scheduling (MS). Remember that the input consists of n jobs with designated processing times $p_1, p_2, \ldots, p_n$ and a positive integer $m \geq 2$. The problem is to schedule these n jobs on m identical machines. A schedule of jobs is an assignment of the jobs to the machines. The objective is to minimize the time after which all jobs are executed, i.e., to minimize the maximum working times of the m machines.

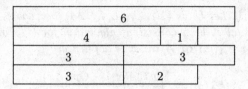

Fig. 4.1.

For instance, for seven jobs of processing times $3, 2, 4, 1, 3, 3, 6$, and 4 machines, an optimal scheduling is depicted in Figure 4.1. The cost of this solution is 6.

Now we present a simple greedy algorithm for MS and show that it is a 2-approximation algorithm. The idea is first to sort the values $p_1, p_2, \ldots, p_n$ in a nonincreasing order and then to assign the largest nonrealized job to one of the first machines that would be free.

Algorithm 4.2.1.3 (GMS (GREEDY MAKESPAN SCHEDULE)).

Input:　$I = (p_1, \ldots, p_n, m)$, n, m, $p_1, \ldots, p_n$ positive integers and $m \geq 2$.
Step 1:　Sort $p_1, \ldots, p_n$.
　　　　To simplify the notation we assume $p_1 \geq p_2 \geq \cdots \geq p_n$ in the rest of the algorithm.
Step 2:　**for** $i = 1$ **to** m **do**
　　　　　　begin　$T_i := \{i\}$;
　　　　　　　　$Time(T_i) := p_i$
　　　　　　end
　　　　{In the initialization step the m largest jobs are distributed to the m machines. At the end, T_i should contain the indices of all jobs assigned to the ith machine for $i = 1, \ldots, m$.}
Step 3:　**for** $i = m + 1$ **to** n **do**
　　　　　　begin　compute an l such that
　　　　　　　　$Time(T_l) := \min\{Time(T_j) | 1 \leq j \leq m\}$;
　　　　　　　　$T_l := T_l \cup \{i\}$;
　　　　　　　　$Time(T_l) := Time(T_l) + p_i$
　　　　　　end
Output:　$(T_1, T_2, \ldots, T_m)$.

Now, let us prove that GMS is a 2-approximation algorithm for MS. Let $I = (p_1, p_2, \ldots, p_n, m)$ with $p_1 \geq p_2 \geq \cdots \geq p_n$ as input. First, we observe the straightforward fact that

$$Opt_{\mathrm{MS}}(I) \geq p_1 \geq p_2 \geq \cdots \geq p_n. \tag{4.1}$$

Clearly,

$$Opt_{\mathrm{MS}}(I) \geq \frac{\sum_{i=1}^{n} p_i}{m} \tag{4.2}$$

because $\left(\sum_{i=1}^{n} p_i\right)/m$ is the average work time over all m machines. Since p_k is the smallest value among $p_1, p_2, \ldots, p_k$ and $\frac{1}{k}\sum_{i=1}^{k} p_i$ is the average value over $p_1, p_2, \ldots, p_k$,

$$p_k \leq \frac{\sum_{i=1}^{k} p_i}{k} \tag{4.3}$$

for every $k = 1, 2, \ldots, n$. Now we distinguish two possibilities according to the relation between n and m.

(1) Let $n \leq m$.

Since $Opt_{MS}(I) \geq p_1$ (4.1) and $cost(\{1\}, \{2\}, \ldots, \{n\}, \emptyset, \ldots, \emptyset) = p_1$, GMS has found an optimal solution and so the approximation ratio is 1.

(2) Let $n > m$.

Let T_l be such that $cost(T_l) = \sum_{r \in T_l} p_r = cost(\mathrm{GMS}(I))$, and let k be the largest index in T_l. If $k \leq m$, then $|T_l| = 1$ and so $Opt_{MS}(I) = p_1 = p_k$ and $\mathrm{GMS}(I)$ is an optimal solution.

Now, assume $m < k$. Following Figure 4.2 we see that

$$Opt_{MS}(I) \geq cost(\mathrm{GMS}(I)) - p_k \tag{4.4}$$

because of $\sum_{i=1}^{k-1} p_i \geq m \cdot [cost(\mathrm{GMS}(I)) - p_k]$ and (4.2).

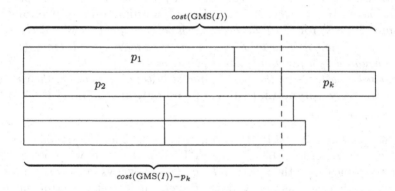

Fig. 4.2.

Combining the inequalities (4.3) and (4.4) we obtain

$$cost(\mathrm{GMS}(I)) - Opt_{MS}(I) \underset{(4.4)}{\leq} p_k \underset{(4.3)}{\leq} \left(\sum_{i=1}^{k} p_i\right)\bigg/k. \tag{4.5}$$

Finally, we bound the relative error by the following calculation:

$$\frac{cost(\mathrm{GMS}(I)) - Opt_{MS}(I)}{Opt_{MS}(I)} \underset{\substack{(4.5)\\(4.2)}}{\leq} \frac{\left(\sum_{i=1}^{k} p_i\right)/k}{\left(\sum_{i=1}^{n} p_i\right)/m} \leq \frac{m}{k} < 1.$$

$\square$

Exercise 4.2.1.4. Change the greedy strategy of the algorithm GMS to an arbitrary choice, i.e., without sorting $p_1, p_2, \ldots, p_n$ (removing Step 1 of GMS), assign the jobs in the on-line manner[2] as described in Step 2 and 3 of GMS. Prove that this simple algorithm, called GRAHAM'S ALGORITHM, is a 2-approximation algorithm for MS, too. $\square$

Exercise 4.2.1.5. Find, for every integer $m \geq 2$, an input instance I_m of MS such that $R_{\text{GMS}}(I) = \frac{cost(\text{GMS}(I))}{Opt_{\text{MS}}(I)}$ is as large as possible. $\square$

In Section 4.3 we present several further approximation algorithms. The simplest examples illustrating Definition 4.2.1.1 are the 2-approximation algorithms for the vertex problem (Algorithm 4.3.2.1) and for the simple knapsack problem (Algorithm 4.3.4.1).

Usually one can be satisfied if one can find a δ-approximation algorithm for a given optimization problem with a conveniently small δ. But for some optimization problems we can do even better. For every input instance x, the user may choose an arbitrarily small relative error ε, and we can provide a feasible solution to x with a relative error at most ε. In such a case we speak about approximation schemes.

Definition 4.2.1.6. *Let* $U = (\Sigma_I, \Sigma_O, L, L_I, \mathcal{M}, cost, goal)$ *be an optimization problem. An algorithm A is called a* **polynomial-time approximation scheme (PTAS) for U**, *if, for every input pair $(x, \varepsilon) \in L_I \times \mathbb{R}^+$, A computes a feasible solution $A(x)$ with a relative error at most ε, and $Time_A(x, \varepsilon^{-1})$ can be bounded by a function[3] that is polynomial in $|x|$. If $Time_A(x, \varepsilon^{-1})$ can be bounded by a function that is polynomial in both $|x|$ and ε^{-1}, then we say that A is a* **fully polynomial-time approximation scheme (FPTAS) for U**.

A simple example of a PTAS is Algorithm 4.3.4.2 for the simple knapsack problem, and Algorithm 4.3.4.11 is a FPTAS for the knapsack problem.

In the typical case the function $Time_A(x, \varepsilon^{-1})$ grows in both $|x|$ and ε^{-1}. This means that one has to pay for the decrease in the relative error (increase of the quality of the output) by an increase in the time complexity (amount of computer work). The advantage of PTASs is that the user has the choice of ε in this tradeoff of the quality of the output and of the amount of computer work $Time_A(x, \varepsilon^{-1})$. FPTASs are very convenient[4] because $Time_A(x, \varepsilon^{-1})$ does not grow too quickly with ε^{-1}.

---

[2] Whenever a machine becomes available (free), the next job on the list is assigned to begin processing on that machine.

[3] Remember that $Time_A(x, \varepsilon^{-1})$ is the time complexity of the computation of the algorithm A on the input (x, ε).

[4] Probably a FPTAS is the best that one can have for a NP-hard optimization problem.

4.2.2 Classification of Optimization Problems

Following the notion of approximability we divide the class NPO of optimization problems into the following five subclasses:

NPO(I):   Contains every optimization problem from NPO for which there exists a FPTAS.
            {In Section 4.3 we show that the knapsack problem belongs to this class.}

NPO(II):   Contains every optimization problem from NPO that has a PTAS.
            {In Section 4.3.4 we show that the makespan scheduling problem belongs to this class.}

NPO(III):   Contains every optimization problem $U \in$ NPO such that
            (i) there is a polynomial-time δ-approximation algorithm for some $\delta > 1$, and
            (ii) there is no polynomial-time d-approximation algorithm for U for some $d < \delta$ (possibly under some reasonable assumption like P $\neq$ NP), i.e., there is no PTAS for U.
            {The minimum vertex cover problem, MAX-SAT, and $\triangle$-TSP are examples of members of this class.}

NPO(IV):   Contains every $U \in NPO$ such that
            (i) there is a polynomial-time $f(n)$-approximation algorithm for U for some $f : \mathbb{N} \to \mathbb{R}^+$, where f is bounded by a polylogarithmic function, and
            (ii) under some reasonable assumption like P $\neq$ NP, there does not exist any polynomial-time δ-approximation algorithm for U for any $\delta \in \mathbb{R}^+$.
            {The set cover problem belongs to this class.}

NPO(V):   Contains every $U \in$ NPO such that if there exists a polynomial-time $f(n)$-approximation algorithm for U, then (under some reasonable assumption like P $\neq$ NP) $f(n)$ is not bounded by any polylogarithmic function.
            {TSP and the maximum clique problem are well-known members of this class.}

Note that assuming P $\neq$ NP none of the classes NPO(I), NPO(II), NPO(III), NPO(IV) or NPO(V) is empty. Thus, we have to deal with all these five types of hardness according to the approximability of problems in NPO.

4.2.3 Stability of Approximation

If one is asking for the hypothetical border between tractability and intractability for optimization problems it is not so easy to give an unambiguous answer. FPTASs and PTASs are usually very practical but we know examples, where the time complexity of a PTAS[5] is around $n^{40 \cdot \varepsilon^{-1}}$ for the input length n and this is obviously not very practical. On the other hand, some $(\log_2 n)$-approximation algorithm may be considered to be practical for some applications. What people usually consider to be intractable are problems in NPO(V). But one has to be careful with this, too. We are working with the worst case complexity, and also the relative error and the approximation ratio of an approximation algorithm are defined in the worst case manner. The problem instances of concrete applications may be much easier than the hardest ones, even much easier than the average ones. It could be helpful to split the set of input instances L_I of a $U \in$ NPO(V) into some (possibly infinitely many) subclasses according to the hardness of their polynomial-time approximability, and to have an efficient algorithm deciding the membership of any input instance to one of the subclasses considered. In order to reach this goal one can use the notion of the stability of approximation.

Informally, one can explain the concept of stability with the following scenario. One has an optimization problem for two sets of input instances L_1 and L_2, $L_1 \subset L_2$. For L_1 there exists a polynomial-time δ-approximation algorithm A for some $\delta > 1$, but for L_2 there is no polynomial-time γ-approximation algorithm for any $\gamma > 1$ (if NP is not equal P). One could pose the following question: "Is the use of the algorithm A really restricted to inputs from L_1?" Let us consider a distance measure d in L_2 determining the distance $d(x)$ between L_1 and any given input $x \in L_2 - L_1$. Now, one can look for how "good" the algorithm A for the inputs $x \in L_2 - L_1$ is. If, for every $k > 0$ and every x with $d(x) \leq k$, A computes a $\gamma_{k,\delta}$-approximation of an optimal solution for x ($\gamma_{k,\delta}$ is considered to be a constant depending on k and δ only), then one can say that A is "(approximation) stable" according to the distance measure d.

Obviously, the idea of this concept is similar to that of the stability of numerical algorithms. But instead of observing the size of the change of the output value according to a small change of the input value, we look for the size of the change of the approximation ratio according to a small change in the specification (some parameters, characteristics) of the set of problem instances considered. If the exchange of the approximation ratio is small for every small change in the specification of the set of problem instances, then we have a stable algorithm. If a small change in the specification of the set of problem instances causes an essential (depending on the size of the input instances) increase of the relative error, then the algorithm is unstable.

The concept of stability enables us to show positive results extending the applicability of known approximation algorithms. As we shall see later, the concept motivates us to modify an unstable algorithm A in order to get a stable algorithm B that achieves the same approximation ratio on the original set of input instances as A has, but B can also be successfully used outside of the original set of input instances. This concept is useful because there are a lot

---

[5] We consider the PTAS for the geometrical (Euclidean) TSP.

of problems for which an additional assumption on the "parameters" of the
input instances leads to an essential decrease in the hardness of the problem.
So, such effects are the starting points for trying to partition the whole set of
input instances into a spectrum of classes according to approximability.

Definition 4.2.3.1. *Let* $U = (\Sigma_I,\ \Sigma_O,\ L,\ L_I,\ \mathcal{M},\ cost,\ goal)$ *and* $\overline{U} = (\Sigma_I, \Sigma_O, L, L, \mathcal{M}, cost, goal)$ *be two optimization problems with* $L_I \subset L$. *A*
distance function for $\overline{U}$ **according to** L_I *is any function* $h_L : L \to \mathbb{R}^{\geq 0}$
satisfying the properties

(i) $h_L(x) = 0$ *for every* $x \in L_I$, *and*
(ii) h *is polynomial-time computable.*

Let h *be a distance function for* $\overline{U}$ *according to* L_I. *We define, for any* $r \in \mathbb{R}^+$,

$$Ball_{r,h}(L_I) = \{w \in L \mid h(w) \leq r\}.^6$$

Let A *be a consistent algorithm for* $\overline{U}$, *and let* A *be an* ε-*approximation al-*
gorithm for U *for some* $\varepsilon \in \mathbb{R}^{>1}$. *Let* p *be a positive real. We say that*
A *is* p-**stable according to** h *if, for every real* $0 < r \leq p$, *there ex-*
ists a $\delta_{r,\varepsilon} \in \mathbb{R}^{>1}$ *such that* A *is a* $\delta_{r,\varepsilon}$-*approximation algorithm for* $U_r = (\Sigma_I, \Sigma_O, L, Ball_{r,h}(L_I), \mathcal{M}, cost, goal)$.

A *is* **stable according to** h *if* A *is* p-*stable according to* h *for every*
$p \in \mathbb{R}^+$. *We say that* A *is* **unstable according to** h *if* A *is not* p-*stable for*
any $p \in \mathbb{R}^+$.

For every positive integer r, *and every function* $f_r : \mathbb{N} \to \mathbb{R}^{>1}$ *we say that*
A *is* $(r, f_r(n))$-**quasistable according to** h *if* A *is an* $f_r(n)$-*approximation*
algorithm for $U_r = (\Sigma_I, \Sigma_O, L, Ball_{r,h}(L_I), \mathcal{M}, cost, goal)$.

Example 4.2.3.2. Consider $TSP = (\Sigma_I, \Sigma_O, L, L, \mathcal{M}, cost, minimum)$ and
$\triangle\text{-}TSP = (\Sigma_I, \Sigma_O, L, L_\triangle, \mathcal{M}, cost, minimum)$ as defined in Section 2.3. Remember that L contains all weighted complete graphs and $L_\triangle$ contains all
weighted complete graphs with a weight function c satisfying the triangle inequality. A natural way to define a distance function h for TSP according
to $L_\triangle$ is to measure the "distance" of any input instance from the triangle
inequality. For instance, this can be done in the following ways:
For every input instance (G, c),

$$dist(G, c) = \max\left\{0, \max\left\{\frac{c(\{u, v\})}{c(\{u, p\}) + c(\{p, v\})} - 1 \,\middle|\, u, v, p \in V(G),\right.\right.$$

$$\left.\left. u \neq v, u \neq p, v \neq p\right\}\right\},$$

$$dist_k(G, c) = \max\left\{0, \max\left\{\frac{c(\{u, v\})}{\sum_{i=1}^m c(\{p_i, p_{i+1}\})} - 1 \,\middle|\, u, v \in V(G) \text{ and}\right.\right.$$

---

[6] $Ball_{r,h}(L_I)$ contains all input instances whose specification "differs" at most r
from the specification of the input instances in L_I.

$$u = p_1, p_2, \ldots, p_m = v \text{ is a simple path between } u \text{ and } v$$

$$\left. \left. \text{of length at most } k \text{ (i.e., } m + 1 \leq k) \right\} \right\}$$

for every integer $k \geq 2$, and

$$distance(G, c) = \max\{ dist_k(G, c) \mid 2 \leq k \leq |V(G)| - 1 \}.$$

We observe that $Ball_{r,dist}(L_\triangle)$ contains exactly those weighted graphs for which

$$c(\{u, v\}) \leq (1 + r)(c(\{u, p\}) + c(\{p, v\}))$$

for all pairwise different vertices u, v, p. This is a natural extension of the triangle inequality specification. The distance function $dist_k$ says that if $dist_k(G, c) \leq r$, then the cost of the direct connection (u, v) between any two vertices u and v is not larger than $(1 + r)$ times the cost of any path between u and v of length at most k.   □

Exercise 4.2.3.3. Prove that the functions $dist$, $dist_k$, and $distance$ (defined in Example 4.2.3.2) are distance functions for TSP according to $L_\triangle$.   □

Some examples of stable and unstable algorithms according to the distance functions $dist$ and $distance$ for TSP are given in Section 4.3.5.

In Definition 4.2.3.1 distance functions were defined in a very general form. It should be clear that the investigation of the stability according to a distance function h is of interest only if h reasonably "partitions" the set of problem instances. The following exercise shows a distance function that is not helpful for the study of approximability of any problem.

Exercise 4.2.3.4. Let $U = (\Sigma_I, \Sigma_O, L, L_I, \mathcal{M}, cost, goal)$ and $\overline{U} = (\Sigma_I, \Sigma_O, L, L, \mathcal{M}, cost, goal) \in \text{NPO}$. Let h_{index} be defined as follows:

(i) $h_{index}(w) = 0$ for every $w \in L_I$, and
(ii) $h_{index}(u)$ is equal to the order of u according to the canonical order of words in Σ_I^*.

Prove that

a) h_{index} is a distance function of $\overline{U}$ according to L_I.
b) For every δ-approximation algorithm A for U, if A is consistent for $\overline{U}$, then A is stable according to h_{index}.   □

Exercise 4.2.3.3 shows that it is not interesting to consider a distance function h with the property

$$|Ball_{r,h}(L_I)| - |Ball_{q,h}(L_I)| \text{ is finite}$$

for every $r > q$. The distance function h' investigated later has the following additional property (called the **property of infinite jumps**):

"If $Ball_{q,h'}(L_I) \subset Ball_{r,h'}(L_I)$ for some $q < r$,
then $|Ball_{r,h'}(L_I)| - |Ball_{q,h'}(L_I)|$ is infinite."

But this additional property on h' also does not need to secure any kind of "reasonability" of h' for the study of approximation stability. The best approach to define a distance function is to take a "natural" function according to the specification of the set L_I of input instances.

Exercise 4.2.3.5. Define two optimization problems $U = (\Sigma_I, \Sigma_O, L, L_I, \mathcal{M},$ $cost, goal)$ and $\overline{U} = (\Sigma_I, \Sigma_O, L, L, \mathcal{M}, cost, goal)$ from NPO with infinite $|L| -$ $|L_I|$, and a distance function h for $\overline{U}$ according to L_I such that:

(i) h has the property of infinite jumps, and
(ii) for every δ-approximation algorithm A for U, if A is consistent for $\overline{U}$, then A is stable according to h.   □

We see that the existence of a stable c-approximation algorithm for U immediately implies the existence of a $\delta_{r,c}$-approximation algorithm for U_r for any $r > 0$. Note that applying the concept of stability to PTASs one can get two different outcomes. Let us consider a PTAS A as a collection of polynomial-time $(1 + \varepsilon)$-approximation algorithms A_ε for every $\varepsilon \in \mathbb{R}^+$. If A_ε is stable according to a distance measure h for every $\varepsilon > 0$, then we can obtain either

(i) a PTAS for $U_r = (\Sigma_I, \Sigma_O, L, Ball_{r,h}(L_I), \mathcal{M}, cost, goal)$ for every $r \in \mathbb{R}^+$ (this happens, for instance, if $\delta_{r,\varepsilon} = 1 + \varepsilon \cdot f(r)$, where f is an arbitrary function), or
(ii) a $\delta_{r,\varepsilon}$-approximation algorithm for U_r for every $r \in \mathbb{R}^+$, but no PTAS for U_r for any $r \in \mathbb{R}^+$ (this happens, for instance, if $\delta_{r,\varepsilon} = 1 + r + \varepsilon$).

To capture these two different situations we introduce the notion of "superstability".

Definition 4.2.3.6. *Let $U = (\Sigma_I, \Sigma_O, L, L_I, \mathcal{M}, cost, goal)$ and $\overline{U} = (\Sigma_I, \Sigma_O,$ $L, L, \mathcal{M}, cost, goal)$ be two optimization problems with $L_I \subset L$. Let h be a distance function for $\overline{U}$ according to L_I, and let $U_r = (\Sigma_I, \Sigma_O, L, Ball_{r,h}(L_I),$ $\mathcal{M}, cost, goal)$ for every $r \in \mathbb{R}^+$. Let $A = \{A_\varepsilon\}_{\varepsilon > 0}$ be a PTAS for U.*

*If, for every $r > 0$ and every $\varepsilon > 0$, A_ε is a $\delta_{r,\varepsilon}$-approximation algorithm for U_r, we say that the PTAS A is **stable according to h**.*

If $\delta_{r,\varepsilon} \leq f(\varepsilon) \cdot g(r)$, where

(i) f and g are some functions from $\mathbb{R}^{\geq 0}$ to $\mathbb{R}^+$, and
(ii) $\lim_{\varepsilon \to 0} f(\varepsilon) = 0$,

*then we say that the PTAS A is **superstable according to h**.*

Observation 4.2.3.7. If A is a superstable (according to a distance function h) PTAS for an optimization problem $U = (\Sigma_I, \Sigma_O, L, L_I, \mathcal{M}, cost, goal)$, then A is a PTAS for the optimization problem $U_r = (\Sigma_I, \Sigma_O, L, Ball_{r,h}(L_I), \mathcal{M},$ $cost, goal)$ for any $r \in \mathbb{R}^+$.

An example of an superstable PTAS according to a distance function h for the knapsack problem is given in Section 4.3.5.

4.2.4 Dual Approximation Algorithms

In the previous approximation concepts we have focused on searching for a feasible solution whose cost well approximates the cost of an optimal solution. Sometimes there are reasons to consider another scenario. Let us consider an optimization problem $U = (\Sigma_I, \Sigma_O, L, L, \mathcal{M}, cost, goal)$, where the constraints determining $\mathcal{M}(x)$ for every input instance $x \in L$ are either not known precisely (i.e., one has an approximation of the constraints only) or not rigid, so that a small violation of the constraints is a natural model of allowed flexibility. An example may be the simple knapsack problem (SKP), where, for any input $I = (a_1, \ldots, a_n, b)$,

$$\mathcal{M}(I) = \left\{ (x_1, \ldots, x_n) \in \{0, 1\}^n \ \bigg| \ \sum_{i=1}^{n} x_i a_i \le b \right\}.$$

Now, it may be possible that some values of $a_1, \ldots, a_n, b$ (especially b) are not precisely determined and a small violation of the constraint $\sum_{i=1}^{n} x_i a_i \le b$ to a constraint $\sum_{i=1}^{n} x_i a_i \le b + \varepsilon$ for small ε could be accepted. In such a case one can prefer to compute a solution S whose cost is at least as good as[7] $Opt_U(I)$ by allowing $S \notin \mathcal{M}(I)$, but where S is not "too far" from the specification of the elements in $\mathcal{M}(I)$.

To formalize this idea one needs a distance function again. But this distance function has to measure the distance of a solution S from the set of feasible solutions $\mathcal{M}(I)$ in contrast to the distance function measuring a distance between input instances in the concept of approximation stability.

Definition 4.2.4.1. *Let* $U = (\Sigma_I, \Sigma_O, L, L_I, \mathcal{M}, cost, goal)$ *be an optimization problem. A* **constraint distance function** *for* U *is any function* $h : L_I \times \Sigma_O^* \to \mathbb{R}^{\ge 0}$ *such that*

(i) $h(x, S) = 0$ *for every* $S \in \mathcal{M}(x)$,
(ii) $h(x, S) > 0$ *for every* $S \notin \mathcal{M}(x)$, *and*
(iii) h *is polynomial-time computable.*

For every $\varepsilon \in \mathbb{R}^+$, *and every* $x \in L_I$, $\mathcal{M}_\varepsilon^h(x) = \{ S \in \Sigma_O^* \mid h(x, S) \le \varepsilon \}$ *is the* ε-*ball of* $\mathcal{M}(x)$ *according to* h.

To illustrate Definition 4.2.4.1 consider the bin-packing problem (BIN-P). Remember that an input instance of BIN-P is $I = (p_1, p_2, \ldots, p_n)$, where $p_i \in [0, 1]$ for $i = 1, \ldots, n$, and the objective is to pack the pieces into bins of unit size 1 in such a way that the minimum number of bins is used. For any

---

[7] If $goal = maximum$ then $cost(S) \ge Opt_U(I)$, and if $goal = minimum$ then $cost(S) \le Opt_U(I)$.

solution $T = T_1, T_2, \ldots, T_m$ $(T_i \subseteq \{1, \ldots, n\}$ for $i = 1, \ldots, m)$, one can define the constraint distance function

$$h(I, T) = \max \left\{ 0, \max \left\{ \sum_{l \in T_i} p_l \,\middle|\, i = 1, 2, \ldots, m \right\} - 1 \right\}.$$

If there exists an ε such that

$$\sum_{l \in T_i} p_l \leq 1 + \varepsilon$$

for all $i = 1, \ldots, m$, then $T \in \mathcal{M}_\varepsilon^h(I)$.

Definition 4.2.4.2. Let $U = (\Sigma_I, \Sigma_O, L, L_I, \mathcal{M}, cost, goal)$ be an optimization problem, and let h be a constraint distance function for U.

An optimization algorithm A for U is called an **h-dual ε-approximation algorithm for U**, if for every $x \in L_I$,

(i) $A(x) \in \mathcal{M}_\varepsilon^h(x)$, and
(ii) $cost(A(x)) \geq Opt_U(x)$ if $goal = maximum$, and
    $cost(A(x)) \leq Opt_U(x)$ if $goal = minimum$.

Definition 4.2.4.3. Let $U = (\Sigma_I, \Sigma_O, L, L_I, \mathcal{M}, cost, goal)$ be an optimization problem, and let h be a constraint distance function for U.

An algorithm A is called **h-dual polynomial-time approximation scheme (h-dual PTAS for U)**, if

(i) for every input $(x, \varepsilon) \in L_I \times \mathbb{R}^+$, $A(x, \varepsilon) \in \mathcal{M}_\varepsilon^h(x)$,
(ii) $cost(A(x, \varepsilon)) \geq Opt_U(x)$ if $goal = maximum$, and
    $cost(A(x, \varepsilon)) \leq Opt_U(x)$ if $goal = minimum$, and
(iii) $Time_A(x, \varepsilon^{-1})$ is bounded by a function that is polynomial in $|x|$.

If $Time_A(x, \varepsilon^{-1})$ can be bounded by a function that is polynomial in both $|x|$ and ε^{-1}, then we say that A is a **h-dual fully polynomial-time approximation scheme (h-dual FPTAS) for U**.

In Section 4.3.6 we give not only an example of a dual approximation algorithm, but we also show how one can use dual approximation algorithms in order to design ordinary approximation algorithms.

The next parts of this chapter show the usefulness of the terminology and classification defined above for the development and design of approximation algorithms for hard optimization problems.

Keywords introduced in Section 4.2

δ-approximation algorithms, PTAS, FPTAS, stability of approximation algorithms, dual approximation algorithms

Summary of Section 4.2

A small change in the specification of an optimization problem may have an essential influence on the hardness of the problem. For instance, the jump from NP-hardness to deterministic polynomial time can be caused by the following changes in the problem specification:

(i) One requires a feasible solution within a given approximation ratio instead of an exact optimal solution, or

(ii) one requires that the input instances have a "reasonable" additional property (that may be typical for inputs from some real applications).

The notion of stability of numerical algorithms can be extended to approximation stability of approximation algorithms in order to get a fine analysis of the hardness of the optimization problem considered. This analysis partially overcomes the drawbacks of the worst case complexity approach by partitioning the set of input instances into subclasses according to the polynomial-time approximability.

Dual approximation algorithms approximate the set of feasible solutions (feasibility) rather than the optimality. They are of practical interest especially if one does not have any exact (but only estimated) constraints on the feasibility of solutions.

4.3 Algorithm Design

4.3.1 Introduction

This section is an introduction to the design of approximation algorithms. The main aim is to present and to illustrate methods and ideas that have been successful in the design of approximation algorithms. We are far from trying to give an overview of the huge area of approximation algorithms here. To achieve our goal without dealing with nontrivial mathematics and too many technical details, we have to prefer the presentation of transparent examples of approximation algorithms over the presentation of the best known ones. We want to show that the fundamental general methods for algorithm design like greedy algorithms (see Algorithm 4.3.2.11 for SCP, and Algorithm 4.3.4.1 for SKP), local search algorithms (see Algorithm 4.3.3.1 for MAX-CUT), dynamic programming (see Algorithm 4.3.4.11 for KP), backtracking (see PTASs for KP in Section 4.3.4), relaxation to linear programming (Algorithm 4.3.2.19 for WEIGHT-VCP), and their combinations and modifications can be successful in the design of approximation algorithms, too. On the other hand, we explain how to develop and apply approaches that are specific for the area of the design of approximation algorithms in combinatorial optimization.

Section 4.3 is organized as follows. Section 4.3.2 is devoted to the set cover problem and its special subproblem – the vertex cover problem. We show there that a very simple efficient heuristic provides a 2-approximation algorithm for the vertex cover problem, and that the greedy approach leads to a $(\ln n)$-approximation algorithm for the set cover problem.[8] Further, we apply the method of relaxation by reduction to linear programming in order to get a 2-approximation algorithm for the weighted minimum vertex cover problem, too.

Section 4.3.3 presents another example of a simple approach that leads to a good approximation. A simple strategy of local improvements (local search) provides a polynomial-time 2-approximation algorithm for the maximum cut problem (Max-Cut). This means that a suitable definition of local improvements can determine such a structure on the set of feasible solutions, that every local optimum is a good approximation of total optima.

Section 4.3.4 is devoted to the knapsack problem and PTASs. First, we show that the greedy approach leads to an efficient 2-approximation algorithm for the simple knapsack problem. Finding some compromise between the greedy approach and the total search we derive a PTAS for the simple knapsack problem. Then, applying the concept of stability of approximation we show that a natural modification of the PTAS for the simple knapsack problem results in a PTAS for the general knapsack problem. Finally, we combine dynamic programming with an approximation of input values in order to get a FPTAS for the knapsack problem.

In Section 4.3.5 we study the traveling salesperson problem that is one of the hardest approximation problems.[9] First, we present two special efficient approximation algorithms that achieves constant approximation ratios for $\triangle$-TSP, i.e., for the input instances that satisfy the triangle inequality. Then, we use the concept of stability of approximation in order to partition the set of input instances of the general TSP into an infinite spectrum of classes according to an achievable polynomial-time approximability. Section 4.3.6 presents the usefulness of the concept of dual approximation algorithms. First, a dual PTAS for the bin-packing problem is designed, and then this dual PTAS is used to develop a PTAS for the makespan scheduling problem. This design method works because the nonrigid constraints of the bin-packing problem can be viewed as the objective of the makespan scheduling problem; i.e., the bin-packing problem is in some sense "dual" to the makespan scheduling problem.

4.3.2 Cover Problems, Greedy Method, and Relaxation to Linear Programming

In this section we consider the minimum vertex cover problem (Min-VCP), the set cover problem (SCP), and the weighted minimum vertex cover problem

---

[8] Observe that assuming $P \neq NP$ one can prove the nonexistence of any polynomial-time $((1 - \varepsilon) \cdot \ln n)$-approximation algorithm for the set cover problem, and so the efficient greedy method is the best possible approximation approach for this problem.

[9] It belongs to NPO(V).

(WEIGHT-VCP). The first algorithm presented is a 2-approximation algorithm for MIN-VCP. The idea of this algorithm is very simple. In fact, one quickly finds a maximal matching of the given graph and considers all vertices adjacent to the edges of the matching as a vertex cover. Since the cardinality of every maximal matching is a lower bound on the cardinality of any vertex cover, the proposed algorithm has a relative error of at most 1.

Algorithm 4.3.2.1.   Input: A graph $G = (V, E)$.

Step 1: $C := \emptyset$ {during the computation $C \subseteq V$, and at the end C should contain a vertex cover};

$A := \emptyset$ {during the computation $A \subseteq E$ is a matching, and at the end A is a maximal matching};

$E' := E$ {during the computation $E' \subseteq E$, E' contains exactly the edges that are not covered by the actual C, and at the end $E' = \emptyset$}.

Step 2:   **while** $E' \neq \emptyset$

do begin choose an arbitrary edge $\{u, v\}$ from E';

$C := C \cup \{u, v\}$;

$A := A \cup \{\{u, v\}\}$;

$E' := E' - \{$all edges incident to u or $v\}$

end

Output: C.

Example 4.3.2.2. We illustrate the work of Algorithm 4.3.2.1 on the input $G = (\{a, b, c, d, e, f, g, h\}, \{\{a, b\}, \{b, c\}, \{c, e\}, \{c, d\}, \{d, e\}, \{d, f\}, \{d, g\}, \{d, h\}, \{e, f\}, \{h, g\}\})$ depicted in Figure 4.3a. An optimal vertex cover is $\{b, e, d, g\}$ with cardinality 4. This output cannot be obtained by any choice of edges in Algorithm 4.3.2.1.

Next we consider the following choice. First $\{b, c\}$ is moved to A, $C := \{b, c\}$ and the restgraph is depicted in Figure 4.3b. Then the edge $\{e, f\}$ is chosen (see Figure 4.3c). After that $C = \{b, c, e, f\}$ and $A = \{\{b, c\}, \{e, f\}\}$. In the last step the edge $\{d, g\}$ is chosen (see Figure 4.3d). This results in the vertex cover $\{b, c, e, f, d, g\}$.   □

Exercise 4.3.2.3. (a) Find a subgraph G' of the graph G in Fig. 4.3 such that Algorithm 4.3.2.1 can output a vertex cover whose cardinality is 6 while the cardinality of an optimal vertex cover for G' is 3.

(b) Find, for every positive integer n, a Graph G_n with a vertex cover of the cardinality n, but where Algorithm 4.3.2.1 can compute a vertex cover of the size $2n$.   □

Now let us analyze the complexity and the approximation ratio of Algorithm 4.3.2.1.

Lemma 4.3.2.4. *Algorithm 4.3.2.1 works in time $O(|E|)$.*

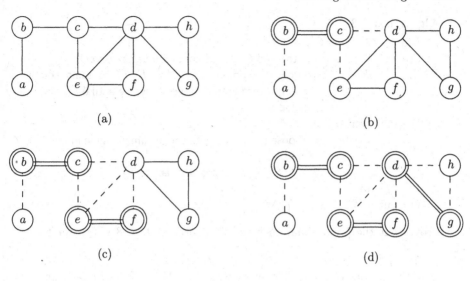

Fig. 4.3.

Proof. Using a convenient data structure this claim is obvious because every edge of the input graph is handled exactly once in the algorithm.   □

Lemma 4.3.2.5. *Algorithm 4.3.2.1 always computes a vertex cover with an approximation ratio at most 2.*

Proof. The fact that Algorithm 4.3.2.1 outputs a vertex cover is obvious because the algorithm halts when $E' = \emptyset$ (i.e., all edges are covered by the vertices in C).

To prove that the approximation ratio is at most 2, we observe that $|C| = 2 \cdot |A|$, and that A is a matching of the input graph $G = (V, E)$. To cover the $|A|$ edges of the matching A one needs at least $|A|$ vertices. Since $A \subseteq E$, the cardinality of every vertex cover is at least $|A|$, i.e., $Opt_{\text{MIN-VCP}}(G) \geq |A|$. Thus, $|C| = 2 \cdot |A| \leq 2 \cdot Opt_{\text{MIN-VCP}}(G)$.   □

Exercise 4.3.2.6. Construct an infinite family of graphs for which Algorithm 4.3.2.1 always (independent of the choice of edges from E) computes an optimal vertex cover.   □

One may say that Algorithm 4.3.2.1 is a very naive heuristic and propose a more involved choice of the vertices for the vertex cover. The most natural idea could be to use the greedy strategy that always chooses a vertex with the maximal degree in the rest graph. The claim of the following exercise shows the (maybe a little bit surprising) fact that this greedy strategy does not lead to a better approximation than the approximation ratio of Algorithm 4.3.2.1.

Exercise 4.3.2.7. [*] Consider the following greedy algorithm for VCP.

Algorithm 4.3.2.8.

Input: A graph $G = (V, E)$.

Step 1: $C := \emptyset$ {at the end C should contain a vertex cover}

$E' := E$ {during the computation $E' \subseteq E$, E' contains exactly the edges that are not covered by the actual C, and at the end $E' = \emptyset$}.

Step 2:   **while** $E' \neq \emptyset$

do begin choose a vertex v with the maximal degree in (V, E');

$C := C \cup \{v\}$;

$E' := E' - \{$all edges adjacent to $v\}$

end

Output: C.

Prove that the approximation ratio of Algorithm 4.3.2.8 is greater than 2. $\square$

MIN-VCP is one of the problems for which no PTAS exists (under the assumption P $\neq$ NP). But, for every such problem, one can search for subclasses of input instances for which MIN-VCP is not that hard. One example in this direction provides the following exercise.

Exercise 4.3.2.9. (a) Design a polynomial-time algorithm for MIN-VCP when the set of input graphs is restricted to the trees.

(b) Design a polynomial-time d-approximation algorithm for MIN-VCP with $d < 2$ when the input graphs have their degree bounded by 3. $\square$

Exercise 4.3.2.10. If $C \subseteq V$ is a vertex cover of a graph $G = (V, E)$, then $V - C$ is a clique in $\overline{G} = (V, \overline{E})$, where $\overline{E} = \{\{u, v\} \mid u, v \in V, u \neq v\} - E$. Is it possible to use Algorithm 4.3.2.1 to get a δ-approximation algorithm for the maximum clique problem for some $\delta \in \mathbb{R}^{>1}$? $\square$

In what follows we consider the set cover problem (SCP). This problem is in the class NPO(IV). Surprisingly, the naive greedy approach provides the best possible approximation ratio.

Algorithm 4.3.2.11.

Input: $(X, \mathcal{F})$, where X is a finite set, $\mathcal{F} \subseteq \mathcal{P}ot(X)$ such that $X = \bigcup_{Q \in \mathcal{F}} Q$.

Step 1: $C := \emptyset$ {during the computation $C \subseteq \mathcal{F}$ and at the end C is a set cover of $(X, \mathcal{F})$};

$U := X$ {during the computation $U \subseteq X$, $U = X - \bigcup_{Q \in C} Q$ for the actual C, and at the end $U = \emptyset$}.

Step 2:   **while** $U \neq \emptyset$

do begin choose an $S \in \mathcal{F}$ such that $|S \cap U|$ is maximal;

$U := U - S$;

$C := C \cup \{S\}$

end

Output: C.

Lemma 4.3.2.12. *Algorithm 4.3.2.11 is a $Har(\max\{|S| \mid S \in \mathcal{F}\})$-approximation algorithm for* SCP.

Proof. Let $C = \{S_1, S_2, \ldots, S_r\}$ be the output of Algorithm 4.3.2.11, where S_i is the ith set chosen in Step 2 of Algorithm 4.3.2.11. Since $U = \emptyset$ it is obvious that C is a set cover for the input $(X, \mathcal{F})$.

For every $x \in X$, we define the **weight of x according to C** as follows:

$$Weight_C(x) = \frac{1}{|S_i - (S_1 \cup S_2 \cup \cdots \cup S_{i-1})|}$$

if $x \in S_i - \bigcup_{j=1}^{i-1} S_j$. Thus, $Weight_C(x)$ is determined when x is covered by Algorithm 4.3.2.11 for the first time. If one defines

$$Weight_C(C) = \sum_{x \in X} Weight_C(x)$$

then obviously $Weight_C(C) = |C| = cost(C)$. Let $C_{Opt} \in Output_{SCP}(X, \mathcal{F})$ be an optimal solution with $cost(C_{Opt}) = Opt_{SCP}(X, \mathcal{F})$. Our goal is to prove that the difference between $Weight_C(C)$ and $|C_{Opt}|$ is not too large relative to $|C_{Opt}| = cost(C_{Opt})$. To achieve this goal we shall prove

$$\text{for all } S \in \mathcal{F} : \sum_{x \in S} Weight_C(x) \leq Har(|S|). \tag{4.6}$$

Let, for every set $S \in \mathcal{F}$ and every $i \in \{0, 1, \ldots, |C|\}$,

$$cov_i(S) = |S - (S_1 \cup S_2 \cup \cdots \cup S_i)|.$$

Obviously, for every $S \in \mathcal{F}$, $cov_0(S) = |S|$, $cov_{|C|}(S) = 0$, and $cov_{i-1}(S) \geq cov_i(S)$ for $i = 1, \ldots, |C|$. Let k be the smallest number with the property $cov_k(S) = 0$. Since $cov_{i-1}(S) - cov_i(S)$ elements from S are covered in the ith run of Step 2,

$$\sum_{x \in S} Weight_C(x) = \sum_{i=1}^{k} \left[(cov_{i-1}(S) - cov_i(S)) \cdot \frac{1}{\left| S_i - \bigcup_{j=1}^{i-1} S_j \right|} \right]. \tag{4.7}$$

Observe that

$$\left| S_i - \bigcup_{j=1}^{i-1} S_j \right| \geq \left| S - \bigcup_{j=1}^{i-1} S_j \right| = cov_{i-1}(S) \tag{4.8}$$

because in the opposite case S would have been chosen instead of S_i in the ith run of Step 2 of Algorithm 4.3.2.11. Following (4.7) and (4.8) we obtain

$$\sum_{x \in S} Weight_C(x) \le \sum_{i=1}^{k} (cov_{i-1}(S) - cov_i(S)) \cdot \frac{1}{cov_{i-1}(S)}. \qquad (4.9)$$

Since

$$Har(b) - Har(a) = \sum_{i=a+1}^{b} \frac{1}{i} \ge (b-a) \cdot \frac{1}{b}$$

we have

$$(cov_{i-1}(S) - cov_i(S)) \cdot \frac{1}{cov_{i-1}(S)} \le Har(cov_{i-1}(S)) - Har(cov_i(S)) \qquad (4.10)$$

for every $i = 1, \ldots, k$.

Inserting (4.10) into (4.9) we obtain the aimed relation (4.6):

$$\sum_{x \in S} Weight_C(x) \le \sum_{i=1}^{k} (Har(cov_{i-1}(S)) - Har(cov_i(S)))$$
$$= Har(cov_0(S)) - Har(cov_k(S)) = Har(|S|) - Har(0)$$
$$= Har(|S|).$$

Now, we are ready to relate $|C| = cost(C) = Weight_C(C)$ to C_{Opt}. Since C_{Opt} is also a set cover of X, one has

$$|C| = \sum_{x \in X} Weight_C(x) \le \sum_{S \in C_{Opt}} \sum_{x \in S} Weight_C(x). \qquad (4.11)$$

Inserting (4.6) into (4.11) we finally obtain

$$|C| \le \sum_{S \in C_{Opt}} \sum_{x \in S} Weight_C(x) \le \sum_{S \in C_{Opt}} Har(|S|)$$
$$\le |C_{Opt}| \cdot Har(\max\{|S| \mid S \in \mathcal{F}\}).$$

$\square$

Corollary 4.3.2.13. *Algorithm 4.3.2.11 is a $\ln(|X|)$-approximation algorithm for* SCP.

Proof. The claim follows immediately from $\max\{|S| \mid S \in \mathcal{F}\} \le |X|$ and $Har(d) \le \ln d + 1$. $\square$

Exercise 4.3.2.14. (*) Find an input instance $(X, \mathcal{F})$ for which Algorithm 4.3.2.11 provides an output with an approximation ratio in $\Omega(\ln n)$. $\square$

Lemma 4.3.2.15. *The time complexity of Algorithm 4.3.2.11 is in $O(n^{3/2})$.*

Proof. Let us encode the inputs in such a way that one can consider $|X| \cdot |\mathcal{F}|$ to be the input size of an input instance $(X, \mathcal{F})$. One run of Step 2 costs $O(|X| \cdot |\mathcal{F}|)$ time. The number of runs of Step 2 is bounded by $\min\{|X|, |\mathcal{F}|\} \leq (|X| \cdot |\mathcal{F}|)^{1/2}$. Thus, the time complexity of Algorithm 4.3.2.11 is in $O(n^{3/2})$.
$\square$

Following Corollary 4.3.2.13 and Lemma 4.3.2.15 we obtain our main result.

Theorem 4.3.2.16. *Algorithm 4.3.2.11 is a polynomial-time $(\ln n)$-approximation algorithm for* SCP.[10]

Exercise 4.3.2.17. Is it possible to implement Algorithm 4.3.2.11 to compute in linear time according to $\sum_{S \in \mathcal{F}} |S|$?
$\square$

One can easily observe that MIN-VCP is a special case of SCP. For a graph $G = (V, E)$, $V = \{v_1, \ldots, v_n\}$, the corresponding input instance of SCP is $(E, \{E_1, \ldots, E_n\})$, where $E_i \subseteq E$ contains all edges adjacent to v_i for $i = 1, \ldots, n$. Algorithm 4.3.2.11, then, is a $(\ln n)$-approximation algorithm for MIN-VCP, too.

Exercise 4.3.2.18. Let $\mathcal{G}_d$ be the set of graphs with the degree bounded by d. For which d does Algorithm 4.3.2.11 ensure a better approximation ratio than Algorithm 4.3.2.1?
$\square$

Algorithm 4.3.2.1 for MIN-VCP does not provide any constant approximation ratio for WEIGHT-VCP. To design a 2-approximation algorithm for WEIGHT-VCP we use the method of relaxation by reduction to linear programming. Remember that the task is, for a given graph $G = (V, E)$ and a function $c : V \to \mathbb{N} - \{0\}$, to find a vertex cover S with the minimal cost $\sum_{v \in S} c(v)$. As already presented in Section 3.7, an instance (G, c) of WEIGHT-VCP can be expressed as the following 0/1-linear program:

$$\text{minimize} \sum c(v_i) \cdot x_i$$
$$\text{subject to } x_r + x_s \geq 1 \text{ for all } \{v_r, v_s\} \in E, and$$
$$x_j \in \{0, 1\} \text{ for } j = 1, \ldots, n$$

if $V = \{v_1, v_2, \ldots, v_n\}$ and $X = (x_1, x_2, \ldots, x_n) \in \{0, 1\}^n$ describes the set S_X of vertices that contains v_i iff $x_i = 1$. Relaxing the constraints $x_j \in \{0, 1\}$ for $j = 1, \ldots, n$ to $x_j \geq 0$ for $j = 1, 2, \ldots, n$, one obtains an instance of LP. The following algorithm solves this instance of LP and rounds the result in order to get a feasible solution to the original instance of WEIGHT-VCP.

---

[10] Note that assuming P $\neq$ NP, the approximation ratio $\ln n$ cannot be improved, so the naive greedy strategy is the best one for SCP.

Algorithm 4.3.2.19.

Input: A graph $G = (V, E)$, and a function $c : V \to \mathbb{N} - \{0\}$. Let $V = \{v_1, v_2, \ldots, v_n\}$.

Step 1:   Express the instance (G, c) as an instance $I_{0/1\text{-LP}}(G, c)$ of 0/1-linear programming and relax it to the corresponding instance $I_{\text{LP}}(G, c)$ of linear programming.

Step 2:   Solve $I_{\text{LP}}(G, c)$.
          Let $X = (x_1, x_2, \ldots, x_n) \in (\mathbb{R}^{\geq 0})^n$ be the computed optimal solution.

Step 3:   Set $S_X := \{v_i \mid x_i \geq 1/2\}$.

Output: S_X.

Theorem 4.3.2.20. *Algorithm 4.3.2.19 is a 2-approximation algorithm for* WEIGHT-VCP.

Proof. First, we show that S_X is a feasible solution to the instance (G, c). Since X satisfies $x_r + x_s \geq 1$ for every $\{v_r, v_s\} \in E$,

$$x_r \geq 1/2 \text{ or } x_s \geq 1/2.$$

Thus, at least one of the vertices v_r and v_s is in S_X and the edge $\{v_r, v_s\}$ is covered by S_X.

Now we prove that $cost(S_X) \leq 2 \cdot Opt_{\text{WEIGHT-VCP}}(G, c)$. First of all observe that

$$Opt_{\text{WEIGHT-VCP}}(G, c) = Opt_{0/1\text{-LP}}(I_{0/1\text{-LP}}(G, c)) \geq Opt_{\text{LP}}(I_{\text{LP}}(G, c)). \quad (4.12)$$

Since

$$cost(S_X) = \sum_{v \in S_X} c(v) = \sum_{x_i \geq 1/2} c(v_i) \leq \sum_{x_i \geq 1/2} 2 \cdot x_i \cdot c(v_i) \leq 2 \cdot \sum_{i=1}^{n} x_i \cdot c(v_i)$$

$$= 2 \cdot Opt_{\text{LP}}(I_{\text{LP}}(G, c)) \underset{(4.12)}{\leq} 2 \cdot Opt_{\text{WEIGHT-VCP}}(G, c),$$

the approximation ratio of Algorithm 4.3.2.19 is at most 2. $\square$

Note that the method of relaxation to linear programming is especially suitable for the weighted versions of optimization problems. This is because of the fact that the weights do not have any influence on the number and the form of the constraints (i.e., the weighted version of a problem has the same constraints as the unweighted version of this problem), and so there is usually no essential difference between the complexity of algorithms designed by the method of relaxation to linear programming for the weighted version of an optimization problem and the simple, unweighted version of this problem.

Summary of Section 4.3.2

Very simple "heuristic" approaches like greedy or "arbitrary choice" may lead to the design of efficient approximation algorithms with a good (sometimes even best possible) approximation ratio.

MIN-VCP is a representative of the class NPO(III).

SCP is a representative of the class NPO(IV).

The method of relaxation by reduction to linear programming is often suitable for the weighted versions of optimization problems. Combining this method with rounding one can design a 2-approximation algorithm for WEIGHT-VCP.

4.3.3 Maximum Cut Problem and Local Search

The maximum cut problem (MAX-CUT) is one of the problems for which a simple local search algorithm provides outputs with a constant approximation ratio. This is due to the property that the value of any local optimum is not too far from the value of the total optima. Obviously, this property can depend on the definition of "locality" in the space of feasible solutions. For MAX-CUT it is sufficient to take a very small locality, which can be determined by the simple transformation of moving one vertex from one side of the cut to the other side of the cut.

Algorithm 4.3.3.1.

Input: A graph $G = (V, E)$.
Step 1: $S = \emptyset$
{the cut is considered to be $(S, V - S)$; in fact S can be chosen arbitrarily in this step};
Step 2:   **while**   there exists such a vertex $v \in V$ that the movement of v from one side of the cut $(S, V - S)$ to the other side of $(S, V - S)$ increases the cost of the cut.
do begin take a $u \in V$ whose movement from one side of $(S, V - S)$ to the other side of $(S, V - S)$ increases the cost of the cut, and move this u to the other side.
end
Output: $(S, V - S)$.

Lemma 4.3.3.2. *Algorithm 4.3.3.1 runs in time* $O(|E| \cdot |V|)$.

Proof. A run of Step 2 costs $O(|V|)$ time. Step 2 can be at most $|E|$ times repeated because each run increases the cost of the current cut at least by 1 and the cost can be at most $|E|$. $\qquad\square$

Theorem 4.3.3.3. *Algorithm 4.3.3.1 is a polynomial-time 2-approximation algorithm for* MAX-CUT.

Proof. It is obvious that Algorithm 4.3.3.1 computes a feasible solution to every given input and Lemma 4.3.3.2 proves that this happens in polynomial time.

It remains to be proven that the approximation ratio is at most 2. There is a very simple way to argue that the approximation ratio is at most 2. Let (Y_1, Y_2) be the output of Algorithm 4.3.3.1. Every vertex in Y_1 (Y_2) has at least as many edges to vertices in Y_2 (Y_1) as edges to vertices in Y_1 (Y_2). Thus, at least half of the edges of the graph is in the $cut(Y_1, Y_2)$. Since the cost of an optimal cut cannot exceed $|E|$, the proof is finished.

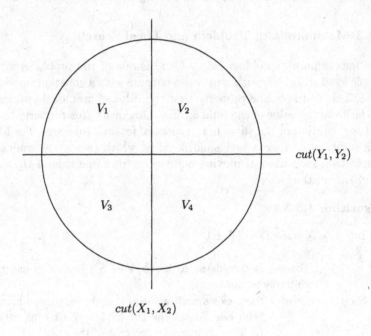

Fig. 4.4.

In what follows we give an alternative proof. The reason for giving it is to show that for many input instances the approximation ratio is clearly better than 2. Let (X_1, X_2) be an optimal cut for an input $G = (V, E)$, and let (Y_1, Y_2) be a cut computed by Algorithm 4.3.3.1, i.e., (Y_1, Y_2) is a local maximum according to the transformation used in Algorithm 4.3.3.1. Let $V_1 = Y_1 \cap X_1$, $V_2 = Y_1 \cap X_2$, $V_3 = Y_2 \cap X_1$, and $V_4 = Y_2 \cap X_2$ (see Figure 4.4).

Thus, $(X_1, X_2) = (V_1 \cup V_3, V_2 \cup V_4)$, and $(Y_1, Y_2) = (V_1 \cup V_2, V_3 \cup V_4)$ because $X_1 \cup X_2 = Y_1 \cup Y_2 = V$. Let $e_{ij} = |E \cap \{\{x, y\} \mid x \in V_i, y \in V_j\}|$ be the number of edges between V_i and V_j for $1 \leq i \leq j \leq 4$. Thus,

$$cost(X_1, X_2) = e_{12} + e_{14} + e_{23} + e_{34},$$

and
$$cost(Y_1, Y_2) = e_{13} + e_{14} + e_{23} + e_{24}.$$

Our aim is to show

$$cost(X_1, X_2) \leq 2 \cdot cost(Y_1, Y_2) - \left(\sum_{i=1}^{4} e_{ii} + e_{13} + e_{24} \right).$$

For every vertex x in V_1, the number of edges between x and the vertices in $V_1 \cup V_2$ is at most the number of edges between x and the vertices in $V_3 \cup V_4$. (In the opposite case, Algorithm 4.3.3.1 would move x from $Y_1 = V_1 \cup V_2$ to $Y_2 = V_3 \cup V_4$). Summing via all vertices in V_1 we obtain

$$2e_{11} + e_{12} \leq e_{13} + e_{14}. \tag{4.13}$$

Because none of the cuts (X_1, X_2) and (Y_1, Y_2) can be improved by the local transformation of Algorithm 4.3.3.1, analogous claims as those for V_1 are true for every vertex of V_2, V_3, and V_4, respectively. This results in the following inequalities:

$$2e_{22} + e_{12} \leq e_{23} + e_{24} \tag{4.14}$$
$$2e_{33} + e_{34} \leq e_{23} + e_{13} \tag{4.15}$$
$$2e_{44} + e_{34} \leq e_{14} + e_{24}. \tag{4.16}$$

Taking $((4.13) + (4.14) + (4.15) + (4.16))/2$ one obtains

$$\sum_{i=1}^{4} e_{ii} + e_{12} + e_{34} \leq e_{13} + e_{14} + e_{23} + e_{24}. \tag{4.17}$$

Adding $e_{14} + e_{23}$ to both sides of the relation (4.17) we get

$$\sum_{i=1}^{4} e_{ii} + e_{12} + e_{34} + e_{14} + e_{23} \leq 2e_{23} + 2e_{14} + e_{13} + e_{24}. \tag{4.18}$$

Now we obtain

$$
\begin{aligned}
cost(X_1, X_2) &= e_{12} + e_{34} + e_{14} + e_{23} \\
&\leq \sum_{i=1}^{4} e_{ii} + e_{12} + e_{14} + e_{34} + e_{23} \\
&\underset{(4.18)}{\leq} 2e_{23} + 2e_{14} + e_{13} + e_{24} \\
&\leq 2(e_{23} + e_{14} + e_{13} + e_{24}) = 2 \cdot cost(Y_1, Y_2)
\end{aligned}
$$

The above calculation directly implies

$$cost(X_1, X_2) \leq 2 \cdot cost(Y_1, Y_2) - e_{13} - e_{24}.$$

$\square$

Exercise 4.3.3.4. Show that one can even prove $cost(X_1, X_2) < 2 \cdot cost(Y_1, Y_2)$ in the proof of Theorem 4.3.3.3.   □

Exercise 4.3.3.5. Find input instances for which the approximation ratio of the solutions computed by Algorithm 4.3.3.1 is almost 2.   □

Exercise 4.3.3.6. Let k-maximum cut problem (k-MAX-CUT) be a restricted version of MAX-CUT, where the degree of the input graphs is bounded by $k \geq 3$. Design a δ_k-approximation algorithm for k-MAX-CUT for every $k \geq 3$ in such a way that $\delta_k < \delta_{k+1} < 2$ for every $k \geq 3$.   □

Summary of Section 4.3.3

There are hard optimization problems whose local optima (according to a reasonable structure of the space of feasible solutions) have costs that well approximate the cost of total optima. In such cases local search algorithms may provide a very efficient solution to the given optimization problem.

4.3.4 Knapsack Problem and PTAS

The knapsack problem (KP) is a representative of the class NPO(I) and so KP is one of the easiest NP-hard optimization problems according to the polynomial-time approximability. We use this problem to illustrate a few concepts for the design of approximation algorithms. First, we use a compromise between total search and greedy algorithms in order to design a PTAS for the simple knapsack problem (SKP). We show that this PTAS is stable according to some reasonable distance measure but that this stability does not suffice to create a PTAS for KP. Then, simply modifying this PTAS for SKP, we obtain a new, superstable PTAS for SKP that provides a PTAS for the general KP. Finally, we "merge" dynamic programming with a kind of approximation of input instances in order to get an FPTAS for KP.

For every input instance $w_1, \ldots, w_n, b$ of SKP, and every $T \subseteq \{1, \ldots, n\}$, remember that $cost(T) = \sum_{i \in T} w_i$. If $cost(T) \leq b$, then T is a feasible solution in $\mathcal{M}(w_1, w_2, \ldots, w_n, b)$.

First, we observe that the simple greedy algorithm for SKP is already a 2-approximation algorithm for SKP.

Algorithm 4.3.4.1. Greedy-SKP

Input: Positive integers $w_1, w_2, \ldots, w_n, b$ for some $n \in \mathbb{N}$.
Step 1:   Sort $w_1, w_2, \ldots, w_n$. For simplicity we may assume $w_1 \geq w_2 \geq \cdots \geq w_n$.
Step 2: $T := \emptyset$; $cost(T) := 0$;
Step 3:   **for** $i = 1$ **to** n **do**
              if $cost(T) + w_i \leq b$ **then**

do begin $T := T \cup \{i\}$;
$$cost(T) := cost(T) + w_i$$
end

Output: T.

Clearly, Algorithm 4.3.4.1 works in the time $O(n \log n)$ because Step 1 takes $O(n \log n)$ time, Step 2 takes $O(1)$ time, and Step 3 takes $O(n)$ time.

To see that Algorithm 4.3.4.1 is a 2-approximation algorithm for SKP it is sufficient to show that $cost(T) \geq b/2$ or T is an optimal solution. Without loss of generality one can assume $b \geq w_1 \geq w_2 \geq \cdots \geq w_n$. Let $j+1$ be the smallest integer not in T (i.e., $\{1, 2, \ldots, j\} \subseteq T$). Note, that $T \neq \emptyset$ (i.e., $j \geq 1$) because $w_1 \leq b$. If $j = 1$, then $w_1 + w_2 > b$. Since $w_1 \geq w_2$ it is obvious that $cost(T) = w_1 > \frac{b}{2}$. Thus, we may assume $j \geq 2$. In general, we have

$$cost(T) + w_{j+1} > b \geq Opt_{\mathrm{SKP}}(w_1, \ldots, w_n, b).$$

Since $w_1 \geq w_2 \geq \cdots \geq w_n$,

$$w_{j+1} \leq w_j \leq \frac{w_1 + w_2 + \cdots + w_j}{j} \leq \frac{b}{j}. \tag{4.19}$$

Thus, $cost(T) > b - w_{j+1} \geq b - \frac{b}{j} \geq b/2$ for every integer $j \geq 2$.

The inequality (4.19) is the kernel of the design idea of the following PTAS for SKP. If $T_{Opt} = \{i_1, i_2, \ldots, i_r\}$, $w_{i_1} \geq w_{i_2} \geq \cdots \geq w_{i_r}$, is an optimal solution and one is able to compute a solution T that contains the j largest weights $w_{i_1}, w_{i_2}, \ldots, w_{i_j}$ of T_{Opt} and $cost(T) + w_{i_{j+1}} > b$, then the difference

$$cost(T_{Opt}) - cost(T) \leq w_{i_{j+1}} \leq \frac{b}{i_{j+1} - 1} \leq \frac{b}{j}$$

is small relative to $cost(T_{Opt})$. To achieve an output T of at least this quality one considers all subsets of $\{1, \ldots, n\}$ of the cardinality at most j extended by the greedy approach, and outputs the best one of them. Obviously, the amount of work grows with j, but the relative error decreases with j.

Algorithm 4.3.4.2.

Input: Positive integers $w_1, w_2, \ldots, w_n, b$ for some $n \in \mathbb{N}$, and a positive real number ε, $0 < \varepsilon < 1$.

Step 1:   Sort $w_1, w_2, \ldots, w_n$. For simplicity, we assume $b \geq w_1 \geq w_2 \geq \cdots \geq w_n$.

Step 2: $k := \lceil 1/\varepsilon \rceil$.

Step 3:   For every set $S \subseteq \{1, 2, \ldots, n\}$ with $|S| \leq k$ and $\sum_{i \in S} w_i \leq b$, extend S to S^* by using the greedy approach described in Step 3 of Algorithm 4.3.4.1.

        {The sets S are created sequentially in the lexicographical order by backtracking, and the up-till-now best S^* is always saved.}

Output: A set S^* with the maximal $cost(S^*)$ among all sets created in Step 3.

Theorem 4.3.4.3. *Algorithm 4.3.4.2 is a* PTAS *for SKP.*

Proof. First, we consider the time complexity $Time(n)$ of Algorithm 4.3.4.2 for an input (I, ε), where $I = w_1, \ldots, w_n, b$. Step 1 can be executed in time $O(n \log n)$ and Step 2 in time $O(1)$. The number of sets $S \subseteq \{1, \ldots, n\}$ with $|S| \leq k$ is

$$\sum_{0 \leq i \leq k} \binom{n}{i} \leq \sum_{0 \leq i \leq k} n^i = \frac{n^{k+1} - 1}{n - 1} = O(n^k).$$

Since these sets can be "created" in lexicographical order, one can construct the next set from the previous one in time $O(1)$. To extend a set S to S^* by the greedy approach costs $O(n)$ time. Thus,

$$Time(n) \leq \left[\sum_{0 \leq i \leq k} \binom{n}{i} \right] \cdot O(n) = O(n^{k+1}) = O(n^{\lceil 1/\varepsilon \rceil + 1}).$$

Now, we show that the approximation ratio

$$R_{\text{Algorithm 4.3.4.2}}(I, \varepsilon) \leq 1 + \frac{1}{k} \leq 1 + \varepsilon$$

for every input $I = w_1, w_2, \ldots, w_n, b$ of SKP. Let $M = \{i_1, i_2, \ldots, i_p\}$, $i_1 < i_2 < \cdots < i_p$ be an optimal solution for I, i.e., $cost(M) = Opt_{\text{SKP}}(I)$. We distinguish two possibilities according to the relation between p and $k = \lceil 1/\varepsilon \rceil$.

If $p \leq k$, then Algorithm 4.3.4.2 considers the set M in Step 3 and so S^* is an optimal solution (i.e., the relative error is 0).

Let $p > k$. Algorithm 4.3.4.2 extends the set $P = \{i_1, i_2, \ldots, i_k\}$ containing the indices of the k greatest weights $w_1, w_2, \ldots, w_{i_k}$ involved in $cost(M) = w_{i_1} + \cdots + w_{i_p}$. If $P^* = M$, then we are ready. If $P^* \neq M$, then there exists $i_q \in M - P^*$ such that $i_q > i_k \geq k$ and

$$cost(P^*) + w_{i_q} > b \geq cost(M). \tag{4.20}$$

Since

$$w_{i_q} \leq \frac{w_{i_1} + w_{i_2} + \cdots + w_{i_k} + w_{i_q}}{k + 1} \leq \frac{cost(M)}{k + 1} \tag{4.21}$$

we obtain

$$R(I, \varepsilon) = \frac{cost(M)}{cost(S^*)} \leq \frac{cost(M)}{cost(P^*)} \underset{(4.20)}{\leq} \frac{cost(M)}{cost(M) - w_{i_q}}$$

$$\underset{(4.21)}{\leq} \frac{cost(M)}{cost(M) - \frac{cost(M)}{k+1}} = \frac{1}{1 - \frac{1}{k+1}} = \frac{k+1}{k}$$

$$= 1 + \frac{1}{k} \leq 1 + \varepsilon.$$

$\square$

We observe that Algorithm 4.3.4.2 is consistent for KP. An input $w_1, \ldots, w_n$, $b, c_1, \ldots, c_n$ of KP is an input of SKP if $w_i = c_i$ for $i = 1, \ldots, n$, so, the relative difference between w_i and c_i seems to be a natural distance measure from the specification $w_i = c_i$. This is the reason for defining the distance function $DIST$ for any input $w_1, w_2, \ldots, w_n, b, c_1, \ldots, c_n$ of KP as follows.

$$DIST(w_1, \ldots, w_n, b, c_1, \ldots, c_n) =$$
$$\max \left\{ \max \left\{ \frac{c_i - w_i}{w_i} \,\middle|\, c_i \geq w_i,\ i \in \{1, \ldots, n\} \right\}, \right.$$
$$\left. \max \left\{ \frac{w_i - c_i}{c_i} \,\middle|\, w_i \geq c_i,\ i \in \{1, \ldots, n\} \right\} \right\}.$$

Let $KP_\delta = (\Sigma_I, \Sigma_O, L, Ball_{\delta,DIST}(L_I), \mathcal{M}, cost, maximum)$ for any $\delta \in \mathbb{R}^+$. Now we show that Algorithm 4.3.4.2 is stable according to $DIST$ but this result does not imply the existence of a PTAS for KP_δ for any $\delta > 0$.

Let us consider $\{ASKP_\varepsilon\}_{\varepsilon > 0}$ as the collection of $(1 + \varepsilon)$-approximation algorithms determined by Algorithm 4.3.4.2.

Lemma 4.3.4.4. *For every $\varepsilon > 0$ and every $\delta > 0$, the algorithm $ASKP_\varepsilon$ is an approximation algorithm for KP_δ with*

$$\varepsilon_A(n) \leq \varepsilon + \delta(2 + \delta)(1 + \varepsilon).$$

Proof. Let $w_1 \geq w_2 \geq \cdots \geq w_n$ for an input $I = w_1, \ldots, w_n, b, c_1, \ldots, c_n$, and let $k = \lceil 1/\varepsilon \rceil$. Let $U = \{i_1, i_2, \ldots, i_l\} \subseteq \{1, 2, \ldots, n\}$ be an optimal solution for I. If $l \leq k$, then $ASKP_\varepsilon$ outputs an optimal solution with $cost(U)$ because $ASKP_\varepsilon$ has considered U as a candidate for the output in Step 3.

Consider the case $l > k$. $ASKP_\varepsilon$ has considered the greedy extension of $T = \{i_1, i_2, \ldots, i_k\}$ in Step 2. Let $T^* = \{i_1, i_2, \ldots, i_k, j_{k+1}, \ldots, j_{k+r}\}$ be the greedy extension of T. Obviously, it is sufficient to show that the difference $cost(U) - cost(T^*)$ is small relative to $cost(U)$, because the cost of the output of SKP_ε is at least $cost(T^*)$. We distinguish the following two possibilities according to the weights of the feasible solutions U and T^*.

(i) Let $\sum_{i \in U} w_i - \sum_{j \in T^*} w_j \leq 0$.
Obviously, for every i, $(1 + \delta)^{-1} \leq \frac{c_i}{w_i} \leq 1 + \delta$. Thus,

$$cost(U) = \sum_{i \in U} c_i \leq (1 + \delta) \cdot \sum_{i \in U} w_i$$

and

$$cost(T^*) = \sum_{j \in T^*} c_j \geq (1 + \delta)^{-1} \cdot \sum_{j \in T^*} w_j.$$

In this way we obtain

$$cost(U) - cost(T^*) \leq (1+\delta) \cdot \sum_{i \in U} w_i - (1+\delta)^{-1} \cdot \sum_{j \in T^*} w_j$$

$$\leq (1+\delta) \cdot \sum_{i \in U} w_i - (1+\delta)^{-1} \cdot \sum_{i \in U} w_i$$

$$= \frac{\delta \cdot (2+\delta)}{1+\delta} \cdot \sum_{i \in U} w_i$$

$$\leq \frac{\delta \cdot (2+\delta)}{1+\delta} \cdot \sum_{i \in U} (1+\delta) \cdot c_i$$

$$= \delta \cdot (2+\delta) \cdot \sum_{i \in U} c_i$$

$$= \delta \cdot (2+\delta) \cdot cost(U).$$

Finally,

$$\frac{cost(U) - cost(T^*)}{cost(U)} \leq \frac{\delta \cdot (2+\delta) \cdot cost(U)}{cost(U)} = \delta \cdot (2+\delta).$$

(ii) Let $d = \sum_{i \in U} w_i - \sum_{j \in T^*} w_j > 0$.

Let c be the cost of the first part of U with the weight $\sum_{j \in T^*} w_j$. Then, in the same way as in (i), one can establish

$$\frac{c - cost(T^*)}{c} \leq \delta \cdot (2+\delta). \tag{4.22}$$

It remains to bound $cost(U) - c$, i.e., the cost of the last part of U with the weight d. Obviously, $d \leq b - \sum_{j \in T^*} w_j \leq w_{i_r}$ for some $r > k$, $i_r \in U$ (if not, then $\mathrm{ASKP}_\varepsilon$ would add i_r to T^* in the greedy procedure). Since $w_{i_1} \geq w_{i_2} \geq \cdots \geq w_{i_l}$,

$$d \leq w_{i_r} \leq \frac{w_{i_1} + w_{i_2} + \cdots + w_{i_r}}{r} \leq \frac{\sum_{i \in U} w_i}{k+1} \leq \varepsilon \cdot \sum_{i \in U} w_i. \tag{4.23}$$

Since $cost(U) \leq c + d \cdot (1+\delta)$ we obtain

$$\frac{cost(U) - cost(T^*)}{cost(U)} \leq \frac{c + d \cdot (1+\delta) - cost(T^*)}{cost(U)}$$

$$\underset{(4.23)}{=} \frac{c - cost(T^*)}{cost(U)} + \frac{(1+\delta) \cdot \varepsilon \cdot \sum_{i \in U} w_i}{cost(U)}$$

$$\underset{(4.22)}{\leq} \delta \cdot (2+\delta) + (1+\delta) \cdot \varepsilon \cdot (1+\delta)$$

$$= 2\delta + \delta^2 + \varepsilon \cdot (1+\delta)^2$$

$$= \varepsilon + \delta \cdot (2+\delta) \cdot (1+\varepsilon). \qquad \square$$

Because KP is a maximization problem Lemma 4.3.4.4 still does not provide any evidence about a constant approximation ratio of the algorithm $\mathrm{ASKP}_\varepsilon$ for KP_δ. The following assertion shows it in a straightforward way[11].

Lemma 4.3.4.5. *For every $\varepsilon > 0$ and every $\delta > 0$, the algorithm $\mathrm{ASKP}_\varepsilon$ is a $(1 + \delta^2 + \varepsilon + \varepsilon \cdot \delta^2)$-approximation algorithm for KP_δ.*

Proof. Let $I = (w_1, w_2, \ldots, w_n, b, c_1, c_2, \ldots, c_n)$ be an input instance of KP_δ, i.e.,

$$(1 + \delta)^{-1} \le \frac{w_i}{c_i} \le 1 + \delta \tag{4.24}$$

for all $i = 1, 2, \ldots, n$.

Let U be an optimal solution of I and let T^* be the output of $\mathrm{ASKP}_\varepsilon$ for I. Since $\mathrm{ASKP}_\varepsilon$ is a $(1 + \varepsilon)$-approximation algorithm for $I' = (w_1, w_2, \ldots, w_n, b)$ of SKP_ε, we have

$$\frac{\sum_{i \in U} w_i}{\sum_{j \in T^*} w_j} \le 1 + \varepsilon. \tag{4.25}$$

Now, we are ready to estimate the approximation ratio of $\mathrm{ASKP}_\varepsilon$ for the input instance I.

$$R(I, \varepsilon) = \frac{cost(U)}{cost(T^*)} \underset{(4.24)}{\le} \frac{\sum_{i \in U} w_i \cdot (1 + \delta)}{\sum_{j \in T^*} w_j \cdot (1 + \delta)^{-1}}$$

$$= (1 + \delta)^2 \cdot \frac{\sum_{i \in U} w_i}{\sum_{j \in T^*} w_j} \underset{(4.25)}{\le} (1 + \delta)^2 \cdot (1 + \varepsilon)$$

$$= 1 + \delta^2 + \varepsilon + \varepsilon \cdot \delta^2.$$

Corollary 4.3.4.6. *The PTAS Algorithm 4.3.4.2 is stable according to DIST, but not superstable according to DIST.*

Proof. The first assertion directly follows from Lemma 4.3.4.4. To see that SKP is not superstable according to *DIST* it is sufficient to consider the following input:[12]

$$w_1, w_2, \ldots, w_m, u_1, u_2, \ldots, u_m, b, c_1, c_2, \ldots, c_{2m},$$

where $w_1 = w_2 = \cdots = w_m$, $u_1 = u_2 = \cdots = u_m$, $w_i = u_i + 1$ for $i = 1, \ldots, m$, $b = \sum_{i=1}^m w_i$, $c_1 = c_2 = \cdots = c_m = (1 - \delta)w_1$ and $c_{m+1} = c_{m+2} = \cdots = c_{2m} = (1 + \delta)u_1$. $\qquad \square$

We see that the PTAS SKP is stable according to *DIST*, but this does not suffice to get a PTAS for KP_δ for any $\delta > 0$. This is because in the approximation ratio we have the additive factor δ^2 that is independent of ε.

---

[11] The main reason for presenting Lemma 4.3.4.4 is that the considerations involved in its proof are essential for improving $\mathrm{ASKP}_\varepsilon$ later.

[12] Note that m should be chosen to be essentially larger than ε^{-1} for a given ε.

In what follows we change Algorithm 4.3.4.2 a little bit in such a way that we obtain a PTAS for every KP_δ, $\delta > 0$. The modification idea is very natural – to sort the input values according to the cost per one weight unit.

Algorithm 4.3.4.7. PTAS MOD-SKP

Input: Positive integers $w_1, w_2, \ldots, w_n, b, c_1, \ldots, c_n$ for some $n \in \mathbb{N}$, and some positive real number ε, $1 > \varepsilon > 0$.

Step 1:   Sort $\frac{c_1}{w_1}, \frac{c_2}{w_2}, \ldots, \frac{c_n}{w_n}$. For simplicity we may assume $\frac{c_i}{w_i} \geq \frac{c_{i+1}}{w_{i+1}}$ for $i = 1, \ldots, n-1$.

Step 2:   Set $k = \lceil 1/\varepsilon \rceil$.

Step 3:   The same as Step 3 of Algorithm 4.3.4.2, but the greedy procedure follows the ordering of the w_is of Step 1.

Output: The best T^* constructed in Step 3.

Let MOD-SKP$_\varepsilon$ denote the algorithm given by PTAS MOD-SKP for any fixed $\varepsilon > 0$.

Lemma 4.3.4.8. *For every ε, $1 > \varepsilon > 0$ and every $\delta \geq 0$, MOD-SKP$_\varepsilon$ is a $(1 + \varepsilon \cdot (1 + \delta) \cdot (1 + \varepsilon))$-approximation algorithm for KP_δ.*

Proof. Let $U = \{i_1, i_2, \ldots, i_l\} \subseteq \{1, 2, \ldots, n\}$, where $w_{i_1} \geq w_{i_2} \geq \cdots \geq w_{i_l}$[13] be an optimal solution for the input $I = w_1, \ldots, w_n, b, c_1, \ldots, c_n$.

If $l \leq k$, then MOD-SKP$_\varepsilon$ provides an optimal solution.

If $l > k$, then we consider a $T^* = \{i_1, i_2, \ldots, i_k, j_{k+1}, \ldots, j_{k+r}\}$ as a greedy extension of $T = \{i_1, i_2, \ldots, i_k\}$. Again, we distinguish two possibilities according to the sizes of $\sum_{i \in U} w_i$ and $\sum_{j \in T^*} w_j$.

(i) Let $\sum_{i \in U} w_i - \sum_{j \in T^*} w_j < 0$.

Now, we show that this is impossible because it contradicts the optimality of U. Both $cost(U)$ and $cost(T^*)$ contain $\sum_{s=1}^{k} c_{i_s}$. For the rest T^* contains the best choice of w_is according to the cost of one weight unit. The choice of U per one weight unit cannot be better. Thus, $cost(U) < cost(T^*)$.

(ii) Let $d = \sum_{i \in U} w_i - \sum_{j \in T^*} w_j \geq 0$.

Because of the optimal choice of T^* according to the cost per one weight unit, the cost c of the first part of U with the weight $\sum_{j \in T^*} w_j$ is at most $cost(T^*)$, i.e.,

$$c - cost(T^*) \leq 0 \tag{4.26}$$

Since U and T^* contain the same k indices $i_1, i_2, \ldots, i_k$, and $w_{i_1}, \ldots, w_{i_k}$ are the largest weights in both U and T^*, the same consideration as in the proof of Lemma 4.3.4.4 yields (see (4.23))

$$d \leq \varepsilon \cdot \sum_{i \in U} w_i, \text{ and } cost(U) \leq c + d \cdot (1 + \delta). \tag{4.27}$$

---

[13] Observe that at this place we consider the order according to the weights and not according to the cost per one weight unit.

Considering U different from T^*, there exists an $m \in \{k+1, \ldots, l\}$ such that

$$i_m \in U - T^* \text{ and } \sum_{j \in T^*} w_j + w_{i_m} > b \geq \sum_{i \in U} w_i.$$

Therefore, MOD-SKP$_\varepsilon$ is also an ε-approximation algorithm[14] for SKP and so

$$\frac{\sum_{i \in U} w_i}{\sum_{j \in T^*} w_j} \leq 1 + \varepsilon. \tag{4.28}$$

Hence,

$$
\begin{aligned}
R(I, \varepsilon) &= \frac{cost(U)}{cost(T^*)} \\
&\underset{(4.27)}{\leq} \frac{c + d \cdot (1 + \delta)}{cost(T^*)} \\
&= \frac{cost(T^*) + c - cost(T^*) + d \cdot (1 + \delta)}{cost(T^*)} \\
&= 1 + \frac{c - cost(T^*) + d \cdot (1 + \delta)}{cost(T^*)} \\
&\underset{(4.26)}{\leq} 1 + \frac{d \cdot (1 + \delta)}{cost(T^*)} \\
&\underset{(4.27)}{\leq} 1 + \varepsilon \cdot (1 + \delta) \cdot \frac{\sum_{i \in U} w_i}{\sum_{j \in T^*} w_j} \\
&\underset{(4.28)}{\leq} 1 + \varepsilon \cdot (1 + \delta) \cdot (1 + \varepsilon).
\end{aligned}
$$

$\square$

Exercise 4.3.4.9. Show, that for every $\varepsilon \geq 0$ and every $\delta \geq 0$, the relative error of MOD-SKP$_\varepsilon$ for KP$_\varepsilon$ is at most $\varepsilon \cdot (1 + \delta)^2$. $\square$

We observe that the collection of MOD-SKP$_\varepsilon$ algorithms is a PTAS for every KP_δ with a constant $\delta \geq 0$ (independent of the size $2n+1$ of the input).

Theorem 4.3.4.10. MOD-SKP *is superstable according to DIST, and so Algorithm 4.3.4.7 is a PTAS for KP.*

To obtain an FPTAS for KP we consider a completely new approach. In Section 3.2 we presented Algorithm 3.2.2.2 (based on dynamic programming) for KP working in time $O(n \cdot Opt_{KP}(I))$ for any input $I = w_1, \ldots, w_n, b, c_1, \ldots, c_n$. The trouble is that $Opt_{KP}(I)$ can be exponential in the input length and

---

[14] see the proof of Theorem 4.3.4.3 for detailed arguments.

so the time complexity of Algorithm 3.2.2.2 has to be considered to be exponential. The idea of our FPTAS for KP is to "approximate" every input $I = w_1, \ldots, w_n, b, c_1, \ldots, c_n$ by another input with $I' = w_1, w_2, \ldots, w_n, b, c'_1, \ldots, c'_n$ with $\sum_{i=1}^{n} c'_i$ polynomial in n, and then to apply Algorithm 3.2.2.2 to the input I' in order to get a feasible solution for I. If one wants to obtain $Opt_{KP}(I')$ to be small in n, then one has to divide the input values $c_1, \ldots, c_n$ by a same large number d. Obviously, the efficiency increases with growing d, but the approximation ratio increases with growing d, too. So, d may be chosen dependent on the user preference.

Algorithm 4.3.4.11. FPTAS for KP

Input: $w_1, \ldots, w_n, b, c_1, \ldots, c_n \in \mathbb{N}, n \in \mathbb{N}, \varepsilon \in \mathbb{R}^+$.
Step 1: $c_{max} := \max\{c_1, \ldots, c_n\}$;
$$t := \left\lfloor \log_2 \frac{\varepsilon \cdot c_{max}}{(1+\varepsilon) \cdot n} \right\rfloor;$$
Step 2: **for** $i = 1$ **to** n
           do $c'_i := \lfloor c_i \cdot 2^{-t} \rfloor$.
Step 3: Compute an optimal solution T' for the input
           $I' = w_1, \ldots, w_n, b, c'_1, \ldots, c'_n$ by Algorithm 3.2.2.2.
Output: T'.

Theorem 4.3.4.12. *Algorithm 4.3.4.11 is an* FPTAS *for KP.*

Proof. First, we show that Algorithm 4.3.4.11 is an approximation scheme. Since the output T' (an optimal solution for I') is a feasible solution for I', and I and I' do not differ in the weights $w_1, \ldots, w_n, b$, T' is a feasible solution for I, too. Let T be an optimal solution for the original input I. Our goal is to show that

$$R(I) = \frac{cost(T, I)}{cost(T', I)} \leq 1 + \varepsilon.$$

We have:

$$cost(T, I) = \sum_{j \in T} c_j$$

$$\geq \sum_{j \in T'} c_j = cost(T', I) \quad \{\text{because } T \text{ is optimal for } I \text{ and } T' \text{ is feasible for } I\}$$

$$\geq 2^t \cdot \sum_{j \in T'} c'_j \quad \{\text{follows from } c'_j = \lfloor c_j \cdot 2^{-t} \rfloor\}$$

$$\geq 2^t \sum_{j \in T} c'_j \quad \{\text{because } T' \text{ is optimal for } I'\}$$

$$= \sum_{j \in T} 2^t \cdot \lfloor c_j \cdot 2^{-t} \rfloor \quad \{\text{because } c'_j = \lfloor c_j \cdot 2^{-t} \rfloor\}$$

$$\geq \sum_{j \in T} 2^t \left(c_j \cdot 2^{-t} - 1 \right) \geq \left(\sum_{j \in T} c_j \right) - n \cdot 2^t = cost(T, I) - n \cdot 2^t.$$

We have, thus, proved

$$cost(T, I) \geq cost(T', I) \geq cost(T, I) - n \cdot 2^t, \text{ i.e.,} \tag{4.29}$$

$$0 \leq cost(T, I) - cost(T', I) \leq n \cdot 2^t$$

$$\leq n \cdot \frac{\varepsilon \cdot c_{max}}{(1 + \varepsilon) \cdot n} = \varepsilon \cdot \frac{c_{max}}{1 + \varepsilon}. \tag{4.30}$$

Since we may assume $cost(T, I) \geq c_{max}$ ($w_i \leq b$ for every $i = 1, \ldots, n$), we obtain from (4.29) and (4.30)

$$cost(T', I) \geq c_{max} - \varepsilon \cdot \frac{c_{max}}{1 + \varepsilon}. \tag{4.31}$$

Finally,

$$R(I) = \frac{cost(T, I)}{cost(T', I)} = \frac{cost(T', I) + cost(T, I) - cost(T', I)}{cost(T', I)}$$

$$\leq 1 + \frac{\varepsilon \cdot \frac{c_{max}}{1 + \varepsilon}}{cost(T', I)} \quad \{\text{because of (4.30)}\}$$

$$\leq 1 + \frac{\varepsilon \cdot \frac{c_{max}}{1 + \varepsilon}}{c_{max} - \varepsilon \cdot \frac{c_{max}}{1 + \varepsilon}} \quad \{\text{because of (4.31)}\}$$

$$= 1 + \frac{\varepsilon}{1 + \varepsilon} \cdot \frac{1}{1 - \frac{\varepsilon}{1 + \varepsilon}} = 1 + \frac{\varepsilon}{1 + \varepsilon} \cdot (1 + \varepsilon) = 1 + \varepsilon.$$

Now, we have to prove that the time complexity of Algorithm 4.3.4.11 is polynomial in n and ε^{-1}. Step 1 and Step 2 can be executed in time $O(n)$. Step 3 is the run of Algorithm 3.2.2.2 on the input I' and this can be performed in time $O(n \cdot Opt_{KP}(I'))$. We bound $Opt_{KP}(I')$ as follows:

$$Opt_{KP}(I') \leq \sum_{i=1}^{n} c_i' = \sum_{i=1}^{n} \left\lfloor c_i \cdot 2^{-\left\lfloor \log_2 \frac{\varepsilon \cdot c_{max}}{(1+\varepsilon) \cdot n} \right\rfloor} \right\rfloor$$

$$\leq \sum_{i=1}^{n} \left(c_i \cdot 2 \cdot \frac{(1 + \varepsilon) \cdot n}{\varepsilon \cdot c_{max}} \right)$$

$$= 2 \cdot (1 + \varepsilon) \cdot \varepsilon^{-1} \cdot \frac{n}{c_{max}} \cdot \sum_{i=1}^{n} c_i$$

$$\leq 2 \cdot (1 + \varepsilon) \cdot \varepsilon^{-1} \cdot n^2 \in O\left(\varepsilon^{-1} \cdot n^2\right).$$

Thus, Algorithm 4.3.4.11 works in time $O\left(\varepsilon^{-1} \cdot n^3\right)$. □

Exercise 4.3.4.13. Consider Algorithm 4.3.4.11 for parameters other than $d = 2^{-t}$. By a suitable choice of d find an FPTAS for KP, whose time complexity is quadratic in n. □

Exercise 4.3.4.14. $^{(*)}$ Design a new FPTAS for KP such that the running time is in

(i) $O\left(n \cdot \log n + n \cdot \varepsilon^{-2}\right)$,
(ii) $O\left(n \cdot \log_2\left(\varepsilon^{-1}\right) + \varepsilon^{-4}\right)$.

$\square$

Summary of Section 4.3.4

We presented the following two methods for the design of polynomial-time approximation algorithms:

(i) A combination of the greedy approach and total search in order to get a PTAS.
(ii) An application of a pseudo-polynomial-time optimization algorithm A that may be of complexity $O(p(n) \cdot F)$, where p is a polynomial in the number n of input variables and F is linear in the size of the values of the input variables. Since the value of a number is exponential in the size of its representation, F can be exponential in the input length. To be able to efficiently apply A, one "reduces" the inputs with F exponential to n to inputs with F polynomial to n. This may lead to a good approximation ratio for every input instance.

We have applied (i) to get a PTAS for SKP, and we have applied (ii) to design an FPTAS for KP. Moreover, we have shown that considering the notion of the superstability of approximation, one can find a natural way to extend the use of a PTAS to a class of problem instances that is essential larger than the original class.

4.3.5 Traveling Salesperson Problem and Stability of Approximation

The traveling salesperson problem (TSP) is a representative of the class NPO(V), i.e., one of the hardest optimization problems according to polynomial-time approximability.[15] Because of this one could consider the approximation approach unsuitable for TSP. We show here that this needs not to be true even for the hardest problems from NPO(V). First, we show that there is a subset $L_\triangle$ of input instances such that

(i) one can design efficient approximation algorithms for inputs in $L_\triangle$,
(ii) several applications produce inputs either from $L_\triangle$ or inputs whose properties are not very far from the specification of $L_\triangle$.

This could already be useful for some applications but we still improve on it. We show that one can modify the above mentioned algorithms to get stable algorithms. In this way one can apply these algorithms for inputs outside $L_\triangle$ and efficiently compute an upper bound on the approximation ratio for every input.

---

[15] More information about the inapproximability of TSP will be given in Section 4.4.

First, we present two algorithms for $\triangle$-TSP, i.e., TSP with the inputs satisfying the triangle inequality.

Algorithm 4.3.5.1.

Input: A complete graph $G = (V, E)$, and a cost function $c : E \to \mathbb{N}^+$ satisfying the triangle inequality

$$c(\{u, v\}) \leq c(\{u, w\}) + c(\{w, v\})$$

for all three different $u, v, w \in V$ {i.e., $(G, c) \in L_\triangle$}.

Step 1:   Construct a minimal spanning tree T of G according to c.

Step 2:   Choose an arbitrary vertex $v \in V$. Perform depth-first-search of T from v, and order the vertices in the order that they are visited. Let H be the resulting sequence.

Output: The Hamiltonian tour $\overline{H} = H, v$.

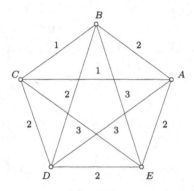

Fig. 4.5.

Example We illustrate the work of Algorithm 4.3.5.1 on the weighted complete graph (K_5, c) in Figure 4.5. One can easily verify that c satisfies the triangle inequality. Figure 4.6 shows the construction of a minimal spanning tree of (K_5, c) by the greedy method. The edges chosen for the spanning tree are depicted as the thick lines in Figure 4.6.

Step 2 of Algorithm 4.3.5.1 is illustrated in Figure 4.7. The depth-first-search of the spanning tree T of Figure 4.6d is begun from the vertex C. The first edge of T chosen is $\{C, D\}$ in Figure 4.7a. Since D has not been visited up till now, the edge $\{C, D\}$ is taken for the Hamiltonian tour. This fact is depicted in Figure 4.7b. In what follows the thick lines are used for edges chosen for the Hamiltonian tour. Since there is no possibility to continue from D in T, we return to C and choose the edge $\{C, A\}$ in the depth-first-search (Figure 4.7c). Since C has been already visited and A was still not visited, we take the edge $\{D, A\}$ for the Hamiltonian tour (Figure 4.7d). We, then, have fixed C, D, A as the beginning of the Hamiltonian tour, and we have visited the vertices C, D, A up till now in the depth-first-search. The only possibility to continue from A in the depth-first-search of T is to take the edge $\{A, E\}$ (Figure 4.7e).

Since the vertex E has not been visited yet, we take the edge $\{A, E\}$ for our Hamiltonian tour. Because there is no possibility to continue from E in T, we return to A. But it is impossible to continue from A, too, and so the depth-first-search returns to vertex C. The only remaining possibility to continue from C (Figure 4.6d) is to take the edge $\{C, B\}$ (Figure 4.7g). Since B is the next new vertex in the depth-first-search, we fix $\{E, B\}$ as the next edge of the Hamiltonian tour (Figure 4.7h). But this completes the depth-first-search of T and so we close the Hamiltonian tour by taking the edge $\{B, C\}$ (Figure 4.7e). The cost of the resulting Hamiltonian tour C, D, A, E, B, C is $2 + 3 + 2 + 3 + 1 = 11$.

Note that the optimal Hamiltonian tour is C, B, D, E, A, C and its cost is 8 (see Figure 4.13).   □

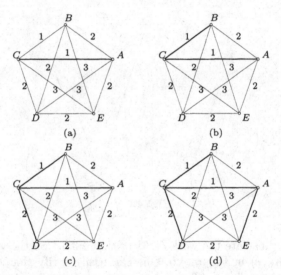

Fig. 4.6.

Theorem 4.3.5.2. *Algorithm 4.3.5.1 is a polynomial-time 2-approximation algorithm for $\triangle$-TSP.*

Proof. First, we analyze the time complexity of Algorithm 4.3.5.1. The first step as well as the second step correspond to the well-known $O(|E|)$-time algorithms. Thus, $Time_A(n) \in O(|E|) \subseteq O(n^2)$.

Let H_{Opt} be an optimal Hamiltonian tour with $cost(H_{Opt}) = Opt_{\triangle\text{-TSP}}(I)$ for an input instance $I = ((V, E), c)$. Let $\overline{H}$ be the output of Algorithm 4.3.5.1.

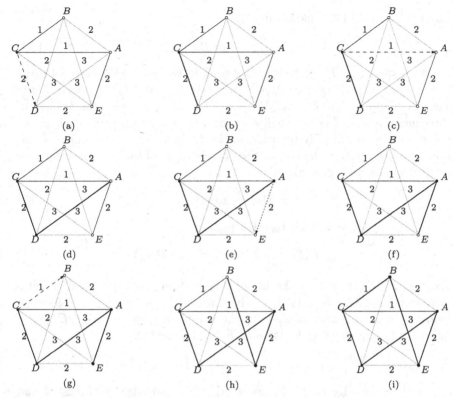

Fig. 4.7. The simulation of Step 2 of Algorithm 4.3.5.1 for the input instance (K_5, c) and T depicted in Figure 4.6d. The thick lines are used to denote edges chosen for the Hamiltonian tour. The dotted lines point out the actual steps of the depth-first-search procedure.

Obviously, $\overline{H} = H, v$ is a feasible solution because H is a permutation of the vertices of V.

First, we observe that

$$cost(T) = \sum_{e \in E(T)} c(e) \leq cost(H_{Opt}) \tag{4.32}$$

because the deletion of any edge of H_{Opt} results in a spanning tree and T is a minimal spanning tree.

Consider W to be the path that corresponds to the depth-first-search of T. W goes twice via every edge of T (once in every direction of every edge[16]). If $cost(W)$ is the sum of the costs of all edges of W, then

$$cost(W) = 2 \cdot cost(T). \tag{4.33}$$

---

[16] Note that W can be considered as an Eulerian tour of a multigraph T_2 that is constructed from T by doubling every edge.

From (4.32) and (4.33) one obtains

$$cost(W) \leq 2 \cdot cost(H_{Opt}). \tag{4.34}$$

We observe that $\overline{H}$ can be obtained from W by exchanging some subpaths $u, v_1, \ldots, v_k, v$ in W by the edges[17] $\{u, v\}$. In fact, this can be done by the successive application of a simple operation that exchanges subsequences of three nodes u, w, v of W by the direct connections u, v (this exchange happens if the node w has already occurred in the prefix of W). This simple operation does not increase the cost of the path because of the triangle inequality $(c\{u, v\} \leq c\{u, w\} + c\{w, v\})$. Thus,

$$cost(\overline{H}) \leq cost(W). \tag{4.35}$$

Finally, (4.34) and (4.35) together provide

$$cost(\overline{H}) \leq cost(W) \leq 2 \cdot cost(H_{Opt}). \qquad \square$$

Example We show that the upper bound 2 on the approximation ratio of Algorithm 4.3.5.1 is tight. Let n be a positive, odd integer. Consider the following input instance of $\triangle$-TSP: the complete graph $G = (V, E)$, $V = \{v_1, \ldots, v_n\}$, and the cost function $c : E \to \{1, 2\}$, where

$$c(\{v_2, v_3\}) = c(\{v_3, v_4\}) = \cdots = c(\{v_{n-1}, v_n\}) = 2 \text{ and } c(e) = 1$$

for every other edge of E (Figure 4.8a), where all edges with cost 2 and some edges of cost 1 are depicted. Obviously, $(V, \{\{v_1, v_i\} \mid i = 1, \ldots, n\})$ is a minimal spanning tree of G according to c, and $cost(T) = n - 1$ (Figure 4.8b). The output of Algorithm 4.3.5.1 may be[18] the Hamiltonian tour[19] $\overline{H} = v_1, v_2, v_3, \ldots, v_n, v_1$ with $cost(\overline{H}) = 2n - 2$ (Figure 4.8c). But an optimal Hamiltonian tour H_{Opt} is (Figure 4.8d)

$$v_1, v_2, v_4, v_6, \ldots, v_{2i}, v_{2i+2}, \ldots, v_{n-1}, v_n, v_3, v_5, v_7, \ldots, v_{2j-1}, v_{2j+1}, \ldots, v_{n-2}, v_1.$$

We observe that $\{v_{n-1}, v_n\}$ is the only edge with cost 2 in H_{Opt} and so $cost(H_{Opt}) = n + 1$. Then

$$\frac{cost(\overline{H})}{cost(H_{Opt})} = \frac{2n - 2}{n + 1},$$

which tends to 2 with growing n. $\qquad \square$

If one looks at Algorithm 4.3.5.1, then one can see the following strategy behind this algorithm.

---

[17] Direct connections

[18] Note that there are several different minimal spanning trees of G, but we cannot remove the possibility that Algorithm 4.3.5.1 computes exactly the tree in Figure 4.8a.

[19] Observe that $\overline{H}$ contains all expensive edges.

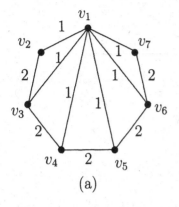

(a)

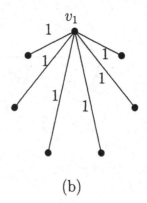

(b)

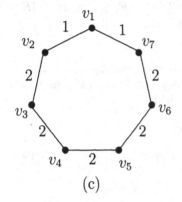

(c)

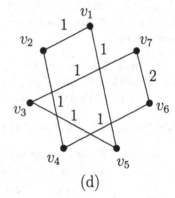

(d)

Fig. 4.8.

(i) Construct a spanning tree T with the *cost* bounded by $Opt_{\triangle\text{-TSP}}(I)$.

(ii) Extend T to a multigraph T^* such that T^* contains an Eulerian tour W and $cost(T^*) = cost(W)$ is as small as possible.

(iii) Shorten the Eulerian tour W in order to get a Hamiltonian tour H. Because of the triangle inequality $cost(H) \leq cost(W)$.

Following this strategy the only part where one can save costs in comparison to Algorithm 4.3.5.1 is part (ii). In Algorithm 4.3.5.1 we get T^* by doubling every edge, but this may be too expensive. In fact we only need to achieve an even degree of all vertices.[20] This can be achieved by taking a "cheap" matching on those vertices of T that have an odd degree. This idea works because the number of vertices with an odd degree is even in any graph.

---

[20] This is a sufficient and necessary condition to get an Eulerian tour.

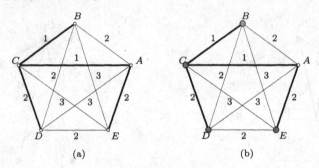

Fig. 4.9.

Lemma 4.3.5.3. *For every graph T, the number of vertices of an odd degree is even.*

Proof. The sum of the degrees of all vertices of $T = (V, E)$ is exactly $2 \cdot |E|$. Since $2 \cdot |E|$ is an even number, the number of vertices with an odd degree must be even.   □

Lemma 4.3.5.3 shows that we can apply our idea of building a multigraph whose vertices all have an even degree by adding matching to the minimal spanning tree. This results in the following algorithm.

Algorithm 4.3.5.4. CHRISTOFIDES ALGORITHM

Input: A complete graph $G = (V, E)$, and a cost function $c : E \to \mathbb{N}^+$ satisfying the triangle inequality.

Step 1:   Construct a minimal spanning tree T of G according to c.

Step 2: $S := \{v \in V \mid deg_T(v) \text{ is odd}\}$.

Step 3:   Compute a minimum-weight[21] perfect[22] matching M on S in G.

Step 4:   Create the multigraph $G' = (V, E(T) \cup M)$ and construct an Eulerian tour ω in G'.

Step 5:   Construct a Hamiltonian tour H of G by shortening ω (i.e., by removing all repetitions of the occurrences of every vertex in ω in one run via ω from the left to the right).

Output: H.

Example We illustrate the work of CHRISTOFIDES ALGORITHM on K_5 with the same weight function c as considered for the illustration of Algorithm 4.3.5.1. In Step 1 a minimal spanning tree T is constructed (Figure 4.9a). The vertices B, C, D, and E have an odd degree in T (Figure 4.9b). The task now is to compute a minimal perfect matching M in K_5 on B, C, D, E. The result

---

[22] A matching M is perfect on a set S of vertices if every vertex of the graph (S, M) has a degree of exactly 1, i.e., every vertex is "included" in the matching.

[22] According to c

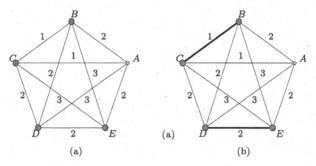

Fig. 4.10.

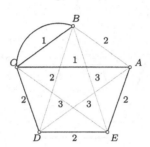

Fig. 4.11.

$\{\{C,B\},\{D,E\}\}$ is depicted in Figure 4.10b. In Step 4 the Eulerian tour $\omega = C, D, E, A, C, B, C$ in the multigraph $G' = (\{A, B, C, D, E\}, \{\{C, D\}, \{D, E\}, \{E, A\}, \{A, C\}, \{C, B\}, \{B, C\}\}$ depicted in Figure 4.11 is constructed.

The shortening of $\omega = C, D, E, A, C, B, C$ to C, D, E, A, B, C in Step 5 is depicted in Figure 4.12. We start at the vertex C. We follow the first edge $\{C, D\}$ of ω (Figure 4.12a) and take it for our Hamiltonian tour H (Figure 4.12b). In what follows the edges fixed for the Hamiltonian tour are depicted by the thick lines in Figure 4.12. The next edge of ω is $\{D, E\}$ (Figure 4.12c) and we take it for H (Figure 4.12d) because E is visited for the first time in ω. The third edge of ω is $\{E, A\}$ (Figure 4.12e) and $\{E, A\}$ is taken to H because A has been reached for the first time in ω. The next edge in ω is $\{A, C\}$ (Figure 4.12g). We do not take it for H because the vertex C has already appeared in ω and there is still the vertex B that has not occurred up till now in ω (Figure 4.12h). The next edge of ω is $\{C, B\}$ (Figure 4.12i). Since B is still not in H, we connect the vertex A (the last visited vertex on H up till now) with B by the edge $\{A, B\}$ (Figure 4.12j). The edge $\{B, C\}$ closes the Hamiltonian tour $H = C, D, E, A, B, C$ (Figure 4.12k) of cost 9.

Observe that H is not optimal, but cheaper than the Hamiltonian tour C, D, A, E, B, C produced by Algorithm 4.3.5.1. An optimal Hamiltonian tour has cost 8 and it is depicted in Figure 4.13.   □

Theorem 4.3.5.5. *The* CHRISTOFIDES ALGORITHM *is a polynomial-time* 1.5-*approximation algorithm for* △-TSP.

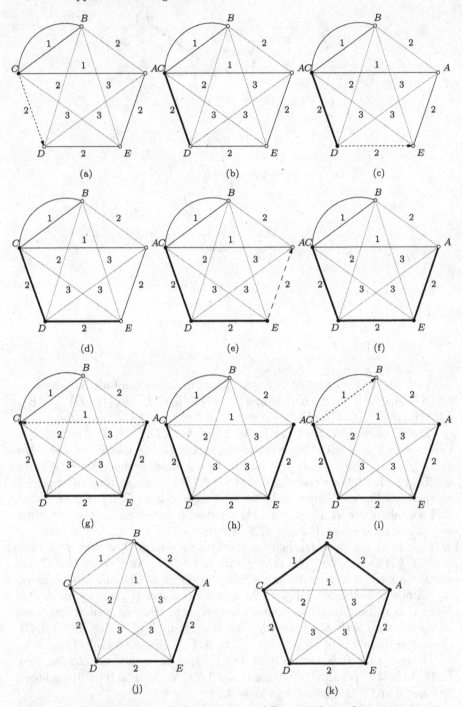

Fig. 4.12. An illustration of the execution of Step 5 of the CHRISTOFIDES AL-GORITHM. The depth-first-search from the vertex C in the multigraph depicted in Figure 4.11 is performed.

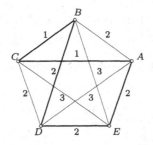

Fig. 4.13.

Proof. First, we analyze the time complexity $Time_{Ch}(n)$ of the CHRISTOFIDES ALGORITHM. The first step can be executed by the standard algorithms that solve the minimal spanning tree problem in $O(|E|) = O(n^2)$ time. Step 2 can be completed in $O(n^2)$ steps. A minimum-weight perfect matching in a graph satisfying the triangle inequality can be computed in $O(n^2 \cdot |E|)$ time. The multigraph G' can be created in $O(n^2)$ time,[23] and an Eulerian tour of G' can be computed in $O(|V(T)|+|M|) = O(n)$ time. Finally, Step 5 can be performed in $O(|V(T)| + |M|) = O(n)$ time. Thus, $Time_{Ch}(n) = O(n^2 \cdot e) = O(n^4)$.

Let H_{Opt} be an optimal Hamiltonian tour in $G = (V, E)$ with $cost(H_{Opt}) = Opt_{\triangle\text{-TSP}}(I)$ for an input instance $I = (G, c)$ of $\triangle$-TSP. Let H be the output of the Christofides algorithm, and let ω be the Eulerian tour constructed in Step 4 of the CHRISTOFIDES ALGORITHM. Obviously,

$$cost(\omega) = cost(G') = \sum_{e \in E(T) \cup M} c(e) = cost(T) + cost(M). \qquad (4.36)$$

Because of the triangle inequality and (4.36) we have

$$cost(H) \leq \underset{(4.36)}{cost(\omega)} = cost(T) + cost(M). \qquad (4.37)$$

Now, we bound $cost(T)$ and $cost(M)$ according to $cost(H_{Opt})$. Since the deletion of any edge of H_{Opt} results in a spanning tree, and T is a minimal spanning tree, we have

$$cost(T) \leq cost(H_{Opt}). \qquad (4.38)$$

Let $S = \{v_1, \ldots, v_{2m}\}$, with m a positive integer, be the set[24] of all vertices of an odd degree in T. Without loss of generality, one can write

$$H_{Opt} = v_1, \alpha_1, v_2, \alpha_2, \ldots, \alpha_{2m-1}, v_{2m}, \alpha_{2m}, v_1,$$

where α_i is a (potentially empty) path for $i = 1, \ldots, 2m$ (Figure 4.14). Consider the following matchings in G:

---

[23] This can be even done in $O(n)$ time by a suitable representation of T and M but this does not matter for our calculations.

[24] Constructed in Step 2 of Christofides algorithm

$$M_1 := \{\{v_1, v_2\}, \{v_3, v_4\}, \ldots, \{v_{2m-1}, v_{2m}\}\}, \text{ and}$$
$$M_2 := \{\{v_2, v_3\}, \{v_4, v_5\}, \ldots, \{v_{2m-2}, v_{2m-1}\}, \{v_{2m}, v_1\}\}$$

depicted in Figure 4.14. Because of the triangle inequality

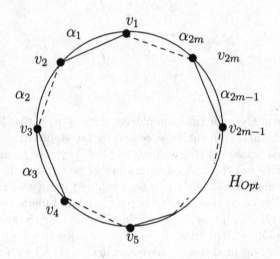

Fig. 4.14.

$$cost(v_i, \alpha_i, v_{i+1}) \geq c(\{v_i, v_{i+1}\}) \tag{4.39}$$

for every $i = 1, \ldots, 2m$.[25] The inequality (4.39) implies

$$cost(H_{Opt}) \geq \sum_{i=1}^{2m} c(\{v_i, v_{i+1}\}) = cost(M_1) + cost(M_2). \tag{4.40}$$

Since M_1 and M_2 are perfect matchings of S

$$cost(M_1) \geq cost(M) \text{ and } cost(M_2) \geq cost(M). \tag{4.41}$$

Statements (4.40) and (4.41) directly imply

$$cost(M) \leq \min\{cost(M_1), cost(M_2)\} \leq \frac{1}{2} cost(H_{Opt}). \tag{4.42}$$

Finally, putting (4.38) and (4.42) into (4.37) we get

$$cost(H) \underset{(4.37)}{\leq} cost(T) + cost(M) \leq cost(H_{Opt}) + \frac{1}{2} cost(H_{Opt}) = \frac{3}{2} cost(H_{Opt}).$$

$$\square$$

---

[25] Set $v_{2m+1} = v_1$.

Example This example shows that the analysis of the approximation ratio of the CHRISTOFIDES ALGORITHM in Theorem 4.3.5.2 is tight. Consider the complete graph G_{2t} whose part is depicted in Figure 4.15a. All depicted edges have the cost 1, but $c(\{v_1, v_{2t}\}) = t$. All other edges have costs that correspond to their Euclidean lengths (distances) in Figure 4.15a.

A minimal spanning tree T is depicted in Figure 4.15b and $cost(T) = 2t-1$. Observe that every node in T has an odd degree. A minimum-weight perfect matching for $V(T)$ is $M = \{\{v_1, v_2\}, \{v_3, v_4\}, \ldots, \{v_{2t-1}, v_{2t}\}\}$ as depicted in Figure 4.15c. Observe that $cost(M) = t$. The output of the CHRISTOFIDES ALGORITHM can be (Figure 4.15d)

$$H = v_1, v_2, v_3, v_4, \ldots, v_{2t-2}, v_{2t-1}, v_{2t}, v_1$$

with $cost(H) = 2t-1+t = 3t-1$ if the CHRISTOFIDES ALGORITHM constructs the Eulerian tour w by starting with $v_1, v_2, v_3, v_4, \ldots, v_{2t-1}, v_{2t}, v_{2t+1}, \ldots$. Since the only optimal Hamiltonian tour (Figure 4.15e)

$$H_{Opt} = v_1, v_2, v_4, v_6, \ldots, v_{2t-4}, v_{2t-2}, v_{2t}, v_{2t-1}, v_{2t-3}, \ldots, v_5, v_3, v_1$$

has the cost 2t, we obtain

$$\frac{cost(H)}{cost(H_{Opt})} = \frac{3t-1}{2t},$$

which tends to $\frac{3}{2}$ with growing t.   □

We have seen that $\triangle$-TSP admits a polynomial-time 1.5-approximation algorithm. On the other hand the general TSP does not admit any polynomial-time $p(n)$-approximation algorithm for any polynomial p (if P $\neq$ NP). The latter fact will be proven in the next section. The next exercise shows that if we restrict $\triangle$-TSP to the geometrical TSP (*Euc*-TSP), then one can even design a PTAS.

Exercise 4.3.5.6. Consider the following modification of $\triangle$-TSP. A problem instance I is a weighted graph $((V, E), c)$ with two special vertices q and s from V, where c satisfies the triangle inequality. A solution for I is any path between q and s that contains each vertex of V exactly once (i.e., a path between q and s of length $n - 1$, that does obtain any vertex twice). The aim is to find a solution with the minimal cost.

(a) Design a 2-approximation algorithm for this minimization problem.
(b) Try to modify the CHRISTOFIDES ALGORITHM in order to get a 5/3-approximation algorithm for this problem.

□

Exercise 4.3.5.7. Consider the following algorithm for $\triangle$-TSP.

Input: A complete graph $G = (V, E)$ and a cost function $c : E \to \mathbb{N}^+$ satisfying the triangle inequality.

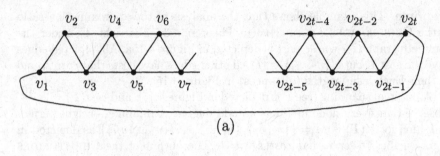

(a)

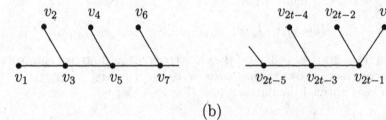

(b)

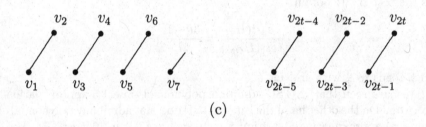

(c)

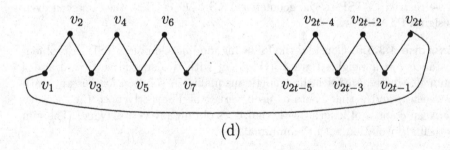

(d)

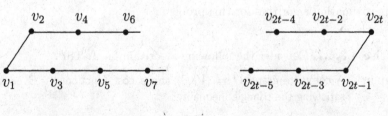

(e)

Step 1:   Find a cheapest edge $\{u, v\}$ of E and start with the tour $w = u, v, u$.

Step 2:   **while** $w = v_1, v_2, \ldots, v_k, v_1$ does not visit all vertices of V **do**
      begin estimate a vertex $w \in V - \{v_1, \ldots, v_k\}$ and an index
      $j \in \{1, \ldots, k\}$ such that
$$cost(\{w, v_j\}) = \min\{cost(\{r, s\}) \mid r \in V - \{v_1, \ldots, v_k\}, \text{ and} $$
$$s \in \{v_1, \ldots, v_k\}\};$$
      set $w = v_1, v_2, \ldots, v_j, w, v_{j+1}, \ldots, v_k, v_1$
      end

Output: w

Show, that this algorithm is a 2-approximation algorithm for $\triangle$-TSP.   □

Exercise 4.3.5.8. $^{(**)}$ Design a PTAS for Euc-TSP.   □

We see that the hardness of TSP is sensible according to the set of problem instances. But to restrict our consideration to the three versions Euc-TSP, $\triangle$-TSP, and TSP is too rough for practical purposes. There are many real inputs that are not in $L_\triangle$ (do not satisfy the triangle inequality), but that do not break the triangle inequality too much. Is it possible to compute efficiently good approximate solutions for such inputs? How does the approximation ratio increase with the "distance" of input instances from the triangle inequality? To answer these questions we use the concept of stability of approximation. In what follows we consider two natural distance measures for TSP. Let TSP $= (\Sigma_I, \Sigma_O, L, L, \mathcal{M}, cost, minimum)$ and $\triangle$-TSP $= (\Sigma_I, \Sigma_O, L, L_\triangle, \mathcal{M}, cost, minimum)$. We observe that Algorithm 4.3.5.1 and CHRISTOFIDES ALGORITHM are consistent for the general TSP. We define for every $x = (G, c) \in L$,

$$dist(x) = \max\left\{0, \max\left\{\frac{c(\{u, v\})}{c(\{u, p\}) + c(\{p, v\})} - 1 \,\middle|\, u, v, p \in V(G)\right\}\right\},$$

and

$$distance(x) = \max\left\{0, \max\left\{\frac{c(\{u, v\})}{\sum_{i=1}^m c(\{p_i, p_{i+1}\})} - 1 \,\middle|\, u, v \in V(G), \right.\right.$$
$$\text{and } u = p_1, p_2, \ldots, p_{m+1} = v$$
$$\left.\left. \text{is a simple path between } u \text{ and } v \text{ in } G\right\}\right\}.$$

For simplicity we consider the size of $x = (G, c)$ as the number of the vertices of G^{26} instead of the real length of the code of x over Σ_I.

We observe that $dist(G, c) \leq r$ implies the so-called **$(1 + r)$-relaxed triangle inequality**

---

[26] Note that the real input length is like $\sum_{e \in E(G)} \lceil \log_2 c(e) \rceil$, because one needs $\lceil \log_2 c(e) \rceil$ bits to code the value $c(e)$.

$$c(\{u, v\}) \leq (1 + r)[c(\{u, w\}) + c(\{w, v\})]$$

for all three different vertices $u, v, w \in V(G)$. The measure *distance* involves a much harder requirement than *dist*. $c(\{u, v\})$ may not be larger than $(1 + r)$ times the cost of any path between u and v. The next results show that these two distance functions are really different because the approximation algorithms for $\triangle$-TSP considered above are stable according to *distance* but not according to *dist*.

Let, for every positive real number r,

$$\triangle\text{-TSP}_r = (\varSigma_I, \varSigma_O, L, Ball_{r, dist}(L_\triangle), \mathcal{M}, cost, minimum).$$

Lemma 4.3.5.9. CHRISTOFIDES ALGORITHM *is stable according to distance.*

Proof. Let $I = (G, c) \in Ball_{r, distance}$ for an $r \in \mathbb{R}^+$. Let ω_I be the Eulerian tour constructed in Step 4 and $H_I = v_1, v_2, v_3, \ldots, v_n, v_{n+1}$, where $v_{n+1} = v_1$, be the Hamiltonian tour constructed by shortening ω_I in Step 5. Clearly,

$$\omega_I = v_1, P_1, v_2, P_2, v_3, \ldots, v_n, P_n, v_{n+1},$$

where P_i is a path between v_i and v_{i+1} for $i = 1, 2, \ldots, n$. Since $c(\{v_i, v_{i+1}\})$ is at most $(1 + r)$ times the cost of v_i, P_i, v_{i+1} for all $i \in \{1, 2, \ldots, n\}$,

$$cost(H_I) \leq (1 + r) \cdot cost(\omega_I). \tag{4.43}$$

Following (4.37) we know that

$$cost(\omega_I) = cost(T_I) + cost(M_I), \tag{4.44}$$

where T_I is the minimal spanning tree constructed in Step 1 and M_I is the matching computed in Step 3. For inputs from $L_\triangle$, (4.42) claims $cost(M) \leq \frac{1}{2} cost(H_{Opt})$, where H_{Opt} is a minimal Hamiltonian tour. Following the arguments of the proof of Theorem 4.3.5.5 for the input I with $distance(I) \leq r$, we obtain

$$cost(H_{Opt}) \geq (1 + r) \cdot [cost(M_1) + cost(M_2)] \tag{4.45}$$

instead of (4.40) for the matchings M_1 and M_2 with the same meaning as in the proof of Theorem 4.3.5.5.

Since M_I is a minimal perfect matching on S, (4.45) implies

$$cost(M_I) \leq \frac{1}{2}(1 + r) \cdot cost(H_{Opt}). \tag{4.46}$$

Following (4.44), (4.46), and (4.38)[27] we obtain

---

[27] That is true for any input instance of TSP.

$$cost(\omega_I) \underset{(4.44)}{=} cost(T_I) + cost(M_I) \underset{\substack{(4.38)\\(4.46)}}{\leq} cost(H_{Opt}) + (1+r) \cdot cost(H_{Opt})$$

$$= (2+r) \cdot cost(H_{Opt}) \tag{4.47}$$

Finally,

$$cost(H_I) \underset{(4.43)}{\leq} (1+r) \cdot cost(\omega_I) \underset{(4.47)}{\leq} (1+r) \cdot (2+r) \cdot cost(H_{Opt}).$$

$\square$

Corollary 4.3.5.10. *For every positive real number r,* CHRISTOFIDES AL-
GORITHM *is a polynomial-time $(1+r) \cdot (2+r)$-approximation algorithm for*
$(\Sigma_I, \Sigma_O, L, Ball_{r,distance}(L_\triangle), \mathcal{M}, cost, minimum)$.

Exercise 4.3.5.11. Prove that Algorithm 4.3.5.1 is a polynomial-time $2 \cdot (1 +$
$r)$-approximation algorithm for $(\Sigma_I, \Sigma_O, L, Ball_{r,distance}(L_\triangle), \mathcal{M}, cost, mini-$
$mum)$. For which input instances from L is Algorithm 4.3.5.1 better than
CHRISTOFIDES ALGORITHM? $\square$

The next results show that Algorithm 4.3.5.1 and CHRISTOFIDES ALGO-
RITHM provide a very weak approximation for input instances of $\triangle$-TSP$_r$ for
any $r \in \mathbb{R}^+$. First, we show a partially positive result and then we prove that
it cannot be essentially improved.

Lemma 4.3.5.12. *For every positive real number r,* CHRISTOFIDES ALGO-
RITHM *is $(r, O(n^{\log_2((1+r)^2)}))$-quasistable for dist.*

Proof. Let $I = (G, c) \in Ball_{r,dist}(L_\triangle)$ for an $r \in \mathbb{R}^+$. Let T_I, ω_I, M_I, and H_I
have the same meaning as in the proof of the previous lemma. To exchange a
path v, P, u of a length m, $m \in \mathbb{N}^+$, for the edge $\{v, u\}$ we proceed as follows.
For any $p, s, t \in V(G)$, one can exchange the path p, s, t for the edge $\{p, t\}$ by
the cost increase bounded by the multiplicative constant $(1+r)$. This means
that reducing the length m of a path to the length $\lceil m/2 \rceil$ increases the cost
of the connection between u and v by at most $(1+r)$ times. After at most
$\lceil \log_2 m \rceil$ such reduction steps one reduces the path v, P, u of length m to the
path v, u, and

$$cost(u, v) = c(\{v, u\}) \leq (1+r)^{\lceil \log_2 m \rceil} \cdot cost(v, P, u). \tag{4.48}$$

Following the arguments of the proof of Theorem 4.3.5.5 we get from (4.48)

$$cost(M_I) \leq \frac{1}{2} \cdot (1+r)^{\lceil \log_2 n \rceil} \cdot cost(H_{Opt}) \tag{4.49}$$

instead of (4.42), and

$$cost(H_I) \leq (1+r)^{\lceil \log_2 n \rceil} cost(\omega_I) \tag{4.50}$$

instead of (4.37). Thus,

$$cost(H_I) \underset{(4.50)}{\leq} (1+r)^{\lceil \log_2 n \rceil} cost(\omega_I) \underset{(4.36)}{=} (1+r)^{\lceil \log_2 \rceil} [cost(T_I) + cost(M_I)]$$

$$\underset{\substack{(4.38) \\ (4.49)}}{\leq} (1+r)^{\lceil \log_2 n \rceil} \left[cost(H_{Opt}) + \frac{1}{2}(1+r)^{\lceil \log_2 n \rceil} \cdot cost(H_{Opt}) \right]$$

$$= (1+r)^{\lceil \log_2 n \rceil} \left(1 + \frac{1}{2}(1+r)^{\lceil \log_2 n \rceil} \right) \cdot cost(H_{Opt})$$

$$= O\left(n^{\log_2((1+r)^2)} \cdot cost(H_{Opt}) \right).$$

$\square$

Exercise 4.3.5.13. Analyze the quasistability of Algorithm 4.3.5.1 according to *dist*. $\square$

Now we show that the result of Lemma 4.3.5.12 cannot be essentially improved. To show this, we construct an input for which the CHRISTOFIDES ALGORITHM provides a very poor approximation.

We construct a weighted complete graph from $Ball_{r,dist}(L_\triangle)$ as follows (Figure 4.16). We start with the path $p_0, p_1, \ldots, p_n$ for $n = 2^k$, $k \in \mathbb{N}$, where every edge $\{p_i, p_{i+1}\}$ has weight 1. Then we add edges $\{p_i, p_{i+2}\}$ for $i = 0, 1, \ldots, n-2$ with weight $2 \cdot (1+r)$. Generally, for every $m \in \{1, \ldots, \log_2 n\}$, we define $weight(\{p_i, p_{i+2^m}\}) = 2^m \cdot (1+r)^m$ for $i = 0, \ldots, n - 2^m$. For all other edges one can take maximal possible weights in such a way that the constructed input is in $Ball_{r,dist}(L_I)$.

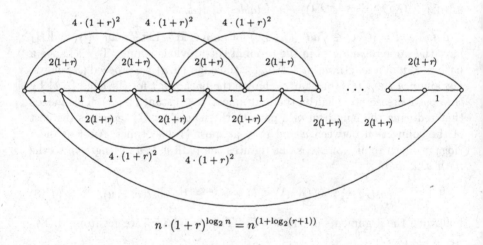

$$n \cdot (1+r)^{\log_2 n} = n^{(1+\log_2(r+1))}$$

Fig. 4.16.

Let us have a look on the work of the CHRISTOFIDES ALGORITHM on the input $(G, weight)$. There is only one minimal spanning tree that corresponds to the path containing all edges of weight 1 (Figure 4.16). Since every path contains exactly two vertices of odd degree, the Eulerian graph constructed in Step 4 is the cycle $D = p_0, p_1, p_2, \ldots, p_n, p_0$ with the n edges of weight 1 and the edge of the maximal weight $n \cdot (1 + r)^{\log_2 n} = n^{1 + \log_2(1+r)}$. Since the Eulerian tour is a Hamiltonian tour (Figure 4.16), the output of the CHRISTOFIDES ALGORITHM is unambiguously the cycle $p_0, p_1, \ldots, p_n, p_0$ with cost $n + n(1+r)^{\log_2 n}$. Analogously, Algorithm 4.3.5.1 outputs the Hamiltonian tour D, too. The optimal tour for this input is $H_{Opt} =$

$$p_0, p_2, p_4, \ldots, p_{2i}, p_{2(i+1)}, \ldots, p_n, p_{n-1}, p_{n-3}, \ldots, p_{2i+1}, p_{2i-1}, \ldots, p_3, p_1, p_0.$$

This tour contains two edges $\{p_0, p_1\}$ and $\{p_{n-1}, p_n\}$ of weight 1 and all $n - 2$ edges of weight $2 \cdot (1 + r)$. Thus, $cost(H_{Opt}) = 2 + 2 \cdot (1 + r) \cdot (n - 2)$ and

$$\frac{cost(D)}{cost(H_{Opt})} = \frac{n + n \cdot (1 + r)^{\log_2 n}}{2 + 2 \cdot (1 + r) \cdot (n - 2)} \geq \frac{n^{1 + \log_2(1+r)}}{2n \cdot (1 + r)} = \frac{n^{\log_2(1+r)}}{2(1 + r)}.$$

Thus, we have proved the following result.

Lemma 4.3.5.14. *For every* $r \in \mathbb{R}^+$, *if the* CHRISTOFIDES ALGORITHM *or Algorithm 4.3.5.1 is* $(r, f_r(n))$-*quasistable for dist, then*

$$f_r(n) \geq n^{\log_2(1+r)}/(2 \cdot (1 + r)).$$

Corollary 4.3.5.15. *Algorithm 4.3.5.1 and the* CHRISTOFIDES ALGORITHM *are unstable for dist.*

The positive results about the stability of Algorithm 4.3.5.1 and the CHRISTOFIDES ALGORITHM according to *distance* show that both algorithms can be useful for a much larger set of inputs than the original set $L_\triangle$ of input instances of $\triangle$-TSP. But the stability according to *dist* would provide low approximation ratios for a substantially larger class of input instances. The key question is whether one can modify the above algorithms to get algorithms that are stable according to *dist*. In what follows, we give a positive answer to this question.

As we have observed by proving the instability of Algorithm 4.3.5.1 and the CHRISTOFIDES ALGORITHM according to *dist*, the main problem is that shortening a path $u_1, u_2, \ldots, u_{m+1}$ to the edge u_1, u_{m+1} can lead to

$$cost(\{u_1, u_{m+1}\}) = (1 + r)^{\lceil \log_2 m \rceil} \cdot cost(u_1, u_2, \ldots, u_{m+1}).$$

This can increase the cost of the constructed Hamiltonian path by the multiplicative factor $(1 + r)^{\lceil \log_2 n \rceil}$ in the comparison with the cost of the Eulerian tour. The rough idea, then, is to construct a Hamiltonian tour by shortening only short paths of the minimal spanning tree constructed in Step 1 of both algorithms.

To realize this idea we shall prove that, for every tree $T = (V, E)$, the graph $T^3 = (V, \{\{x, y\} \mid x, y \in V$, there is a path x, P, y in T of a length at most 3$\})$

contains a Hamiltonian tour H. This means that every edge $\{u, v\}$ of H has a corresponding unique path $u, P_{u,v}, v$ in T of a length at most 3. This is a positive development, but it still does not suffice for our purposes. The remaining problem is that we need to estimate a good upper bound on the cost of the path $P(H) = u_1, P_{u_1,u_2}, u_2, P_{u_2,u_3}, u_3, \ldots, u_{n-1}P_{u_{n-1},u_n}, u_n, P_{u_n,u_1}, u_1$ (in T) that corresponds to the Hamiltonian tour $u_1, u_2, \ldots, u_n, u_1$ in T^3. Note that in Algorithm 4.3.5.1 the resulting Hamiltonian tour can be viewed as a shortening of the Eulerian tour[28] with a cost at most twice of the cost of T. But, we do not know the frequency of the occurrences of particular edges of T in $P(H)$. It may happen that the most expensive edges of T occur more frequently in $P(H)$ than the cheap edges. Observe also that $cost(T^3)$ cannot be bounded by $c \cdot cost(T)$ for any constant c independent on T, because T^3 may be even a complete graph for some trees T. Thus, we need the following technical lemma proving that T^3 contains a Hamiltonian tour H such that each edge of T occurs at most twice in $P(H)$.

Definition 4.3.5.16. *Let T be a tree. For every edge $\{u, v\} \in E(T)$, let $u, P_{u,v}, v$ be the unique simple path between u and v in T.*

*Let k be a positive integer. Let $U = u_1, u_2, \ldots, u_m$ be any simple path in T^k. Then, we define the **U-path in T** as*

$$P_T(U) = u_1, P_{u_1,u_2}, u_2, P_{u_2,u_3}, \ldots, u_{m-1}, P_{u_{m-1},u_m}, u_m.$$

Lemma 4.3.5.17. *Let T be a tree with $n \geq 3$ vertices, and let $\{p, q\}$ be an edge of T. Then, T^3 contains a Hamiltonian path $U = v_1, v_2, \ldots, v_n$, $p = v_1$, $v_n = q$, such that every edge of $E(T)$ occurs exactly twice in $P_T(H)$, where $H = U, p$ is a Hamiltonian tour in T^3.*

Proof. We prove this assertion by induction on the number of vertices of T.

(1) Let $n = 3$. The only tree of three vertices is

$$T = (\{v_1, v_2, v_3\}, \{\{v_1, v_2\}, \{v_2, v_3\}\})$$

and the corresponding T^3 is the complete graph of three vertices

$$(\{v_1, v_2, v_3\}, \{\{v_1, v_2\}, \{v_2, v_3\}, \{v_1, v_3\}\}).$$

Thus, the only Hamiltonian tour in T^3 is v_1, v_2, v_3, v_1. The claim of Lemma 4.3.5.17 is true, since $P_T(v_1, v_2, v_3, v_4) = v_1, v_2, v_3, v_2, v_1$.

(2) Let $n \geq 4$ and assume that Lemma 4.3.5.17 is true for trees with fewer than n vertices. Let $T = (V, E)$ be a tree, $|V| = n$. Let $\{p, q\}$ be an arbitrary edge of T. Consider the graph $T' = (V, E - \{\{p, q\}\})$ that consists of two trees T_p and T_q, where T_p $[T_q]$ is the component of T' containing the vertex p $[q]$. Obviously, $|V(T_p)| \leq n - 1$ and $|V(T_q)| \leq n - 1$. Let p' and q', respectively, be a neighbor of p and q, if any, in T_p and T_q, respectively.

---

[28] The Eulerian tour uses every edge of T exactly twice.

Now, we fix some Hamiltonian paths U_p and U_q in T_p^3 and T_q^3, respectively. To do it, we distinguish three possibilities according to the cardinalities of T_p and T_q.

(i)   If $|V(T_p)| = 1$, then set $U_p = p = p'$.

(ii) If $|V(T_p)| = 2$, then set $U_p = p, p'$.

(iii) If $3 \le |V(T_p)| \le n-1$, then we can apply the induction hypothesis. We set U_p to be a Hamiltonian path from p to p' in T_p^3 such that $P(U_p, p)$ contains every edge of T_p exactly twice.

A Hamiltonian path U_q in T_q^3 can be fixed in the same way as U_p was fixed above (Figure 4.17).

Now, consider the path U_p, U_q^R obtained by connecting U_p and the reverse of U_q by the edge $\{p', q'\}$. Observe, that $\{p', q'\} \in T^3$, because p', p, q, q' is a path in T. Following Figure 4.17, it is obvious that U_p, U_q^R is a Hamiltonian path in T^3, and that U_p, U_q^R, p is a Hamiltonian tour in T^3.

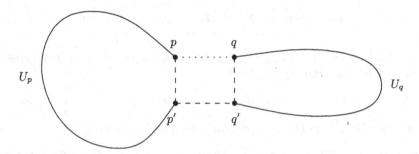

Fig. 4.17.

Observe (by the induction hypothesis or the trivial cases with $|V(T_p)| \le 2$) that $P_{T_p}(U_p, p')$ the Hamiltonian tour U_p, p' in T^3 contains every edge of T_p exactly twice. Thus, $P_{T_p}(U_p)$ contains every edge, but the edge $\{p, p'\}$ of T_p exactly twice. The edge $\{p, p'\}$ is contained exactly once in $P_{T_p}(U_p)$. Similarly, $P_{T_q}(U_q)$ contains every edge of T_q twice, but the edge $\{q, q'\}$ once. Finally, $P_T(U_p, U_q^R, p)$ contains every edge of T exactly twice, because

(i)   this is clear from the properties of U_p and U_q^R for every edge from $E - \{\{p, q\}, \{p, p'\}, \{q, q'\}\}$,

(ii) the edge $\{p', q'\} \in T^3$ connecting U_p and U_q (Figure 4.17) is realized by the path p', p, q, q' containing edges $\{p, p'\}$, $\{p, q\}$, and $\{q, q'\}$ of E, and

(iii) the connection of U_p, U_q^R with p is realized directly by the edge $\{p, q\}$.
                                                                                    □

Algorithm 4.3.5.18. SEKANINA'S ALGORITHM

Input: A complete graph $G = (V, E)$, and a cost function $c : E \to \mathbb{N}^+$.

Step 1: Construct a minimal spanning tree T of G according to c.

Step 2:   Construct T^3.

Step 3:   Find a Hamiltonian tour H in T^3 such that $P_T(H)$ contains every
edge of T exactly twice.

Output: H.

Theorem 4.3.5.19. SEKANINA'S ALGORITHM *is a polynomial-time* 2-*approximation algorithm for* $\triangle$-TSP.

Proof. Obviously, Step 1 and 2 of SEKANINA'S ALGORITHM can be performed in time $O(n^2)$. Using Lemma 4.3.5.17 one can implement Step 3 in time $O(n)$. Thus, the time complexity of SEKANINA'S ALGORITHM is in $O(n^2)$.

Let H_{Opt} be an optimal solution for an input instance (G, c) of $\triangle$-TSP. Following the inequality (4.32) we have $cost(T) \leq cost(H_{Opt})$. The output H of SEKANINA'S ALGORITHM can be viewed as shortening the path $P_T(H)$ by removing repetitions of vertices in $P_T(H)$. Since $P_T(H)$ contains every edge of T exactly twice,

$$cost(P_T(H)) = 2 \cdot cost(T) \underset{(4.32)}{\leq} 2 \cdot cost(H_{Opt}). \qquad (4.51)$$

Since H is obtained from $P_T(H)$ by exchanging simple subpaths by an edge, and c satisfies the triangle inequality,

$$cost(H) \leq cost(P_T(H)). \qquad (4.52)$$

Combining (4.51) and (4.52) we obtain $cost(H) \leq 2 \cdot cost(H_{Opt})$. $\square$

Theorem 4.3.5.20. *For every positive real number* r, SEKANINA'S ALGORITHM *is a polynomial-time* $2(1+r)^2$-*approximation algorithm for* $\triangle$-TSP$_r$.

Proof. Since SEKANINA'S ALGORITHM always outputs a Hamiltonian tour, it is consistent for TSP. Obviously, the inequality (4.51) is also true for any input instance of the general TSP.

Let (G, c) be an input instance of $\triangle$-TSP$_r$. Since $(G, c) \in Ball_{r,dist}(L_{\triangle})$,

$$c(\{v_1, v_4\}) \leq (1+r)^2 \cdot cost(v_1, v_2, v_3, v_4), \text{ and}$$
$$c(\{u_1, u_3\}) \leq (1+r) \cdot cost(u_1, u_2, u_3)$$

for all edges $\{u_1, u_3\}$, $\{v_1, v_4\} \in E(G)$ and every path v_1, v_2, v_3, v_4 between v_1 and v_4 and every path u_1, u_2, u_3 between u_1 and u_3. Since H is obtained from $P_T(H)$ by exchanging a simple subpath of $P_T(H)$ of length at most 3,

$$cost(H) \leq (1+r)^2 \cdot cost(P_T(H)). \qquad (4.53)$$

Combining (4.51) and (4.53) we finally obtain

$$cost(H) \leq 2 \cdot (1+r)^2 \cdot cost(P_T(H)).$$

$\square$

Corollary 4.3.5.21. SEKANINA'S ALGORITHM *is stable according to* dist.

Exercise 4.3.5.22. [*] Prove that SEKANINA'S ALGORITHM is a $\left(\frac{3}{2} \cdot (1+r)^2 + \frac{1}{2}(1+r)\right)$-approximation algorithm for $\triangle$-TSP$_r$.   □

Exercise 4.3.5.23. [*] Modify CHRISTOFIDES ALGORITHM in order to get a stable algorithm according to *dist*.   □

Thus, we have reached our final aim to divide the set of all instances of TSP into an infinite spectrum in such a way that the sets of this spectrum have upper bounds on the polynomial-time approximability of their input instances. The above analysis of TSP shows that it is reasonable to measure the hardness of the TSP instances by the distance function *dist*, i.e., by the degree of violation of the triangle inequality.

Now we pose a natural question. Our partition of the set of all instances (of TSP) has the set of the instances of $\triangle$-TSP as the kernel, and the best known approximation ratio for $\triangle$-TSP is 3/2 due to the CHRISTOFIDES ALGORITHM. Is there a possibility to partition the set of all instances in such a way that the kernel of this partition would be better approximable than $\triangle$-TSP? The answer to this question is positive because we can even start with a kernel that is solvable in polynomial time (i.e., with the best possible approximation ratio 1). The idea is simply to partition the set of all instances of $\triangle$-TSP into a spectrum with respect to their hardness, and so we do not need to make any change of the above partition of instances of TSP outside $\triangle$-TSP. Then, these two partitions together provide a partition of all input instances of TSP that start with a kernel that is solvable in polynomial time. The main point is that this new partition of $\triangle$-TSP is also based on the triangle inequality.

Definition 4.3.5.24. *Let* (G, c) *be an instance of* TSP *and let* $1 > p \geq 1/2$ *be a real number. We say that* (G, c) *satisfies the so-called* **p-strengthen triangle inequality** *if*

$$c(\{u, v\}) \leq p \cdot [c(\{u, w\}) + c(\{w, v\})]$$

for every three different vertices u, v, w *of* G.

We observe that $p = 1/2$ forces that all edges of G have the same weight, and so all Hamiltonian tours have the same costs.[29] If one sets $p = 1$ then one obtains the ordinary triangle inequality. Now we can extend our notation $\triangle$-TSP$_r$ for negative rs. Let, for every real p, $1/2 \leq p < 1$, $L_{str(p)}$ contain all instances of TSP that satisfy the p-strengthen triangle inequality. We define

$$\triangle\text{-TSP}_{p-1} = (\Sigma_I, \Sigma_O, L, L_{str(p)}, \mathcal{M}, cost, minimum)$$

for all p, $1/2 \leq p < 1$. Taking $\triangle$-TSP$_{-1/2}$ as the kernel of $\triangle$-TSP, one can define, for every $(G, c) \in L_\triangle$, the distance between (G, c) and $L_{str(1/2)}$ as

---

[29] Note that to define a q-strengthen triangle inequality for a $q < 1/2$ does not make any sense because there is no instance of TSP satisfying it.

$p - 1/2$, where p is the smallest real number such that $(G, c) \in L_{str(p)}$. Then considering the concept of stability of approximation one could investigate the approximability of $\triangle$-TSP_{1-p} according to $p - 1/2$. Since it is more transparent to measure the approximation ratio with respect to p than with respect to $p - 1/2$, we use the first possibility here. Our aim is to show that, for all instances of $\triangle$-TSP, the CHRISTOFIDES ALGORITHM guarantees an approximation ratio that ranges from 1 to 3/2 in dependency of p. We start with the following simple observation[30]. Let, for every instance (G, c) of TSP, $G = (V, E)$, $c_{min}(G) = \min\{c(e) \,|\, e \in E\}$ and $c_{max}(G) = \max\{c(e) \,|\, e \in E\}$.

Lemma 4.3.5.25. *Let* $1/2 \leq p < 1$, *and let* (G, c) *be an instance of* $\triangle$-TSP_{p-1}.

(i) For any two edges e_1, e_2 *of* G *with a common endpoint*

$$cost(e_1) \leq \frac{p}{1-p} \cdot cost(e_2).$$

(ii) $c_{max}(G) \leq \dfrac{2p^2}{1-p} \cdot c_{min}(G)$.

Proof. (i) For any triangle consisting of edges e_1, e_2, e_3, the p-strengthen triangle inequality implies

$$c(e_1) \leq p \cdot (c(e_2) + c(e_3)) \text{ and } c(e_3) \leq p \cdot (c(e_1) + c(e_2)). \qquad (4.54)$$

(4.54) directly implies

$$c(e_1) \leq p \cdot [c(e_2) + p(c(e_1) + c(e_2))] = pc(e_2) + p^2 c(e_2) + p^2 c(e_1),$$

and so

$$c(e_1) \leq \frac{p + p^2}{1 - p^2} \cdot c(e_2) = \frac{p}{1-p} \cdot c(e_2).$$

(ii) Let $\{a, b\}$ be an edge with $c(\{a, b\}) = c_{min}(G)$, and let $\{c, d\}$ be an edge with $c(\{c, d\}) = c_{max}(G)$. If these edges have a common endpoint (i.e., $|\{a, b, c, d\}| < 4$), then (i) implies

$$c_{max}(G) \leq \frac{p}{1-p} \cdot c_{min}(G) \leq \frac{2p^2}{1-p} \cdot c_{min}(G).$$

Let $|\{a, b, c, d\}| = 4$. Consider now the edges $\{a, c\}$ and $\{a, d\}$ that have the common endpoint a with $\{a, b\}$. Applying (i) we obtain

$$c(\{a, c\}) \leq \frac{p}{1-p} \cdot c(\{a, b\}) = \frac{p}{1-p} \cdot c_{min}(G), \text{ and} \qquad (4.55)$$

$$c(\{a, d\}) \leq \frac{p}{1-p} \cdot c_{min}(G).$$

---

[30] Note that this does not hold for $\triangle$-TSP.

Since $\{a,c\}$, $\{a,d\}$, and $\{c,d\}$ build a triangle, (4.55) together with the p-strengthen triangle inequality imply

$$
\begin{aligned}
c_{max} \;=\;\; & c(\{c,d\}) \le p(c(\{a,c\}) + c(\{a,d\})) \\
\le \;\; & p \cdot 2 \cdot \frac{p}{1-p} \cdot c_{min}(G) = \frac{2p^2}{1-p} \cdot c_{min}(G) \qquad \square
\end{aligned}
$$

The assertion (ii) of Lemma 4.3.5.25 directly implies that, for every Hamiltonian tour H of an instance (G,c) of $\triangle\text{-TSP}_{p-1}$,

$$
\frac{cost(H)}{Opt_{\mathrm{TSP}}(G,c)} \le \frac{2p^2}{1-p},
$$

and so every consistent algorithm for $\triangle\text{-TSP}$ is a $(2p^2/(1-p))$-approximation algorithm for $\triangle\text{-TSP}_{p-1}$. In what follows we improve this trivial approximation ratio.

Theorem 4.3.5.26. *For every* $p \in [1/2, 1)$, *the* CHRISTOFIDES ALGORITHM *is a* $\left(1 + \frac{2p-1}{3p^2-2p+1}\right)$*-approximation algorithm for* $\triangle\text{-TSP}_{p-1}$.

Proof. The CHRISTOFIDES ALGORITHM is a $3/2$-approximation algorithm for $\triangle\text{-TSP}$. The rough idea of the proof is to show that due to the p-strengthen triangle inequality shortening[31] of paths improves the approximation ratio $3/2$ to $\delta(p) = 1 + (2p-1)/(3p^2 - 2p + 1)$. The crucial point is that shortening is performed in two different parts of the CHRISTOFIDES ALGORITHM. The obvious part is Step 5 where a Hamiltonian tour is constructed by shortening the Eulerian tour ω created in Step 4. A little bit hidden shortening is involved in Step 3 where a minimum-weight perfect matching M is computed. In some sense[32] M can be viewed as shortening of some subpaths of an optimal Hamiltonian tour (Figure 4.14). Thus, if M contains a small number of edges and consequently we cannot save any non-negligible part of costs by shortening the Eulerian tour ω in Step 5, then we saved a non-negligible part of the cost when building the matching M. On the other hand, if $|M|$ is large then we cannot guarantee saving substantial costs by building M, but by shortening ω in Step 5. The following formal proof is nothing else than the calculation of the tradeoff between the amounts of costs saved due to these two shortening possibilities. The crucial point in the below calculation is that due to p-strengthen triangle inequality shortening a path v, w, u to the path v, u saves cost of at least

$$
(1-p) \cdot (c(\{v,w\} + c(\{w,u\})) \ge (1-p) \cdot 2 \cdot c_{min}(G).
$$

Let (G,c) be an instance of $\triangle\text{-TSP}_{p-1}$. Let H be the output of the CHRISTOFIDES ALGORITHM for the input (G,c), and let H_{Opt} be an optimal Hamiltonian tour to (G,c). Let T be the minimal spanning tree of (G,c)

---

[31] Shortening means exchanging a path for an edge.

[32] See the analysis of CHRISTOFIDES ALGORITHM in Theorem 4.3.5.5.

constructed in Step 1 of the CHRISTOFIDES ALGORITHM, and let $v_1, v_2, \ldots, v_k$, $k = 2m$, $m \in \mathbb{N} - \{0\}$, be the vertices of odd degree in T in the order as they appear in H_{Opt}. As in the proof of Theorem 4.3.5.5 (Figure 4.14) we consider two matchings $M_1 = \{\{v_1, v_2\}, \{v_3, v_4\}, \ldots, \{v_{k-1}, v_k\}\}$ and $M_2 = \{\{v_2, v_3\}, \{v_4, v_5\}, \ldots, \{v_{k-2}, v_{k-1}\}, \{v_k, v_1\}\}$. Due to the p-strengthen triangle inequality[33]

$$cost(H_{Opt}) \geq cost(M_1) + cost(M_2) + (n - k) \cdot (1 - p) \cdot 2c_{min}(G). \quad (4.56)$$

Since M is an optimal matching

$$
\begin{aligned}
cost(M) &\leq \frac{1}{2}\left(cost(M_1) + cost(M_2)\right) \\
&\leq \frac{1}{2}cost(H_{Opt}) - (n - k) \cdot (1 - p) \cdot c_{min}(G) \quad (4.57)
\end{aligned}
$$

Since one can construct a spanning tree from H_{Opt} by deleting an arbitrary edge of H_{Opt}, we have

$$
\begin{aligned}
cost(T) &\leq cost(H_{Opt}) - c_{min}(G) \\
&\leq cost(H_{Opt}) - (1 - p) \cdot 2 \cdot c_{min}(G). \quad (4.58)
\end{aligned}
$$

Let ω be the Eulerian tour constructed in Step 4 of the CHRISTOFIDES ALGORITHM. Clearly,

$$cost(\omega) = cost(T) + cost(M). \quad (4.59)$$

Substituting (4.57) and (4.58) in (4.59) we obtain

$$cost(\omega) \leq \frac{3}{2}cost(H_{Opt}) - (n - k) \cdot (1 - p) \cdot c_{min}(G) - (1 - p) \cdot 2c_{min}(G). \quad (4.60)$$

Since ω consists of $n - 1 + (k/2)$ edges and H has n edges, the application of the p-strengthen triangle inequality provides

$$
\begin{aligned}
cost(H) \quad &\leq \quad cost(\omega) - \left(\frac{k}{2} - 1\right) \cdot (1 - p) \cdot 2 \cdot c_{min}(G) \\
&\underset{(4.60)}{\leq} \quad \frac{3}{2}cost(H_{Opt}) - (n - k) \cdot (1 - p) \cdot c_{min}(G) \\
&\qquad - \frac{k}{2}(1 - p) \cdot 2 \cdot c_{min}(G) \\
&= \quad \frac{3}{2}cost(H_{Opt}) - n \cdot (1 - p) \cdot c_{min}(G) \\
&\underset{L.4.3.5.25}{\leq} \quad \frac{3}{2}cost(H_{Opt}) - \frac{(1 - p)^2}{2p^2} \cdot n \cdot c_{max}(G).
\end{aligned}
$$

---

[33] The construction of $M_1 \cup M_2$ of k edges from H_{Opt} of n edges can be done in $(n - k)$ consecutive steps where a path consisting of two edges is shortened by an edge in each of these steps.

Let $\Gamma = \{\gamma \geq 1 \mid cost(H) \leq \gamma cost(H_{Opt}) \leq n \cdot c_{max}\}$. Then, for any $\gamma \in \Gamma$,

$$cost(H) \leq \frac{3}{2} cost(H_{Opt}) - \frac{(1-p)^2}{2p^2} \cdot \gamma cost(H_{Opt})$$

$$= \left(\frac{3}{2} - \gamma \cdot \frac{(1-p)^2}{2p^2}\right) \cdot cost(H_{Opt}).$$

As a consequence

$$cost(H) \leq \min\left\{\max\left\{\gamma, \frac{3}{2} - \gamma \cdot \frac{(1-p)^2}{2p^2}\right\} \,\middle|\, \gamma \in \Gamma\right\} \cdot cost(H_{Opt}).$$

This minimum is achieved for $\gamma = \frac{3}{2} - \gamma \cdot \frac{(1-p)^2}{2p^2}$. This leads to

$$\gamma = \frac{3p^2}{3p^2 - 2p + 1} = 1 + \frac{2p - 1}{3p^2 - 2p + 1}$$

which completes our proof. $\square$

Observe that

$$\delta(p) = 1 + (2p - 1)/(3p^2 - 2p + 1)$$

is equal to 1 for $p = 1/2$, and that $\delta(p) = \frac{3}{2}$ for $p = 1$. Thus, the approximation ratio $\delta(p)$ starts with $\frac{3}{2}$ for instances of $\triangle$-TSP, and continuously converges to 1 with the degree of sharpening of the triangle inequality.

Exercise 4.3.5.27. Let A be an α-approximation algorithm for $\triangle$-TSP. Consider the following algorithm $\text{Red}(A)_\beta$ for $\triangle_\beta$-TSP with $1/2 < \beta < 1$.

$\text{Red}(A)_\beta$

Input: A complete graph $G = (V, E)$, and a cost function $c : E \rightarrow \mathbb{N}^+$ that satisfies the β-strengthen triangle inequality.
Step 1:   For every $c \in E$ compute $c'(e) := c(e) - (1 - \beta) \cdot 2c_{\min}$.
Step 2:   Execute the work of A on the problem instance $((V, E), c')$.
Output: $A((V, E), c')$

Estimate the approximation ration of $\text{Red}(A)_\beta$ with respect to α and β. To which values does $\text{Red}(A)_\beta$ converge for β approaching $1/2$ and β approaching 1? $\square$

Summary of Section 4.3.5

TSP is in class NPO(V), i.e., one of the hardest optimization problems from NPO with respect to the polynomial-time approximability. But it is very sensible according to some restrictions of the set of input instances that are connected with metrics. If the input (G, c) can be interpreted as a set of points in the Euclidean space, then one even has a PTAS for such inputs. For problem instances satisfying the triangle inequality one still has constant approximation algorithms.

The crucial point is that one can split the set of all instances of TSP into infinitely many subclasses according to the degree of violation of the triangle inequality, and, for each such subclass, one can guarantee an upper bound on the approximation ratio. These upper bounds on the approximation ratio grows slowly with the degree of the triangle inequality violation (distance from $\triangle$-TSP), and they are independent of the size of the instances. This splitting has been achieved by the concept of the stability of approximation, so one has an algorithm working on all instances of the general TSP within the upper bounds on the approximation ratio in each of the problem instance subclasses.

A similar approach is used to partition the set of all instances of $\triangle$-TSP into infinitely many subclasses according to the strengthened triangle inequality. The upper bound on the approximation ratio of these subclasses ranges from 1 to 3/2.

4.3.6 Bin-Packing, Scheduling, and Dual Approximation Algorithms

The aims of this section are:

(i) to design a dual polynomial-time approximation scheme[34] (dual PTAS) for the bin-packing problem (BIN-P), and
(ii) to use the dual PTAS for the BIN-P to design a PTAS for the makespan scheduling problem (MS).

Thus, we do not only illustrate the concept of dual approximation algorithms, but we present the design of dual approximation algorithms as a method for the design of ordinary approximation algorithms.

Remember that an input instance of BIN-P is a vector $I = (r_1, r_2, \ldots, r_n)$ over rationals from $[0, 1]$. A feasible solution of I is any set S of vectors from $\{0, 1\}^n$ such that, for every $Y \in S$,

$$Y^T \cdot (r_1, r_2, \ldots, r_n) \leq 1,$$

and

$$\sum_{Y \in S} Y = (1, 1, \ldots, 1).$$

The objective is to minimize the cardinality of S.

We define a function $h : L_I \times \Sigma_O^* \to \mathbb{R}^{\geq 0}$ as follows. For every input instance $I = (r_1, r_2, \ldots, r_n)$ and every set $S \subseteq \{0, 1\}^m$ with $\sum_{Y \in S} Y = (1, 1, \ldots, 1)$,

$$h(I, S) = \max \left\{ \max \left\{ Y^T \cdot (r_1, r_2, \ldots, r_n) | Y \in S \right\} - 1, 0 \right\}.$$

---

[34] Remember that the concept of dual approximation algorithms was introduced in Section 4.2.4 that contains all formal definitions needed for reading this section.

Obviously, h is a constraint-distance function for BIN-P.

The design of an h-dual PTAS for BIN-P is realized in the following three steps:

Step 1: Use the method of dynamic programming to design a polynomial-time algorithm DPB-P for input instances of BIN-P that contain a constant number of different values of r_is (i.e., the input involves a lot of multiple occurrences of some values r_i).

Step 2: Apply DPB-P (in a similar way as for the knapsack problem in Section 4.3.4) to obtain an h-dual PTAS for the input instances of BIN-P that do not contain "very small" r_is.

Step 3: Use the above h-dual PTAS to design an h-dual PTAS for the general BIN-P.

We start with Step 1. For every positive integer s, s-bin-packing problem briefly s-BIN-P, is a restricted BIN-P with at most s different values among all input values $r_1, \ldots, r_n$, $n \in \mathbb{N}$. To simplify the matter, we consider an input instance of an s-BIN-P as $I = (q_1, \ldots, q_s, n_1, \ldots, n_s)$, where $q_i \in (0, 1]$ for $i = 1, \ldots, s$, $n = \sum_{i=1}^{s} n_i$, $n_i \in \mathbb{N} - \{0\}$ for $i = 1, \ldots, n$. This means that we have n_i values q_i in the input for every $i = 1, 2, \ldots, s$. To solve s-BIN-P in polynomial-time according to n we use dynamic programming. Let, for every m_i, $0 \leq m_i \leq n_i$, $i = 1, \ldots, m$, BIN-P$(m_1, m_2, \ldots, m_s)$ denote the value $Opt_{s\text{-Bin-P}}(q_1, \ldots, q_s, m_1, \ldots, m_s)$. The dynamic approach means to compute BIN-P$(m_1, \ldots, m_s)$ for all $(m_1, \ldots, m_s) \in \{0, 1, \ldots, n_1\} \times \cdots \times \{0, 1, \ldots, n_s\}$ by the recurrence

$$\text{Bin-P}(m_1, \ldots, m_s) =$$
$$1 + \min_{x_1, \ldots, x_s} \left\{ \text{Bin-P}(m_1 - x_1, \ldots, m_s - x_s) \,\middle|\, \sum_{i=1}^{s} x_i q_i \leq 1 \right\}. \quad (4.61)$$

Algorithm 4.3.6.1 (DPB-P$_s$).

Input: $q_1, \ldots, q_s$, $n_1, \ldots, n_s$, where $q_i \in (0, 1]$ for $i = 1, \ldots, s$, and $n_1, \ldots, n_s$ are positive integers.

Step 1: BIN-P$(0, \ldots, 0) := 0$;
Bin-P$(h_1, \ldots, h_s) := 1$ for all $(h_1, \ldots, h_s) \in \{0, \ldots, n_1\} \times \cdots \times \{0, \ldots, n_s\}$ such that $\sum_{i=1}^{s} h_i q_i \leq 1$ and $\sum_{i=1}^{s} h_i \geq 1$.

Step 2: Compute Bin-P$(m_1, \ldots, m_s)$ with the corresponding optimal solution $T(m_1, \ldots, m_s)$ by the recurrence (4.61) for all $(m_1, \ldots, m_s) \in \{0, \ldots, n_1\} \times \cdots \times \{0, \ldots, n_s\}$.

Output: BIN-P$(n_1, \ldots, n_s)$, $T(m_1, \ldots, m_s)$.

One can easily observe that the number of different subproblems Bin-P$(m_1, \ldots, m_s)$ of Bin-P$(n_1, \ldots, n_s)$ is

$$n_1 \cdot n_2 \cdots \cdot n_s \leq \left(\frac{\sum_{i=1}^{s} n_i}{s} \right)^s = \left(\frac{n}{s} \right)^s.$$

Thus, the time complexity of Algorithm 4.3.6.1 (DPB-P) is in $O((\frac{n}{s})^{2s})$, i.e., it is polynomial in n.

Now, we realize the second step of our design of an h-dual PTAS for BIN-P. Let, for every ε, $0 < \varepsilon < 1$, Bin-P$_\varepsilon$ be a subproblem of BIN-P, where every input value is larger than ε. The idea is to approximate every input instance $I = (q_1, \ldots, q_n)$ by rounding (approximating) all values $q_1, \ldots, q_n$ to some $s = \frac{\log_2(1/\varepsilon)}{\varepsilon}$ fixed values, and to use DPB-P$_s$ to work on this rounded input.

Algorithm 4.3.6.2 (BP-PTA$_\varepsilon$).

Input: $(q_1, q_2, \ldots, q_n)$, where $\varepsilon < q_1 \leq \cdots \leq q_n \leq 1$.
Step 1:   Set $s := \lceil \log_2(1/\varepsilon)/\varepsilon \rceil$;
$l_1 := \varepsilon$, and
$l_j := l_{j-1} \cdot (1 + \varepsilon)$ for $j = 2, 3, \ldots, s$;
$l_{s+1} = 1$.
{This corresponds to the partitioning of the interval $(\varepsilon, 1]$ into s subintervals $(l_1, l_2]$, $(l_2, l_3]$, $\ldots$, $(l_s, l_{s+1}]$.}
Step 2:   **for** $i = 1$ **to** s **do**
do begin $L_i := \{q_1, \ldots, q_n\} \cap (l_i, l_{i+1}]$;
$n_i := |L_i|$;
end
{We consider that every value of L_i is rounded to the value l_i in what follows.}
Step 3:   Apply DPB-P$_s$ on the input $(l_1, l_2, \ldots, l_s, n_1, n_2, \ldots, n_s)$.

Lemma 4.3.6.3. *For every $\varepsilon \in (0, 1)$, BP-PTA$_\varepsilon$ is an h-dual ε-approximation algorithm for Bin-P$_\varepsilon$, and*

$$Time_{\text{BP-PTA}_\varepsilon}(n) = O\left(\left(\frac{\varepsilon \cdot n}{\log_2(1/\varepsilon)} \right)^{\frac{2}{\varepsilon} \cdot \log_2 \left(\frac{1}{\varepsilon} \right)} \right).$$

Proof. We first analyze the time complexity BP-PTA$_\varepsilon$. Since $s \leq n$ it is obvious that Step 1 and Step 2 can be executed in $O(n^2)$ time. Since the time complexity of DPB-P$_s$ is in $O((\frac{n}{s})^{2s})$ and $s = \frac{1}{\varepsilon} \cdot \log_2\left(\frac{1}{\varepsilon}\right)$, the proof is completed.

To prove that BP-PTA$_\varepsilon$ is an h-dual ε-approximation algorithm for Bin-P$_\varepsilon$, we have to prove that, for every input $I = (q_1, q_2, \ldots, q_n)$, $\varepsilon < q_1 \leq \cdots \leq q_n \leq 1$, the following two facts hold:

(i) $r = cost(T_1, \ldots, T_r) = \text{Bin-P}(n_1, \ldots, n_s) \leq Opt_{\text{Bin-P}}(I)$, where $(T_1, \ldots, T_r)$ is the optimal solution for the input $Round(I) = (l_1, \ldots, l_s, n_1, \ldots, n_s)$ computed by BP-PTA$_\varepsilon$ [T_i is the set of the multiplicatives of the indices of the values packed in the ith bin], and

(ii) for every $j = 1, \ldots, r$, $\sum_{a \in T_j} q_a \leq 1 + \varepsilon$.

The fact (i) is obvious because $Round(I)$ can be considered as $(p_1, \ldots, p_n)$, where $p_i \leq q_i$ for every $i \in \{1, \ldots, n\}$.

Since $\mathrm{DPB\text{-}P}_s(Round(I)) \leq Opt_{\mathrm{Bin\text{-}P}}(I)$, we obtain

$$\mathrm{Bin\text{-}P}(n_1, \ldots, n_s) = Opt_{\mathrm{Bin\text{-}P}}(Round(I)) \leq Opt_{\mathrm{Bin\text{-}P}}(I).$$

To prove (ii), consider an arbitrary set of indices $T \in \{T_1, T_2, \ldots, T_r\}$. Let $x_T = (x_1, \ldots, x_n)$ be the corresponding description of the set of indices assigned to this bin for $Round(I)$. We can bound $\sum_{j \in T} q_j$ as follows:

$$\sum_{j \in T} q_j \leq \sum_{i=1}^{s} x_i l_{i+1} = \sum_{i=1}^{s} x_i l_i + \sum_{i=1}^{s} x_i (l_{i+1} - l_i) \leq 1 + \sum_{i=1}^{s} x_i (l_{i+1} - l_i). \quad (4.62)$$

Since $l_i > \varepsilon$ for every $i \in \{1, \ldots, s\}$, the number of pieces in a bin is at most $\lfloor \frac{1}{\varepsilon} \rfloor$, i.e.,

$$\sum_{i=1}^{s} x_i \leq \left\lfloor \frac{1}{\varepsilon} \right\rfloor. \quad (4.63)$$

Let, for $i = 1, \ldots, s$, a_i be the fraction of the bin T filled by values of size l_i. Obviously,

$$x_i \leq \frac{a_i}{l_i} \quad (4.64)$$

for every $i \in \{1, 2, \ldots, s\}$. Inserting (4.64) into (4.62) we obtain

$$
\begin{aligned}
\sum_{j \in T} q_j \;&\underset{(4.62)}{\leq}\; 1 + \sum_{i=1}^{s} x_i (l_{i+1} - l_i) \\
&\underset{(4.64)}{\leq}\; 1 + \sum_{i=1}^{s} \frac{a_i}{l_i}(l_{i+1} - l_i) \\
&=\; 1 + \sum_{i=1}^{s} \left[a_i \cdot \frac{l_{i+1}}{l_i} - a_i \right] \\
&=\; 1 + \sum_{i=1}^{s} a_i \cdot \left(\frac{l_{i+1}}{l_i} - 1 \right) \\
&=\; 1 + \sum_{i=1}^{s} a_i \cdot \varepsilon = 1 + \varepsilon \cdot \sum_{i=1}^{s} a_i = 1 + \varepsilon
\end{aligned}
$$

$\square$

Now, we are prepared to present the dual PTAS for the general BIN-P. The idea is to use, for a given ε, the algorithm $\mathrm{BP\text{-}PTA}_\varepsilon$ for the input part that consists of values that are larger than ε, and then to pack the remaining small pieces in bins of content size smaller or equal to 1 if possible. If all bins are overfilled, then one takes a new bin for the rest of the small pieces.

Algorithm 4.3.6.4 (Bin-PTAS).

Input: (I, ε), where $I = (q_1, q_2, \ldots, q_n)$, $0 \leq q_1 \leq q_2 \leq \cdots \leq q_n \leq 1$, $\varepsilon \in (0, 1)$.

Step 1:   Find i such that $q_1 \leq q_2 \leq \cdots \leq q_i \leq \varepsilon \leq q_{i+1} \leq q_{i+2} \leq \cdots \leq q_n$.

Step 2:   Apply BP-PTA$_\varepsilon$ on the input $(q_{i+1}, \ldots, q_n)$. Let $T = (T_1, \ldots, T_m)$ be the output BP-PTA$_\varepsilon(q_{i+1}, \ldots, q_n)$.

Step 3:   For every i such that $\sum_{j \in T_i} q_j \leq 1$ pack one of the small pieces from $\{q_1, \ldots, q_i\}$ into T_i until $\sum_{j \in T_i} q_j > 1$ for all $j \in \{1, 2, \ldots, n\}$.
If there are still some small pieces to be assigned, take a new bin and pack the pieces there until this bin is overfilled. Repeat this last step several times, if necessary.

Theorem 4.3.6.5. Bin-PTAS *is an* h-*dual polynomial-time approximation scheme for* BIN-P.

Proof. First, we analyze the time complexity of Bin-PTAS. Step 1 can be executed in linear time. (If one needs to sort the input values, then it takes $O(n \log n)$ time.) Following Lemma 4.3.6.3 the application of BP-PTA$_\varepsilon$ on the input values larger than ε runs in time polynomial according to n. Step 3 can be implemented in linear time.

Now we have to prove that for every input (I, ε), $I = (q_1, \ldots, q_n)$, $\varepsilon \in (0, 1)$,

(i) $cost(\text{Bin-PTAS}(I, \varepsilon)) \leq Opt_{\text{Bin-P}}(I)$, and

(ii) every bin of Bin-PTAS(I, ε) has a size of at most $1 + \varepsilon$.

The condition (ii) is obviously fulfilled because BP-PTA$_\varepsilon$ is an h-dual ε-approximation algorithm, i.e., the bins of BP-PTA$_\varepsilon(q_{i+1}, \ldots, q_n)$ have a size of at most $1 + \varepsilon$. One can easily observe that the small pieces $q_1, \ldots, q_i$ are added to BP-PTA$_\varepsilon(q_{i+1}, \ldots, q_n)$ in Step 3 in such a way that no bin has a size greater than $1 + \varepsilon$.

To prove (i) we first observe that (Lemma 4.3.6.3)

$$Opt_{\text{Bin-P}}(q_{i+1}, \ldots, q_n) \geq cost(\text{BP-PTA}_\varepsilon(q_{i+1}, \ldots, q_n)).$$

Now, if one adds a new bin in Step 3 of Bin-PTAS, then it means that all bins have sizes larger than 1. Thus, the sum of the capacities (sizes) of these bins is larger than its number and so any optimal solution must contain one bin more. □

Exercise 4.3.6.6. Design a dual PTAS for the knapsack problem. □

In what follows we shall show that Bin-PTAS can be used to design a polynomial-time approximation scheme for the makespan scheduling problem (MS). Let Bin-PTAS$_\varepsilon$ denote the version of Bin-PTAS for a fixed $\varepsilon > 0$. The idea is that Bin-P and MS are dual in the sense that the constraints of BIN-P may be viewed as the objective of MS. More precisely, for any input instance (I, m), $I = (p_1, \ldots, p_n)$, of MS,

$$Opt_{\text{Bin-P}} \left(\frac{p_1}{d}, \frac{p_2}{d}, \ldots, \frac{p_n}{d} \right) \leq m \Leftrightarrow Opt_{\text{MS}}(I, m) \leq d. \qquad (4.65)$$

The equivalence (4.65) is obvious. If one can schedule jobs $p_1, \ldots, p_n$ in time d on m machines, then one can pack pieces $\frac{p_1}{d}, \frac{p_2}{d}, \ldots, \frac{p_n}{d}$ into m bins of the unit size 1. Vice versa, if one can pack pieces $\frac{p_1}{d}, \ldots, \frac{p_n}{d}$ into m unit bins, then one can schedule jobs $p_1, \ldots, p_n$ on m machines in time d.

If one knows the value $Opt_{\text{MS}}(I, m)$ [but not necessarily an optimal solution for (I, m)], then one can apply

$$\text{Bin-PTAS}_\varepsilon \left(\frac{p_1}{Opt_{\text{MS}}(I, m)}, \ldots, \frac{p_n}{Opt_{\text{MS}}(I, m)} \right)$$

to get a feasible solution for (I, m) of MS with the approximation ratio $1 + \varepsilon$. The solution $\text{Bin-PTAS}_\varepsilon \left(\frac{p_1}{d^*}, \ldots, \frac{p_n}{d^*} \right)$ for $d^* = Opt_{\text{MS}}(I, m)$ is feasible for MS, because Theorem 4.3.6.5 and the equivalence (4.65) imply

$$cost \left(\text{Bin-PTAS}_\varepsilon \left(\frac{p_1}{d^*}, \ldots, \frac{p_n}{d^*} \right) \right) \leq Opt_{\text{Bin-P}} \left(\frac{p_1}{d^*}, \ldots, \frac{p_n}{d^*} \right) \underset{(4.65)}{\leq} m.$$

Theorem 4.3.6.5 claims that every bin of $\text{Bin-PTAS}_\varepsilon \left(\frac{p_1}{d^*}, \ldots, \frac{p_n}{d^*} \right)$ has a size of at most $1 + \varepsilon$. Thus, the time of the corresponding scheduling is at most

$$(1 + \varepsilon) \cdot d^* = (1 + \varepsilon) \cdot Opt_{\text{MS}}(I, m).$$

The remaining problem is to estimate $Opt_{\text{MS}}(I, m)$. This can be solved by binary search (iteratively using Bin-PTAS in every step of the binary search algorithm), where (4.65) shows us in which direction one has to continue. To be able to start the binary search, we need some initial lower and upper bounds on $Opt_{\text{MS}}(I, m)$. Consider the value

$$ATLEAST(I, m) = \max \left\{ \frac{1}{m} \sum_{i=1}^{n} p_i, \max \{p_1, \ldots, p_n\} \right\}$$

for every input instance (I, m), $I = (p_1, \ldots, p_n)$, of MS. Since $Opt_{\text{MS}}(I, m) \geq \max\{p_1, \ldots, p_n\}$ and the optimal scheduling with m machines has time at least $\frac{1}{m} \cdot \sum_{i=1}^{n} p_i$, we obtain

$$Opt_{\text{MS}}(I, m) \geq ATLEAST(I, m). \qquad (4.66)$$

Following the proof of the fact that the approximation ratio of the greedy approach for MS is at most 2 in Example 4.3.4.14, one immediately sees that

$$Opt_{\text{MS}}(I, m) \leq \frac{1}{m} \sum_{i=1}^{n} p_i + \max\{p_1, \ldots, p_n\} \leq 2 \cdot ATLEAST(I, m). \quad (4.67)$$

Now we can describe the resulting PTAS for MS.

Algorithm 4.3.6.7.

Input: $((I,m),\varepsilon)$, where $I = (p_1,\dots,p_n)$, for some $n \in \mathbb{N}$, $p_1,\dots,p_n$, m are positive integers, and $\varepsilon > 0$.

Step 1:   Compute $ATLEAST := \max\left\{\frac{1}{m}\sum_{i=1}^n p_i, \max\{p_1,\dots,p_n\}\right\}$;

            Set $LOWER := ATLEAST$;

            $UPPER := 2 \cdot ATLEAST$;

            $k := \lceil \log_2(4/\varepsilon) \rceil$.

Step 2:   **for** $i = 1$ **to** k **do**

            do begin $d := \frac{1}{2}(UPPER + LOWER)$;

                  call Bin-PTAS$_{\varepsilon/2}$ on the input $\left(\frac{p_1}{d}, \frac{p_2}{d}, \dots, \frac{p_n}{d}\right)$;

                  $c := cost\left(\text{Bin-PTAS}_{\varepsilon/2}\left(\frac{p_1}{d}, \dots, \frac{p_n}{d}\right)\right)$

                  if $c > m$ **then** $LOWER := d$

                          else $UPPER := d$

            end

Step 3:   Set $d^* := UPPER$;

            call Bin-PTAS$_{\varepsilon/2}$ on the input $\left(\frac{p_1}{d^*}, \dots, \frac{p_n}{d^*}\right)$.

Output: Bin-PTAS$_{\varepsilon/2}\left(\frac{p_1}{d^*}, \dots, \frac{p_n}{d^*}\right)$.

Theorem 4.3.6.8. *Algorithm 4.3.6.7 is a PTAS for MS.*

Proof. First, we analyze the time complexity of Algorithm 4.3.6.7. Step 1 can be performed in linear time. Step 2 consists of k calls of Bin-PTAS$_{\varepsilon/2}$, where k is a constant according to the size of (I,m). Step 3 contains one call of Bin-PTAS$_{\varepsilon/2}$. Since Bin-PTAS$_{\varepsilon/2}$ runs in polynomial-time according to n, Algorithm 4.3.6.7 is a polynomial-time algorithm, too.

Now, we prove that Bin-PTAS$_{\varepsilon/2}\left(\frac{p_1}{d^*}, \dots, \frac{p_n}{d^*}\right)$ is a feasible solution for the input instance $(p_1,\dots,p_n)$, m of MS (i.e., that the jobs $p_1,\dots,p_n$ are assigned to at most m machines). Following the eqivalence (4.65) and the inequality (4.67) we see that after adjusting $UPPER$ in Step 1,

$$\text{Bin-PTAS}_{\varepsilon/2}\left(\frac{p_1}{UPPER}, \dots, \frac{p_n}{UPPER}\right)$$

is a feasible solution for $(p_1,\dots,p_n)$, m.

Since we change $UPPER$ for

$$d = \frac{1}{2}(UPPER + LOWER) \text{ only if } cost\left(\text{Bin-PTAS}_{\varepsilon/2}\left(\frac{p_1}{d}, \dots, \frac{p_n}{d}\right)\right) \le m,$$

Bin-PTAS$_{\varepsilon/2}\left(\frac{p_1}{UPPER}, \dots, \frac{p_n}{UPPER}\right)$ is a feasible solution to the instance I, m of MS for every value assigned to the variable $UPPER$ during the execution of Step   2. Thus, Bin-PTAS$_{\varepsilon/2}\left(\frac{p_1}{d^*}, \dots, \frac{p_n}{d^*}\right)$ must be also a feasible solution for the input instance I, m of MS.

Since every bin of the solution $S = \text{Bin-PTAS}_{\varepsilon/2}\left(\frac{p_1}{d^*}, \dots, \frac{p_n}{d^*}\right)$ of the input $\frac{p_1}{d^*}, \dots, \frac{p_n}{d^*}$ of Bin-P has a size of at most $1 + \varepsilon/2$,

$$cost_{MS}\left(\text{Bin-PTAS}_{\varepsilon/2}\left(\frac{p_1}{d^*},\ldots,\frac{p_n}{d^*}\right)\right) \le \left(1+\frac{\varepsilon}{2}\right)\cdot d^*, \qquad (4.68)$$

when S is interpreted as a feasible solution for (I,m) of MS. The remaining question is: "How well is $Opt_{MS}(I,m)$ approximated by d^*?" At the beginning we know

$$LOWER = ATLEAST \le Opt_{MS}(I,m) \le 2 \cdot ATLEAST = UPPER.$$

In every execution of the cycle of Step 2 the difference between $UPPER$ and $LOWER$ is halved. After $k = \lceil \log_2(4/\varepsilon) \rceil$ reduction steps

$$UPPER - LOWER \le \frac{1}{2^k}\cdot ATLEAST.$$

Since $Opt_{MS}(I,m) \in [LOWER, UPPER]$ during the whole run of Algorithm 4.3.6.7,

$$d^* - Opt_{MS}(I,m) \le \frac{1}{2^k}\cdot ATLEAST \le \frac{1}{2^k}Opt_{MS}(I,m).$$

Thus,

$$\frac{d^* - Opt_{MS}(I,m)}{Opt_{MS}(I,m)} \le \frac{1}{2^k} \le \frac{\varepsilon}{4}. \qquad (4.69)$$

The inequality (4.68) and (4.69) imply together

$$\begin{aligned}
cost\left(\text{Bin-PTAS}_{\varepsilon/2}\left(\frac{p_1}{d^*},\ldots,\frac{p_n}{d^*}\right)\right) &\underset{(4.68)}{\le} \left(1+\frac{\varepsilon}{2}\right)\cdot d^* \\
&\underset{(4.69)}{\le} \left(1+\frac{\varepsilon}{2}\right)\cdot\left(1+\frac{\varepsilon}{4}\right)\cdot Opt_{MS}(I,m) \\
&\le (1+\varepsilon)\cdot Opt_{MS}(I,m)
\end{aligned}$$

for every $\varepsilon \in (0,1]$. $\qquad\qquad\qquad\qquad\qquad\qquad\qquad\qquad\square$

Summary of Section 4.3.6

A dual approximation algorithm approximates the feasibility of a problem instead of its optimality. This approach can be of practical interest if the constraints on the feasibility are not precisely determined, or simply not rigid, so that their small violation is a natural model of an existing feasibility.

The bin-packing problem admits a dual PTAS. This dual PTAS can be designed by using dynamic programming for some subclass of the input instances of S of BIN-P and then by approximating every input instance of BIN-P with an input instance of S.

The dual approximation algorithms are not only an approach to solve optimization problems with nonrigid constraints, but also a tool for the design of ordinary approximation algorithms. This can work especially well if the nonrigid constraints of an optimization problem can be considered as an objective of another ("dual") optimization problem. The bin-packing problem and the makespan scheduling problem have such a relationship. Thus, the dual PTAS for BIN-P can be "transformed" into a PTAS for MS.

4.4 Inapproximability

4.4.1 Introduction

The aim of this section is to present some techniques that, under the assumption P $\neq$ NP (or a similar one), prove lower bounds on polynomial-time approximability for concrete problems. These lower bounds may be very different, ranging from the nonexistence of a PTAS for a given problem to n^a, $a > 0$, lower bounds on the approximation ratio of any polynomial-time approximation algorithm for the given problem. Here, we present the following three methods for proving lower bounds on polynomial-time approximability.

(i) **Reduction to NP-hard decision problems**
   This is the classical reduction method as used in the theory of NP-completeness. First, one considers an optimization problem U as an approximation problem, i.e., as a problem of finding a feasible solution within a fixed approximation ratio d. Then one reduces a NP-hard problem to this approximation problem. The direct consequence is that if P $\neq$ NP, then U does not admit any polynomial-time d-approximation algorithm. In Section 4.4.2 we illustrate this approach to prove that if P $\neq$ NP, then there is no polynomial-time $p(n)$-approximation algorithm for TSP for any polynomial p.

(ii) **Approximation-preserving reductions**
   Assume that we have already a proof that an optimization problem U does not admit any polynomial-time d-approximation algorithm for a fixed constant d (under the assumption P $\neq$ NP). This clearly means that U does not admit any PTAS. If one wants to prove that another optimization problem W does not admit any PTAS, then it is sufficient to reduce U to W by a reduction that somehow preserves the approximation ratio.
   In Section 4.4.3 we present two of the possible approximation-preserving reductions and so-called APX-completeness.

(iii) **Application of the PCP-Theorem**
   This method is based on one of the deepest and hardest results of computer science – the so-called PCP-(Probabilistically Checkable Proofs) Theorem. This theorem provides a completely new, surprising characterization of the class NP, and thus a new representation of languages from NP. Starting from this representation one can reduce any decision problem (language) from NP to some approximation problems. In this way, assuming P $\neq$ NP, one obtains the evidence that the approximation problems are hard.
   In Section 4.4.4 we apply this concept in order to show that there does not exist any PTAS for MAX-SAT.[35]

Before presenting the above three methods in the subsequent sections, we provide a useful definition.

---

[35] Note that this crucial result was the starting point that made the use of the approximation-preserving reduction method possible.

Definition 4.4.1.1. *Let* $U = (\Sigma_I, \Sigma_O, L, L_I, \mathcal{M}, cost, goal)$ *be an optimization problem. For every* $c \in \mathbb{R}^{>1}$, *we define a* **c-approximation problem** *to* U, *shortly* **c-App(U)**, *as the problem of finding, for every* $x \in L_I$, *a feasible solution* $S(x) \in \mathcal{M}(x)$ *such that*

$$\max\left\{ \frac{cost(S(x))}{Opt_U(x)}, \frac{Opt_U(x)}{cost(S(x))} \right\} \le c.$$

For every function $f : \mathbb{N} \to \mathbb{N}$, *we define an* $f(n)$**-approximation problem** *to* U, *shortly* $f(n)$**-App(U)**, *as the problem of finding, for every* $x \in L_I$, *a feasible solution* $S(x) \in \mathcal{M}(x)$ *such that*

$$\max\left\{ \frac{cost(S(x))}{Opt_U(x)}, \frac{Opt_U(x)}{cost(S(x))} \right\} \le f(|x|).$$

We simply observe that the existence of an c-approximation algorithm for an optimization problem U directly implies that c-App(U) can be solved in polynomial time. On the other hand, if one has a polynomial-time algorithm for c-App(U), then this algorithm is a polynomial-time c-approximation algorithm for U. The only reason for presenting Definition 4.4.1.1 is to express the aim for searching for an approximate solution as a computing problem.

4.4.2 Reduction to NP-Hard Problems

In this section we present the oldest and simplest technique for proving that approximation problems to some optimization problems are hard. The idea is to reduce an NP-hard problem (language) to an approximation problem c-App(U) and so to show that c-App(U) does not admit any polynomial-time algorithm if P $\neq$ NP. This directly implies the nonexistence of any polynomial-time c-approximation algorithm for U under the assumption P $\neq$ NP.

The general idea of such a classical reduction is the following one. Let $L \subseteq \Sigma^*$ be NP-hard, and let c-App(U) be an approximation problem to a minimization problem $U = (\Sigma_I, \Sigma_O, L, L_I, \mathcal{M}, cost, goal) \in$ NPO, $c \in \mathbb{R}^{>1}$. One has to find a polynomial-time transformation F from Σ^* to L_I such that there exists an efficiently computable function $f_F : \mathbb{N} \to \mathbb{N}$ with the following properties:

(i) for every $x \in L$, $\mathcal{M}(F(x))$ contains a solution y with $cost(y) \le f_F(|x|)$,
(ii) for every $x \in \Sigma^* - L$, all solutions $z \in \mathcal{M}(F(x))$ have $cost(z) > c \cdot f_F(|x|)$.

Now, it should be clear how the transformation F can be used to decide the membership of a given $x \in \Sigma^*$ to L if there is a polynomial-time c-approximation algorithm A for U. One transforms x to $F(x)$ and looks on $A(F(x))$, and $cost(A(F(x)))$.

If $cost(A(F(x)) \le c \cdot f_F(|x|)$, then surely[36] $x \in L$.

---
[36] Because of (ii)

If $cost(A(F(x))) > c \cdot f_F(|x|)$, then $x \notin L$. This is because $x \in L$ implies

$$Opt_U(F(x)) \underset{(i)}{\leq} f_F(|x|) < \frac{1}{c} cost(A(F(x))), \qquad (4.70)$$

and (4.70) contradicts the fact that A is a c-approximation algorithm for U.

In what follows we illustrate this technique on the traveling salesperson problem (TSP).

Lemma 4.4.2.1. *For every positive integer d, the Hamiltonian tour (HT) problem is polynomial-time reducible to d-App(TSP).*

Proof. Let $G = (V, E)$ be an input instance of the HT problem. Remember that $G \in L_{HT}$ if and only if G contains a Hamiltonian tour. We propose to consider the polynomial-time transformation F that assigns the complete (undirected) graph $K_{|V|} = (V, E')$ and the cost function $c : E' \to \mathbb{N}^+$ to G as follows:

$c(e) = 1$ if $e \in E$, and
$c(e) = (d-1)|V| + 2$ if $e \notin E$ (i.e., if $e \in E' - E$).

We set $f_F(|G|) = |V|$. We observe that

(i) if G contains a Hamiltonian tour, then $K_{|V|}$ contains a Hamiltonian tour of cost $|V|$, i.e., $Opt_{TSP}(K_{|V|}, c) = |V| = f_F(|G|)$.

(ii) if G does not contain any Hamiltonian tour, then each Hamiltonian tour in $K_{|V|}$ contains at least one edge of $E' - E$. Thus, every Hamiltonian tour in $K_{|V|}$ has a cost of at least

$$|V| - 1 + (d-1)|V| + 2 = d \cdot |V| + 1 > d \cdot |V| = d \cdot f_F(|G|).$$

Clearly, if d-App(TSP) can be solved in polynomial time, then the HT problem can be solved in polynomial time, too. □

Corollary 4.4.2.2. *If $P \neq NP$, then there is no polynomial-time d-approximation algorithm for TSP for any constant d.* □

The proof of Lemma 4.4.2.1 shows that the cost function c can be chosen in such a way that the difference between $Opt_{TSP}(F(x))$ and $Opt_{TSP}(F(y))$ may be arbitrarily large if $x \in L_{HT}$ and $y \notin L_{HT}$ ($|x| = |y|$). We can even prove that the TSP problem is in NPO(V) because it does not admit any polynomial-time $p(n)$-approximation algorithm for any polynomial p.

Theorem 4.4.2.3. *If $P \neq NP$, then there is no polynomial-time, $p(n)$-approximation algorithm for TSP for any polynomial p.*

Proof. Consider the reduction of the HT problem to TSP as done in the proof of Lemma 4.4.2.1. We change the definition of the cost of edges of E' as follows:

$c(e) = 1$ if $e \in E$, and

$$c(e) = |V| \cdot 2^{|V|} + 1 \text{ if } e \notin E.$$

Observe that F is still polynomial-time computable, because the numbers of the size $|V| \cdot 2^{|V|}$ can be saved in $|V| + \lceil \log_2 |V| \rceil$ space and so the representation of $(K_{|V|}, c)$ is of length at most $O(|V|^3)$. One can easily observe that $F((V, E))$ can be derived in time $O(|V|^3)$, too.

Now, if $G = (V, E)$ contains a Hamiltonian tour, then $Opt_{\text{TSP}}(F(G)) = |V|$. But if G does not contain any Hamiltonian tour, then every Hamiltonian tour in $F(G) = (K_{|V|}, c)$ has cost at least

$$|V| - 1 + 2^{|V|} \cdot |V| + 1 > 2^{|V|} |V| = 2^{|V|} \cdot f_F(|G|).$$

So, if $L_{\text{HT}} \notin$ P, then there does not exist any polynomial-time $2^{\Omega(\sqrt[3]{n})}$-approximation algorithm for the TSP problem.[37]   □

Exercise 4.4.2.4. For which exponential functions g can the lower bound g on the polynomial-time inapproximability of the TSP problem be proved?   □

4.4.3 Approximation-Preserving Reductions

The concept of approximation-preserving reductions is similar to that of NP-completeness. While the concept of NP-completeness serves as an argument for claiming[38] that a given problem is not in P if P $\neq$ NP, the concept of approximation-preserving reductions primarily provides a method for proving that an optimization problem does not admit any PTAS under the assumption P $\neq$ NP. For this purpose the analogy of the class NP in the NP-completeness concept is the following class APX:

$$\mathbf{APX} = \{U \in \text{NPO} \,|\, \text{there exists a polynomial-time}$$
$$c\text{-approximation algorithm for } U, c \in \mathbb{R}^{>1}\}.$$

Our goal is to prove that some specific problems from APX do not admit any PTAS, or even, for a specific constant d, that there does not exist any polynomial-time d-approximation algorithm for them. Analogous to NP-completeness, we try to define a subclass of APX that contains the hardest optimization problems of APX. The hardness has to be measured by the existence or nonexistence of a PTAS for the given problem here.

What we need is a reduction R that has the following property. If an optimization problem U_1 can be reduced by R to an optimization problem U_2, and U_1 does not admit any PTAS, then U_2 may not admit any PTAS.

---

[37] Note that we cannot automatically conclude the nonexistence of any polynomial-time 2^n-approximation algorithm, because the size of $F(G)$ is $|V|^3 + |V|^2 \cdot \lceil \log_2 |V| \rceil$ and so $|V|$ is approximately the third root of the size of the input instance $F(G)$ of TSP.

[38] As a method for proving

There are several ways of defining a reduction having this property. Here, we present two different kinds of reductions.

We start with the so-called approximation-preserving reduction, or AP-reduction.

Definition 4.4.3.1. *Let $U_1 = (\Sigma_{I,1}, \Sigma_{O,1}, L_1, L_{I,1}, \mathcal{M}_1, cost_1, goal_1)$ and $U_2 = (\Sigma_{I,2}, \Sigma_{O,2}, L_2, L_{I,2}, \mathcal{M}_2, cost_2, goal_2)$ be optimization problems. We say that U_1 is* **AP-reducible** *to U_2, $U_1 \leq_{AP} U_2$, if there exist functions*

$$F : \Sigma_{I,1}^* \times \mathbb{Q}^+ \to \Sigma_{I,2}^*,$$
$$H : \Sigma_{I,1}^* \times \mathbb{Q}^+ \times \Sigma_{O,2}^* \to \Sigma_{O,1}^*,$$

and a constant $\alpha > 0$ such that

(i) for all $x \in L_{I,1}$ with $\mathcal{M}_1(x) \neq \emptyset$ and all $\varepsilon \in \mathbb{Q}^+$,

$$F(x,\varepsilon) \in L_{I,2} \text{ and } \mathcal{M}_2(F(x,\varepsilon)) \neq \emptyset,$$

(ii) for all $x \in L_{I,1}$, all $\varepsilon \in \mathbb{Q}^+$, and all $y \in \mathcal{M}_2(F(x,\varepsilon))$,

$$H(x,\varepsilon,y) \in \mathcal{M}_1(x),$$

(iii) F and H are computable in a time that is polynomial in $|x|$ and $|y|$ for any fixed number ε,

(iv) the time complexity of F and H is nonincreasing with ε for all fixed input sizes $|x|$ and $|y|$, and

(v) for all $x \in L_{I,1}$, all $\varepsilon \in \mathbb{Q}^+$, and all $y \in \mathcal{M}_2(F(x,\varepsilon))$:

$$\max\left\{ \frac{Opt_{U_2}(F(x,\varepsilon))}{cost_2(y)}, \frac{cost_2(y)}{Opt_{U_2}(F(x,\varepsilon))} \right\} \leq 1 + \varepsilon \text{ implies}$$
$$\max\left\{ \frac{cost_1(H(x,\varepsilon,y))}{Opt_{U_1}(x)}, \frac{Opt_{U_1}(x)}{cost_1(H(x,\varepsilon,y))} \right\} \leq 1 + \alpha \cdot \varepsilon.$$

Now, let us look at Definition 4.4.3.1 to see whether it satisfies our requirements. We wanted to have a reduction from U_1 to U_2 such that $U_1 \leq_{AP} U_2$ would imply the existence of a PTAS for U_1 if there is a PTAS for U_2. In another direction, if there is no PTAS for U_1, there should be no PTAS for U_2.

Lemma 4.4.3.2. *Let U_1 and U_2 be optimization problems as described in Definition 4.4.3.1. If there exists a PTAS for U_2, and $U_1 \leq_{AP} U_2$, then there exists a PTAS for U_1.*

Proof. Let $U_1 \leq_{AP} U_2$, i.e., there exist functions F, H and a constant $\alpha > 0$ with the properties (i), (ii), (iii), (iv), and (v) of Definition 4.4.3.1. We construct a PTAS A_1 for U_1 provided there exists a PTAS A_2 for U_2. The input for our PTAS A_1 is $(x,\delta) \in L_{I,1} \times \mathbb{Q}^+$, where x is the input instance of U_1

and δ is the maximal relative error allowed. Now we apply the transformation (function) F on (x, δ) to get an input instance $F(x, \delta)$ of U_2.

We choose $\varepsilon = \frac{\delta}{\alpha}$ and execute the PTAS A_2 on the input $(F(x, \delta), \varepsilon) \in L_{I,2} \times Q^+$. The output of A_2 is an $y \in \mathcal{M}_2(F(x, \delta))$ such that

$$\max \left\{ \frac{Opt_{U_2}(F(x, \varepsilon))}{cost_2(y)}, \frac{cost_2(y)}{Opt_{U_2}(F(x, \varepsilon))} \right\} \le 1 + \varepsilon.$$

Following (ii) and (v) the transformation H computes, for the input (x, δ, y), a feasible solution $H(x, \varepsilon, y) \in \mathcal{M}_1(x)$ and

$$\max \left\{ \frac{cost_1(H(x, \varepsilon, y))}{Opt_{U_1}(x)}, \frac{Opt_{U_1}(x)}{cost_1(H(x, \varepsilon, y))} \right\} \le 1 + \alpha \cdot \varepsilon = 1 + \alpha \cdot \frac{\delta}{\alpha} = 1 + \delta.$$

Thus, A_1 computes a feasible solution for x with a relative error of at most δ.

To show that A_1 is a PTAS it remains to show that the time complexity of A_1 is polynomial in the input size $|x|$, for any fixed constant δ. Since F is computable in polynomial time according to $|x|$ for any fixed δ [the condition (iii) of Definition 4.4.3.1], $|F(x, \delta)|$ is polynomial in $|x|$. The PTAS A_2 works on $(F(x, \delta), \varepsilon)$ in polynomial time according to $|(F(x, \delta))|$ for any fixed ε. The result $y = A_2(F(x, \delta), \varepsilon)$, then, has a length polynomial in $|F(x, \delta)|$, and so in $|x|$, too.

The last step of A_1 is to compute $H(x, \delta, y) \in \mathcal{M}_1(x)$. Following (iii) this can be performed in polynomial time according to $|x|$ and $|y|$ for any fixed δ. Thus, we can conclude that A_2 works in polynomial time according to $|x|$ for any relative error δ, i.e., A_2 is a PTAS for U_1. $\square$

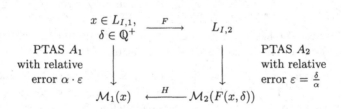

Fig. 4.18.

Observe that we did not need the condition (iv) of Definition 4.4.3.1 to prove Lemma 4.4.3.2. This condition says only that it is not reasonable to use more time to find a feasible solution with large relative error than to find a feasible solution with a small relative error. The scheme illustrating Definition 4.4.3.1 and Lemma 4.4.3.2 is given in Figure 4.18.

Observe that to reach the required relative error δ one has to suitably choose the relative error ε of the PTAS A_2 in order to compensate for the deterioration (by a constant multiplicative factor α) caused by the transformations F and H.

Exercise 4.4.3.3. Prove that the relation $\leq_{\text{AP}}$ is transitive. $\square$

Definition 4.4.3.4. *An optimization problem* $U \in$ NPO *is called* APX-complete, if

(i) $U \in$ APX, *and*
(ii) for every $W \in$ APX, $W \leq_{\text{AP}} U$.

Currently, we do not really deal with APX-completeness, because due to the PCP-Theorem (see the next section) we can (under the assumption $P \neq NP$) prove for some optimization problems that they do not admit any polynomial-time c-approximation algorithm for a fixed constant c. From the practical point of view, it is sufficient to reduce[39] one of these problems to the considered optimization problem U in order to show that there is no PTAS for U if $P \neq NP$.

In what follows, we present a simple example of an AP-reduction.

Lemma 4.4.3.5. MAX-SAT$\leq_{\text{AP}}$MAX-CL.

Proof. To prove Lemma 4.4.3.5 we have to find F, H, α with the properties as described in Definition 4.4.3.1. We set $\alpha = 1$. Now we describe the work of the transformation F for an input C, δ, where $\delta \in \mathbb{Q}^+$ and $C = C_1 \wedge C_2 \wedge \cdots \wedge C_m$, $m \in \mathbb{N}$, is a formula in CNF over a set of variables $\{x_1, \ldots, x_n\}$, $n \in \mathbb{N}$. Let $C_i = l_{i1} \vee l_{i2} \vee \cdots \vee l_{ij_i}$ for $i = 1, \ldots, m$.

The transformation F works independently on δ and constructs an undirected graph $G_C = (V, E)$ defined as follows:

(i) $V = \{(i, p) \mid 1 \leq i \leq m, 1 \leq p \leq j_i\}$
     {Every occurrence of a literal in C corresponds to a vertex of G.}
(ii) $E = \{\{(r, s), (p, q)\} \mid r \neq p \text{ and } l_{r,s} \not\equiv \bar{l}_{p,q}\}$
     {We connect vertices corresponding to literals from different clauses only, provided one of the literals is not the negation of the other one.}

The transformation H maps cliques of $G_C = F(C, \delta)$ to assignments γ: $\{x_1, \ldots, x_n\} \to \{0, 1\}^n$. Since there is no edge between two vertices labeled by x_i and $\bar{x}_i$ for any $i \in \{1, \ldots, n\}$, we can define H as follows. Let $Q = (V_Q, E_Q)$ be a clique of G_C. If a vertex (r, s) of G_C is in V_Q and $l_{r,s} = x_j$ for some $j \in \{1, \ldots, n\}$, then set $x_j = 1$. If $(r, s) \in V_Q$ and $l_{rs} = \bar{x}_z$ for some $z \in \{1, \ldots, n\}$, then set $x_z = 0$. If this procedure does not assign the Boolean values to all variables from $\{x_1, \ldots, x_n\}$, choose the values for the remaining variables in an arbitrary way. Let $H(G_C)$ denote the resulting assignment to $\{x_1, \ldots, x_n\}$.

Obviously, F and H satisfy the conditions (i), (ii), (iii), and (iv) of Definition 4.4.3.1. To see (v) it is sufficient to recognize that every clique

---

[39] By AP-reduction

$Q = (V_Q, E_Q)$ of G_C determines an assignment $F(Q)$ that satisfies at least[40] $|V_Q|$ clauses, and that every assignment β to $\{x_1, \ldots, x_n\}$ satisfying r clauses determines a clique of G_C of exactly r vertices. Thus, $Opt_{\text{MAX-SAT}}(C) = Opt_{\text{MAX-CL}}(G_C)$, and

$$cost_{\text{MAX-SAT}}(H(G_C)) \geq cost_{\text{Max-CL}}(Q) = |V_Q|$$

for every $Q = (V_Q, E_Q) \in \mathcal{M}_{\text{MAX-CL}}(G_C)$.   □

Exercise 4.4.3.6. [*] Is MAX-CL AP-reducible to MAX-SAT?   □

Exercise 4.4.3.7. Prove MAX-3SAT$\leq_{\text{AP}}$MAX-CL.   □

Note that the AP-reduction may also have another application other than to prove the nonexistence of a PTAS for an optimization problem. If $U_1 \leq_{\text{AP}} U_2$ with parameters F, H, and α, and there is no polynomial-time $(1 + c)$-approximation algorithm for U_1 and a given constant c, then there is no polynomial-time $\left(1 + \frac{c}{\alpha}\right)$-approximation algorithm for U_2. One can, then, use the AP-reduction to prove specific lower bounds on the polynomial-time approximability. The AP-reduction of Lemma 4.4.3.5 with $\alpha = 1$ claims that every lower bound on the inapproximability of MAX-SAT is also a lower bound on the inapproximability of MAX-CL.

Exercise 4.4.3.8. Prove that if $U_1 \leq_{\text{AP}} U_2$ and U_1 does not admit any polynomial-time d-approximation algorithm for any $d \in \mathbb{Q}^{\geq 1}$, then U_2 does not admit any polynomial-time approximation algorithm with any constant approximation ratio either.   □

Exercise 4.4.3.9. Define a class of optimization problems (distinct from APX), and a reduction R, such that the hardest problems from $\mathcal{A}$ according to the reduction R do not have any polynomial-time approximation algorithm with a constant approximation ratio if P $\neq$ NP.   □

Currently, the most successful reduction for proving inapproximability results is the so-called gap-preserving reduction (GP-reduction). The goal is to transform an already known fact that an optimization problem does not admit a polynomial-time c-approximation algorithm for a specific c to a proof that another optimization problem does not admit a polynomial-time d-approximation algorithm for some d. c and d may be considered to be constant or functions of the input size. The crucial idea is to consider gap problems that are special problems corresponding to optimization problems.

---

[40] If $(r, s) \in V_Q$, then $H(G_C)(l_{r,s}) = 1$ and so the clause C_r is satisfied. Because of the definition of G_C there is no edge between two literals of the same clause, i.e., V_Q contains at most one vertex for each clause. This means that at least $|V_Q|$ different clauses must be satisfied. Some additional clauses may be satisfied by the assignment of the values to variables, whose values were not determined by V_Q.

Definition 4.4.3.10. *Let s, c be positive reals, $0 < s \leq c$. Let $U = (\Sigma_I, \Sigma_O, L, L_I, \mathcal{M}, cost, goal)$ be an optimization problem from NPO. We define the* **GAP$_{s,c}$-U** *decision problem as follows:*

Input: An $x \in L_I$ such that either $Opt_U(x)/|x| \geq c$ or $Opt_U(x)/|x| < s$.
Output: "yes" if $Opt_U(x)/|x| \geq c$;
         "no" if $Opt_U(x)/|x| < s$.

The usefulness of Definition 4.4.3.4 is explained by the following lemma.

Lemma 4.4.3.11. *Let $U \in$ NPO, and let $0 < s \leq c$ be some constants. If* GAP$_{s,c}$-U *is NP-hard, then, assuming* P $\neq$ NP, *there is no polynomial-time $\frac{c}{s}$-approximation algorithm for U.*

Proof. If there would be a polynomial-time (c/s)-approximation algorithm A for U, then one could use A to decide GAP$_{s,c}$-U as follows. Without loss of generality assume that U is a maximization problem. For every allowed input x of GAP$_{s,c}$-U one computes $A(x) \in \mathcal{M}$ and looks for $cost(A(x))$. We show that

$$cost(A(x)) < s \cdot |x| \text{ iff } Opt_U(x) < s \cdot |x|.$$

If $Opt_U(x) < s \cdot |x|$, then obviously $cost(A(x)) \leq Opt_U(x) < s \cdot |x|$.

If $Opt_U(x)$ is not smaller than $s \cdot |x|$, then $Opt_U(x) \geq c \cdot |x|$. Since A is an (c/s)-approximation algorithm,

$$\frac{Opt_U(x)}{cost(A(x))} \leq \frac{c}{s}.$$

This directly implies

$$cost(A(x)) \geq \frac{s}{c} \cdot Opt_U(x) \geq \frac{s}{c} \cdot c \cdot |x| \geq s \cdot |x|.$$

Thus, using A one can decide GAP$_{c,s}$-U. $\square$

Observe that we have already implicitly considered a gap problem for TSP in Lemma 4.4.2.1, where we reduced the HT problem to GAP$_{a,d}$-TSP for any $a > 1$ and any $d > 1$.

The basic idea of gap-preserving reduction is to realize the reduction of the NP-hardness of polynomial-time approximability of an optimization problem on the level of the corresponding gap problems.

Definition 4.4.3.12. *Let $U_1 = (\Sigma_{I,1}, \Sigma_{O,1}, L_1, L_{I,1}, \mathcal{M}_1, cost_1, maximum)$ and $U_2 = (\Sigma_{I,2}, \Sigma_{O,2}, L_2, L_{I,2}, \mathcal{M}_2, cost_2, maximum)$ be two maximization problems. A* **gap-preserving reduction (GP-reduction)** *from U_1 to U_2* **with parameters (c, s)** *and (c', s'), $0 \leq s \leq c \leq 1$, $0 \leq s' \leq c' \leq 1$, is a polynomial-time algorithm A that satisfies the following properties:*

(i) for every input instance $x \in L_{I,1}$, $A(x) \in L_{I,2}$,

(ii) $\dfrac{Opt_{U_1}(x)}{|x|} \geq c$ *implies* $\dfrac{Opt_{U_2}(A(x))}{|A(x)|} \geq c'$, *and*

(iii) $\dfrac{Opt_{U_1}(x)}{|x|} < s$ *implies* $\dfrac{Opt_{U_2}(A(x))}{|A(x)|} < s'$.

Observation 4.4.3.13. Let U_1 and U_2 be two maximization problems. If there is a GP-reduction from U_1 to U_2 with some parameters (c, s) and (c', s'), $0 \leq s \leq c \leq 1$, $0 \leq s' \leq c' \leq 1$, and $\mathrm{GAP}_{c,s}$-U_1 is NP-hard, then $\mathrm{GAP}_{c',s'}$-U_2 is NP-hard (i.e., following Lemma 4.4.3.11 there is no polynomial-time $\frac{c'}{s'}$-approximation algorithm for U_2 if P $\neq$ NP).

The advantage of the GP-reduction concept over the concept of the AP-reduction is its simplicity. A GP-reduction is simply a polynomial-time algorithm reducing a gap to another gap, while an AP-reduction consists of two polynomial-time transformations related by a set of properties.

Now, we present two simple examples of GP-reductions. In what follows we consider the number of clauses as the size of an input instance of MAX-E3SAT, and the number of equalities mod 2 as the size of MAX-E3LINMOD2.

Lemma 4.4.3.14. *For every $h \in [1/2, 1)$ and every small $\varepsilon \in (0, (1 - h)/2)$, there is a GP-reduction from MAX-E3LINMOD2 to MAX-E3SAT with parameters $(h + \varepsilon, 1 - \varepsilon)$ and $(\frac{3}{4} + \frac{h}{4} + \frac{\varepsilon}{4}, 1 - \frac{\varepsilon}{4})$.*

Proof. We give a GP-reduction from MAX-E3LINMOD2 to MAX-E3SAT. Remember that an input instance of MAX-E3LINMOD2 is a set S of equations

$$x + y + z \equiv a \,(\mathrm{mod}\ 2), \tag{4.71}$$

where x, y, z are Boolean variables and $a \in \{0, 1\}$. The objective is to maximize the number of satisfied equations.

For every S we construct a formula Φ_S over the same set of variables in the following way. For every equation $x + y + z \equiv 0 \,(\mathrm{mod}\ 2)$ we add

$$\Phi_0(x, y, z) = (\overline{x} \lor y \lor z) \land (x \lor \overline{y} \lor z) \land (x \lor y \lor \overline{z}) \land (\overline{x} \lor \overline{y} \lor \overline{z})$$

to Φ_S. For every equation $x + y + z \equiv 1 \,(\mathrm{mod}\ 2)$ we add

$$\Phi_1(x, y, z) = (x \lor \overline{y} \lor \overline{z}) \land (\overline{x} \lor y \lor \overline{z}) \land (\overline{x} \lor \overline{y} \lor z) \land (x \lor y \lor z)$$

to Φ_S. We observe that $\Phi_0(x, y, z)$ and $\Phi_1(x, y, z)$ together contain all 8 clauses over the set of variables $\{x, y, z\}$, i.e., no clause over $\{x, y, z\}$ is in both $\Phi_0(x, y, z)$ and $\Phi_1(x, y, z)$.

Obviously, for any assignment to $\{x, y, z\}$ exactly 7 of 8 clauses are satisfied and one clause is not satisfied. We immediately see that $\Phi_0(x, y, z)$ and $\Phi_1(x, y, z)$ have been chosen in such a way that, for every $a \in \{0, 1\}$, and every $\alpha : \{x, y, z\} \to \{0, 1\}$,

(i) if $x + y + z \equiv a \,(\,\mathrm{mod}\ 2)$ is satisfied by α, then all four clauses of $\Phi_a(x, y, z)$ are satisfied, and

(ii) if $x + y + z \equiv a \pmod 2$ is not satisfied by α, then exactly three clauses of $\Phi_a(x, y, z)$ are satisfied (i.e., exactly one clause of $\Phi_a(x, y, z)$ is not satisfied).

Now, we show that the above construction of Φ_S is a GP-reduction with parameters $(h + \varepsilon, 1 - \varepsilon)$ and $(\frac{3}{4} + \frac{h}{4} + \frac{\varepsilon}{4}, 1 - \frac{\varepsilon}{4})$. Let S be an input instance of MAX-E3LINMOD2 of $|S|$ clauses. If $Opt(S) \geq (1 - \varepsilon) \cdot |S|$, then the number of satisfied clauses of Φ_S is at least

$$4 \cdot (1 - \varepsilon) \cdot |S| + 3 \cdot \varepsilon \cdot |S| = (4 - \varepsilon) \cdot |S|.$$

Since Φ_S contains exactly $4 \cdot |S|$ clauses ($|\Phi_S| = 4 \cdot |S|$),

$$\frac{Opt(\Phi_S)}{|\Phi_S|} \geq \frac{(4 - \varepsilon) \cdot |S|}{4 \cdot |S|} \geq 1 - \frac{\varepsilon}{4}.$$

Let S be an input instance of MAX-E3LINMOD2 with $Opt(S) < (h + \varepsilon) \cdot |S|$. Then the number of satisfied clauses of Φ_S is smaller than

$$4 \cdot (h + \varepsilon) \cdot |S| + 3 \cdot (1 - h - \varepsilon) \cdot |S| = (3 + h + \varepsilon) \cdot |S|.$$

Thus,

$$\frac{Opt(\Phi_S)}{|\Phi_s|} < \frac{(3 + h + \varepsilon) \cdot |S|}{4 \cdot |S|} = \frac{3}{4} + \frac{h}{4} + \frac{\varepsilon}{4}.$$

$\square$

The usefulness of Lemma 4.4.3.14 can especially be seen in connection with the PCP-Theorem, which will be presented in the next section. A nontrivial application of the PCP-Theorem provides the NP-hardness of GAP$_{1/2+\varepsilon, 1-\varepsilon^-}$ MAX-E3LINMOD2. Applying Lemma 4.4.3.14 for $h = \frac{1}{2}$ and $\varepsilon = 4\delta$, one obtains the NP-hardness of GAP$_{7/8+\delta, 1-\delta}$-MAX-E3SAT for every small $\delta > 0$. Thus, assuming P $\neq$ NP, there is no polynomial-time $(\frac{8}{7} - \varepsilon')$-approximation algorithm for MAX-E3SAT for any $\varepsilon' > 0$. In Section 5.3 we shall design a polynomial-time $\frac{8}{7}$-approximation algorithm for MAX-E3SAT, and so the above lower bound on the polynomial-time inapproximability of MAX-E3SAT is optimal.

Lemma 4.4.3.15. *For all a, b, $0 < a \leq b \leq 1 - \frac{6}{10}$, there is a GP-reduction from MAX-E3SAT to MAX-2SAT with parameters (b, a) and $\left(\frac{6}{10} + \frac{b}{10}, \frac{6}{10} + \frac{a}{10}\right)$.*

Proof. Let $C = C_1 \wedge C_2 \wedge \cdots \wedge C_m$ be an input instance of MAX-E3SAT, where $C_i = l_{i1} \vee l_{i2} \vee l_{i3}$ for $i = 1, \ldots, m$. Let the 3-CNF C be over a set of variables $\{x_1, \ldots, x_n\}$.

Now we construct a formula Φ_C in 2-CNF over the extended set of variables $\{x_1, \ldots, x_n, y_1, \ldots, y_m\}$, where $y_1, \ldots, y_m$ are new variables. We replace each clause $C_i = l_{i1} \vee l_{i2} \vee l_{i3}$ by the formula

$$\Phi(C_i) = (l_{i1}) \wedge (l_{i2}) \wedge (l_{i3})$$
$$\wedge (\bar{l}_{i1} \vee \bar{l}_{i2}) \wedge (\bar{l}_{i2} \vee \bar{l}_{i3}) \wedge (\bar{l}_{i1} \vee \bar{l}_{i3})$$
$$\wedge (y_i) \wedge (l_{i1} \vee \bar{y}_i) \wedge (l_{i2} \vee \bar{y}_i) \wedge (l_{i3} \vee \bar{y}_i).$$

Thus, $\Phi_C = \bigwedge_{i=1}^{m} \Phi(C_i)$, and it is clear that Φ_C is an input instance of MAX-2SAT.

The crucial point is the following observation. For every assignment α : $\{x_1, \ldots, x_n\} \to \{0, 1\}$, and all $i = 1, 2, \ldots, m$,

(i) if C_i is satisfied by α, then there exists a choice for y_i such that exactly 7 clauses of the 10 clauses of $\Phi(C_i)$ are satisfied,

(ii) if C_i is not satisfied by α, then, for every choice of a Boolean value for y_i, at most 6 clauses of the 10 clauses of $\Phi(C_i)$ are satisfied, and

(iii) there is no choice for y_i that leads to more than 7 satisfied clauses.

If, for an assignment α, $\alpha(l_{i1}) = \alpha(l_{i2}) = \alpha(l_{i3}) = 1$, then the choice $y_i = 1$ leads to 7 satisfied clauses of $\Phi(C_i)$. In any other case[41] of a satisfying assignment the choice $y_i = 0$ provides 7 satisfied clauses. If $\alpha(l_{i1}) = \alpha(l_{i2}) = \alpha(l_{i3}) = 0$ (i.e., C_i is not satisfied by α), then the first three clauses of $\Phi(C_i)$ and at least one of the last four clauses of $\Phi(C_i)$ cannot be satisfied. One can easily verify that $Opt(\Phi(C_i)) = 7$ if $\Phi(C_i)$ is considered as a formula over $\{x_1, \ldots, x_n, y_i\}$.

It remains to use (i), (ii), and (iii) to show that the transformation from C to Φ_C is a GP-reduction with parameters

$$(a, b) \text{ and } \left(\frac{6}{10} + \frac{a}{10}, \frac{6}{10} + \frac{b}{10} \right).$$

Let C be a formula with $Opt(C) < a \cdot m = a \cdot |C|$. Then the number of satisfied clauses of Φ_C is smaller than

$$7 \cdot a \cdot m + 6 \cdot (1 - a) \cdot m = (6 + a) \cdot m.$$

Since Φ_C consists of $10 \cdot m$ clauses (i.e., we consider $|\Phi_C| = 10m$)

$$\frac{Opt(\Phi_C)}{|\Phi_C|} < \frac{(6 + a) \cdot m}{10m} = \frac{6}{10} + \frac{a}{10}.$$

Let C be a formula with $Opt(C) \geq b \cdot m = b \cdot |C|$. Then the number of satisfied clauses in the optimal solution for Φ_C is at least

$$7 \cdot b \cdot m + 6 \cdot (1 - b) \cdot m = (6 + b) \cdot m.$$

Thus,

$$\frac{Opt(\Phi_C)}{|\Phi_C|} \geq \frac{(6 + b) \cdot m}{10 \cdot m} = \frac{6}{10} + \frac{b}{10}.$$

$\square$

---

[41] If one or two of the literals are 1 by the given assignment α.

The fact that $\text{GAP}_{7/8+\delta,1-\delta}$-MAX-E3SAT is NP-hard for every small $\delta > 0$ together with Lemma 4.4.3.15 imply that $\text{GAP}_{55/80+\varepsilon,7/10-\varepsilon}$-MAX-2SAT is NP-hard. Since $\frac{7/10}{55/80} = \frac{56}{55}$, assuming P $\neq$ NP, there is no polynomial-time $(\frac{56}{55} - \varepsilon)$-approximation algorithm for MAX-2SAT. Thus, MAX-2SAT is also in the class NPO(III).

The next result shows that the MAX-CL problem can be reduced to itself in such a way that the gap grows.

Lemma 4.4.3.16. MAX-CL *can be reduced to* MAX-CL *by a GP-reduction with parameters* $(a, 1 - \varepsilon)$, $(a^2, (1 - \varepsilon)^2)$ *for every* $a \in (0, 1 - \varepsilon)$ *and any small* $\varepsilon \in (0, \frac{1}{2})$.

Proof. Let $G = (V, E)$ be an input instance of MAX-CL. Consider the size of G, $|G|$, equal to $|V|$. We construct the graph $G \times G = (V_{G \times G}, E_{G \times G})$ determined by[42]

$V_{G \times G} = \{< v, u > \, | \, v, u \in V\} = V \times V$, and

$E_{G \times G} = \{\{< v, u >, < r, s >\} \, | \, v, u, r, s \in V \text{ and } \{v, r\}, \{u, s\} \in E\}$.

One can easily observe that

$$|V_{G \times G}| = |V|^2 \text{ and } Opt_{\text{MAX-CL}}(G \times G) = (Opt_{\text{MAX-CL}}(G))^2.$$

If $Opt_{\text{MAX-CL}}(G) \geq (1 - \varepsilon) \cdot |G|$, then

$$Opt_{\text{MAX-CL}}(G \times G) \geq (1 - \varepsilon)^2 \cdot |G|^2 = (1 - \varepsilon)^2 \cdot |G \times G|.$$

If $Opt_{\text{MAX-CL}}(G) < a \cdot |G|$, then

$$Opt_{\text{MAX-CL}}(G \times G) < a^2 \cdot |G|^2 = a^2 \cdot |G \times G|.$$

$\square$

Corollary 4.4.3.17. *(a) If* $\text{GAP}_{(a,1-\varepsilon)}$-MAX-CL *is* NP-*hard for some* $a \in (0,1)$ *and for any small* ε, $1 - a > \varepsilon > 0$, *then* $\text{GAP}_{(a^2,1-\delta)}$-MAX-CL *is* NP-*hard for every small* δ, $1 - a^2 > \delta > 0$.

(b) For every $c > 1$, *if there is a polynomial-time c-approximation algorithm for* MAX-CL, *then there is a polynomial-time* $\sqrt{c}$-*approximation algorithm for* MAX-CL.

We see that (a) of Corollary 4.4.3.17 implies that if $\text{GAP}(a, 1-\varepsilon)$-MAX-CL is NP-hard for some $a \in (0,1)$ and every small $\varepsilon > 0$, then $\text{GAP}(b, 1 - \delta)$-MAX-CL is NP-hard for all $b \in (0,1)$ and every small δ, $0 < \delta < 1-b$. Thus, if there is no polynomial-time c-approximation algorithm for MAX-CL for some $c > 1$, then there is no polynomial-time d-approximation algorithm for any $d > 1$. On the other hand, (b) of Corollary 4.4.3.17 implies that the existence

---

[42] Informally, $G \times G$ is constructed by replacing each vertex of G by a copy of G and joining two such copies completely if the corresponding vertices in G are adjacent.

of a polynomial-time c-approximation algorithm for MAX-CL for some $c > 1$ forces the existence of a PTAS for MAX-CL. Thus, we have either a PTAS for MAX-CL, or no polynomial-time constant-approximation algorithm for MAX-CL. Since MAX-SAT is APX-complete[43] and MAX-SAT $\leq_{AP}$ MAX-CL by Lemma 4.4.3.5, the second alternative is true.

4.4.4 Probabilistic Proof Checking and Inapproximability

In this section we explain the concept of probabilistic proof checking, present the PCP-Theorem, and apply the PCP-Theorem to show that there is no PTAS for MAX-E3SAT. The crucial point is that the PCP-Theorem provides a completely new characterization of languages in NP that is suitable for reducing any language (decision problem) from NP to GAP$_{(a,1)}$-MAX-SAT for an $a < 1$.

First of all we define probabilistic verifiers that are a generalization of the (deterministic) polynomial-time verifiers introduced in Section 2.3.3. A **probabilistic verifier** V is a polynomial-time Turing machine with four tapes (see Figure 4.19):

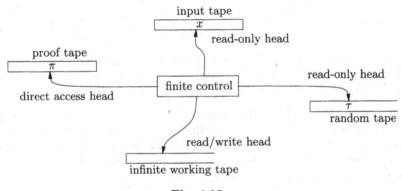

Fig. 4.19.

(a) a usual input tape containing an input string x,
(b) a usual working tape (internal memory of V),
(c) a random tape containing a random string $\tau \in \{0,1\}^*$,
(d) a proof tape that contains a string $\pi \in \{0,1\}^*$, (candidate for being a proof of the fact $x \in L$).

V uses read-only heads on the input tape and on the random tape, and it has a direct access to read any position of the proof tape. As usual, V has a read/write head on the working tape.

---

[43] Note that we did not prove it here.

A probabilistic verifier works as follows. Reading the input tape (containing an input x) and the random tape (containing a random string τ) V computes a list of indices $i_1, i_2, \ldots, i_c$ for some positive integer c. Then V looks at the bits $\pi_{i_1}, \pi_{i_2}, \ldots, \pi_{i_c}$ of π. Depending on $\pi_{i_1}, \pi_{i_2}, \ldots, \pi_{i_c}$, x, and τ, V decides either to accept x or to reject x. V works in time polynomial in $|x|$, and so c and $|\tau|$ are considered to be polynomial in $|x|$, too. There is no restriction on the length of π.

Let $V(x, \tau, \pi) \in \{\text{accept}, \text{reject}\}$ be the decision of V for given x, τ, and π. Obviously, $V(x, \tau, \pi)$ is unambiguously determined by x, τ, and π, because V computes deterministically if τ is given.

Let r and q be functions from $\mathbb{N}$ to $\mathbb{N}$. An $(r(n), q(n))$-**restricted probabilistic verifier** is a verifier that, for any input x, uses random strings of length at most $r(|x|)$ and queries at most $q(|x|)$ bits from the proof.

In what follows we consider that every $(r(n), q(n))$-restricted probabilistic verifier uses random strings of length exactly $r(n)$ on the input length n. Further, we assume the uniform probability distribution $Prob$ over the strings, i.e., $Prob(\tau) = \frac{1}{2^{|\tau|}}$. For any $(r(n), q(n))$-restricted probabilistic verifier, any input x, and any proof π, we define

$$Prob_\tau[V(x, \tau, \pi) = \text{accept}] = \sum_{\substack{\tau \\ V(x, \tau, \pi) = \text{accept}}} Prob(\tau).$$

We say that a probabilistic verifier V accepts a language L if

(i) $x \in L$ implies $\exists \pi \in \{0,1\}^*$ such that $V(x, \tau, \pi) = \text{accept}$ for every τ (i.e., $Prob_\tau[V(x, \tau, \pi) = \text{accept}] = 1$), and

(ii) $x \notin L$ implies $\forall \pi \in \{0,1\}^* : Prob_\tau[V(x, \tau, \pi) = \text{accept}] \leq 1/2$.

The condition (i) is called **completeness** and means that, for every input $x \in L$, there must exist a proof π_x of the fact $x \in L$ such that V accepts x for every random string τ. The condition (ii) is called **soundness**. It assures that, for every $x \notin L$, every string π is recognized to be no proof of the fact $x \in L$ with a probability of at least $1/2$.

Example 4.4.4.1. Consider the following $(0, n)$-restricted probabilistic verifier V_S that accepts SAT. For any satisfiable formula Φ over a set of variables $X = \{x_1, \ldots, x_m\}$, any assignment to X that satisfies Φ is considered to be a proof of the fact $\Phi \in$ SAT. The verifier V_S computes deterministically, i.e., the random tape is empty. V_S reads the first m bits of the proof tape and interprets them as an assignment α to X. V_S checks whether α satisfies Φ, and accepts Φ if and only if α satisfies Φ. So, for a correct proof, V_S always accepts the formula Φ. If Φ is not satisfiable, V_S rejects Φ for any content of the proof tape. Since m is always smaller than $n = |\Phi|$ and V_S works in polynomial time, V_S is a $(0, n)$-restricted probabilistic verifier for SAT.   □

Example 4.4.4.1 only shows the well-known fact (Theorem 2.3.3.9) that the problems from NP have a deterministic polynomial-time verifier.[44] The

---

[44] Remember that SAT is NP-complete.

principle question is whether one can use randomization in order to decrease the amount of queried bits from the proof. The following example shows that it is possible.

Example 4.4.4.2. Consider the $\text{GAP}_{1-\varepsilon,1}\text{-E3SAT}$ problem for some $\varepsilon \in (0,1)$. Remember that an input instance is a formula Φ in the 3CNF that is either satisfiable or at most a $(1-\varepsilon)$ fraction of clauses is satisfiable (i.e., $Opt_{\text{MAX-3SAT}}(\Phi) \leq (1-\varepsilon) \cdot m$, where Φ consists of m clauses). We design a $\left(\log_{1-\varepsilon}(1/2) \cdot \log_2 n, 3\lceil \log_{1-\varepsilon}(1/2)\rceil\right)$-restricted probabilistic verifier V that accepts Φ if and only if Φ is satisfiable.

Let $\Phi = F_1 \wedge F_2 \wedge \cdots \wedge F_m$, $m \in \mathbb{N}$, be a formula in 3CNF over a set of variables $\{x_1,\ldots,x_d\}$, $d \leq 3m$. Let $n = |\Phi|$, and $k = \lceil \log_{1-\varepsilon}(1/2)\rceil$. Obviously, $3m \leq n$ and $d \leq n$. First, the verifier V chooses uniformly k indices $i_1, i_2, \ldots, i_k \in \{1,\ldots,m\}$ at random. Formally, it means that the random tape contains a random string τ of length $k \cdot \lceil \log_2 m\rceil$ that binary codes $i_1, i_2, \ldots, i_k$. Then V looks up the bits corresponding to the indices of all variables that appear in $F_{i_1} \wedge F_{i_2} \wedge \cdots \wedge F_{i_k}$. Thus, the number of queried bits is at most $3k$.

V accepts Φ if these at most $3k$ bits determine a satisfying assignment for $F_{i_1} \wedge F_{i_2} \wedge \cdots \wedge F_{i_k}$. In the opposite case V rejects Φ.

Let us prove that V is a probabilistic verifier for $\text{GAP}_{1-\varepsilon,1}\text{-E3SAT}$. If Φ is satisfiable, then there exists an assignment π to $\{x_1,\ldots,x_d\}$ that satisfies Φ. If π is the content of the proof tape, then V always accepts Φ, i.e., $Prob_\tau[V(\Phi,\tau,\pi) = \text{accept}] = 1$.

Let Φ be not satisfiable. This means that fewer than $(1-\varepsilon) \cdot m$ clauses of Φ are satisfied for any assignment to $\{x_1,\ldots,x_d\}$. Let π be a fixed binary string that is considered to be the content of the proof tape. V interprets the prefix of π of the length d as an assignment α to $\{x_1,\ldots,x_d\}$. Since α does not satisfy at least $\varepsilon \cdot m$ clauses of Φ, the probability of uniformly choosing at random a clause Φ_i, which is not satisfied by α, is at least ε. Remember that V rejects Φ if α does not satisfy at least one of the k randomly chosen clauses $F_{i_1}, F_{i_2}, \ldots, F_{i_k}$. The probability that all these clauses are satisfied by α is at most

$$(1-\varepsilon)^k.$$

Thus, the probability that at least one of the k clauses is not satisfied by α is at least

$$1 - (1-\varepsilon)^k = 1 - (1-\varepsilon)^{\lceil \log_{1-\varepsilon}(\frac{1}{2})\rceil} \geq \frac{1}{2},$$

i.e.,

$$Prob_\tau[V(\Phi,\tau,\pi) = \text{reject}] \geq \frac{1}{2}.$$

$\square$

Example 4.4.4.2 shows that a hard problem can admit a probabilistic verifier that uses $O(\log_2 n)$ random bits and queries $O(1)$ bits of the proof tape only.

We define, for all functions r, q from $\mathbb{N}$ to $\mathbb{N}$,

$$\mathbf{PCP}(r, q) = \{L \mid L \text{ is accepted by an } (r(n), q(n))\text{-restricted}$$
$$\text{probabilistic verifier}\}.$$

Let $\mathcal{W}$ and $\mathcal{V}$ be subclasses of the class of all nondecreasing functions from $\mathbb{N}$ to $\mathbb{N}$. We define

$$\mathbf{PCP}(\mathcal{W}, \mathcal{V}) = \bigcup_{\substack{r \in \mathcal{W} \\ q \in \mathcal{V}}} \mathrm{PCP}(r, q).$$

Observation 4.4.4.3. For every $\varepsilon \in (0, 1)$,

$$\mathrm{GAP}_{(1-\varepsilon),1}\text{-Max-E3Sat} \in \mathrm{PCP}(O(\log_2 n), O(1)).$$

Observation 4.4.4.4. Let $Poly(n)$ be the class of all polynomials over $\mathbb{N}$.

(i) $\mathrm{P} = \mathrm{PCP}(0, 0)$, and
(ii) $\mathrm{NP} = \mathrm{PCP}(0, Poly(n))$.

Proof. The claim (i) is obvious because a verifier may simply simulate a polynomial-time algorithm for a given language L without making any use of the random and proof tapes. On the other hand, a verifier without any access to its random and proof tapes is a usual polynomial-time Turing machine.

We already know (see Section 2.3.3) that, for any x from a language L from NP, there exists a proof π_x of a polynomial length in the size of x such that a deterministic verifier (reading the whole π_x) can verify whether π_x is a proof of the fact $x \in L$ in polynomial time. On the other hand a probabilistic verifier without any access to its random tape can be easily simulated by a polynomial-time nondeterministic Turing machine.[45]

$\square$

In what follows we present the PCP-Theorem that is probably the deepest result of theoretical computer science. The proof is based on some very nontrivial techniques (including arithmetization) from complexity theory and so it is beyond the elementary level of this book. Fortunately, it is not necessary to learn the proof in order to prove inapproximability of optimization problems. The assertion of the PCP-Theorem can be directly applied to prove lower bounds on the inapproximability of some optimization problems, and using appropriate reductions, these lower bounds can be extended to a variety of optimization problems.

Theorem 4.4.4.5 (PCP-Theorem).

$$\mathrm{NP} = \mathrm{PCP}(O(\log_2 n), 11).$$

---

[45] This Turing machine starts to work by nondeterministically generating (guessing) a word π that is a candidate for a proof.

Idea of the proof. The relation $\mathrm{PCP}(O(\log_2 n), O(1)) \subseteq \mathrm{NP}$ can be easily proved. Let $L \in \mathrm{PCP}(O(\log_2 n), O(1))$. Then there exists an $(c \cdot \log_2 n, d)$-restricted probabilistic verifier V for L for some positive integers c and d. A nondeterministic Turing machine starts to simulate the probabilistic verifier V by nondeterministically guessing (generating) a queried part of a word π that candidates for being a proof of the fact $x \in L$. Since the length of the content of the random tape is bounded by $c \cdot \log_2 n$, there are at most n^c different random strings appearing on the random tape. One can remove randomness (the use of the random tape) by generating all n^c random strings in the lexicographical order and deterministically simulate the work of V for every one of these random strings generated. If all n^c simulations finish with acceptance, then one accepts the input. In the opposite case the input is rejected. Since every simulation takes polynomial time and we execute n^c simulations, the whole deterministic computation runs in polynomial time.

To remove the use of the proof tape we use nondeterminism. In any computation with a fixed random string τ, at most d bits of the proof are queried. In all the n^c simulated computations the number of queried bits is bounded by $d \cdot n^c$. A nondeterministic Turing machine A can, at the beginning of its computation, nondeterministically guess all these bits as well as their positions in the proof. Since V works in polynomial time, the index of any queried bit can be represented by a binary string of a polynomial length. Thus, the common length of the binary string nondeterministically generated by A is bounded by a polynomial in the input length. The rest of the computation consists of the n^c deterministic simulations as described above. Thus, A is a polynomial time nondeterministic Turing machine accepting L.

The hard direction of the proof is to prove $\mathrm{NP} \subseteq \mathrm{PCP}(O(\log_2 n), 11)$. A very rough idea of the proof is to construct a special proof $\pi(x, L)$ of the fact $x \in L$ for any language $L \in \mathrm{NP}$ and any input $x \in L$. This proof must have the property that every local part of $\pi(x, L)$ contains an essential global information of the proof. Looking at 11 randomly chosen bits of $\pi(x, L)$ is sufficient to verify the correctness of $\pi(x, L)$ for the fact $x \in L$. $\square$

Now, we apply the PCP-Theorem to prove that there is no PTAS for MAX-E3SAT.

Theorem 4.4.4.6. *Let* $\varepsilon = (9 \cdot 2^{12})^{-1}$. $\mathrm{GAP}_{1-\varepsilon,1}$-3SAT *is NP-hard.*

Proof. We show that every language L from the class NP is reducible to $\mathrm{GAP}_{1-\varepsilon,1}$-3SAT. Since $\mathrm{NP} = \mathrm{PCP}(O(\log_2 n), 11)$ there exists a constant $c \in \mathbb{N}$ and a $(c \cdot \log_2 n, 11)$-restricted probabilistic verifier V_L that accepts L. The idea is to construct, for every $x \in \Sigma_L^*$, a formula Φ_x in 3CNF such that

(i) if $x \in L$, then Φ_x is satisfiable,
(ii) if $x \notin L$, then, for any assignment to the variables of Φ_x, at least an ε fraction of the clauses of Φ_x is not satisfied, and
(iii) Φ_x can be constructed from V_L and x in polynomial time.

In this way one reduces the recognition of any language L from NP to $\text{GAP}_{1-\varepsilon,1}\text{-3SAT}$.

We now describe how to construct the 3SAT problem instance Φ_x for a given input word x and a given verifier V_L. We interpret the proof π queried by V_L as a binary string $\pi = \pi_1\pi_2\pi_3\ldots$, where the ith bit is represented by the variable x_i. First, we construct a formula F_x over a subset of $X = \{x_1, x_2, x_3, \ldots\}$ such that F_x is in 11CNF and satisfies the above properties (i), (ii), and (iii). We set

$$F_x = \bigwedge_\tau F_{x,\tau}$$

where $\bigwedge_\tau$ runs over all n^c random strings τ used by V_L. For a fixed τ, $F_{x,\tau}$ is determined as follows. If, for the input x and for the random string τ, V_L queried the bits of π on the positions $i_1, i_2, \ldots, i_{11}$, then $F_{x,\tau}$ is over the set of variables $X_\tau = \{x_{i_1}, x_{i_2}, \ldots, x_{i_{11}}\}$. Since x and τ are fixed, V_L unambiguously determines a Boolean function $f_{x,\tau} : \{0,1\}^{11} \to \{0,1\}$ over X_τ such that

$$f_{x,\tau}(a_{i_1}, a_{i_2}, \ldots, a_{i_{11}}) = 1 \text{ iff } V_L(x, \tau, \pi_1\ldots a_{i_1}\pi_{i_1+1}\ldots a_{i_2}\pi_{i_2+1}\ldots a_{i_{11}}\pi_{i_{11}+1}\ldots)$$
$$= \text{accept.}$$

Since $f_{x,\tau}$ is over 11 variables, it can be unambiguously represented by a formula $F_{x,\tau}$ in 11CNF. Obviously, $F_{x,\tau}$ contains at most 2^{11} clauses.

If $x \in L$, there exists a proof π such that $V_L(x, \tau, \pi) = \text{accept}$ for every random string τ. Thus, π determines an assignment to X such that all $F_{x,\tau}$ are satisfied, i.e., F_x is satisfiable.

If $x \notin L$, then, for every proof π, V_L accepts x for at most one half of all n^c random strings. Thus, for every assignment to the set of variables of F_x, at least one half of the formulae $F_{x,\tau}$ are not satisfied. Since we have n^c formulae $F_{x,\tau}$, there are at least $\frac{n^c}{2}$ unsatisfied clauses in F_x for every assignment to the variables of F_x. Since the number of clauses of F_x is at most $2^{11} \cdot n^c$, at least a $\frac{1}{2^{12}}$ fraction of the clauses are not satisfied by any assignment.

To determine $f_{x,\tau}$ for a fixed x and τ one has to simulate the work of V_L on x and τ for all 2^{11} assignments to $x_{i_1}, \ldots, x_{i_{11}}$ (i.e., for all possible proofs). Since, for a given τ, V_L works deterministically in polynomial time, and 2^{11} is a constant, every function $f_{x,\tau}$ (and so its formula $F_{x,\tau}$) can be generated in polynomial time. Since the number of formulae $F_{x,\tau}$ is n^c for the fixed constant c, the whole formula F_x can be generated in polynomial time.

Now it remains to transform F_x into an equivalent formula Φ_x in 3CNF. This can be simply realized by the standard reduction from SAT to 3SAT as presented in Section 2.3. Every clause of $F_{x,\tau}$ can be transformed into the conjunction of 9 clauses of $\Phi_{x,\tau}$ in 3CNF by using 8 new variables. If $F_{x,\tau}$ is satisfied by an assignment α, then there exists an extension of α to the set of new variables such that $\Phi_{x,\tau}$ is satisfied, too. If there is at least one clause of $F_{x,\tau}$ not satisfied by an assignment β, then any extension of β to an assignment including the new variables cannot satisfy $\Phi_{x,\tau}$ (i.e., at least one clause of $\Phi_{x,\tau}$ is not satisfied for any assignment to its variables). Taking $\Phi_x = \bigvee_\tau \Phi_{x,\tau}$, we see that

(1) $F_{x,\tau}$ is satisfiable if and only if $\Phi_{x,\tau}$ is satisfiable, and
(2) if m subformulae $F_{x,\tau}$ of F_x are not satisfied by an assignment β, then exactly m subformulae $\Phi_{x,\tau}$ of Φ_x are not satisfied by any extension of the assignment β to the set of all variables of Φ_x.

Because of (1), and the fact "$x \in L \Leftrightarrow F_x$ is satisfiable", the condition (i) holds for Φ_x. Because of (2) and the fact that, for $x \notin L$, at least $\frac{n^c}{2}$ subformulae $F_{x,\tau}$ of F_x are not satisfied by any assignment, $x \notin L$ implies that at least $\frac{n^c}{2}$ subformulae $\Phi_{x,\tau}$ of Φ_x are not satisfied by any assignment to the set of the input variables of Φ_x. Thus, if $x \notin L$, at least $\frac{n^c}{2}$ clauses of Φ_x are not satisfied for any assignment. Since the number of clauses of Φ_x is at most $9 \cdot 2^{11} \cdot n^c$, at least a

$$\frac{1}{9 \cdot 2^{12}}$$

fraction of the clauses of Φ_x are not satisfied for any assignment. Thus, (ii) holds for Φ_x. Since we proved that F_x is computable in polynomial time, and the transformation from F_x to Φ_x is a standard polynomial-time reduction, the condition (iii) is true for Φ_x, too. This completes the proof. $\quad\square$

Corollary 4.4.4.7. *Unless* $P = NP$, *there is no* PTAS *for* MAX-3SAT.

Proof. A direct consequence of Theorem 4.4.4.6 is even that if $P \neq NP$, then there is no polynomial-time $\frac{1}{1-\varepsilon}$-approximation algorithm for MAX-3SAT. $\quad\square$

Using AP-reductions or GP-reductions from MAX-3SAT to other optimization problems one can prove the nonexistence of PTAS for these optimization problems. The result of Lemma 4.4.3.16 (self-reduction of MAX-CL) and Lemma 4.4.3.5 (MAX-SAT$\leq_{AP}$MAX-CL) even imply that there is no polynomial-time c-approximation algorithm[46] for MAX-CL for any positive constant c.

Exercise 4.4.4.8. Assume that one could prove $NP = PCP(O(\log_2 n), d)$ for a positive integer d. Find the largest possible $\varepsilon > 0$ such that one can prove $GAP_{1-\varepsilon,1}$-3SAT is NP-hard by the proof method of Theorem 4.4.4.6. $\quad\square$

Exercise 4.4.4.9. $^{(**)}$ Use the PCP-Theorem to prove that $GAP_{\frac{1}{2}+\varepsilon,1-\varepsilon}$-MAX-E3LINMOD2 is NP-hard for every $\varepsilon \in (0, \frac{1}{2})$. $\quad\square$

The result of Exercise 4.4.4.9 can help to essentially improve the result of Corollary 4.4.4.7.

Theorem 4.4.4.10. *Unless* $P \neq NP$, *there exists no polynomial-time* $\left(\frac{8}{7} - \delta\right)$-*approximation algorithm for* MAX-3SAT *for any small constant* $\delta > 0$.

---

[46] See also Corollary 4.4.3.17

Proof. In Lemma 4.4.3.14 we proved that there is a GP-reduction from MAX-E3LINMOD2 to MAX-E3SAT with parameters

$$(h + \varepsilon, 1 - \varepsilon) \text{ and } \left(\frac{3}{4} + \frac{h}{4} + \frac{\varepsilon}{4}, 1 - \frac{\varepsilon}{4}\right)$$

for every small $\varepsilon \in (0, (1 - h)/2)$ and every $h \in [1/2, 1)$. Choosing $h = \frac{1}{2}$, the parameters of this reduction are

$$\left(\frac{1}{2} + \varepsilon, 1 - \varepsilon\right) \text{ and } \left(\frac{7}{8} + \frac{\varepsilon}{4}, 1 - \frac{\varepsilon}{4}\right).$$

Exercise 4.4.4.9 claims that $\text{GAP}_{1/2+\varepsilon, 1-\varepsilon}$-MAX-E3LINMOD2 is NP-hard for any small $\varepsilon > 0$. Applying the above GP-reduction with parameters

$$\left(\frac{1}{2} + \varepsilon, 1 - \varepsilon\right) \text{ and } \left(\frac{7}{8} + \frac{\varepsilon}{4}, 1 - \frac{\varepsilon}{4}\right)$$

we obtain that $\text{GAP}_{7/8+\varepsilon', 1-\varepsilon'}$-MAX-3SAT is NP-hard for any ε'. Thus, there is no polynomial-time $\left(\frac{8}{7} - \delta\right)$-approximation algorithm for MAX-3SAT for any small $\delta > 0$, unless $\text{P} \neq \text{NP}$.   □

Exercise 4.4.4.11. (**) Prove that for any integer $k \geq 3$ and any small $\varepsilon > 0$, $\text{GAP}_{1-2^{-k}+\varepsilon, 1}$-MAX-EkSAT is NP-hard.   □

Exercise 4.4.4.12. (*) Prove, for some some $\varepsilon > 0$, that $\text{P} \neq \text{NP}$ implies that there is no polynomial-time n^ε-approximation algorithm for MAX-CL.   □

Keywords introduced in Section 4.4

approximation-preserving reduction, the class APX, APX-complete problems, gap problems, gap-preserving reduction, probabilistic verifier

Summary of Section 4.4

There are three basic possibilities for proving lower bounds on the polynomial-time inapproximability of optimization problems. The first possibility is to reduce an NP-hard problem to an approximation problem. This approach is suitable for proving the nonexistence of any polynomial time $p(n)$-approximation algorithm for TSP for any polynomial p, unless $\text{P} \neq \text{NP}$.

If one has lower bounds on the inapproximability of some optimization problems, then one can extend them to other optimization problems by suitable reductions. We have presented two kinds of such reductions: the approximation-preserving reduction (AP-reduction) and the gap-preserving reduction (GP-reduction). Defining the class APX as the class of optimization problems that are polynomial-time approximable with a constant approximation ratio, one can introduce APX-completeness by the

AP-reduction. This concept is analogous to the concept of NP-completeness and if $P \neq NP$, then the APX-complete optimization problems do not admit any PTAS.

The principal possibility for proving lower bounds on inapproximability is based on the PCP-Theorem. The PCP-Theorem gives a new, fundamental characterization of languages in NP by restricted probabilistic verifiers. The restrictions are related to two computational sources of verifiers – the number of random bits and the number of queried bits of the proof. Any language in NP can be accepted by a probabilistic verifier with $O(\log_2 n)$ random bits and 11 queried bits. Starting from this representation of languages in L by $(O(\log_2 n), 11)$-restricted probabilistic verifiers, one can find a polynomial-time reduction from any language in NP to some specific approximation problem. We showed such a reduction to the gap problem of MAX-3SAT. The nonexistence of any PTAS for MAX-3SAT (if $P \neq NP$) follows.

4.5 Bibliographical Remarks

The Graham's algorithm [Gra66] for the makespan scheduling problem was probably the first analyzed approximation algorithm for an optimization problem from NPO. The question of approximability started receiving attention immediately after the concept of NP-completeness was introduced by Cook [Coo71] in 1971. In the next few years several efficient approximation algorithms were designed (see, for instance, [Joh74, SG76, Chr76, IK75, Lov75]). Excellent early references for overviews on approximation algorithms are Garey and Johnson [GJ79], Sahni and Horowitz [SH78], and Papadimitriou and Steiglitz [PS82].

Since the design and analysis of approximation algorithms is one of the central topics of current computer science, there is wealth of literature on approximation algorithms. Recent bestsellers are the books [Hoc97] edited by Hochbaum, and the textbook [ACG⁺99] by Ausiello, Crescenzi, Gambosi, Kann, Marchetti-Spaccamela, and Protasi. The book [Hoc97], written by leading researchers in this field, contains an excellent comprehensive survey of basic concepts and ideas connected with approximation algorithms. Textbook [ACG⁺99] is devoted to almost all aspects related to approximation algorithms, and contains transparent introductory parts as well as complex parts for advanced readers. The concept of the stability of approximation has been recently introduced and investigated in [Hro98, BHK⁺99, Hro99a, Hro99b]. The concept of dual approximation algorithms is due to Hochbaum and Shmoys [HS87].

Papadimitriou and Steiglitz [PS82] attribute the 2-approximation algorithm VCP for the vertex cover problem to Gavril and Yannakakis. The first analysis of the greedy approach for the set cover problem is due to Johnson [Joh74] and Lovász [Lov75]. The analysis presented here is based on a more general result about greedy algorithms proved by Chvátal [Chv79]. The 2-approximation local algorithm for MAX-CUT has been considered for a long time to be the best approximation approach for MAX-CUT. Using the method based on the relaxation of semidefinite programs and rounding, Goemans and Williamson [GW95] essentially improved the approximation factor to 1.139 . . . . The concept initialized a large progress in the application of semidefinite programming in the design of approximation application algorithms (see, for instance, [FG95, AOH96, AHO$^+$96, Asa97]). Early approximation algorithms for the knapsack problems are given by Sahni [Sah75]. The FPTAS for the knapsack problem is due to Ibarra and Kim [IK75]. Some other approximation schemes for the knapsack problem are given by Lawler [Law79]. The 2-approximation algorithm for $\triangle$-TSP was apparently known to researchers for some time. An elaborate version was published by Rosenkrantz, Stearns, and Lewis [RSL77]. Christofides' algorithm is due to Christofides [Chr76]. The investigation of the stability of approximation of these two algorithms for $\triangle$-TSP as presented here was proposed by Böckenhauer, Hromkovič, Klasing, Seibert, and Unger [BHK$^+$99]. The idea of violating the triangle inequality was already investigated by Andreae and Bandelt [AB95], who designed a $\left(\frac{3}{2}\beta^2 + \frac{1}{2}\beta\right)$-approximation algorithm for $\triangle$-TSP$_\beta$. Recently, Böckenhauer et al. [BHK$^+$99] achieved a $\frac{3}{2}\beta^2$-approximation algorithm for $\triangle$-TSP$_\beta$, and Bender and Chekuri [BC99] designed a 4β-approximation algorithm for $\triangle$-TSP$_\beta$. Moreover, if P $\neq$ NP, then $\triangle$-TSP$_\beta$ is not approximable within a factor $1 + \varepsilon \cdot \beta$ for some $\varepsilon < 1$ [BC99]. The possibilities to improve the approximation ratio for $\triangle$-TSP instances with sharpened triangle inequality were discovered by Böckenhauer, Hromkovič, Klasing, Seibert, and Unger [BHK$^+$00]. Böckenhauer and Seibert [BS00] proved explicit lower bounds on the polynomial-time approximability of $\triangle$-TSP with sharpened triangle inequality. The first PTASs for the geometrical TSP were discovered by Arora [Aro97] and Mitchell [Mit96]. They are not very practical because their time complexity is around $O(n^{30 \cdot \frac{1}{\varepsilon}})$ for the approximation ratio $1 + \varepsilon$. Arora [Aro97] achieved a significant improvement by designing a randomized PTAS working in time $O(n \cdot (\log n)^{40 \cdot \frac{1}{\varepsilon}})$. The PTAS for the makespan scheduling problem presented here is due to Hochbaum and Shmoys [HS87]. An excellent, comprehensive, and systematically updated overview on the approximability of optimization problems is given by Crescenzi and Kann [CK99].

The interest in classifying optimization problems with respect to their polynomial-time approximability started immediately after the concept of NP-completeness was discovered by Cook [Coo71]. While a large progress in the design of approximation algorithms was achieved in the 1970s and the 1980s, the attempts to prove inapproximability results were not very successful. One of the few known explicit lower bounds on inapproximability was that of Theorem 4.4.2.3, claiming the nonexistence of any polynomial-time $p(n)$-approximation algorithm for TSP. Sahni and Gonzales [SG76] proved it by the presented reduction to the Hamiltonian tour problem. Similar results for some maximum-subgraph problems can be found in Yannakakis [Yan79]. Much of the work attempted to develop a classification framework for optimization problems according to approximability followed ideas analogous to the concept of NP-completeness (see, for instance, [ADP77, ADP80, AMS$^+$80]). The most successful approach was due to Papadimitriou and Yannakakis [PY91], who introduced the class MAX-SNP by a logical characterization and the notion of completeness for this class by using the so-called L-reduction. The idea behind this concept was that every MAX-SNP-complete optimization problem does not admit any PTAS if and only if MAX-3SAT does not admit any PTAS. Several optimization problems were proved to be MAX-SNP-complete.

The PCP-Theorem is a consequence of very intensive research on so-called interactive proof systems. The concepts of these proof systems were invented by Goldwasser, Micali, and Rackoff [GMR89] and Babai [Bab85]. The notion of probabilistically checkable proofs was introduced by Fortnow, Rompel, and Sipser [FRS88]. The intensive investigation of the power of restricted probabilistic verifiers ([BFL91, BFLS91, FGL91, AS92]) culminated in [ALM92], where Arora, Lund, Motwani, Sudan, and Szegedy proved the PCP-Theorem. The connection between probabilistically checkable proofs and approximability of optimization problems was first discovered by Feige, Goldwasser, Lovász, Safra, and Szegedy [FGL91], who observed that NP $\subseteq$ PCP($\log n, \log n$) and P $\neq$ NP imply that there is no constant factor approximation algorithm for MAX-CL. Arora, Lund, Motwani, Sudan, and Szegedy [ALM92] not only proved the PCP-Theorem but at the same time also proved as a consequence that no APX-complete problem has a PTAS. A self-contained proof of the PCP-Theorem is approximately 50 pages long, and one can find it in [Aro94, HPS94, MPS98, ACG$^+$99]. The *Lecture Notes Tutorial* edited by Mayr, Prömel, and Steger [MPS98] is an excellent, comprehensive survey on the concept of proof verifications and its application for proving lower bounds on inapproximability. Another concise survey on inapproximability results is given by Arora and Lund [AL97] in [Hoc97]. Using this approach, many lower bounds on the approximation problems were established. In what follows we mention only those, that are connected to approximation algorithms presented in this book. Håstad [Hås97b], assuming P $\neq$ NP, proved the lower bound $2-\varepsilon$ and $\frac{2^k}{2^k-1}$, respectively, on the approximation ratio of any polynomial-time approximation algorithm for MAX-E3LINMOD2 and MAX-EkSAT, respectively. Feige [Fei96] proved that, for any $\varepsilon > 0$, a $(1-\varepsilon) \cdot \ln |S|$ upper bound on the approximation ratio of any polynomial-time approximation algorithm for the set cover problem would imply NP $\subseteq$ $DTIME(n^{\log \log n})$. Thus, the presented greedy approach is the best possible one from an approximation point of view. In 1997 Håstad [Hås97b] proved that MAX-CL is not polynomial-time approximable within the approximation ratio 1.0624, and that the minimum vertex cover is not approximable within the approximation ratio 1.1666. Engebretsen [Eng99], Böckenhauer et al. [BHK$^+$99], and Papadimitriou and Vempala [PVe00] proved explicit lower bounds on the approximation ratio for $\triangle$-TSP.

Currently, the most efficient way to prove lower bounds on inapproximability is to extend the known lower bounds by reductions. The idea of approximation-preserving reductions is due to Paz and Moran [PM81] who presented it already in 1981. The problem was that it turned out to be very difficult to prove the completeness of an optimization problem under considered reductions. In 1995 Crescenzi, Kann, Silvestri, and Trevisan [CKS+95] introduced the AP-reductions presented here. Trevisan [Tre97] proved that MAX-3SAT is APX-complete under AP-reduction.

5

Randomized Algorithms

*"For him
who seeks the truth,
an error is nothing unknown."*

JOHANN WOLFGANG VON GOETHE

5.1 Introduction

A randomized algorithm can be viewed as a nondeterministic algorithm that
has a probability distribution for every nondeterministic choice. To simplify
the matter one usually considers only the random choices from two possibilities, each with the probability 1/2. Another possibility is to consider a
randomized algorithm as a deterministic algorithm with an additional input that consists of a sequence of random bits. In other words, a randomized
algorithm may be seen as a set of deterministic algorithms, from which one
algorithm is randomly chosen for the given input.

For a fixed input instance x of the problem considered, the runs (computations) of a randomized algorithm on x may differ in the dependence on
the actual sequence of random bits. This difference can be projected into the
complexity as well as to the outputs of the particular runs. Thus, for instance,
running time or output can be considered as random variables. In general,
randomized algorithms whose outputs can be considered as random variables
are called Monte Carlo algorithms. A randomized algorithm that always (independent of the random choice) computes the correct output and only the
complexity is considered as a random variable is called a Las Vegas algorithm.[1]

Randomized algorithms may be more efficient (faster, more space-efficient,
etc.) and simpler to handle (implement) than their best known deterministic counterparts. Unfortunately, the typical case is that it is not at all clear
whether randomization is really helpful, i.e., whether or not it is possible
to convert a randomized algorithm to a deterministic one without paying

---

[1] More details about this classification are given in Section 5.2.

any significant penalty in the computer resources.[2] This problem is open for polynomial-time computations for any kind of reasonable randomization. Still worse, we do not know any randomized-polynomial time algorithm for an NP-hard computing problem. On the other hand, we know several important problems, like tests on primality, where

(i) we have an efficient polynomial-time randomized algorithm for the problem,
(ii) there is no known polynomial-time deterministic algorithm for the problem, and
(iii) it is unknown whether the problem is in P or not (or whether it is NP-complete or not).

Thus, from the point of view of this book[3] randomized algorithms are extremely helpful because they enable us to efficiently solve problems for which we do not know any other approach making them tractable from the practical point of view. Because a few of these problems are so important and appear so often in every day applications such as primality testing (that is crucial for modern cryptography), randomized approaches have become standard tools in almost all areas of computer science. Moreover, one can combine approximation with randomization in order to efficiently solve NP-hard optimization problems. In this case the approximation ratio can be considered as a random variable and the aim is to produce a feasible solution that is a good approximation of an optimal solution with high probability. In this way one can make an optimization problem tractable even if the approximation approach alone was not able to do it.

Monte Carlo randomized algorithms give probabilistic outputs that are not necessarily correct or not necessarily exact. Somebody may consider that because of this, randomized algorithms are not suitable for critical applications for which a false or uncertain answer cannot be tolerated. This does not need to be true. The error probability can be brought down significantly by independent[4] repetitions of the runs of the randomized algorithms on the same input. The designers of randomized algorithms like to call attention to the fact that often the error probability may even be decreased so that it is essentially smaller than the probability of the occurrence of a hardware error during the significantly longer running time of the best known deterministic algorithm.

---

[2] Theorems proving that randomized computation are strictly more efficient than deterministic ones are known only for some restricted computing models like two-party communication protocols or finite automata. There is more about this in the bibliographical remarks.

[3] Remember that here we consider a problem to be hard even if the best known deterministic algorithm runs in polynomial time but the polynomial has a high degree.

[4] An independent repetition means a run with a new random sequence.

This chapter is organized as follows. Section 5.2 provides the definition of the basic classification of randomized algorithms as well as some of the fundamental paradigms of randomization that intuitively explain the reasons for the possible success for solving hard problems by randomization. In Section 5.3 we apply these paradigms in order to illustrate their power. More precisely, in Section 5.3.2 we present a polynomial-time Las Vegas algorithm that finds a quadratic nonresidue in $\mathbb{Z}_p$ for every prime p. This problem is not known to be in P. In Section 5.3.3 we present Monte Carlo algorithms for primality testing. In Section 5.3.4 we show how to design a Monte Carlo algorithm solving an equivalence problem that is not known to be in P. Section 5.3.5 gives a randomized optimization algorithm for MAX-CUT. Section 5.3.6 is devoted to the design of randomized approximation algorithms for distinct satisfiability problems. Section 5.4 is devoted to derandomization, i.e., to methods that convert randomized algorithms to equivalent deterministic ones. To read this chapter it is reasonable to have some elementary knowledge of algebra, number theory, and probability theory as presented in Sections 2.2.4 and 2.2.5.

5.2 Classification of Randomized Algorithms and Design Paradigms

5.2.1 Fundamentals

If one wants to formalize the notion of randomized algorithms, then one can take deterministic Turing machines A with an infinite additional tape. This additional tape is read-only; it contains an infinite random sequence of 0s and 1s, and A may move on it only from the left to the right. Another possibility is to take a nondeterministic Turing machine with nondeterministic guesses over at most two possibilities and to assign the probability $\frac{1}{2}$ to every such possibility. For our purposes we do not need to present any complete formal definition of randomized algorithms. For us it is sufficient to view randomized algorithms as algorithms that sometimes may ask for some random bits and then may continue to compute in different ways depending on the values of these random bits. The probability theory point of view is that a randomized algorithm A together with a fixed input x determines a probabilistic experiment. This experiment can be described by the probability space $(S_{A,x}, Prob)$, where $S_{A,x} = \{C \mid C$ is a computation (random run) of A on $x\}$, and $Prob$ is a probability distribution over $S_{A,x}$.

For every randomized algorithm A we consider a new complexity measure – the number of random bits used. Let $Random_A(x)$ be the maximum number of random bits used over all random runs (computations) of A on x. Then, for every $n \in \mathbb{N}$,

$$Random_A(n) = \max\left\{Random_A(x) \mid x \text{ is an input of size } n\right\}.$$

This complexity measure is important because of the following two reasons:

(1) It costs a lot of work to produce sequences that may be considered a reasonable "substitute" for truly random sequences, and this cost increases with the length of the random sequences required.[5]

(2) If $Random_A(x)$ of a randomized algorithm A is bounded by a logarithmic function,[6] then the number of distinct runs on any fixed input of size n is bounded by $2^{Random_A(n)} \leq p(n)$ for a polynomial p.[7] This means that one can deterministically simulate A by following all runs of A (one after another). If A computes in polynomial time in every run, then this deterministic simulation runs in polynomial time, too. The above-mentioned simulation is called "**derandomization**".[8] Obviously, if $Random_A(n)$ is large (e.g., linear or even polynomial), then the number of different runs for one input may be too large (e.g., exponential). In this case we do not know any efficient, general method for "converting" randomized algorithms into deterministic ones.

For every run (random computation) C of a randomized algorithm A on an input x, one can consider the probability of the execution[9] of this run on x. This probability is the multiplication over the probabilities of all random choices done in this run C of A on x, i.e., the probability of the corresponding random sequence. In what follows, we denote this probability by $\boldsymbol{Prob_{A,x}(C)}$. In our simplified approach to randomized algorithms it is $\frac{1}{2}$ to the power of the number of random bits asked in C.

Because a randomized algorithm A may produce different results on a fixed input x, the output of a result y is considered an (random) event. The probability that $\boldsymbol{A}$ **outputs** $\boldsymbol{y}$ **for an input** $\boldsymbol{x}$, $\boldsymbol{Prob(A(x) = y)}$, is the sum of all $Prob_{A,x}(C)$, where C outputs y. Obviously, the aim of the randomized algorithm designer is to achieve high $Prob(A(x) = y)$ if y is the correct output for the input x.

Different runs of A on an input x may also have different lengths, i.e., they may have different costs. So, the time complexity becomes a random variable. Let $\boldsymbol{Time(C)}^{10}$ be the time complexity of the run C of A on x. Then the **expected time complexity of** $\boldsymbol{A}$ **on** $\boldsymbol{x}$ is

$$Exp\text{-}Time_A(x) = E[Time] = \sum_C Prob_{A,x}(C) \cdot Time(C),$$

---

[5] We do not want to delve into a deep discussion on this topic, because an essentially nontrivial theory is needed to understand and to solve the production of "random" sequences. A little bit more about this is given in the Bibliographic Notes.

[6] For instance, $c \cdot \log_2 n$

[7] In our example, n^c

[8] Note that there exist also other approaches to derandomization (to efficient deterministic simulation of randomized computations).

[9] Thus, the execution of a run is an elementary event.

[10] Note that $Time$ is a random variable.

where the sum is taken over all runs C of A on x.[11] If one would like to consider $Exp\text{-}Time_A$ as a function of the input size, then one uses the worst case approach, i.e., **the expected time complexity of A is**

$$Exp\text{-}Time_A(n) = \max\left\{Exp\text{-}Time_A(x) \mid x \text{ is an input of size } n\right\}$$

for every $n \in \mathbb{N}$. It is often not easy to analyze $Exp\text{-}Time_A(n)$ for a given randomized algorithm A. To overcome this difficulty one also uses the worst case approach from the beginning. This means:

$$Time_A(x) = \max\left\{Time(C) \mid C \text{ is a run of } A \text{ on } x\right\}.$$

Then, the **(worst case) time complexity of A** is

$$Time_A(n) = \max\left\{Time_A(x) \mid x \text{ is an input of size } n\right\}.$$

Unfortunately, this definition may be misleading[12] in some cases. This is because randomized algorithms may allow infinite runs provided that they occur with a reasonably small probability on any given input. Obviously, a deterministic algorithm is never allowed to take an infinite computation on an input x, because if it does so, this algorithm never solves the given problem for the input x. If this deterministic algorithm has bounded space complexity (which is always the case in reality), then an infinite computation means an infinite loop. By contrast, infinite computations (runs) are acceptable for probabilistic algorithms because an infinite run does not mean an infinite loop. New random decisions may always lead to success resulting in a computation providing the right output. One possibility of simplifying things is to determine the time complexity $Time_A$ from outside and stop the randomized algorithm if it does not halt in this time. One interprets this situation as the output "I was not able to solve the problem in time" (the output "?"). This approach works very well if the probability to be ready in $Time_A(n)$ is reasonably large for every input. In such a case the output "?" is not bad because in the application it only means that the randomized algorithm has to be restarted on the same input[13] from the initial configuration.

5.2.2 Classification of Randomized Algorithms

In what follows we introduce the following fundamental classes of randomized algorithms: Las Vegas, one-sided-error Monte Carlo, two-sided-error Monte Carlo, and unbounded-error Monte Carlo. This classification is natural and convenient for decision problems or the evaluation of functions. For optimization problems this fine classification does not make sense and we introduce the concepts of a randomized optimization algorithm and a randomized approximation algorithm only.

---

[11] Observe that $Exp\text{-}Time_A(x)$ is the expectation $E[Time]$ of the random variable $Time$ in the experiment of the work of A on x.

[12] Even inconsistent

[13] Obviously with a new sequence of random bits

LAS VEGAS ALGORITHMS.

There are two different possibilities for how people define Las Vegas algorithms and their time complexity. We present both these approaches because in different situations one may prefer a particular one. But the main common point for any definition of Las Vegas algorithms is that they never give a wrong output.[14]

The first approach says that a randomized algorithm A is a **Las Vegas algorithm computing a problem** F if for any input instance x of F,

$$Prob(A(x) = F(x)) = 1,$$

where $F(x)$ is the solution of F for the input instance x. For this definition the time complexity of A is always considered to be the expected time complexity $Exp\text{-}Time_A(n)$.

In the second approach to defining Las Vegas algorithms we allow the answer "?".[15] We say that a randomized algorithm A is a **Las Vegas algorithm computing a problem** F if for every input instance x of F,

$$Prob(A(x) = F(x)) \geq \frac{1}{2},$$

and

$$Prob(A(x) = \text{"?"}) = 1 - Prob(A(x) = F(x)) \leq \frac{1}{2}.$$

In this second approach one may consider $Time_A(n)$ as the time complexity of A because the nature of this second approach is to stop after $Time_A(|x|)$ computing steps on x, and to consider a new start (run) on x from the initial configuration if the output was not computed in $Time_A(|x|)$ steps.

The first approach to defining Las Vegas algorithms is usually used for computing a function, while the second one is often used for decision problems.

Example 5.2.2.1. Probably the best known Las Vegas algorithm is the randomized QUICKSORT. Let S be a set of elements with a linear ordering.

Algorithm 5.2.2.2. RQS (RANDOMIZED QUICKSORT)

Input: $a_1, \ldots, a_n,\ a_i \in S$ for $i = 1, \ldots, n,\ n \in \mathbb{N}$.

Step 1:   Choose an $i \in \{1, \ldots, n\}$ uniformly at random.

        {Every $i \in \{1, \ldots, n\}$ has equal probability to be chosen.}

Step 2:   Let A be the multiset $\{a_1, \ldots, a_n\}$.

        if $n = 1$ **output**(S)

        else the multisets $S_<, S_=, S_>$ are created.

$$S_< := \{b \in A \mid b < a_i\};$$
$$S_= := \{b \in A \mid b = a_i\};$$
$$S_> := \{b \in A \mid b > a_i\}.$$

---

[14] That is, the output is not a random variable.

[15] In this case the informal meaning of "?" is more or less "I was not able to solve the problem in one random attempt".

Step 3:   Recursively sort $S_<$ and $S_>$.
Output: $\mathrm{RQS}(S_<)$, $S_=$, $\mathrm{RQS}(S_>)$.

Without any doubt RQS is a randomized algorithm[16] that provides the correct output with probability 1. So, to evaluate the "quality" of RQS, one has to analyze the function $Exp\text{-}Time_{\mathrm{RQS}}$. We omit this analysis here because we gave it already in Example 2.2.5.28 for the case when S is a set.   □

Example 5.2.2.3. Here we consider the problem of finding the kth smallest element of a given set of elements. The idea of the randomized algorithm for this problem is similar to RQS.

Algorithm 5.2.2.4. RANDOM-SELECT(S, k)

Input: $S = \{a_1, a_2, \ldots, a_n\}$, $n \in \mathbb{N}$, and a positive integer $k \leq n$.
Step 1: **if** $n = 1$ **then return** a_1
          else choose an $i \in \{1, 2, \ldots, n\}$ randomly.
Step 2: $S_< = \{b \in S \mid b < a_i\}$;
          $S_> = \{c \in S \mid c > a_i\}$.
Step 3: **if** $|S_<| > k$ **then** RANDOM-SELECT$(S_<, k)$
          else if $|S_<| = k - 1$ **then return** a_i
              else RANDOM-SELECT$(S_>, k - |S_<| - 1)$.
Output: the kth smallest element of S
        (i.e., an a_l such that $|\{b \in S \mid b < a_l\}| = k - 1$).

We can easily observe that, for every input (S, k), the worst case running time is $\Theta(n^2)$ (i.e., there exists a sequence of random bits leading to $\Theta(n^2)$ comparisons). Intuitively, the best random choice in Step 1 is the choice for which $|S_<| = k-1$. A reasonable choice is a choice for which $|S_<|$ and $|S_>|$ have approximately the same size (i.e., after one repetition of Steps 1, 2, and 3 the input instance (S, k) reduces to (S', j) where $|S'|$ is approximately half of $|S|$). A bad choice is a choice for which one of $|S_<|$ and $|S_>|$ is very small and the kth smallest element is in the larger of the sets $S_<$ and $S_>$. In what follows we prove an upper bound on $E[T_{S,k}]$, where $T_{S,k}$ is the random variable counting the number of comparisons of RANDOM-SELECT on the input instance (S, k). Let $|S| = n$. Since $E(T_{S,k})$ depends on $|S| = n$ but not on the particular elements of S, we shall use also the notation $T_{n,k}$ instead of $T_{S,k}$. Let $T_n = \max\{T_{n,k} \mid 1 \leq k \leq n\}$.

The first repetition of Steps 1, 2, and 3 costs $n - 1$ comparisons between elements of S. Since i is uniformly chosen in Step 1, the probability that a_i is the jth smallest number in S is $1/n$ for every $j = 1, \ldots, n$. In that case

$$T_{|S|,k} \leq n - 1 + \max\{T_{|S_<|,k}, T_{|S_>|,k-|S_<|-1}\}$$
$$= n - 1 + \max\{T_{j-1,k}, T_{n-j,k-j}\}$$
$$\leq n - 1 + \max\{T_{j-1}, T_{n-j}\}.$$

---

[16] Because of Step 1

Thus, we obtain the recurrence

$$E[T_n] \leq n - 1 + \frac{1}{n} \sum_{j=1}^{n-1} \max\{E[T_{j-1}], E[T_{n-j}]\}$$

$$\leq n - 1 + \frac{1}{n} \sum_{j=1}^{n-1} E\left[T_{max\{j-1,n-j\}}\right]$$

$$\leq n - 1 + \frac{2}{n} \sum_{l=\lceil n/2 \rceil}^{n-1} E[T_l].$$

Now, we prove $E[T_n] \leq 5 \cdot n$ for every n by induction. Obviously, this is true for $n = 1$. Consider that it is true for all $m < n$. Then

$$E[T_n] \underset{ind.}{\leq} n - 1 + \frac{2}{n} \sum_{l=\lceil n/2 \rceil}^{n-1} 5 \cdot l$$

$$\leq n - 1 + \frac{10}{n} \left(\sum_{l=1}^{n-1} l - \sum_{l=1}^{\lceil n/2 \rceil - 1} l \right)$$

$$= n - 1 + \frac{10}{n} \left(\frac{n \cdot (n-1)}{2} - \frac{(\lceil n/2 \rceil - 1) \cdot \lceil n/2 \rceil}{2} \right)$$

$$\leq n - 1 + 5(n-1) - 5 \cdot (\lceil n/2 \rceil - 1) \cdot \frac{1}{2}$$

$$\leq 5 \cdot n.$$

$\square$

Another example of a Las Vegas algorithm that sometimes outputs "?" is given in what follows. We call attention to the fact that there is one more reason to present Example 5.2.2.5. For the computing model considered there one can even prove that the complexity of every deterministic algorithm[17] computing the given task is higher than the complexity of the designed Las Vegas algorithm.

Example 5.2.2.5. Let us consider the following scenario. One has a computing model, called **one-way (communication) protocol** that consists of two computers C_I and C_{II} connected by a communication link. One considers that C_I and C_{II} have an unbounded computation power. In general, the system has to compute a function $F : A_I \times A_{II} \to \{0, 1\}$ by the following rules:

(i) At the beginning C_I gets an input $x \in A_I$ and C_{II} gets an input $y \in A_{II}$.
(ii) C_I computes a message $\overline{C}_I(x) \in \{0, 1\}^*$.
   C_I is considered as a function $\overline{C}_I : A_I \to \{0, 1\}^*$ with the prefix-freeness property: $\overline{C}_I(x_1)$ is not a proper prefix of $\overline{C}_I(x_2)$ if $x_1 \neq x_2$, $x_1, x_2 \in A_I$.

---

[17] From the restricted class of algorithms defined by the computing model

(iii) C_{II} with the arguments y and $\overline{C}_I(x)$ computes $F(x,y)$;
i.e., C_{II} is considered to be a function $\overline{C}_{II} : A_{II} \times \{0,1\}^* \to \{0,1\}$.

The cost of a computation on an input (x,y) is the length of the message $\overline{C}_I(x)$. The **communication complexity** of a one-way protocol (C_I, C_{II}) is

$$\max \left\{ |\overline{C}_I(x)| \mid x \in A_I \right\}.$$

Now, consider the following function

$$Choice_n : \{0,1\}^n \times \{1,2,\ldots,n\} \to \{0,1\}$$

defined by

$$Choice_n(x_1 x_2 \ldots x_n, i) = x_i.$$

First, we observe that every deterministic one-way protocol (C_I, C_{II}) computing $Choice_n$ has a communication complexity of at least n. This is because C_I has to use at least 2^n different messages, a special one for every input from $\{0,1\}^n$. Let us assume the opposite, i.e., $\overline{C}_I(u) = \overline{C}_I(v)$ for two different $u = u_1 u_2 \ldots u_n$ and $v = v_1 v_2 \ldots v_n$ from $\{0,1\}^n$. Since $u \neq v$, there exists a $j \in \{1,2,\ldots,n\}$ such that $u_j \neq v_j$. Now consider the computation of (C_I, C_{II}) on the inputs (u,j) and (v,j). Since $\overline{C}_I(u) = \overline{C}_{II}(v)$ we get $\overline{C}_{II}(j, \overline{C}_I(u)) = \overline{C}_{II}(j, C_{II}(v))$, i.e., the protocol (C_I, C_{II}) produces the same output for the inputs (u,j) and (v,j). But this cannot be true because

$$Choice_n(u_1 u_2 \ldots u_n, j) = u_j \neq v_j = Choice_n(v_1 v_2, \ldots, v_n, j).$$

So, (C_I, C_{II}) has to use at least 2^n different messages. Due to the prefix-freeness property of $\overline{C}_I$, the communication complexity of (C_I, C_{II}) is at least n.

Note that the communication complexity n is also sufficient, because one can simply define $\overline{C}_I(x) = x$ for every $x \in \{0,1\}^n$ and $\overline{C}_{II}(j, \overline{C}_I(x)) = \overline{C}_{II}(j,x) = Choice_n(x,j)$.

Now, for every even n we give a Las Vegas one-way protocol (D_I, D_{II}) computing $Choice_n$ with the communication complexity $n/2 + 1$.

Las Vegas One-Way Protocol (D_I, D_{II})

Input: (x,j), $x = x_1 \ldots x_n \in \{0,1\}^n$, $j \in \{1,\ldots,n\}$.

Step 1: D_I chooses a random bit $r \in \{0,1\}$.

Step 2: D_I sends the message $c_1 c_2 \ldots c_{n/2+1} = 0 x_1 \ldots x_{n/2} \in \{0,1\}^{n/2+1}$ if $r = 0$, and
D_I sends the message $c_1 c_2 \ldots c_{n/2+1} = 1 x_{n/2+1} \ldots x_n \in \{0,1\}^{n/2+1}$ if $r = 1$.

Step 3: If $r = 0$ and $j \in \{1,2,\ldots,n/2\}$ then D_{II} outputs $c_{j+1} = x_j$.
If $r = 1$ and $j \in \{n/2+1,\ldots,n\}$ then D_{II} outputs $c_{j-n/2+1} = x_j = Choice(x,j)$.
Else, D_{II} outputs "?".

$\square$

ONE-SIDED-ERROR MONTE CARLO ALGORITHMS.

This type of randomized algorithm is considered for decision problems only. Let L be a language, and let A be a randomized algorithm. We say that A is a **one-sided-error Monte Carlo algorithm** recognizing L if[18]

(i) for every $x \in L$, $Prob(A(x) = 1) \geq 1/2$, and
(ii) for every $x \notin L$, $Prob(A(x) = 0) = 1$.

Thus, a one-sided-error Monte Carlo algorithm never says "yes" if the input does not have the required property. It may err in one direction only. Namely, it can sometimes conclude that the input does not have the property if it has.

One-sided-error Monte Carlo algorithms are very practical because the probability of getting the right answer grows exponentially with the number of repetitions. Let $a_1, a_2, \ldots, a_k$ be k answers of k independent[19] runs of a one-sided-error Monte Carlo algorithm A on the same input instance x. If there exists $i \in \{1, \ldots, k\}$ such that $a_i = 1$, then we know with certainty that $x \in L$. If $a_1 = a_2 = \cdots = a_k = 0$ (i.e., in the complementary case), the probability that $x \in L$ is $(1/2)^k$. So, we decide to consider $x \notin L$, and this is true with probability $1 - 1/2^k$.

In the following we show that, for some special computing models, one-sided-error Monte Carlo randomization may drastically decrease the deterministic complexity.

Example 5.2.2.6. Consider the one-way (communication) protocol model introduced in Example 5.2.2.5 and the function

$$Non\text{-}Eq_n : \{0,1\}^n \times \{0,1\}^n \to \{0,1\}$$

defined by

$$Non\text{-}Eq_n(x,y) = 1 \text{ iff } x \neq y.$$

Similarly as in Example 5.2.2.5, we observe that every deterministic one-way protocol computing $Non\text{-}Eq_n$ has a communication complexity of at least n. The argument for this is that for all $u, v \in \{0,1\}^n$, $u \neq v$ implies $\overline{C}_I(u) \neq \overline{C}_I(v)$. Let $\overline{C}_I(u) = \overline{C}_I(v)$ for two different $u, v \in \{0,1\}^n$. Then, assuming that (C_I, C_{II}) computes $Non\text{-}Eq_n$, we obtain the following contradiction:

$$0 = Non\text{-}Eq_n(u,u) = \overline{C}_{II}\left(u, \overline{C}_I(u)\right) = \overline{C}_{II}\left(u, \overline{C}_I(v)\right) = Non\text{-}Eq_n(v,u) = 1.$$

Now, we give a randomized one-way protocol for $Non\text{-}Eq_n$ with communication complexity in $O(\log_2 n)$. Note, that the comparison of x and y is equivalent to the comparison of $Number(x)$ and $Number(y)$[20].

---

[18] Remember that the output 1 means acceptance (yes) and the output 0 represents rejection (no).

[19] This means that $a_1, a_2, \ldots, a_k$ can be considered as independent events.

[20] Remember that $Number(x_1 \ldots x_n) = \sum_{i=1}^{n} x_i \cdot 2^{n-i}$ for every $x_1 \ldots x_n \in \{0,1\}^n$.

Random Inequality (R_I, R_{II})

Input: $x, y \in \{0, 1\}^n$

Step 1: R_I chooses uniformly a prime p from the interval $[2, n^2]$ at random.
{Note that there are approximately $n^2 / \ln n^2$ primes in this interval and so $2\lceil \log_2 n \rceil$ random bits are enough to realize this random choice.}

Step 2: R_I computes $s = Number(x) \bmod p$ and sends p and s to R_{II}.
{The length of the message is $4\lceil \log_2 n \rceil$ ($2\lceil \log_2 n \rceil$ bits for each of p and s). This is possible because $s \leq p \leq n^2$.}

Step 3: R_{II} computes $q = Number(y) \bmod p$.
If $q \neq s$, then R_{II} outputs 1 ("accept").
If $q = s$, then R_{II} outputs 0 ("reject").

The communication complexity of the randomized protocol (R_I, R_{II}) is $4\lceil \log_2 n \rceil$. To prove that (R_I, R_{II}) is a one-sided-error Monte Carlo one-way protocol for $Non\text{-}Eq_n$ we need a deep result of number theory – the Prime Number Theorem that claims that the number of primes in the set $\{1, 2, 3, \ldots, m\}$ is approximately $m / \ln m$, and at least $m / \ln m$ for $m \geq 100$.

If $x = y$, then $Number(x) \bmod p = Number(y) \bmod p$ for every prime p. So,

$$Prob\left((R_I, R_{II}) \text{ rejects } (x, y)\right) = 1.$$

Let $x \neq y$. If [21] $Number(x) \bmod p = Number(y) \bmod p$ for a prime p, then p divides $h = |Number(x) - Number(y)|$. Since $h < 2^n$, h has fewer than n different prime divisors. This means that at most $n - 1$ primes l from the at least $n^2 / \ln n^2$ primes from $\{2, 3, \ldots, n^2\}$ have the property

$$Number(x) \bmod l = Number(y) \bmod l. \tag{5.1}$$

Thus, the probability that R_I randomly chooses a prime with the property (5.1) for the given input (x, y) is at most

$$\frac{n - 1}{n^2 / \ln n^2} \leq \frac{\ln n^2}{n}.$$

Then,

$$Prob\left((R_I, R_{II}) \text{ accepts } (x, y)\right) \geq 1 - \frac{\ln n^2}{n},$$

which is greater than $9/10$ for every $n \geq 100$.

Note, that the protocol (R_I, R_{II}) can be improved by a random choice of k primes [22] $p_1, p_2, \ldots, p_k$ from $\{2, 3, \ldots, n^2\}$. Then, the protocol rejects (x, y) if and only if

$$Number(x) \bmod p_i = Number(y) \bmod p_i$$

for all $i = 1, 2, \ldots, k$. $\square$

---

[21] Despite the fact $x \neq y$
[22] Instead of one

Exercise 5.2.2.7. Change the first step of the one-way protocol $(R_{\mathrm{I}}, R_{\mathrm{II}})$ by choosing randomly a prime from the set $\{2, 3, \ldots, n^c\}$ for a positive integer $c \geq 3$. Analyze

(i) the communication complexity of this protocol, and
(ii) $Prob\,((R_{\mathrm{I}}, R_{\mathrm{II}})$ accepts $(x, y))$ for every input (x, y) with $x \neq y$. $\qquad$ $\square$

TWO-SIDED-ERROR MONTE CARLO ALGORITHMS.

Let F be a computing problem. We say that a randomized algorithm A is a **two-sided-error Monte Carlo algorithm computing F** if there exists a real number ε, $0 < \varepsilon \leq 1/2$, such that for every input x of F

$$Prob(A(x) = F(x)) \geq \frac{1}{2} + \varepsilon.$$

The strategy is to let the algorithm run t times on the given input and to take as the output the result which appears at least $\lceil t/2 \rceil$ times, if any. Let $p = p(x) \geq 1/2 + \varepsilon$ be the probability that A computes the correct result on a given input x in one run. The probability that A gives the correct answer on the input x exactly $i \leq \lfloor t/2 \rfloor$ times in t runs is

$$pr_i(x) = \binom{t}{i} p^i (1-p)^{t-i} = \binom{t}{i} (p(1-p))^i (1-p)^{2(\frac{t}{2}-i)}$$

$$\leq \binom{t}{i} \cdot \left(\frac{1}{4} - \varepsilon^2 \right)^i \left(\frac{1}{4} - \varepsilon^2 \right)^{\frac{t}{2}-i} = \binom{t}{i} \cdot \left(\frac{1}{4} - \varepsilon^2 \right)^{\frac{t}{2}}.$$

Now, consider the following algorithm A_t.

Algorithm A_t

Input: x
Step 1:   Run the algorithm A on x t times independently and save the t outputs
          $y_1, y_2, \ldots, y_t$.
Step 2: $y :=$ an output from the multiset $\{y_1, \ldots, y_t\}$ with the property $y = y_i$ for at least $\lceil t/2 \rceil$ different is from $\{1, \ldots, t\}$, if any.
          $\{y = ?$ if there is no output with at least $\lceil t/2 \rceil$ occurrences in the sequence $y_1, \ldots, y_t.\}$
Output: y

Since A_t computes $F(x)$ if and only if at least $\lceil t/2 \rceil$ runs of A finish with the output $F(x)$, one obtains

$$Prob(A_t(x) = F(x)) \geq 1 - \sum_{i=0}^{\lfloor t/2 \rfloor} pr_i(x)$$

$$> 1 - \sum_{i=0}^{\lfloor t/2 \rfloor} \binom{t}{i} \cdot \left(\frac{1}{4} - \varepsilon^2 \right)^{t/2} > 1 - 2^{t-1} \left(\frac{1}{4} - \varepsilon^2 \right)^{t/2}$$

$$= 1 - \frac{1}{2}(1 - 4\varepsilon^2)^{t/2}. \tag{5.2}$$

Thus, if one looks for a k such that $Prob(A_k(x) = F(x)) \geq 1 - \delta$ for a chosen constant δ and any input x, then it is sufficient to take

$$k \geq \frac{2 \ln 2\delta}{\ln(1 - 4\varepsilon^2)}. \tag{5.3}$$

Obviously, if δ and ε are assumed to be constants[23] then k is a constant, too. Thus, $Time_{A_k}(n) \in O(Time_A(n))$. This means that if A is asymptotically better than the best known deterministic algorithm for F, then A_k is too.

Exercise 5.2.2.8. Consider the one-way protocol model used in Examples 5.2.2.5 and 5.2.2.6, and, for every $n \in \mathbb{N}^+$, the function

$$Equality_n : \{0,1\}^n \times \{0,1\}^n \to \{0,1\},$$

defined by

$$Equality_n(x,y) = 1 \text{ iff } x \equiv y.$$

(i) Prove, for every $n \in \mathbb{N}^+$, that every deterministic one-way protocol computing $Equality_n$ has a communication complexity of at least n.
(ii) Design a two-sided-error Monte Carlo one-way protocol that computes $Equality_n$ within $O(\log_2 n)$ communication complexity.
(iii) Prove that every one-sided-error Monte Carlo one-way protocol that computes $Equality_n$ must have a communication complexity of at least n.   □

UNBOUNDED-ERROR MONTE CARLO ALGORITHMS.

These are the general randomized algorithms also called Monte Carlo algorithms. Let F be a computing problem. We say that a randomized algorithm A is a **(unbounded-error) Monte Carlo algorithm computing A** if, for every input x of F,

$$Prob(A(x) = F(x)) > \frac{1}{2}.$$

To see the essential difference between two-sided-error Monte Carlo algorithms and unbounded-error ones one has to analyze the number of necessary repetitions k (see (5.3)) of an unbounded-error Monte Carlo algorithm A in order to get a two-sided error Monte Carlo algorithm A_k with

$$Prob(A_k(x) = F(x)) \geq 1 - \delta$$

---

[23] Independent of the input size

for some constant δ, $0 \le \delta \le 1/2$. The drawback of A is that the difference between $Prob(A(x) = F(x))$ and $1/2$ tends to 0 with the growth of the input length. For an input x, A can have $2^{Random_A(|x|)}$ different computations, each with probability $2^{-Random_A(|x|)}$.[24] It may happen that

$$Prob(A(x) = F(x)) = \frac{1}{2} + 2^{-Random_A(|x|)} > \frac{1}{2}.$$

Thus, following (5.2) and $\ln(1 + x) \le x$ for $-1 < x < 1$,

$$k = k(|x|) \underset{(5.3)}{=} \frac{2 \ln 2\delta}{\ln(1 - 4 \cdot 2^{-2Random_A(|x|)})} \ge \frac{2 \ln 2\delta}{-4 \cdot 2^{-2Random_A(|x|)}}$$

$$= (-\ln 2\delta) \cdot 2^{2Random_A(|x|)-1} \qquad (5.4)$$

Since $Random_A(|x|) \le Time_A(|x|)$ and one considers $Time_A(n)$ to be bounded by a polynomial, k may be exponential in $|x|$. Summarizing the results we have obtained: in order to get a randomized algorithm $A_{k(n)}$ with

$$Prob(A_{k(|x|)}(x) = F(x)) \ge 1 - \delta$$

from an unbounded-error Monte Carlo algorithm A, one is forced to accept

$$Time_{A_{k(n)}}(n) = O(2^{2Random_A(n)} \cdot Time_A(n)).$$

Exercise 5.2.2.9. Let **ZPP** (zero-error probabilistic polynomial time) be the class of languages (decision problems) that can be accepted (decided) by Las Vegas polynomial-time algorithms (Turing machines). Analogously, let **RP** (randomized polynomial time), **BPP** (bounded-error probabilistic polynomial time), and **PP** (probabilistic polynomial time), respectively, be the classes of languages accepted by one-sided-error, two-sided-error, and unbounded-error Monte Carlo algorithms in polynomial time, respectively. Prove the following relations:

(i) P $\subseteq$ ZPP $\subseteq$ RP $\subseteq$ BPP $\subseteq$ PP
(ii) RP $\subseteq$ NP
(iii) ZPP = RP $\cap$ co-RP
(iv) NP $\subseteq$ co-RP implies NP = ZPP
(v) NP $\cup$ co-NP $\subseteq$ PP. □

RANDOMIZED OPTIMIZATION ALGORITHMS.

If one considers optimization problems, then there is no essential reason to classify randomized algorithms as done above. This is because if one executes k runs of a randomized algorithm A for an optimization problem U, then the

---

[24] Note that $2^{-Random_A(|x|)}$ is not a constant here; it depends on the length of the input.

final output is considered as the best output (according to the cost function) of the k outputs computed in the particular runs. One does not need to try to obtain an output that appears as the result of at least half of the numbers of runs. So, we do not identify the right output with the frequency of its occurrence in a sequence of runs of the randomized algorithm,[25] but we simply take the best one. Then, the probability of obtaining an optimal solution in k rounds is

$$Prob\left(A_k(x) \in Output_U(x)\right) = 1 - \left[Prob\left(A(x) \notin Output_U(x)\right)\right]^k,$$

where A_k is the randomized algorithm consisting of k runs of A.

If $Prob(A(x) \notin Output_U(x)) \le \varepsilon$ for a positive constant $\varepsilon < 1$, then A is as practical as one-sided-error Monte Carlo algorithms. A may also be practical if $Prob(A(x) \in Output_U(x)) \ge 1/p(|x|)$ for a polynomial p, because a polynomial number of runs of A is sufficient to get an optimal solution with high probability.

One can also consider randomized algorithms as approximation algorithms that produce, with high probability, solutions whose cost (quality) is not very far from the cost of the optimal solutions. In this case the approximation ratio is considered to be a random variable. There are two different concepts of randomized approximation algorithms. The first concept requires to achieve an approximation ratio δ with a probability of at least $1/2$. The second concept requires an upper bound δ on the expected approximation ratio.

Definition 5.2.2.10. *Let $U = (\Sigma_I, \Sigma_O, L, L_I, \mathcal{M}, cost, goal)$ be an optimization problem. For any positive real $\delta > 1$ a randomized algorithm A is called a* **randomized δ-approximation algorithm for U** *if*

(i) $Prob(A(x) \in \mathcal{M}(x)) = 1$, and
(ii) $Prob(R_A(x) \le \delta) \ge 1/2$

for every $x \in L_I$.

For every function $f : \mathbb{N} \to \mathbb{R}^+$, we say that A is a **randomized $f(n)$-approximation algorithm for U** *if*

$$Prob(A(x) \in \mathcal{M}(x)) = 1 \text{ and } Prob(R_A(x) \le f(|x|)) \ge \frac{1}{2}$$

for every $x \in L_I$.

A randomized algorithm A is called a **randomized polynomial-time approximation scheme (RPTAS) for U** *if there exists a function $p : L_I \times \mathbb{R}^+ \to \mathbb{N}$ such that for every input $(x, \delta) \in L_I \times \mathbb{R}^+$,*

(i) $Prob(A(x, \delta) \in \mathcal{M}(x)) = 1$ {for every random choice A computes a feasible solution of U},
(ii) $Prob(\varepsilon_A(x, \delta) \le \delta) \ge 1/2$ {a feasible solution, whose relative error is at most δ, is produced with the probability at least $1/2$}, and

---

[25] As by the two-sided-error and unbounded-error Monte Carlo algorithms

(iii) $Time_A(x, \delta^{-1}) \leq p(|x|, \delta^{-1})$ and p is a polynomial in $|x|$.

If $p(|x|, \delta^{-1})$ is polynomial in both its arguments $|x|$ and δ^{-1}, then we say that A is a **randomized fully polynomial-time approximation scheme (RFPTAS) for U**.

Now, we present the second concept of randomized approximation algorithms that focuses on the expected approximation ratio.

Definition 5.2.2.11. *Let $U = (\Sigma_I, \Sigma_O, L, L_I, \mathcal{M}, cost, goal)$ be an optimization problem. For any positive real $\delta > 1$ a randomized algorithm A is called a* **randomized δ-expected approximation algorithm for U** *if*

(i) $Prob(A(x) \in \mathcal{M}(x)) = 1$, and
(ii) $E[R_A(x)] \leq \delta$

for every $x \in L_I$.

Observe that the above concepts of randomized approximation algorithms are different. A randomized δ-approximation algorithm does not need to be a randomized δ-expected approximation algorithm, and vice versa, a randomized δ-expected approximation algorithm does not need to be a randomized δ-approximation algorithm. To see it, consider all computations (runs) of a randomized algorithm A on a problem instance x. Assume that there exist 12 different equiprobable computations where 10 computations output feasible solutions to x with approximation ratio exactly 2, and 2 computations provide feasible solutions with approximation ratio 50. Then $E[R_A(x)] = 1/12 \cdot (10 \cdot 2 + 2 \cdot 50) = 10$. If this happens for all x, A is a randomized 10-expected approximation algorithm. But A is a randomized 2-approximation algorithm because

$$Prob\left(R_A(x) \leq 2\right) = 10 \cdot \frac{1}{12} = \frac{5}{6} > \frac{1}{2}.$$

On the other hand, consider that a randomized algorithm B has 1999 different runs on an input, 1000 runs with approximation ratio 11 and 999 runs with approximation ratio 1. Then $E[R_B(x)]$ is a little bit above 6, but B is not a randomized δ-approximation algorithm for any $\delta < 11$. If one considers the algorithm B_2 that consists of two independent runs of B and the output of B_2 is the best solution of the two computed solutions, then B_2 is even a randomized 1-approximation algorithm (i.e., B_2 computes an optimal solution with a probability greater than $1/2$).

The following example does not only illustrate the above definitions, but it shows a technique for proving that an algorithm A is a randomized δ-approximation algorithm for a reasonable constant δ if an upper bound ε on $E[R_A(x)]$ for every $x \in L_I$ is known.

Example 5.2.2.12. Consider the MAX-EkSAT problem for an integer $k \geq 3$ and Algorithm 2.2.5.26 (RANDOM ASSIGNMENT) from Example 2.2.5.25. For a given formula $F = F_1 \wedge F_2 \wedge \cdots \wedge F_m$ over n variables, the algorithm RANDOM ASSIGNMENT simply generates a random assignment and takes it as the output. We have observed that every clause F_i of k literals is satisfied with probability $1 - 2^{-k}$ by a randomly chosen assignment, i.e., $E[Z_i] = 1 - 2^{-k}$ where Z_i is the random variable defined by $Z_i = 1$ if F_i is satisfied and $Z_i = 0$ otherwise. Let $Z = \sum_{i=1}^{m} Z_i$ be a random variable that counts the number of satisfied clauses of F. Since F is an instance of MAX-EkSAT, all clauses of F consists of exactly k literals. Because of the linearity of expectation

$$E[Z] = E\left[\sum_{i=1}^{m} Z_i\right] = \sum_{i=1}^{m} E[Z_i] = \sum_{i=1}^{m} \left(1 - \frac{1}{2^k}\right) = m \cdot \left(1 - \frac{1}{2^k}\right). \quad (5.5)$$

Since the cost of an optimal solution is at most m,

$$E\left[R_{\text{RANDOM ASSIGNMENT}}(F)\right] \leq \frac{m}{E[Z]} \leq \frac{m}{m \cdot (1 - 2^{-k})} = \frac{2^k}{2^k - 1},$$

i.e., RANDOM ASSIGNMENT is a randomized $(2^k/(2^k - 1))$-expected approximation algorithm for MAX-EkSAT.

Applying (5.5) we prove that RANDOM ASSIGNMENT is a randomized $(2^{k-1}/(2^{k-1} - 1))$-approximation algorithm for MAX-EkSAT. First of all observe that $E[Z] = m(1 - 1/2^k)$ is nothing else than the average[26] of the number of satisfied clauses over all 2^n assignments for F. The number $m(1 - 1/2^k)$ is the average of m and $m(1 - 2/2^k)$, i.e., $m(1 - 1/2^k)$ lies in the middle of m and $m(1 - 2/2^k)$ on the axis (Figure 5.1). To complete the proof it is sufficient

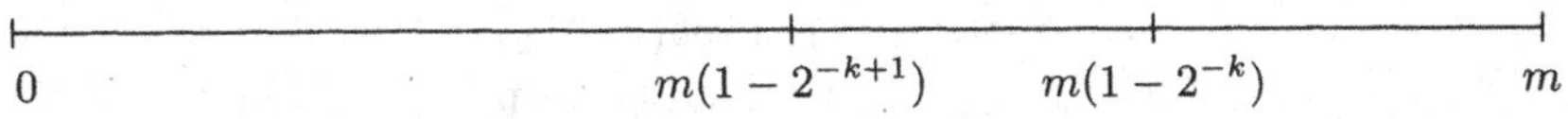

$$\vdash\!\dashv$$

$$0 \qquad\qquad\qquad m(1 - 2^{-k+1}) \qquad m(1 - 2^{-k}) \qquad\qquad m$$

Fig. 5.1. A transparent argument for the fact that at least half of assignments satisfies at least $m(1 - 2^{-k+1})$ clauses.

to show that at least half[27] of the assignments satisfies at least $m(1 - 2^{-k+1})$ clauses because $m/(m(1 - 2^{-k+1})) = 2^{k-1}/(2^{k-1} - 1)$. Let l be the number of assignments satisfying less than $m(1 - 2^{-k+1})$ clauses, and let u be the number of assignments satisfying at least $m(1 - 2^{-k+1})$ assignments. Clearly, $l + u = 2^n$. Then,

$$E[Z] \leq \frac{1}{2^n} \cdot \left(l \cdot \left[m\left(1 - 2^{-k+1}\right) - 1\right] + u \cdot m\right). \quad (5.6)$$

(5.5) and (5.6) together imply

---

[26] This is because all assignments to the variables of F have the same probability $1/2^n$ to be chosen.

[27] 2^{n-1} many

$$2^n \cdot m(1 - 2^{-k}) \leq l \cdot [m(1 - 2^{-k+1}) - 1] + u \cdot m$$

which is possible only if $u > l$.   □

The next example presents a randomized 2-expected approximation algorithm for MAX-CUT.

Example 5.2.2.13. Let us consider the following simple algorithm RANSAM for MAX-CUT.

Algorithm 5.2.2.14. RANSAM

Input: An undirected graph $G = (V, E)$.
Step 1: $V_1 = \emptyset$; $V_2 = \emptyset$;
Step 2:   **for** every vertex $v \in V$ **do**
            put v equiprobably to either V_1 or V_2.
Output: The cut (V_1, V_2).

Obviously, the output (V_1, V_2) of the algorithm RANSAM is a cut of G, and so (V_1, V_2) is a feasible solution for the input instance $G = (V, E)$. Now we prove that RANSAM is a randomized 2-expected approximation algorithm for MAX-CUT. Clearly,

$$Opt_{\text{MAX-CUT}}(G) \leq |E|.$$

Let, for every $e = \{u, v\} \in E$, X_e be a random variable defined as follows:

$$X_e = \begin{cases} 0 \text{ if both } u, v \in V_1 \text{ or both } u, v \in V_2 \\ 1 \text{ else (i.e., if } e \text{ is in the cut } (V_1, V_2)). \end{cases}$$

We easily observe that

$$\begin{aligned} Prob(X_{\{u,v\}} = 1) &= Prob(u \in V_1 \wedge v \in V_2) + Prob(u \in V_2 \wedge v \in V_1) \\ &= Prob(u \in V_1) \cdot Prob(v \in V_2) \\ &\quad + Prob(u \in V_2) \cdot Prob(v \in V_1) \\ &= \frac{1}{2} \cdot \frac{1}{2} + \frac{1}{2} \cdot \frac{1}{2} = \frac{1}{2}, \end{aligned}$$

and $Prob(X_{\{u,v\}} = 0) = 1/2$. So, for every $e \in E$,

$$E[X_e] = 1 \cdot Prob(X_e = 1) + 0 \cdot Prob(X_e = 0) = \frac{1}{2}.$$

By linearity of expectation we obtain, for the random variable $X = \sum_{e \in E} X_e$,

$$E[X] = \sum_{e \in E} E[X_e] = \frac{1}{2} \cdot |E| \geq \frac{1}{2} Opt_{\text{MAX-CUT}}(G).$$

This directly implies

$$E[R_{\text{RANSAM}}(G)] \leq \frac{|E|}{E[X]} = 2$$

for every graph G.   □

5.2.3 Paradigms of Design of Randomized Algorithms

As we already mentioned, randomized algorithms may be more efficient than their best known deterministic counterparts. We have even seen that for some special models like communication protocols the randomized algorithms are provably more powerful than the deterministic ones. What is the nature of the success (power) of randomization? In what follows we list some of the paradigms[28] that partially explain the power of randomized algorithms. These paradigms contain crucial, general ideas that may be very useful for every designer of randomized algorithms.

FOILING AN ADVERSARY.

The classical adversary argument for a deterministic algorithm establishes a lower bound on the running time of the algorithm by constructing an input instance on which the algorithm fares poorly.[29] The hard input instance thus constructed may be different for each deterministic algorithm. A randomized algorithm can be viewed as a probability distribution on a set of deterministic algorithms. While an adversary may be able to construct an input that foils one (or a small fraction) of the deterministic algorithms in the set, it is difficult to devise a single input that is likely to defeat a randomly chosen algorithm.

This paradigm is common for all randomized algorithms because (as mentioned earlier) a randomized algorithm can be viewed as a set of deterministic algorithms, each one corresponding to one sequence of random bits.

ABUNDANCE OF WITNESSES.

This principle is used for randomized solutions of decision problems. Solving a decision problem may be viewed as the requirement for determining whether an input x has a certain property $(x \in L)$ or not $(x \notin L)$. A witness for x (for the fact $x \in L$) is any string y, such that there is an efficient polynomial-time procedure B that having the input (x, y) proves $x \in L$. For instance, if x is a number and L is the set of composite numbers, then any factor (nontrivial divisor) y of x is a witness for the membership of x in L. To check whether $x \bmod y \equiv 0$ is much easier than to prove $x \in L$ without any additional information. Thus, in general, we consider witnesses only if they essentially decrease the complexity of the decision problem. For many problems the difficulty with finding a witness deterministically is that the witness lies in a search space that is too large to be searched exhaustively[30].

---

[28] Fundamental principles

[29] It has a high time complexity.

[30] Moreover, the witnesses are usually distributed in the search space so irregularly that no deterministic search method that would be more efficient than the exhaustive search is available.

However, by establishing that the space contains a large number of witnesses, it often suffices to choose an element at random from the space. The randomly chosen item is likely to be a witness. If this probability is not high enough, an independent random choice of several items reduces the probability that no witness is found.

Usually, this approach is handled very simple. One has for every given input x a set $CanW(x)$, which contains all items candidating to be a witness for the input x. The set $CanW(x)$ can be efficiently enumerated.[31] Let $Witness(x)$ contain all witnesses of $x \in L$ from $CanW(x)$. This approach works very well if for every $x \in L$ the set $Witness(x) \subseteq CanW(x)$ has a cardinality proportional to the cardinality of $CanW(x)$.

To illustrate this paradigm consider Example 5.2.2.6. A prime p is a witness for the fact $x \neq y$ if

$$Number(x) \bmod p \neq Number(y) \bmod p.$$

For every input $(x, y) \in \{0,1\}^n \times \{0,1\}^n$, $CanW(x)$ is the set of all primes from $\{2, 3, \ldots, n^2\}$, where $n = |x| = |y|$. We know $|CanW(x)|$ is approximately $n^2/\ln n^2$. For every x, y, $x \neq y$, at most $n-1$ primes from $CanW(x)$ are not witnesses of the fact $x \neq y$ (i.e., $Number(x) \bmod p = Number(y) \bmod p$). So, at least

$$|CanW(x)| - n + 1$$

primes from $CanW(x)$ are witnesses for $x \neq y$. Because of this we have obtained that the randomized protocol gives the right answer with the probability

$$\frac{|CanW(x)| - n + 1}{|CanW(x)|}$$

for every input (x, y) with $x \neq y$.

A more elaborated application of the method of abundance of witnesses is given for primality testing in Section 5.3.3.

FINGERPRINTING.

Fingerprinting, called also Freivalds' technique, is used for deciding equivalence problems. Typically, an equivalence problem is defined over a class of objects (Boolean functions, numbers, words, languages) and their representation. The crucial point is that there are (possibly infinitely) many correct representations[32] of an object and an equivalence problem is to decide whether two given representations R_1 and R_2 represent the same object. Fingerprinting works in the following three steps.

---

[31] Often, $CanW(x)$ is the same set for all inputs of size $|x|$, i.e., $CanW(x) = CanW(y)$ for all inputs x, y with $|x| = |y|$.

[32] For instance, for any Boolean function f, there are infinitely many branching programs that represent f.

Step 1.   Choose randomly a mapping h from a class M.
Step 2.   Compute $h(R_1)$ and $h(R_2)$.
          {$h(R_i)$ is called a **fingerprint** of R_i for $i = 1, 2$.}
Step 3.   **if** $h(R_1) = h(R_2)$ **then** **output** "yes"
          else **output** "no".

The main idea is that a fingerprint $h(R_i)$ of R_i is substantially shorter than R_i, but $h(R_i)$ is not a representation of an object O_i because $h(R_i)$ contains a partial information about O_i only. Thus, the gain is that the comparsion between the fingerprints $h(R_1)$ and $h(R_2)$ is simpler to execute than the direct comparison of R_1 and R_2. The drawback is that representations of distinct objects may have the same image (fingerprint), i.e., we lose the guarantee to always give a correct answer. The set M of mappings is chosen in such a way that if R_1 and R_2 represent the same object, then the fingerprint $h(R_1)$ and $h(R_2)$ are equivalent for every $h \in M$ (i.e., we obtain the correct answer with certainty in this case). If R_1 and R_2 represent distinct objects, then we require $Prob_{h \in M}(h(R_1) \neq h(R_2)) \geq 1/2$. Thus, fingerprinting provides a one-sided-error Monte Carlo algorithm for the given nonequivalence problem, and so a two-sided-error Monte Carlo algorithm for the corresponding equivalence problem.

Fingerprinting can be viewed as a special version of the method of abundance of witnesses. A mapping $h \in M$ is considered as a witness of the nonequivalence of R_1 and R_2 if $h(R_1) \neq h(R_2)$. Thus, the comparison of two strings in Example 5.2.2.6 is a typical example of the fingerprinting technique. More elaborated applications of the fingerprinting technique are given in Section 5.3.4, where nonequivalence problems for polynomials over finite fields and for one-time-only branching programs are studied.

RANDOM SAMPLING.

A random sample from a population is often a representative of the whole population. A special case of this paradigm is the so-called **probabilistic method**. This method is based on the following two facts.

Claim 1. Any random variable   assumes at least one value that is not
         smaller than its expectation, and at least one value that is not
         greater than its expectation.
Claim 2. If an object chosen randomly from a given universe satisfies a
         property with a positive probability, then there must exist an
         object in the universe that satisfies that property.

Despite the fact that these claims may seem too obvious, they turn out to provide a surprising amount of power. Sometimes they lead to the best known approaches for solving hard computing problems. A nice example is the algorithm RANDOM ASSIGNMENT for MAX-EkSAT presented in Example 5.2.2.12. Another example is the algorithm RANSAM given in Example 5.2.2.13.

RELAXATION AND RANDOM ROUNDING.

This paradigm may also be considered as a special case of random sampling. The idea is to relax a given optimization problem in such a way that the relaxed problem can be solved in polynomial time. The typical realization is based on reducing the original optimization problem to an integer (or Boolean) programming problem, and then to relax it to a linear programming problem over reals. The last problem can be solved efficiently. Usually, the solution of the relaxed problem is not a feasible solution to the original problem. But this solution is used to compute probabilities that are used to randomly round the real values of the solution in order to get a feasible solution of the original problem. A successful application of this approach is presented in Section 5.3.6 for the MAX-SAT problem.

Keywords introduced in Section 5.2

randomized algorithms, derandomization, expected complexity, Las Vegas algorithm, one-sided-error Monte Carlo algorithm, two-sided-error Monte Carlo algorithm, unbounded-error Monte Carlo algorithm, randomized approximation algorithm, randomized polynomial-time approximation schemes, foiling an adversary, abundance of witnesses, random sampling, relaxation and random rounding

Summary of Section 5.2

A randomized algorithm can be viewed as a probability distribution over a set of deterministic algorithms. Randomized algorithms are mainly classified into Las Vegas and Monte Carlo algorithms. Las Vegas algorithms never err, and Monte Carlo algorithms may produce a wrong output with a bounded probability.

For a Las Vegas algorithm that returns the right output in each computation, one usually investigates the expected time complexity $Exp\text{-}Time$. $Exp\text{-}Time$ is the expectation of the random variable equal to the time complexity of the particular computations. For a Las Vegas algorithm that may output "?", one may consider the usual worst case complexity.

A constant number of repetitions of one-sided-error or two-sided-error bounded Monte Carlo algorithms is sufficient to increase the success of these algorithms to $1 - \delta$ for arbitrarily small δ. This does not need to be the case for unbounded-error Monte Carlo algorithms that may require an exponential number of repetitions to essentially increase the probability of success. Already for one-sided-error Monte Carlo randomization, there are special computing models where this randomization proves to be much more powerful than determinism.

Some of the fundamental principles of the design of randomized algorithms are the abundance of witnesses, fingerprinting, random sampling, and relaxation with random rounding. The paradigm of the abundance of witnesses may be useful for designing randomized algorithms for decision problems. It works very well if one can randomly find a witness for every input with a high probability. The known deterministic methods are not efficient because the space containing witnesses is too large to be searched exhaustively and the distribution of witnesses in the space is too irregular for any efficient search strategy. The idea of the paradigm of random sampling is that a random sample from a population is a representative of the population as a whole. The method of relaxation and random rounding is a method for solving optimization problems by the reduction to linear programming and relaxation.

5.3 Design of Randomized Algorithms

5.3.1 Introduction

The aim of Section 5.3 is to present concrete, transparent examples of the design of randomized algorithms for hard computing problems. In Section 5.3.2 we present an efficient Las Vegas algorithm that finds a quadratic nonresidue mod p for any given prime p. This problem is not known to be in P. The designed Las Vegas algorithm is based on the method of random sampling. One uniformly chooses an element a from $\{1, \ldots, p-1\}$ at random and verifies whether a is a quadratic nonresidue (mod p). This simple idea works because exactly half of the elements from $\{1, 2, \ldots, p - 1\}$ are quadratic nonresidues (mod p). The reason why we do not have a deterministic polynomial-time algorithm for this task is that we do not have any idea how the quadratic nonresidues (mod p) are distributed in the set $\{1, \ldots, p - 1\}$.

In Section 5.3.3 we present one of the most famous randomized algorithms – the Solovay-Strassen algorithm for primality testing. This algorithm is a one-sided Monte Carlo algorithm and it is based on the method of the abundance of witnesses. The whole design of this algorithm may be viewed as a search for a suitable notion of a witness for primality. This is probably one of the most instructive applications of the paradigm of the abundance of witnesses. We close Section 5.3.3 by showing how this algorithm can be applied to searching for large primes.

In Section 5.3.4 we present fingerprinting as a special version of the method of abundance of witnesses. To decide the nonequivalence of two one-time-only branching programs we reduce this problem to an equivalence problem for some more general objects than one-time-only branching programs. Surprisingly, this generalization may be helpful because there are many suitable witnesses in the generalized framework.

Section 5.3.5 is devoted to the presentation of a randomized optimization algorithm for the MIN-CUT problem. The MIN-CUT problem is in P, but the designed randomized algorithm is more efficient than the best known deterministic algorithm. The approach is to start with a naive randomized approach (similar to random sampling) and improve its efficiency by a clever organization of run repetitions. The main idea is that one does not need to realize complete independent runs of a randomized algorithm in order to increase the probability of success (in finding an optimal solution in this case). It is sufficient to repeat only such critical parts of the computation in which the probability of success is essentially decreased.

In Section 5.3.6 we design two randomized approximation algorithms for the MAX-SAT problem. One algorithm is based on random sampling and the second one is based on relaxation and random rounding. These algorithms are incomparable according to the achievable expected approximation ratio (for some inputs, the first algorithm provides a better approximation ratio than the second one, and, for other inputs, it is vice versa). Combining these two algorithms we obtain a randomized $(4/3)$-expected approximation algorithm for MAX-SAT.

Section 5.3.7 is devoted to the concept of lowering worst case complexity of exponential algorithms (introduced in Section 3.5). A randomization of multistart local search is used to design a one-sided-error Monte Carlo algorithm for 3SAT. This algorithm runs in $O(|F| \cdot n^{3/2}(4/3)^n)$ time for any 3CNF F over n variables.

5.3.2 Quadratic Residues, Random Sampling, and Las Vegas

In this section we consider the following problem. Given a prime p, find an $a \in \mathbb{Z}_p$ such that a is a quadratic nonresidue[33] (mod p). No deterministic polynomial-time algorithm is known for this problem. On the other hand, the complementary problem – to find a quadratic residue modulo p – is very simple. One takes an arbitrary $a \in \{1, 2, \ldots, p-1\}$, computes $b = a^2 \bmod p$, and gives the output b.

In what follows we give a Las Vegas polynomial-time algorithm for the problem of finding a quadratic nonresidue. The idea of this algorithm is based on the method of random sampling. More precisely, it is based on the following two facts:

(A) For a given prime p and an $a \in \mathbb{Z}_p$, it is possible to decide whether a is a quadratic residue (mod p) in polynomial time.

(B) For every prime p, exactly half of the elements of $\mathbb{Z}_p$ are quadratic residues.

So, to design a Las Vegas algorithm for this problem it is sufficient to choose an element $a \in Z_p$ randomly[34] and then to verify whether a is a quadratic residue in polynomial time.[35]

Before proving the facts (A) and (B), we show that one can efficiently compute $a^b \bmod p$. The idea is very simple. To compute $a^b \bmod p$ if $b = 2^k$, it is enough to use $k \odot_p$ operations by so-called "repeated squaring":

---

[33] Remember that a is a quadratic residue (mod p) if there exists an $x \in \mathbb{Z}_p$ such that $x^2 = x \cdot x = a$. In the opposite case, a is a quadratic nonresidue (mod p).

[34] In the uniform way

[35] This is possible because of (A).

$$a^2 \bmod p = a \odot_p a, \; a^4 \bmod p = a^2 \odot_p a^2, \; a^8 \bmod p = a^4 \odot_p a^4, \dots,$$
$$a^{2^k} \bmod p = a^{2^{k-1}} \odot_p a^{2^{k-1}}.$$

In general $1 < b = \sum_{i=0}^k b_i \cdot 2^i$, i.e., $b = Number(b_k b_{k-1} \dots b_0)$ for some positive integer k. To compute a^b is the same as to compute

$$a^{b_0 \cdot 2^0} \cdot a^{b_1 \cdot 2^1} \cdot a^{b_2 \cdot 2^2} \cdot \dots \cdot a^{b_k \cdot 2^k},$$

where $a^{b_i \cdot 2^i}$ is either a^{2^i} if $b_i = 1$ or 1 if $b_i = 0$. Since a^{2^i} can be computed by the method of repeated squaring, one can write the algorithm for the modular exponentiation as follows.

Algorithm 5.3.2.1. REPEATED SQUARING

Input: Positive integers a, b, p, where $b = Number(b_k b_{k-1} \dots b_0)$.
Step 1: $C := a; \; D := 1$.
Step 2: **for** $I := 0$ **to** k **do**
         begin if $b_I = 1$ **then** $D := D \cdot C \bmod p$;
                 $C := C \cdot C \bmod p$
         end
Step 3: **return** D
Output: $D = a^b \bmod p$.

One can easily observe that the algorithm REPEATED SQUARING uses only $2(k+1)$ operations over $\mathbb{Z}_p$, while the input length[36] is $3(k+1)$. $O(k^2)$ binary operations are sufficient to realize any multiplication (modulo p) of two numbers of length $k+1$.

Now we are ready to prove the facts (A) and (B).

Theorem 5.3.2.2 (Euler's Criterion). *For every* $a \in \mathbb{Z}_p$,

(i) if a is a quadratic residue modulo p, then $a^{(p-1)/2} \equiv 1 (\bmod \; p)$, and
(ii) if a is a quadratic nonresidue modulo p, then $a^{(p-1)/2} \equiv -1 (\bmod \; p)$.

Proof. The above statement is trivial for $\mathbb{Z}_2$, therefore assume from now on that $p > 2$. Referring to Fermat's Theorem,

$$a^{p-1} \equiv 1 (\bmod \; p), \text{i.e.}, \; a^{p-1} - 1 \equiv 0 (\bmod \; p).$$

Since $p > 2$, one can write $p = 2p' + 1$ (i.e., $p' = (p-1)/2$) and so

$$a^{p-1} - 1 = a^{2p'} - 1 = (a^{p'} - 1) \cdot (a^{p'} + 1) \equiv 0 (\bmod \; p).$$

Since a product is divisible by p only if one of its factors is, it immediately appears that either $a^{p'} - 1$ or $a^{p'} + 1$ must be divisible by p. So,[37]

---

[36] Note that $a, b \in \{1, \dots, p-1\}$ and that $k + 1 = \lceil \log_2 p \rceil$.
[37] Note that this fact was already proved in Theorem 2.2.4.32.

$$a^{(p-1)/2} \equiv 1 (\bmod\ p) \text{ or } a^{(p-1)/2} \equiv -1 (\bmod\ p). \tag{5.7}$$

Now, suppose that a is a quadratic residue modulo p. Then there exists an $x \in \mathbb{Z}_p$ such that

$$a \equiv x^2 (\bmod\ p).$$

Due to the facts

$$a^{(p-1)/2} = (x^2)^{(p-1)/2} \equiv x^{p-1} (\bmod\ p)$$

and $x^{p-1} \equiv 1 (\bmod\ p)$ according to Fermat's Theorem, we obtain

$$a^{(p-1)/2} \equiv x^{p-1} \equiv 1 (\bmod\ p).$$

Now, we suppose that a is a quadratic nonresidue modulo p. Because of (5.7) it is sufficient to show that $a^{(p-1)/2} \not\equiv 1 (\bmod\ p)$. Since $(\mathbb{Z}_p^*, \odot_p)$ is a cyclic group there exists a generator g of $\mathbb{Z}_p^*$. Because a is a quadratic nonresidue modulo p, a is an odd power of g. Let $a = g^{2l+1}$ for some non-negative integer l. Then

$$a^{(p-1)/2} = \left(g^{2l+1}\right)^{(p-1)/2} = g^{l \cdot (p-1)} \cdot g^{(p-1)/2}. \tag{5.8}$$

Due to Fermat's Theorem $g^{p-1} \equiv 1 (\bmod\ p)$ and so

$$g^{l \cdot (p-1)} = \left(g^{p-1}\right)^l \equiv 1 (\bmod\ p).$$

Inserting it in (5.8) we obtain

$$a^{(p-1)/2} \equiv g^{(p-1)/2} (\bmod\ p).$$

But $g^{(p-1)/2}$ cannot be congruent to 1 modulo p because then

$$g^{(p-1)/2+i} = g^{(p-1)/2} \cdot g^i \equiv g^i (\bmod\ p)$$

for every positive integer i, i.e., g generates at most $(p-1)/2$ different elements. Obviously, this would contradict the assumption that g is a generator of $\mathbb{Z}_p^*$.

$\square$

Using Euler's Criterion of Theorem 5.3.2.2 and the algorithm REPEATED SQUARING we see that one can efficiently decide whether a is a quadratic residue modulo p for any $a \in \mathbb{Z}_p$ and any prime p.[38] The following theorem proves the fact (B).

Theorem 5.3.2.3. *For every odd prime p, exactly half [39] of the nonzero elements of $\mathbb{Z}_p$ are quadratic residues modulo p.*

---

[38] This is exactly the claim (A) presented at the very beginning of Section 5.3.2.
[39] $(p-1)/2$

Proof. We have to prove that

$$|\{1^2 \bmod p, 2^2 \bmod p, \dots, (p-1)^2 \bmod p\}| = (p-1)/2. \qquad (5.9)$$

We observe that for every $x \in \{1, \dots, p-1\}$,

$$(p-x)^2 = p^2 - 2px + x^2 = p(p-2x) + x^2 \equiv x^2 (\bmod p).$$

Thus, we have proved that the number of quadratic residues modulo p is at most $(p-1)/2$.

Now it is sufficient to prove that for every $x \in \{1, \dots, p-1\}$, the congruence $x^2 \equiv y^2 \bmod p$ has at most one solution $y \in \{1, 2, \dots, p-1\}$ different from x.

Without loss of generality we assume $y > x$, i.e., $y = x + i$ for some $i \in \{1, 2, \dots, p-2\}$. Thus,

$$x^2 \equiv (x+i)^2 \equiv x^2 + 2ix + i^2 (\bmod p).$$

This directly implies

$$2ix + i^2 = i(2x + i) \equiv 0 (\bmod p).$$

Since $\mathbb{Z}_p$ is a field[40] and $i \in \{1, 2, \dots, p-1\}$,[41]

$$2x + i \equiv 0 (\bmod p). \qquad (5.10)$$

Since the congruence (5.10) has exactly one solution[42] $i \in \{1, \dots, p-1\}$, the proof is completed.[43] $\qquad \square$

Now, we are ready to give the Las Vegas algorithm for finding a quadratic nonresidue in $\mathbb{Z}_p$ for any given prime p.

Algorithm 5.3.2.4. QUADRATIC NONRESIDUE

Input: A prime p.

Step 1: Choose randomly an $a \in \{1, \dots, p-1\}$ in the uniform way.

Step 2: Compute $X := a^{(p-1)/2} \bmod p$ by the algorithm REPEATED SQUARING.

Step 3: if $X = p - 1$ then return a
else return "I was not successful in this attempt"

Output: a number $a \in \{1, \dots, p-1\}$ such that a is a quadratic nonresidue modulo p, if any.

---

[40] Particularly it means that if $a \cdot b = 0$, then either $a = 0$ or $b = 0$.

[41] $i \not\equiv 0 \pmod p$

[42] $i = p - 2x$ if $2x < p$ and $i = -(2x - p)$ if $2x > p$

[43] That is, there do not exist three pairwise different $x, y, z \in \{1, \dots, p-1\}$ such that $x^2 \equiv y^2 \equiv z^2 \pmod p$.

Following Theorem 5.3.2.2 the algorithm QUADRATIC NONRESIDUE never errs because $a^{(p-1)/2} \equiv -1 = p-1 \ (\bmod\ p)$ if and only if a is a quadratic non-residue modulo p. Due to Theorem 5.3.2.3 half of the elements of $\{1, \ldots, p-1\}$ are quadratic nonresidues and so the algorithm QUADRATIC NONRESIDUE finds a quadratic nonresidue with the probability $1/2$, i.e., it is a Las Vegas algorithm. Remember, k repetitions[44] of any Las Vegas algorithm provide a correct result with the probability at least $1 - 1/2^k$.

The only computational part of the algorithm QUADRATIC NONRESIDUE is the computation of $a^{(p-1)/2} \bmod p$ by the algorithm REPEATED SQUARING, and so the time complexity is in $O((\log_2 p)^3)$.

Exercise 5.3.2.5. Find an implementation of the method repeated squaring such that the algorithm QUADRATIC NONRESIDUE will work in $O((\log_2 p)^2)$ time.   □

Now, one can ask: "Why are we not able to solve the problem of quadratic nonresidues deterministically in polynomial time?" We know that half of the elements of $\{1, 2, \ldots, p-1\}$ are quadratic nonresidues. But we do not have any idea how they are distributed in this set. This means that any deterministic algorithm may result in $(p-1)/2$ unsuccessful attempts to find a quadratic nonresidue. But this means that the time complexity $\Omega(p)$ is exponential in $\log_2 p$. Whether there exists a deterministic polynomial-time algorithm for this task depends on the famous mathematical hypothesis known as the *Extended Riemann Hypothesis*. If this hypothesis holds, then $\mathbb{Z}_p$ must contain a quadratic nonresidue among its $O((\log_2 p)^2)$ smallest elements, i.e., a quadratic nonresidue can be easily identified by trying all these numbers.

Summary of Section 5.3.2

The method of random sampling can be successfully applied to searching for objects with some given properties. It provides a randomized algorithm that can be more efficient than any deterministic algorithm if

(i) there are many objects with the given property relative to the cardinality of the set of all objects considered,

(ii) for a given object, one can efficiently verify whether it has the required property or not, and

(iii) the distribution of the "right" objects among all objects is unknown and cannot be efficiently computed (or at least one does not know how to determine it efficiently).

The distribution of quadratic residues $(\bmod\ p)$ and quadratic nonresidues $(\bmod\ p)$ among the numbers of $\{1, \ldots, p-1\}$ seems to be extremely irregular. Thus, there is no known deterministic polynomial-time algorithm for finding a quadratic nonresidue modulo p for a given prime p. Since the number of quadratic residues

---

[44] Independent runs

(mod p) is equal to the number of quadratic nonresidues (mod p), the design of an efficient Las Vegas algorithm is the most natural approach to solving this problem.

5.3.3 Primality Testing, Abundance of Witnesses, and One-Sided-Error Monte Carlo

To decide whether a given odd integer is prime or composite is one of the most famous[45] computing problems. So far we do not know of any polynomial-time algorithm[46] that can solve it. On the other hand, one needs large primes that consist of one- or two-hundred decimal digits. If a number n is an input, then $\lceil \log_2 n \rceil$ – the length of the binary code of n – is considered to be the size of the input. So, if one uses the trivial algorithm that checks, for every $a \in \{1, \ldots, n-1\}$, whether a is a factor of n, then the time complexity is at least n which is exponential in $\log_2 n$. Already for $\log_2 n = 100$ this algorithm is not executable in the known universe. Also the improvement looking for numbers from $\{2, 3, \ldots, \lfloor \sqrt{n} \rfloor\}$ as candidates for a factor of n does not essentially help to solve the problem.

The idea is to test primality by a randomized algorithm where the main approach is based on the concept of abundance of witnesses. In fact, we are searching for witnesses of the compositeness. Obviously, any factor a of a number n is a witness of the compositeness of n. But if $n = p \cdot q$ for two primes p and q, then p and q are the only witnesses of the compositeness of n among all $\Omega(n)$ candidates. So, we need an equivalent definition of primality. Fermat's Theorem claims:

"For every prime p and every $a \in \{1, \ldots, p-1\}$, $a^{p-1} \equiv 1$ (mod p)."

The subsequent Theorem 2.2.4.32 provides an equivalent definition of primality:

"p is a prime $\iff a^{(p-1)/2}$ mod $p \in \{1, -1\}$ for all $a \in \{1, \ldots, p-1\}$".

If, for a composite number p, there exists $b \in \{1, \ldots, p-1\}$ such that $b^{(p-1)/2}$ mod $p \notin \{1, -1\}$, then b is a witness of the compositeness of p. To be able to apply the abundance of witnesses method, we need a lot of witnesses for every composite p. The next theorem shows that we have enough such witnesses for every odd composite number n such that $(n-1)/2$ is odd, too.

---

[45] This problem is of large theoretical as well as practical interest, especially because of the applications in cryptography.

[46] During the last week of the work on the second edition of this book Agrawal, Kayal and Saxena [AKS02] announced a deterministic polynomial-time algorithm for primality testing. But this does not decrease the interest in designing efficient randomized algorithms for primality testing, because the time complexity of the announced deterministic algorithm is in $O((\log_2 n)^{12})$, and so (in the sense of our terminology) primality testing is a hard problem.

Theorem 5.3.3.1. *For every odd n such that $(n-1)/2$ is odd (i.e., $n \equiv 3 \pmod 4$)* ,

(i) if n is a prime, then $a^{(n-1)/2}$ mod $n \in \{1, -1\}$ for all $a \in \{1, \ldots, n-1\}$,
(ii) if n is composite, then $a^{(n-1)/2}$ mod $n \notin \{1, -1\}$ for at least one half of the a's from $\{1, 2, \ldots, n-1\}$.

Proof. Fact (i) is a direct consequence of Theorem 2.2.4.32.

To prove (ii) we consider the following strategy. Let n be composite. A number $a \in \mathbb{Z}_n$ is called **Eulerian** if $a^{(n-1)/2}$ mod $n \in \{1, -1\}$. We claim that to prove (ii) it is sufficient to find a number $b \in \mathbb{Z}_n - \{0\}$ such that b is not Eulerian and there exists a multiplicative inverse b^{-1} to b. Let us prove this claim. Let $Eu_n = \{a \in \mathbb{Z}_n \,|\, a \text{ is Eulerian}\}$. The idea of the proof is that the multiplication of elements of Eu_n by b is an injective mapping into $\mathbb{Z}_n - Eu_n$. For every $a \in Eu_n$, $a \cdot b$ is not Eulerian because

$$(a \cdot b)^{\frac{n-1}{2}} \text{ mod } n = \left(a^{\frac{n-1}{2}} \text{ mod } n\right) \cdot \left(b^{\frac{n-1}{2}} \text{ mod } n\right) = \pm b^{\frac{n-1}{2}} \text{ mod } n \notin \{1, -1\}.$$

Now it remains to prove that $a_1 \cdot b \not\equiv a_2 \cdot b \pmod n$ if $a_1 \neq a_2$, $a_1, a_2 \in Eu_n$. Let $a_1 \cdot b \equiv a_2 \cdot b \pmod n$. Then by multiplying the congruence with b^{-1} we obtain

$$a_1 = a_1 \cdot b \cdot b^{-1} \text{ mod } n = a_2 \cdot b \cdot b^{-1} \text{ mod } n = a_2.$$

So, $|\mathbb{Z}_n - Eu_n| \geq |Eu_n|$.

Now, we have to search for a $b \in \{1, 2, \ldots, n-1\}$ that is not Eulerian and has an inverse element b^{-1}. To avoid too many technicalities we explain how to search for b in the case $n = p \cdot q$ for two distinct primes p and q only. To do it, consider the Chinese Remainder Theorem that claims

$$\mathbb{Z}_n \text{ is isomorphic to } \mathbb{Z}_p \times \mathbb{Z}_q,$$

where, for every $a \in \mathbb{Z}_n$, the representation of a in $\mathbb{Z}_p \times \mathbb{Z}_q$ is the pair $(a \text{ mod } p, a \text{ mod } q)$. Let a be Eulerian. Then the representation of $a^{(n-1)/2}$ mod $n \in \{1, -1\}$ in $\mathbb{Z}_p \times \mathbb{Z}_q$ is $\left(a^{(n-1)/2} \text{ mod } p, a^{(n-1)/2} \text{ mod } q\right)$ that is equal either to $(1, 1)$ or to $(-1, -1)$[47]. Now, we choose $b \in \mathbb{Z}_n$ as the number corresponding to $(1, -1) = (1, n-1)$. Let us show that b has the required properties. The representation of $b^{\frac{n-1}{2}}$ mod n in $\mathbb{Z}_p \times \mathbb{Z}_q$ is

$$\left(b^{\frac{n-1}{2}} \text{ mod } p, b^{\frac{n-1}{2}} \text{ mod } q\right) = \left(1^{\frac{n-1}{2}} \text{ mod } p, (-1)^{\frac{n-1}{2}} \text{ mod } q\right) = (1, -1)$$

because $(n-1)/2$ is odd. So, b is not Eulerian.[48] Finally, we observe that $b^{-1} = b$. Obviously, $(1, 1)$ is the neutral element according to the multiplication in $\mathbb{Z}_p \times \mathbb{Z}_q$. Since $(1, -1) \odot_{p,q} (1, -1) = (1, 1)$, b is inverse to itself.[49]   □

---

[47] If $a^{(n-1)/2}$ mod $pq = 1$, then $a^{(n-1)/2} = k \cdot p \cdot q + 1$ for some k and so $a^{(n-1)/2}$ mod $p = a^{(n-1)/2}$ mod $q = 1$.

[48] Remember that we proved above that for every Eulerian number a, $a^{(n-1)/2}$ must have the representation (r, r), $r \in \{1, -1\}$ in $\mathbb{Z}_p \times \mathbb{Z}_q$.

[49] If one wants to have an argumentation without the use of the embedded operation $\odot_{p,q}$, one can follow the next consideration. Since $b = k \cdot p + 1$ for some k,

Exercise 5.3.3.2. Consider $n = p \cdot q \cdot r = 3 \cdot 3 \cdot 7 = 63$. Find $b \in \mathbb{Z}_{63}$ such that b is no Eulerian number and there exists b^{-1} such that $b \cdot b^{-1} \equiv 1 \pmod{63}$.   □

Exercise 5.3.3.3. Let $n = p^e$ for some positive integer $e > 1$ and a prime $p > 2$. Prove that $-(p-1)$ is no Eulerian number and possesses a multiplicative inverse in $\mathbb{Z}_n^*$.   □

Exercise 5.3.3.4. (*) Give the complete proof of (ii) of Theorem 5.3.3.1, i.e., find a suitable b for every composite number n with odd $(n-1)/2$.   □

Now, we have all that we need to randomly test primality for positive integers $n > 2$ with odd $(n-1)/2$.

Algorithm 5.3.3.5 (SSSA SIMPLIFIED SOLOVAY-STRASSEN ALGORITHM).

Input: An odd number n with odd $(n-1)/2$.
Step 1:   Choose uniformly an $a \in \{1, 2, \ldots, n-1\}$
Step 2:   Compute $A := a^{\frac{n-1}{2}} \bmod n$
Step 3:   **if** $A \in \{1, -1\}$
          then return ("PRIME") {reject}
          else return ("COMPOSITE") {accept}.

Theorem 5.3.3.6. SSSA *is a polynomial-time one-sided-error Monte Carlo algorithm for the recognition of composite numbers* n *with odd* $(n-1)/2$.

Proof. The only computation of SSSA is in Step 2 and we know that it can be realized efficiently by repeated squaring.

Now, assume that n is a prime. Then due to Theorem 2.2.4.32,[50] $a^{(n-1)/2} \equiv \pm 1 \pmod{n}$, and for every $a \in \{1, \ldots, n-1\}$ the output of SSSA is "PRIME". Thus,

$$Prob(\text{SSSA rejects } n) = 1.$$

If n is composite, then due to the claim (ii) of Theorem 5.3.3.1

$$Prob(\text{SSSA accepts } n) \geq \frac{1}{2}.$$

So, SSSA is a one-sided-error Monte Carlo algorithm for the recognition of composite numbers with odd $(n-1)/2$.   □

Remember the good property of one-sided-error Monte Carlo algorithms yielding the probability $1 - 1/2^k$ of being correct after k independent runs of the algorithm. So, despite of the input restriction, SSSA is very practical,

---

$b^2 = k^2 \cdot p^2 + 2k \cdot p + 1 = p \cdot (k^2 p + 2k) + 1$, i.e., $b^2 \bmod p = 1$. Since $b = l \cdot q + pq - 1$ for some l, $b^2 = l^2 q^2 + 2l \cdot q(pq-1) + (pq-1)^2$. So $b^2 \bmod q = (pq-1)^2 \bmod q = (p^2 q^2 - 2pq + 1) \bmod q = 1$.

[50] And also due to (i) of Theorem 5.3.3.1.

because in several applications one only needs to generate large primes (and not to test a given number).

A natural question is whether this algorithm can be used for all odd n (maybe with a not-so-good probability of success but still a reasonable one). Unfortunately, the answer is negative. There exist so-called **Carmichael numbers** that are composite numbers n with the property

$$a^{n-1} \equiv 1 (\bmod\ n) \text{ for all } a \in \{1, 2, \ldots, n-1\} \text{ with } gcd(a, n) = 1.$$

The first three Carmichael numbers are $561 = 3 \cdot 11 \cdot 17$, $1105 = 5 \cdot 13 \cdot 17$, and $1729 = 7 \cdot 13 \cdot 19$. Another piece of negative news is that there are infinitely many Carmichael numbers.[51]

The existence of Carmichel numbers implies that the attempt to test primality by Fermat's Theorem (i.e., by Fermat's test that asks whether $a^{n-1} \bmod n = 1$) does not work for infinitely many n. The test whether "$a^{(n-1)/2} \bmod n \in \{1, -1\}$" of the SSSA is more elaborated than Fermat's test but, unfortunately, there exist also infinitely many composite numbers that have too few such witnesses.

To overcome this difficulty we have to search for another kind of witness of the compositeness of numbers.

For any prime $p > 2$ and any integer a with $gcd(a, p) = 1$, the **Legendre symbol for a and p** is

$$Leg[\tfrac{a}{p}] = \begin{cases} 1 \text{ if } a \text{ is a quadratic residue mod } p \\ -1 \text{ if } a \text{ is a quadratic nonresidue mod } p. \end{cases}$$

Following Euler's Criterion (Theorem 5.3.2.2), we obtain

$$Leg[\tfrac{a}{p}] = a^{(p-1)/2} \bmod p.$$

A direct consequence of this equality is the fact that, for given a and p, $Leg[\tfrac{a}{p}]$ can be efficiently computed. To extend the SSSA algorithm for primality testing to all odd numbers, we need the following generalization of the Legendre symbol.

Definition 5.3.3.7 (Jacobi Symbol). *Let* $n = p_1^{k_1} \cdot p_2^{k_2} \cdot \cdots \cdot p_l^{k_l}$ *be an odd number for primes* $p_1 < p_2 < \cdots < p_l$ *and positive integers* $k_1, k_2, \ldots, k_l$. *For all positive integers* a *such that* $gcd(a, n) = 1$, *the* **Jacobi symbol of a and n** *is defined by*

$$\boldsymbol{Jac}\left[\tfrac{a}{n}\right] = \prod_{i=1}^{l} \left(Leg[\tfrac{a}{p_i}]\right)^{k_i} = \prod_{i=1}^{l} \left(a^{(p_i-1)/2} \bmod\ p_i\right)^{k_i}.$$

Observation 5.3.3.8. For all positive integers a, n satisfying the assumptions of Definition 5.3.3.7, $Jac\left[\tfrac{a}{n}\right] \in \{1, -1\}$.

---

[51] On the other hand, Carmichael numbers are extremely rare; there are only 255 of them among the first 10^8 integers.

Proof. For every prime p_i, $a^{(p_i-1)/2} \bmod p_i$ is either 1 or -1 (see Theorem 2.2.4.32.) $\square$

Clearly, if one knows the factorization of n, then one can efficiently compute the Jacobi symbol $Jac\left[\frac{a}{n}\right]$ for every a with $gcd(a,n) = 1$. The rules presented in the following exercise provide an efficient way to compute $Jac\left[\frac{a}{n}\right]$ when the factorization of n is unknown.

Exercise 5.3.3.9. Let a, b, n satisfy the assumptions of Definition 5.3.3.7. Prove

(i) $Jac\left[\frac{a \cdot b}{n}\right] = Jac\left[\frac{a}{n}\right] \cdot Jac\left[\frac{b}{n}\right]$,

(ii) if $a \equiv b \pmod{n}$, then $Jac\left[\frac{a}{n}\right] = Jac\left[\frac{b}{n}\right]$,

(iii) for odd coprimes a and n, $Jac\left[\frac{a}{n}\right] = (-1)^{\frac{a-1}{2} \cdot \frac{n-1}{2}} \cdot Jac\left[\frac{n}{a}\right]$,

(iv) $Jac\left[\frac{1}{n}\right] = 1$, and

(v) $Jac\left[\frac{2}{n}\right] = -1$, for $n \equiv 3$ or $5 \pmod 8$
    $Jac\left[\frac{2}{n}\right] = \ \ 1$, for $n \equiv 1$ or $7 \pmod 8$

$\square$

Applying the five properties of Exercise 5.3.3.9 we show how to compute the Jacobi number of the coprimes $n = 245$ and 3861.

$$Jac\left[\tfrac{245}{3861}\right] \underset{(iii)}{=} (-1)^{244 \cdot 3860/4} \cdot Jac\left[\tfrac{3861}{245}\right] = Jac\left[\tfrac{3861}{245}\right]$$

$$\underset{(ii)}{=} Jac\left[\tfrac{186}{245}\right] \underset{(i)}{=} Jac\left[\tfrac{2}{245}\right] \cdot Jac\left[\tfrac{93}{245}\right] \underset{(v)}{=} (-1) \cdot Jac\left[\tfrac{93}{245}\right]$$

$$\underset{(iii)}{=} (-1) \cdot (-1)^{92 \cdot 244/4} \cdot Jac\left[\tfrac{245}{93}\right] \underset{(ii)}{=} (-1) \cdot Jac\left[\tfrac{59}{93}\right]$$

$$\underset{(iii)}{=} (-1) \cdot (-1)^{58 \cdot 92/4} \cdot Jac\left[\tfrac{93}{59}\right] \underset{(ii)}{=} (-1) \cdot Jac\left[\tfrac{34}{59}\right]$$

$$\underset{(i)}{=} (-1) \cdot Jac\left[\tfrac{2}{59}\right] \cdot Jac\left[\tfrac{17}{59}\right] \underset{(v)}{=} (-1) \cdot (-1) \cdot Jac\left[\tfrac{17}{59}\right]$$

$$\underset{(iii)}{=} (-1)^{16 \cdot 58/4} \cdot Jac\left[\tfrac{59}{17}\right] \underset{(ii)}{=} Jac\left[\tfrac{8}{17}\right] \underset{(i)}{=} \left(Jac\left[\tfrac{2}{17}\right]\right)^3$$

$$\underset{(v)}{=} (1)^3 = 1.$$

Exercise 5.3.3.10. Prove the following assertion. For every odd n and every $a \in \{1, 2, \ldots, n-1\}$ with $gcd(a,n) = 1$, $Jac\left[\frac{a}{n}\right]$ can be computed in polynomial time according to $\log_2 n$.
Hint: Use properties (i)-(v) of Exercise 5.3.3.9 to prove it. $\square$

The following theorem provides new, useful definitions of witnesses for the compositeness.

Theorem 5.3.3.11. *For every odd n,*

(i) if n is a prime, then $Jac\left[\frac{a}{n}\right] \equiv a^{(n-1)/2} \pmod{n}$ for all $a \in \{1, 2, \ldots, n-1\}$,

(ii) if n is composite, then $Jac\left[\frac{a}{n}\right] \not\equiv a^{(n-1)/2} \pmod{n}$ for at least half of the numbers a with $gcd(a,n) = 1$.

Proof. First, we prove (i). If n is a prime, then $Jac\left[\frac{a}{n}\right] = Leg\left[\frac{a}{n}\right]$ for all $a \in \mathbb{Z}_n$. Following Euler's Criterion [Theorem 5.3.2.2] and the definition of the Legendre symbols, we obtain

$$Leg\left[\frac{a}{n}\right] = a^{(n-1)/2} \bmod n,$$

which completes the proof of (i).

Now, we proof (ii). Let, for every odd n,

$$\overline{Wit}_n = \{a \in \mathbb{Z}_n^* \mid Jac\left[\frac{a}{n}\right] \equiv a^{(n-1)/2} \pmod{n}\}.$$

Clearly, in our framework, $\mathbb{Z}_n^*$ is the set of all numbers candidating to be a witness of n's compositeness, $\mathbb{Z}_n^* - \overline{Wit}_n$ is the set of all witnesses of n's compositeness and $\overline{Wit}_n$ contains candidates that are no witnesses. The first part (i) of Theorem 5.3.3.11 claims $\mathbb{Z}_n^* = \overline{Wit}_n$ for all odd primes. Here, we have to prove that for all composite odd numbers n

$$|\overline{Wit}_n| \leq |\mathbb{Z}_n^*|/2.$$

Following the property (i) of Jacobi Symbols (Exercise 5.3.3.9) one can easily observe, that $(\overline{Wit}_n, \odot_{\bmod n})$ is a group. Since $\overline{Wit}_n \subseteq \mathbb{Z}_n^*$, Theorem 2.2.4.45 implies that $(\overline{Wit}_n, \odot_{\bmod n})$ is a subgroup of $(\mathbb{Z}_n^*, \odot_{\bmod n})$. Since Theorem 2.2.4.53 (Lagrange's Theorem) claims that, for any finite group $(G, \circ)$ and its subgroup $(H, \circ)$, $|H|$ divides $|G|$, we have

$$|\overline{Wit}_n| \text{ divides } |\mathbb{Z}_n^*|, \text{ i.e., } |\mathbb{Z}_n^*|/|\overline{Wit}_n| \in \mathbb{N} - \{0\}.$$

Thus, it is sufficient to prove that

$$(\overline{Wit}_n, \odot_{\bmod n}) \text{ is a proper subgroup of } (\mathbb{Z}_n^*, \odot_{\bmod n}).$$

To prove $\overline{Wit}_n \subset \mathbb{Z}_n^*$ it suffices to find an element $a \in \mathbb{Z}_n^* - \overline{Wit}_n$. Let

$$n = p_1^{i_1} \cdot p_2^{i_2} \cdot \ldots \cdot p_k^{i_k}$$

be the prime factorization of n ($i_j \in \mathbb{N} - \{0\}$ for $j = 1, \ldots, k$). To simplify the notation we set

$$q = p_1^{i_1} \text{ and } m = p_2^{i_2} \cdot \ldots \cdot p_k^{i_k}.$$

Following Exercise 2.2.4.28 $(\mathbb{Z}_q^*, \odot_{\bmod q})$ is a cyclic group. Let g be the generator of $\mathbb{Z}_q^*$.

Now, we fix a by the following congruences:

$$a \equiv g \pmod{q}$$
$$a \equiv 1 \pmod{m},$$

i.e., the representation of a in $\mathbb{Z}_q \times \mathbb{Z}_m$ is $(g, 1)$. If $m = 1$ (i.e., $n = q = p_1^{i_1}$ with $i_1 \geq 2$), then a is determined by the first congruence only.

First, we have to show that $a \in \mathbb{Z}_n^*$, i.e., that $gcd(a, n) = 1$. So, we have to prove that none of the primes $p_1, \ldots, p_k$ divides a. If p_1 divides a, then $g \equiv a \bmod p_1^{i_1}$ could not be a generator of the cyclic group $(\mathbb{Z}_q^*, \odot \bmod q)$. If p_r divides a for some $r \in \{2, \ldots, k\}$, then $a = p_r b$ for some $b \in \mathbb{N} - \{0\}$. The congruence $a \equiv 1 \bmod m$ implies $a = m \cdot x + 1$ for an $x \in \mathbb{N}$. Hence $a = p_r b = mx + 1 = p_r(m/p_r) \cdot x + 1$, i.e., $p_r(b - (m/p_r)) = 1$, i.e., p_r divides 1.

Finally, we have to prove that $a \notin \overline{Wit}_n$. We distinguish two possibilities with respect to i_1.

(i) Let $i_1 = 1$.
We compute the Jacobi symbol for a and n, and the number $a^{(n-1)/2} \bmod n$ in order to show that they are different.
Remember that $n = p_1 \cdot m$, $m \neq 1$, and $gcd(p_1, m) = 1$.

$$
\begin{aligned}
Jac\left[\tfrac{a}{n}\right] &= \prod_{j=1}^{k} Jac\left[\tfrac{a}{p_j}\right]^{i_j} &&\{\text{by Definition 5.3.3.7}\} \\
&\overset{\cdot}{=} Jac\left[\tfrac{a}{p_1}\right] \cdot \prod_{j=2}^{k}\left(Jac\left[\tfrac{a}{p_j}\right]\right)^{i_j} &&\{i_1 = 1\} \\
&= Jac\left[\tfrac{a}{p_1}\right] \cdot \prod_{j=2}^{k}\left(Jac\left[\tfrac{1}{p_j}\right]\right)^{i_j} &&\{\text{by (ii) of Exercise 5.3.3.9}\} \\
&&&\text{and } a \equiv 1 \ (\bmod\ m)\} \\
&= Jac\left[\tfrac{a}{p_1}\right] &&\{\text{by (iv) of Exercise 5.3.3.9}\} \\
&= Jac\left[\tfrac{g}{p_1}\right] &&\{\text{by (ii) of Exercise 5.3.3.9}\} \\
&&&\text{and } a \equiv g \ (\bmod\ p_1)\} \\
&= Leg\left[\tfrac{g}{p_1}\right] &&\{\text{since } q = p_1 \text{ is a prime}\} \\
&= -1 &&\{\text{since the generator } g \text{ of}\} \\
&&&\mathbb{Z}_{p_1}^* \text{ cannot be a quadratic} \\
&&&\text{residue}\}
\end{aligned}
$$

Thus, $Jac\left[\tfrac{a}{n}\right] = -1$.

Since $a \equiv 1 \ (\bmod\ m)$, we obtain

$$a^{(n-1)/2} \bmod m = (a \bmod m)^{(n-1)/2} \bmod m = 1^{(n-1)/2} \bmod m = 1.$$

Thus, $a^{(n-1)/2} \bmod n = -1 \ (= n - 1 \text{ in } \mathbb{Z}_n^*)$ is impossible because it would imply[52]

---

[52] The representation of $-1 = n-1$ from $\mathbb{Z}_n$ in $\mathbb{Z}_q \times \mathbb{Z}_m$ is $(-1, -1) = (q-1, m-1)$.

$$a^{(n-1)/2} \bmod m = -1 \; (= m - 1 \text{ in } \mathbb{Z}_m^*).$$

Thus, we have obtained

$$-1 = Jac\left[\tfrac{a}{n}\right] \neq a^{(n-1)/2} \bmod n,$$

i.e., $a \in \mathbb{Z}_n^* - \overline{Wit}_n$.

(ii) Let $i_1 \geq 2$.

We prove $a \notin \overline{Wit}_n$ by contradiction. Let $a \in \overline{Wit}_n$. This implies

$$a^{(n-1)/2} \bmod n = Jac\left[\tfrac{a}{n}\right] \in \{-1, 1\},$$

and so

$$a^{n-1} \bmod n = 1.$$

Since $n = q \cdot m$, we also have

$$a^{n-1} \bmod q = 1.$$

Thus,

$$1 = a^{n-1} \bmod q = (a \bmod q)^{n-1} \bmod q = g^{n-1} \bmod q.$$

Since g is a generator of the cyclic group $(\mathbb{Z}_q^*, \odot_{\bmod q})$, its order is $|\mathbb{Z}_q^*|$. The above proved fact $g^{n-1} \bmod q = 1$ implies that the order of g must divide $n-1$, i.e., that

$$|\mathbb{Z}_q^*| \text{ divides } n-1.$$

Since $q = p_1^{i_1}$, $i_1 \geq 2$, we have that[53]

$$p_1 \text{ divides } |\mathbb{Z}_q^*|,$$

and so p_1 divides $n-1$. But $n = p_1^{i_1} \cdot m$ which is a contradiction because no prime p_1 can divide both n and $n-1$.

$\square$

The use of Theorem 5.3.3.11 results in the following general, randomized algorithm for primality testing.

Algorithm 5.3.3.12. SOLOVAY-STRASSEN ALGORITHM

Input: An odd number n.
Step 1: Choose a uniformly at random from $\{1, 2, \ldots, n-1\}$.
Step 2: Compute $gcd(a, n)$.
Step 3: if $gcd(a, n) \neq 1$ then return ("COMPOSITE") {accept}.
Step 4: Compute $Jac\left[\tfrac{a}{n}\right]$ and $a^{(n-1)/2} \bmod n$.

---

[53] Remember, that $\mathbb{Z}_q^* = \{a \in \mathbb{Z}_q \mid gcd(a, q) = 1\} = \{a \in \mathbb{Z}_q \mid p_1 \text{ does not divide } a\}$. Since the number of elements from $\mathbb{Z}_q$ that are divisible by p_1 is exactly $|\mathbb{Z}_q|/p_1$, we obtain $|\mathbb{Z}_q^*| = |\mathbb{Z}_q| - |\mathbb{Z}_q|/p_1 = p_1^{i_1} - p_1^{i_1-1} = p_1 \cdot (p_1^{i_1-1} - p_1^{i_1-2})$.

Step 5:   **if** $Jac\left[\frac{a}{n}\right] \equiv a^{(n-1)/2} \pmod{n}$ **then return** ("PRIME") {reject}
else **return** ("COMPOSITE") {accept}.

Following Theorem 5.3.3.11 we see that the SOLOVAY-STRASSEN ALGO-
RITHM is a one-sided-error Monte Carlo algorithm for primality testing.[54]
Starting again from the key assertion of Theorem 2.2.4.32 that for any odd
$p > 2$

"*p is a prime* $\Leftrightarrow a^{(p-1)/2} \bmod p \in \{1, -1\}$ *for all* $a \in \{1, \ldots, p-1\}$"

we will develop another concept of a witness for compositeness than the
concept used in the SOLOVAY-STRASSEN ALGORITHM. Fermat's Theorem
says that $a^{(p-1)} \bmod p$ is the number 1 for every prime p and any $a \in$
$\{1, 2, \ldots, p-1\}$. The proof of Theorem 2.2.4.32 says that for any prime p,
the number 1 has exactly two square roots 1 and $-1 \bmod p$. But, for a composite number p, the number 1 can have four or more square roots mod p.
For instance,

$$1 \equiv 1^2 \equiv 20^2 \equiv 8^2 \equiv 13^2 \pmod{21},$$

and so $1, -1 = 20, 8, -8 = 13$ are the four square roots of 1 mod 21. Unfortunately, we do not know any polynomial-time algorithm for computing all
square roots of 1 mod n for a given number n. The idea of SSSA was that if
$a^{p-1} \equiv 1 \pmod{p}$, then $a^{(p-1)/2} \bmod p$ is a square root of 1, and so we can get
at least one square root of 1. We extend this idea in the following way. Since
we consider only odd numbers n as inputs for primality testing, we have that
$n - 1$ is even. This means that $n - 1$ can be expressed as follows:

$$n - 1 = s \cdot 2^m$$

for an $m \geq 1$ and an odd $s \in \mathbb{N} - \{0\}$. If one chooses an $a \in \mathbb{Z}_n$, then
$a^{n-1} \bmod n \neq 1$ ensures that n is a composite number. If $a^{n-1} \bmod n = 1$,
then we can compute a root

$$a^{(n-1)/2} \bmod n = a^{s \cdot 2^{m-1}} \bmod n$$

of the number 1. If $a^{(n-1)/2} \bmod n \notin \{1, -1\}$, then we are sure that n is a
composite number. If $a^{s \cdot 2^{m-1}} \bmod n = 1$ and $m \geq 2$, then we again have
the possibility to compute a square root $a^{s \cdot 2^{m-2}} \bmod n$ of the number 1. In
general, if

$$a^{s \cdot 2^{m-1}} \equiv a^{s \cdot 2^{m-2}} \equiv \ldots \equiv a^{s \cdot 2^{m-j}} \equiv 1 \bmod n, \text{ and } m > j,$$

then we can compute another square root $a^{s \cdot 2^{m-j-1}} \bmod n$ of the number 1. If
$a^{s \cdot 2^{m-j-1}} \bmod n \notin \{1, -1\}$, we have proved that n is a composite number. If
$a^{s \cdot 2^{m-j-1}} = 1 \bmod n$ and $m > j+1$, then we can continue in this procedure in
order to compute a next square root of 1. If $m = j + 1$ or $a^{s \cdot 2^{m-j-1}} \bmod n =$

---

[54] To be exact, we have to say "for the recognition of compositeness".

-1, we have to stop because we are not able to use a to prove that n is a composite number.

Let us consider the following example. Let $n = 561 = 3 \cdot 11 \cdot 17$ be the smallest Carmichael number. Thus, $n - 1 = 560 = 2^4 \cdot 35$. The number 7 is a witness of the compositeness of 561, because

$$7^{560} \bmod 561 = 1 \text{ and } 7^{280} \bmod 561 = 67;$$

i.e., 67 is a root of 1 mod 561. This consideration leads to the following definition of a compositeness witness.

Let n be a composite odd number. Let $n - 1 = s \cdot 2^m$ for an odd s and an integer $m \geq 1$. We say that a number $a \in \{1, \ldots, n - 1\}$ is a **root-witness of the compositeness of n** if

(1) $a^{n-1} \bmod n \neq 1$, or
(2) there exists $j \in \{0, 1, \ldots, m - 1\}$, such that

$$a^{s \cdot 2^{m-j}} \bmod n = 1 \text{ and } a^{s \cdot 2^{m-j-1}} \bmod n \notin \{1, -1\}.$$

The following assertion shows that the concept of root-witnesses is suitable for the method of abundance of witnesses.

Theorem 5.3.3.13. *Let $n > 2$ be an odd integer. Then*

(i) if n is a prime, then for all $a \in \{1, \ldots, n - 1\}$, a is no root-witness of the compositeness of n
{i.e., our definition of root-witnesses is a correct definition of witnesses of the compositeness},

(ii) if n is composite, then at least half of the numbers $a \in \{1, \ldots, n - 1\}$ are root-witnesses of the compositeness of n
{i.e., there are many root-witnesses of the compositeness}.

Proof. First, we prove (i). Let n be a prime. Following Fermat's Theorem we have

$$a^{n-1} \bmod n = 1 \text{ for all } a \in \{1, \ldots, n - 1\}$$

and so the condition (1) of the definition of the root-witnesses cannot be fulfilled for any $a \in \{1, \ldots, n - 1\}$.

Now, we have to prove that also (2) cannot be satisfied. Let us assume the contrary. Let $n - 1 = s \cdot 2^m$ for an odd s and an integer $m \geq 1$. Let there exist an $a \in \{0, 1, \ldots, n - 1\}$ and a $j \in \{0, 1, \ldots, m - 1\}$ such that

$$a^{s \cdot 2^{m-j}} \bmod n = 1 \text{ and } a^{s \cdot 2^{m-j-1}} \bmod n \notin \{1, -1\},$$

i.e., there exist a and j such that (2) is satisfied.

Since $a^{s \cdot 2^{m-j}} \bmod n = 1$, we have

$$a^{s \cdot 2^{m-j}} - 1 \equiv 0 \pmod{n}.$$

Factoring $a^{s \cdot 2^{m-j}} - 1$ yields

$$(a^{s \cdot 2^{m-j-1}} - 1) \cdot (a^{s \cdot 2^{m-j-1}} + 1) = a^{s \cdot 2^{m-j}} - 1 \equiv 0 \ (\text{mod } n).$$

Since n is a prime, $(\mathbb{Z}_n, \odot_{\text{mod } n})$ is a field, and so

$$(a^{s \cdot 2^{m-j-1}} - 1) \ \text{mod } n = 0 \ \text{or} \ (a^{s \cdot 2^{m-j-1}} + 1) \ \text{mod } n = 0.$$

Thus, $a^{s \cdot 2^{m-j-1}} \ \text{mod } n \in \{-1, 1\}$, which contradicts (2).

Now, we prove the assertion (ii) claiming that the number of root-witnesses of n is at least $(n-1)/2$ for every odd integer $n > 2$. We consider

$$\mathbb{Z}_n - \{0\} = \{1, 2, \ldots, n-1\} = Wit_n \cup NonWit_n,$$

where $Wit_n = \{a \in \mathbb{Z}_n - \{0\} \mid a$ is a root-witness of the compositeness of $n\}$, and $NonWit_n = (\mathbb{Z}_n - \{0\}) - Wit_n$.

First we show that $NonWit_n \subseteq \mathbb{Z}_n^* = \{a \in \mathbb{Z}_n \mid gcd(a, n) = 1\}$. Because of (1) of the definition of the root-witnesses,

$$a^{n-1} \ \text{mod } n = 1$$

for every non-witness $a \in \mathbb{Z}_n$. This implies $a \cdot a^{n-2} \equiv 1 (\text{mod } n)$, which expresses the fact that a^{n-2} is the multiplicative inverse element to a. Thus, follwing Lemma 2.2.4.57 $gcd(a, n) = 1$ and $a \in \mathbb{Z}_n^*$.

To complete the proof, it is sufficient to show that there is a $B \subseteq \mathbb{Z}_n^*$ such that $NonWit_n \subseteq B$ and $(B, \odot_{\text{mod } n})$ is a proper subgroup of $(\mathbb{Z}_n^*, \odot_{\text{mod } n})$, because these facts imply (Theorem 4.3.5.13 and Corollary 4.3.5.13)

$$|NonWit_n| \le |B| \le |\mathbb{Z}_n^*|/c \le |\mathbb{Z}_n|/c$$

for an integer $c \ge 2$.

We distinguish two cases with respect to whether n is a Carmichael number or not.

(a) Let n be no Carmichael number, i.e., there exists a $b \in \mathbb{Z}_n$ such that

$$b^{n-1} \ \text{mod } n \ne 1.$$

In this case we set

$$B = \{a \in \mathbb{Z}_n^* \mid a^{n-1} \equiv 1 \ (\text{mod } n)\}.$$

B is nonempty, because $1 \in B$. Since for all $a, d \in B$

$$(a \cdot d)^{n-1} \ \text{mod } n = (a^{n-1} \ \text{mod } n) \cdot (d^{n-1} \ \text{mod } n) \ \text{mod } n = 1 \cdot 1 \ (\text{mod } n) = 1,$$

B is closed under $\odot_{\text{mod } n}$, and so, following Theorem 2.2.4.45, B is a subgroup of $\mathbb{Z}_n^*$. Since $a^{n-1} \ \text{mod } n = 1$ for every nonwitness a,

$$NonWit_n \subseteq B.$$

Following our assumption that there exists a $b \in \mathbb{Z}_n^*$ with $b^{n-1} \bmod n \neq 1$ we have

$$b \in \mathbb{Z}_n^* - B,$$

and so $(B, \odot_{\bmod n})$ is a proper subgroup of $(\mathbb{Z}_n^*, \odot_{\bmod n})$.

(b) Let n be a Carmichael number, i.e.,

$$a^{n-1} \bmod n = 1 \text{ for all } a \in \mathbb{Z}_n^*. \tag{5.11}$$

Obviously n is either a prime power $n = p^r$ or n is divisible by two different primes. We consider these two possibilities separately.

(b.1) Let $n = p^r$ for a prime p and an integer $r \geq 2$. Then we prove that n is no Carmichael number and so this case cannot appear. Since n is odd, p must be odd, too. Following Exercise 2.2.4.28 (ii), $(\mathbb{Z}_n^*, \odot_{\bmod n})$ is a cyclic group. Since

$$\mathbb{Z}_n^* = \mathbb{Z}_n - \{h \in \mathbb{Z}_n - \{0\} \mid p \text{ divides } h\}$$

we have

$$|\mathbb{Z}_n^*| = p^r - p^{r-1} = p^{r-1}(p-1).$$

Let g be a generator of $\mathbb{Z}_n^*$. By the property (5.11), we have

$$g^{n-1} \bmod n = 1 = g^0. \tag{5.12}$$

Since the sequence $g^0, g^1, g^2, \ldots$ is periodic with the period $|\mathbb{Z}_n^*|$, (5.12) implies that the number $n - 1 - 0 = n - 1$ is divisible by $|\mathbb{Z}_n^*|$, i.e., that

$$|\mathbb{Z}_n^*| = p^{r-1}(p-1) \text{ divides } n - 1 = p^r - 1.$$

But this is impossible because $r \geq 2$ and p does not divide $p^r - 1$.

(b.2) Let n be no prime power. Since n is odd, we can find odd positive integers p, q such that

$$n = p \cdot q \text{ and } gcd(p, q) = 1.$$

Since $n - 1$ is even, we can express it as

$$n - 1 = s \cdot 2^m,$$

where s is an odd integer and $m \geq 1$. Let

$$Seq_n(a) = \{a^s \bmod n, \ldots, a^{s \cdot 2^{m-1}} \bmod n, a^{s \cdot 2^m} \bmod n\}.$$

For every $a \in \mathbb{Z}_n^*$ define $j_a \in \{0, 1, \ldots, m, Empty\}$ as follows

$$j_a = \begin{cases} Empty & \text{if } -1 \notin Seq_n(a) \\ t & \text{if } a^{s \cdot 2^t} \bmod n = -1 \text{ and } a^{s \cdot 2^i} \bmod n \neq -1 \\ & \text{for all } i \in \{t+1, \ldots, m\}. \end{cases}$$

Since $n - 1 = -1 \in \mathbb{Z}_n^*$ and $(-1)^s = -1$, there must exist an $a \in \mathbb{Z}_n^*$ with $j_a \neq Empty$. So, we can define

$$j = \max\{j_a \in \{0, 1, \ldots, m\} \mid a \in \mathbb{Z}_n^*\}.$$

Now, we set

$$B = \{a \in \mathbb{Z}_n^* \mid a^{s \cdot 2^j} \bmod n \in \{1, -1\}\}.$$

For all $a, b \in B$,

$$(a \cdot b)^{s \cdot 2^j} \bmod n = (a^{s \cdot 2^j} \bmod n) \cdot (b^{s \cdot 2^j} \bmod n) \in \{1, -1\}$$

and so B is closed under $\odot_{\bmod n}$, i.e., following Theorem 2.2.4.45 $(B, \odot_{\bmod n})$ is a subgroup of $(\mathbb{Z}_n^*, \odot_{\bmod n})$.

$$NonWit_n \subseteq B,$$

because for all $a \in NonWit_n$
(i) either $Seq_n(a) = \{1\}$ and then $a^{s \cdot 2^j} = 1 \in \{1, -1\}$, or
(ii) $Seq_n(a) \neq \{1\}$ which means
$$a^{s \cdot 2^m} \equiv a^{s \cdot 2^{m-1}} \equiv \ldots \equiv a^{s \cdot 2^{j_a+1}} \equiv 1 \bmod n \text{ and } a^{s \cdot 2^{j_a}} \equiv -1 \bmod n.$$

Since $j_a \leq j$, $a^{s \cdot 2^j} \in \{1, -1\}$ and so $a \in B$.
It remains to show that B is a proper subset of $\mathbb{Z}_n^*$. We fix an $h \in \mathbb{Z}_n^*$ such that

$$j_h = j.$$

Now, consider the representation of elements of $\mathbb{Z}_n$ in $\mathbb{Z}_p \times \mathbb{Z}_q$ and fix $w \in \mathbb{Z}_n$ by

$$w \equiv h \bmod p$$
$$w \equiv 1 \bmod q.$$

Following $n = p \cdot q$ and $gcd(p, q) = 1$, the Chinese Remainder Theorem claims that such a w exists and that w is unambiguously determined by these congruences.
We show that $w \in \mathbb{Z}_n^* - B$. First we show that $w \notin B$, i.e., that $w^{s \cdot 2^j} \bmod n \notin \{1, -1\}$. Since $h^{s \cdot 2^j} \bmod n = -1 = n - 1$, we have $h^{s \cdot 2^j} \bmod p = p - 1$. Note that $p - 1$ is -1 in $\mathbb{Z}_p$. Therefore[55]

$$w^{s \cdot 2^j} \bmod p = (w \bmod p)^{s \cdot 2^j} \bmod p = (h)^{s \cdot 2^j} \bmod p = p - 1,$$

and

---

[55] Remember that s is odd.

$$w^{s \cdot 2^j} \bmod q = (w \bmod q)^{s \cdot 2^j} \bmod q = 1^{s \cdot 2^j} \bmod q = 1,$$

and so the representation of $w^{s \cdot 2^j} \bmod n$ in $\mathbb{Z}_p \times \mathbb{Z}_q$ is $(p-1, 1) = (-1, 1)$. The representation of 1 in $\mathbb{Z}_p \times \mathbb{Z}_q$ is $(1, 1)$ and the representation of $n - 1 = -1$ in $\mathbb{Z}_p \times \mathbb{Z}_q$ is $(p - 1, q - 1) = (-1, -1)$. Thus $w^{s \cdot 2^j} \bmod n \notin \{1, -1\}$, i.e., $w \notin B$.

To prove $w \in \mathbb{Z}_n^*$ (i.e., to prove $gcd(w, n) = 1$) it is sufficient to prove $gcd(p, w) = 1 = gcd(q, w)$, because $n = p \cdot q$ and $gcd(p, q) = 1$.

Since $h \in \mathbb{Z}_n^*$, we have $gcd(h, n) = 1$. If h does not have any common divisor with $n = p \cdot q$, then h cannot have any common divisor with p, i.e.,

$$gcd(p, h) = 1.$$

Since $w \equiv h \pmod{p}$ means $w = p \cdot k + h$, the fact $gcd(p, w) = b > 1$ would mean that b divides h and so $gcd(p, h) \geq b$. Since we have proved $gcd(p, h) = 1$, b must be equal to 1, and so

$$gcd(p, w) = 1.$$

We determined w by the congruence $w \equiv 1 \bmod q$, which directly implies

$$gcd(q, w) = 1.$$

Thus $gcd(n, w) = 1$ and $w \in \mathbb{Z}_n^* - B$, which completes the proof.   □

Based on the concept of the root-witnesses of compositeness and on the assertion of Theorem 5.3.3.13, we obtain the following randomized algorithm for primality testing.

Algorithm 5.3.3.14. MILLER-RABIN ALGORITHM

Input: An odd number n.
Step 1:   Choose a uniformly at random from $\{1, 2, \ldots, n - 1\}$.
Step 2:   Compute $a^{n-1} \bmod n$.
Step 3:   **if** $a^{n-1} \bmod n \neq 1$ **then**
              return ("COMPOSITE") –accept"
          else begin
              compute s and m such that $n - 1 = s \cdot 2^m$;
              for $i := 0$ **to** $m - 1$ **do**
                  $r[i] := a^{s \cdot 2^i} \bmod n$ –by repeated squaring";
              $r[m] := a^{n-1} \bmod n$;
              if there exists $j \in \{0, 1, \ldots, m - 1\}$, such that
                  $r[m - j] = 1$ and $r[m - j - 1] \notin \{1, -1\}$,
                  then return ("COMPOSITE") –accept"
                  else return ("PRIME") –reject"
          end

Theorem 5.3.3.15. *The* MILLER-RABIN ALGORITHM *is a polynomial-time one-sided-error Monte Carlo algorithm for the recognition of the set of odd composite numbers.*

Proof. The fact that the MILLER-RABIN ALGORITHM is a one-sided-error Monte Carlo algorithm is a direct consequence of Theorem 5.3.3.13 and the fact that this algorithm is an application of the method of abundance of witnesses with respect to root-witnesses of compositeness.

We observe that all numbers $r[i]$ for $i = 0, \ldots, m$ can be computed in one application of the algorithm REPEATED SQUARING that can be executed by $O((\log_2 n)^3)$ binary operations. The rest can be performed in time $O(m)$, i.e., in $O(\log_2 n)$.   □

The last part of this section is devoted to the typical task in public key cryptography:

"Generate a prime of the binary length l for some given positive integer l."

This task is a little bit more complicated than to decide whether a given number is a prime. Due to the Prime Number Theorem[56], we know that, for a randomly chosen number n of the length l, the probability of being a prime is approximately $\frac{1}{\ln n} \approx \frac{1}{l}$. So, several runs of the SOLOVAY-STRASSEN ALGORITHM (or of the MILLER-RABIN ALGORITHM) on different inputs are needed to find a prime. But one has to realize this carefully because the probability of outputting a composite number instead of a prime grows with the number of repetitions (runs) of the SOLOVAY-STRASSEN ALGORITHM on different inputs. Obviously, the solution is to let the SOLOVAY-STRASSEN ALGORITHM run k times on every randomly generated number of the length l for some suitable k.

Algorithm 5.3.3.16. PRIME GENERATION(l, k) (PG(l, k))

Input: l, k.

Step 1:   Set $X :=$ "still not found";
           $I := 0$

Step 2:   **while** $X =$ "still not found" and $I < 2l^2$
           do begin generate randomly a bit sequence $a_1, \ldots, a_{l-2}$ and set
                 $n = 2^{l-1} + \sum_{i=1}^{l-2} a_i 2^i + 1$;
             perform k runs of SOLOVAY-STRASSEN ALGORITHM on n;
             if at least one of the k outputs is "Composite"
                then $I := I + 1$
                else do begin $X :=$ "already found";
                             output(n)
                end

      end

---

[56] Theorem 2.2.4.16

Step 3:   **if** $I = 2l^2$ **output**("I did not find any prime").

First, we analyze the probability that the randomized algorithm PRIME GENERATION(l, k) gives an integer as the output. We know that the probability of choosing a prime from the set of all integers of l bits is at least $1/2l$ for sufficiently large ls. PG(l, k) outputs "I did not find any prime" only if all $2l^2$ randomly generated numbers are composite and for all these numbers their compositeness has been proved in the k runs of the SOLOVAY-STRASSEN ALGORITHM. This is because if n is a prime then SOLOVAY-STRASSEN ALGORITHM always outputs "PRIME". So the probability of the output "I did not find any prime" is at most

$$\left[\left(1 - \frac{1}{2l}\right) \cdot \left(1 - \frac{1}{2^k}\right)\right]^{2l^2} < \left(1 - \frac{1}{2l}\right)^{2l^2} = \left[\left(1 - \frac{1}{2l}\right)^{2l}\right]^l < \left(\frac{1}{e}\right)^l = e^{-l}.$$

Obviously, e^{-l} tends to 0 with growing l and $e^{-l} < \frac{1}{4}$ for every $l \geq 2$. For instance, $e^{-l} < 0.00005$ for $l \geq 10$. Since we usually apply this algorithm for l with $100 \leq l \leq 600$, the probability of the output "I did not find any prime" is negligible.

Now, let us bound the probability of the event that PG(l, k) gives a wrong answer, i.e. that it outputs a composite number n. This obviously depends on the relation between l and k. In what follows we set $k = l$. This choice is "fair" from the complexity point of view because it means that PG(l, k) runs in polynomial time according to l. PG(l, k) outputs a composite number n only if all up to n generated numbers were composite, and the k runs of SOLOVAY-STRASSEN ALGORITHM proved their compositeness but not the compositeness of n. The probability of this event is at most

$$\left(1 - \frac{1}{2l}\right) \cdot \frac{1}{2^l} + \sum_{i=1}^{2l^2-1} \left[\left(1 - \frac{1}{2l}\right) \cdot \left(1 - \frac{1}{2^l}\right)\right]^i \cdot \left(1 - \frac{1}{2l}\right) \cdot \frac{1}{2^l}$$

$$\leq \left(1 - \frac{1}{2l}\right) \cdot \frac{1}{2^l} \cdot \left(\sum_{i=1}^{2l^2-1} \left(1 - \frac{1}{2l}\right)^i + 1\right)$$

$$\leq \left(1 - \frac{1}{2l}\right) \cdot \frac{1}{2^l} \cdot 2l^2 \leq \frac{l^2}{2^{l-1}}.$$

Obviously, $l^2 \cdot 2^{-(l-1)}$ tends to 0 with growing l and $l^2 \cdot 2^{-(l-1)} \leq 1/5$ for $l \geq 10$. For instance, for $l \geq 100$,

$$l^2 \cdot 2^{-(l-1)} \leq 7.9 \cdot 10^{-27},$$

i.e., the probability that the output of PG(l, l) is wrong is smaller than 1 divided by the number of microseconds after the "Big Bang". So this probability is so small that it would be a fascinating experience to see such a rare event.

Exercise 5.3.3.17. Analyze the behavior of $PG(l, k)$ for

(i) $k = 2 \cdot \lceil \log_2 l \rceil$, and
(ii) $k = 2 \cdot (\lceil \log_2 l \rceil)^2$. $\qquad \Box$

Summary of Section 5.3.3

Primality testing is one of the fundamental computing problems. It is not known to be in P nor is it known whether it is NP-hard. The method of abundance of witnesses leads to the design of efficient one-sided-error Monte Carlo algorithms for primality testing. The crucial idea is to find such a characterization (an equivalent definition) of primality that if p is composite, then there is a lot of witnesses of ps compositeness (more precisely at least half of the possible candidates for the testimony (for instance, all numbers a with $gcd(a, p) = 1$) are witnesses of p's compositeness).

A one-sided-error Monte Carlo algorithm for primality testing can be successfully used to design a randomized generator of large primes.

5.3.4 Equivalence Tests, Fingerprinting, and Monte Carlo

In Section 5.2 we used the one-way communication protocol model in order to show that one-sided-error Monte Carlo may essentially be more powerful than determinism. Here, we want to show that the problem of the nonequivalence of two one-time-only branching programs (1BPs), NEq-1BP, can be efficiently solved by a one-sided-error Monte Carlo algorithm. This problem is not easy. Currently we do not know any polynomial time algorithm for this problem, but this problem is not known to be NP-complete. To design a randomized algorithm for NEq-1BP we reduce this problem to deciding the nonequivalence of two polynomials over the field $\mathbb{Z}_p$ for a prime p. First of all, we need a randomized algorithm for the equivalence of two polynomials. We observe that to decide the equivalence of two polynomials $p_1(x_1, \ldots, x_n)$ and $p_2(x_1, \ldots, x_n)$ is the same as to decide whether the polynomial $p_1(x_1, \ldots, x_n) - p_2(x_1, \ldots, x_n)$ is identical to 0.

We start with two technical lemmata.

Lemma 5.3.4.1. *Let d be a non-negative integer. Every polynomial $p(x)$ of a single variable x and of degree d has either at most d roots or is everywhere equal to 0.*

Proof. We perform the proof by an induction on the degree d.

(1) Let $d = 0$. Then $p(x) = c$ for a constant c. If $c \neq 0$, then $p(x)$ has no root.
(2) Let the assertion of Lemma 5.3.4.1 hold for $d - 1$. Let $p(x) \not\equiv 0$, and let a be a root of p. Then $p(x) = (x - a) \cdot p_1(x)$, where $p_1(x) = p(x)/(x - a)$ is a polynomial of degree $d - 1$. Following the induction hypothesis, $p_1(x)$ has at most $d - 1$ roots and so $p(x)$ has at most d roots. $\qquad \Box$

Now we consider the method of the abundance of witnesses for the fact that a polynomial $p(x_1, \ldots, x_n)$ over a finite field is not identical to 0. Obviously, any input $a_1, a_2, \ldots, a_n$ such that $p(a_1, a_2, \ldots, a_n) \neq 0$ is a witness of the fact $p(x_1, x_2, \ldots, x_n) \not\equiv 0$. The aim is to show that there are many such witnesses.

Lemma 5.3.4.2. *Let n be a prime and let d, m be positive integers. Let $p(x_1, \ldots, x_m)$ be a nonzero polynomial over $\mathbb{Z}_n$ in the variables $x_1, x_2, \ldots, x_m$, where each variable has at most degree d in $p(x_1, \ldots, x_m)$. If $a_1, a_2, \ldots, a_m$ are selected uniformly at random from $\mathbb{Z}_n$, then[57]*

$$Prob(p(a_1, a_2, \ldots, a_m) = 0) \leq m \cdot d/n.$$

Proof. We perform the proof by the induction on the number of input variables m.

(1) Let $m = 1$. By Lemma 5.3.4.2 the polynomial $p(x_1)$ has at most d roots. Since the number of distinct elements in $\mathbb{Z}_n$ is n,

$$Prob(p(a_1) = 0) \leq \frac{d}{n} = \frac{d \cdot m}{n}$$

for every randomly chosen $a_1 \in \mathbb{Z}_n$.

(2) Assume that the assertion of Lemma 5.3.4.2 is true for $m - 1$ and let us prove it for m. Obviously, there are polynomials $p_0(x_2, \ldots, x_m)$, $p_1(x_2, \ldots, x_m)$, $\ldots$, $p_d(x_2, \ldots, x_m)$ such that

$$\begin{aligned} p(x_1, x_2, \ldots, x_m) = \qquad\qquad\qquad\qquad\qquad\qquad &(5.13) \\ p_0(x_2, \ldots, x_m) + x_1 \cdot p_1(x_2, \ldots, x_m) + \cdots + x_1^d \cdot p_d(x_2, \ldots, x_m) \end{aligned}$$

If $p(a_1, a_2, \ldots, a_m) = 0$ for some $a_1, a_2, \ldots, a_m$, then either
 (i) $p_i(a_2, \ldots, a_m) = 0$ for all $i = 0, 1, \ldots, d$, or
 (ii) there exists $j \in \{0, 1, \ldots, d\}$ such that $p_j(a_2, \ldots, a_m) \neq 0$ and a_1 is a root of the single variable polynomial

$$\overline{p}(x_1) = p_0(a_2, \ldots, a_m) + p_1(a_2, \ldots, a_m) \cdot x_1 + \cdots + p_d(a_2, \ldots, a_m) \cdot x_1^d.$$

Now, we bound the probabilities of the cases (i) and (ii) separately.
 (i) Since $p(x_1, \ldots, x_m)$ is a nonzero polynomial, (5.13) implies the existence of a $k \in \{0, 1, \ldots, d\}$ such that $p_k(x_2, \ldots, x_m)$ is a nonzero polynomial.[58] The probability that, for randomly chosen $a_2, \ldots, a_m$, $p_j(a_2, \ldots, a_m) = 0$ for all $j = 0, 1, \ldots, d$, is at most the probability that $p_k(a_2, \ldots, a_m) = 0$. Summarizing the consideration above, and applying the induction hypothesis we obtain

---

[57] Observe that $p(x_1, \ldots, x_m)$ can be considered as a random variable in this experiment.

[58] Note that $p_k(x_2, \ldots, x_m)$ has at most $m - 1$ variables.

$Prob$(the event (i) happens) =

$$Prob\big(p_0(a_2, \ldots, a_m) = \cdots = p_d(a_2, \ldots, a_m) = 0\big)$$
$$\leq Prob(p_k(a_2, \ldots, a_m) = 0)$$
$$\underset{(i.h.)}{\leq} (m-1) \cdot d/n.$$

(ii) If $\overline{p}(x_1)$ is a nonzero polynomial, then the induction hypothesis directly implies

$$Prob(\overline{p}(a_1) = 0) \leq d/n,$$

and it is obvious that

$$Prob(\text{the event (ii) happens}) \leq Prob(\overline{p}(a_1) = 0).$$

Finally, for randomly chosen $a_1, \ldots, a_n$,

$$Prob(p(a_1, \ldots, a_m) = 0) = Prob(\text{event (i)}) + Prob(\text{event (ii)})$$
$$\leq (m-1) \cdot d/n + d/n = m \cdot d/n.$$

$\square$

Corollary 5.3.4.3. *Let n be a prime, and let m, d be positive integers. For every nonzero polynomial $p(x_1, \ldots, x_m)$ over $\mathbb{Z}_n$ with degree at most d, there are at least*

$$\left(1 - \frac{m \cdot d}{n}\right) \cdot n^m$$

distinct witnesses from $(\mathbb{Z}_n)^m$ of the nonzeroness of $p(x_1, \ldots, x_m)$. $\square$

If $n \geq 2md$, then Lemma 5.3.4.2 provides the following randomized equivalence test for two polynomials.

Algorithm 5.3.4.4. NEQ-POL

Input: Two polynomials $p_1(x_1, \ldots, x_m)$ and $p_2(x_1, \ldots, x_m)$ over $\mathbb{Z}_n$ with at most degree d, where n is a prime and $n > 2dm$.

Step 1: Choose uniformly $a_1, a_2, \ldots, a_m \in \mathbb{Z}_n$ at random.

Step 2: Evaluate $I := p_1(a_1, a_2, \ldots, a_m) - p_2(a_1, a_2, \ldots, a_m)$.

Step 3: **if** $I \neq 0$ **then output**($p_1 \not\equiv p_2$) {accept}
else **output**($p_1 \equiv p_2$) {reject}.

Observe that NEQ-POL is an application of fingerprinting because in fact we test whether the fingerprint $p_1(a_1, \ldots, a_n)$ of p_1 is identical to the fingerprint $p_2(a_1, \ldots, a_2)$ of p_2 for random $a_1, \ldots, a_n$.

Theorem 5.3.4.5. *Algorithm NEQ-POL is a polynomial time one-sided-error Monte Carlo algorithm that decides the nonequivalence of two polynomials.*

Proof. Since the only computation part of NEQ-POL is the evaluation of a polynomial in Step 2, it is obvious that NEQ-POL is a polynomial-time algorithm.

If $p_1 \equiv p_2$, then $p_1(a_1, \ldots, a_m) = p_2(a_1, \ldots, a_m)$ for all $a_1, a_2, \ldots, a_m$ from $\mathbb{Z}_n$ and so $I = 0$. So,

$$Prob(\text{NEQ-POL rejects } (p_1, p_2)) = 1.$$

If $p_1 \not\equiv p_2$, then $p_1(x_1, \ldots, x_m) - p_2(x_1, \ldots, x_m)$ is a nonzero polynomial. Following Lemma 5.3.4.2 and the fact $n \geq 2dm$,

$$Prob\left(p_1(a_1, a_2, \ldots, a_m) - p_2(a_1, a_2, \ldots, a_m) = 0\right) \underset{\text{Lemma 5.3.4.2}}{\leq} m \cdot d/n \leq \frac{1}{2}.$$

Thus,

$$Prob(\text{NEQ-POL accepts } (p_1, p_2)) =$$
$$Prob(p_1(a_1, \ldots, a_m) - p_2(a_1, \ldots, a_m) \neq 0) \geq 1 - \frac{m \cdot d}{n} \geq \frac{1}{2}.$$

$\square$

Observe, that NEQ-POL can be also viewed as a two-sided-error Monte Carlo algorithm for the problem of the equivalence of two polynomials.

Now we move back to our primary decision problem about the equivalence of two one-time-only branching programs (1BPs). The idea used above for polynomials cannot be straightforwardly applied to the equivalence of branching programs. Consider two 1BPs A_1 and A_2 over the set of variables $\{x_1, \ldots, x_m\}$ that compute two Boolean functions f_1 and f_2 respectively. If f_1 and f_2 differ on only one of the 2^m inputs, then a random choice of the values of $x_1, \ldots, x_m$ can distinguish A_1 and A_2 only with the probability $1/2^m$. So, if the polynomials p_1 and p_2 would be exchanged by the 1BPs A_1 and A_2 in NEQ-POL, then it could happen that the answer "equivalent" is given with the high probability of $1 - 1/2^m$ despite the fact that A_1 and A_2 compute different functions.

The way to overcome this difficulty is to construct a polynomial for every 1BP, and then to test the equivalence of the constructed polynomials instead of the original 1BPs. The kernel of this idea is that the polynomial is over $\mathbb{Z}_p$ for some sufficiently large prime p and that it is equivalent with the given 1BP on all Boolean inputs.

Now we describe the construction of the polynomial $p_A(x_1, \ldots, x_m)$ to a given 1BP A. The idea is to assign a polynomial to every node of A and to every edge of A by the following rules:

(i) The constant polynomial 1 is assigned to the start node.
(ii) If a node v is labelled by the variable x, and the polynomial assigned to v is $p_v(x_1, \ldots, x_m)$, then the polynomial $x \cdot p_v(x_1, \ldots, x_m)$ is assigned to the outgoing edge labelled by 1, and the polynomial $(1 - x) \cdot p_v(x_1, \ldots, x_m)$ is assigned to the outgoing edge labelled by 0.

(iii) If all incoming edges of a node u have already assigned polynomials, then the sum of all these polynomials is assigned to u.

(iv) $p_A(x_1, \ldots, x_m)$ is the polynomial assigned to the sink that is labelled by "1".

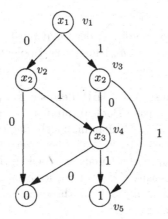

Fig. 5.2.

Example 5.3.4.6. Consider the 1BP A depicted in Figure 5.2. A computes the Boolean function $f_A : \{0,1\}^3 \to \{0,1\}$, such that $f_A(a_1, a_2, a_3) = 1$ if and only if $a_1 + a_2 + a_3 \geq 2$. Following our construction we have

$$p_{v_1}(x_1, x_2, x_3) = 1$$
$$p_{v_2}(x_1, x_2, x_3) = 1 \cdot (1 - x_1) = 1 - x_1$$
$$p_{v_3}(x_1, x_2, x_3) = 1 \cdot x_1 = x_1$$
$$p_{v_4}(x_1, x_2, x_3) = x_2 \cdot p_{v_2}(x_1, x_2, x_3) + (1 - x_2) \cdot p_{v_3}(x_1, x_2, x_3)$$
$$= x_2 \cdot (1 - x_1) + (1 - x_2) \cdot x_1$$
$$p_A(x_1, x_2, x_3) = p_{v_5}(x_1, x_2, x_3) = x_3 \cdot p_{v_4}(x_1, x_2, x_3) + x_2 \cdot p_{v_3}(x_1, x_2, x_3)$$
$$= x_3(x_2 \cdot (1 - x_1) + (1 - x_2) \cdot x_1) + x_2 \cdot x_1.$$

Observe that we can transform $p_A(x_1, x_2, x_3)$ in a form that is very similar to DNF as follows:

$$p_A(x_1, x_2, x_3) = x_3 \cdot x_2 \cdot (1 - x_1) + x_3 \cdot (1 - x_2) \cdot x_1 + x_2 \cdot x_1$$
$$= (1 - x_1) \cdot x_2 \cdot x_3 + x_1 \cdot (1 - x_2) \cdot x_3$$
$$+ (x_3 + (1 - x_3)) \cdot x_2 \cdot x_1$$
$$= (1 - x_1) \cdot x_2 \cdot x_3 + x_1 \cdot (1 - x_2) \cdot x_3$$
$$+ x_1 \cdot x_2 \cdot x_3 + x_1 \cdot x_2 \cdot (1 - x_3).$$

As in the case of DNF the four elementary "conjunctions" $(1 - x_1)x_2x_3$, $x_1(1 - x_2)x_3$, $x_1x_2x_3$ and $x_1x_2(1 - x_3)$ correspond exactly to the four input assignments $(0,1,1)$, $(1,0,1)$, $(1,1,1)$, and $(1,1,0)$, respectively. Obviously, these are exactly those assignments for which A evaluates the output "1". □

Observation 5.3.4.7. For every 1BP A over the set of variables $\{x_1, x_2, \ldots, x_m\}$,

(i) $p_A(x_1, \ldots, x_m)$ is a polynomial of degree at most 1 for every variable,
(ii) $p_A(a_1, \ldots, a_m) = A(a_1, \ldots, a_m)$ for every Boolean input $(a_1, \ldots, a_m) \in \{0, 1\}^m$.

Proof. (i) is obvious because A is a 1BP. Let us prove (ii). Since A is a deterministic computing model, each Boolean input $a = a_1, a_2, \ldots, a_m \in \{0, 1\}^m$ with $A(a_1, a_2, \ldots, a_m) = 1$ [$= 0$] unambiguously determines a path P_A from the source to the sink labeled by "1" ["0"]. For every Boolean assignment to the variables $x_1, \ldots, x_m$, all polynomials assigned to the nodes of A evaluate a Boolean value. One can easily observe that, for every Boolean assignment a, the only polynomials evaluating to 1 are those on the computation path P_A that corresponds to this assignment. □

Lemma 5.3.4.8. *For every two 1BPs A and B,*
    A and B are equivalent if and only if p_A and p_B are identical.

Proof. To see this we transform every polynomial of degree at most 1 into a special "normal" form similar to DNF for the representation of Boolean functions.[59] This normal form is the sum of "elementary multiplications" $y_1y_2 \ldots y_m$, where either $y_i = x_i$ or $y_i = (1 - x_i)$ for every $i = 1, 2, \ldots, m$. Obviously, two polynomials of degree 1 are equivalent if and only if they have their normal forms identical. Moreover, every elementary multiplication of this normal form corresponds to one input assignment on which the corresponding 1BP computes "1". So, A and B are equivalent if and only if the normal forms of p_A and p_B are identical.

It remains to show that one can unambiguously assign the normal form to every polynomial[60] p_A of degree 1. First, one applies the distributive rules to get a sum of elementary multiplications. If an elementary multiplication y does not contain a variable x, then we exchange y by two elementary multiplications $x \cdot y$ and $(1 - x) \cdot y$. Obviously, an iterative application of this rule results in the normal form. □

Now we are ready to present our randomized algorithm for deciding the nonequivalence of two 1BPs.

---

[59] See the transformation of $p_A(x_1, x_2, x_3)$ in Example 5.3.4.6.
[60] Note that the following way to construct the normal form of p_A does not provide any efficient deterministic algorithm for deciding the equivalence of two 1BPs. The reason for this claim is that the normal form can have an exponential length in the length of the polynomial p_A assigned to the 1BP A.

Algorithm 5.3.4.9. NEQ-1BP

Input: Two 1BPs A and B over the set of variables $\{x_1, x_2, \ldots, x_m\}$, $m \in \mathbb{N}$.

Step 1:   Construct the polynomials p_A and p_B.

Step 2:   Apply the algorithm NEQ-POL on $p_A(x_1, \ldots, x_m)$ and $p_B(x_1, \ldots, x_m)$ over some $\mathbb{Z}_n$, where n is a prime that is larger than $2m$.

Output: The output of NEQ-POL.

Theorem 5.3.4.10. NEQ-1BP *is a polynomial-time one-sided-error Monte Carlo algorithm for the problem of nonequivalence of two 1BPs.*

Proof. The construction of p_A and p_B in Step 1 can be done in a time that is quadratic in the input size (representation of 1BPs). Since NEQ-POL works in polynomial time and the sizes of p_A and p_B (as inputs of NEQ-POL) are polynomial in the size of the input of NEQ-1BP, Step 2 is also done in polynomial time.

Due to Lemma 5.3.4.8 we know that A and B are equivalent if and only if p_A and p_B are equivalent. If A and B are equivalent (i.e., when p_A and p_B are equivalent), then NEQ-POL rejects (p_A, p_B) with a probability of 1, i.e., we have no error on this side. If A and B are not equivalent, then NEQ-POL accepts (p_A, p_B) with the probability of at least

$$1 - m/n.$$

Since $n > 2m$, this probability is at least $1/2$. $\square$

Corollary 5.3.4.11. *Exchanging the output "accept" and "reject" in the algorithm* NEQ-1BP, *one obtains a two-sided-error Monte Carlo algorithm for the problem of equivalence of two 1BPs.*

Exercise 5.3.4.12. Find an efficient implementation of NEQ-POL and analyze precisely its time complexity. $\square$

Summary of Section 5.3.4

Section 5.3.4 introduces an interesting, powerful technique for the application of fingerprinting as a special version of the method of abundance of witnesses for equivalence problems. If one is not able to find enough witnesses for the equivalence (or nonequivalence) of two given objects, then one can transform these objects having meaning in a restricted "world" to a more general world. In our example we have moved from Boolean algebra to finite fields $\mathbb{Z}_n$ for a sufficiently large prime n. Surprisingly, one can find enough witnesses for the nonequivalence of the objects in this generalized context and thus can randomly decide the original problem in the restricted world.

5.3.5 Randomized Optimization Algorithms for MIN-CUT

To give a transparent example of a simple randomized optimization algorithm we consider the MIN-CUT problem for multigraphs.[61] Note that this problem is known to be in P, because one can solve it by $|V|^2$ applications of some network flow algorithm. Since the best deterministic flow algorithm runs in time $O(|V| \cdot |E| \cdot \log(|V|^2/|E|))$ for a network (V, E), one may consider the MIN-CUT problem to be hard in some applications.

In what follows we design efficient randomized algorithms for this optimization task. These algorithms are based on a simple operation called contraction. For a given multigraph G and an edge $e = \{x, y\}$ of G, **contraction(G, e)** replaces the vertices x and y by a new vertex $ver(x, y)$, and every edge $\{r, s\}$ with $r \in \{x, y\}$ and $s \notin \{x, y\}$ is replaced by the edge $\{ver(x, y), s\}$. The edge $\{x, y\}$ is removed (we do not allow any self-loop) and the rest of the graph remains unchanged. The resulting multigraph is denoted by $G/\{e\}$. Figure 5.3 illustrates a sequence of such contractions.

Observe that given a collection of edges $F \subseteq E$, the effect of contracting the edges in F does not depend on the order of contractions. So, we denote the resulting multigraph by G/F. Obviously, each vertex of a contracted multigraph corresponds to a set of vertices of the original multigraph $G = (V, E)$, and the edges on G/F are exactly those edges in E whose endpoints are not collapsed into the same vertex in G/F.

The concept of our first randomized algorithm is very simple. We realize a random sequence of contractions on a given multigraph G until a multigraph of two vertices is reached. The two sets of vertices of G corresponding to these two vertices determine a cut of G. This concept works because of the crucial observation that

"any edge contraction does not reduce the size of a minimal cut in G."

This is obviously true because every cut of G/F corresponds to a cut of G for any F.

Algorithm 5.3.5.1. RANDOM CONTRACTION

Input: A connected multigraph $G = (V, E)$.
Output: a cut (V_1, V_2) of G.
Step 1:   Set, for every $v \in V$, $label(v) = \{v\}$.
Step 2:   **while** G has more than 2 vertices
          do begin choose uniformly at random an edge $e = \{x, y\} \in E(G)$;
                    $G := G/\{e\}$;
                    set $label(z) = label(x) \cup label(y)$ for the new vertex z of G,
                    and replace the edges as described above

---

[61] Remember that the MIN-CUT problem is to find a cut (V_1, V_2) of a multigraph $G = (V, E)$ such that $Edge(V_1, V_2) = E \cap \{\{x, y\} \mid x \in V_1, y \in V_2\}$ has the minimal cardinality.

     end
Step 3: **if** $G = (\{u, v\}, E')$ for a multiset E',
     then output($label(u), label(v)$).

Figure 5.3 illustrates the work of the algorithm RANDOM CONTRAC-TION on a multigraph of 5 vertices. So, after 3 contractions $G/\{\{x, v\}\}$, $(G/\{\{x, v\}\})/\{\{z, y\}\}$, and $(G/\{\{x, v\}, \{z, y\}\})/\{\{ver(x, v), ver(y, z)\}\}$ a cut of three edges is produced.

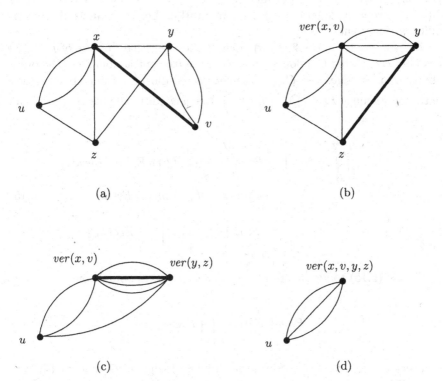

Fig. 5.3. (a)A multigraph G; (b) The multigraph $G/\{\{x, v\}\}$; (c) The multigraph $G/\{\{x, v\}, \{z, y\}\}$; (d) The multigraph $G/\{\{x, v\}, \{z, y\}, \{ver(x, v), ver(y, z)\}\}$.

Theorem 5.3.5.2. RANDOM CONTRACTION *is a polynomial-time random-ized optimization algorithm that finds an optimal cut with a probability of at least* $\frac{2}{n \cdot (n-1)}$ *for any multigraph of n vertices.*

Proof. If one represents multigraphs by giving weights on single edges to ex-press the multiplicity, then one can easily see that RANDOM CONTRACTION can be implemented in $O(n^2)$ time.

  As already mentioned above, it is clear that RANDOM CONTRACTION com-putes a cut of the given multigraph G. We have to prove that the probability of the optimality of this cut is at least $\frac{2}{n \cdot (n-1)}$.

Let $G = (V, E)$ be a multigraph, and let $C_{min} = (V_1, V_2)$ be a minimal cut of G. Let $|E(C_{min})| = k$ be a positive integer.[62]

In what follows we show that the probability that RANDOM CONTRACTION computes this specific minimal cut C_{min} is at least $\frac{2}{n \cdot (n-1)}$. First, we observe that

$$\text{``the total number of edges of } G \text{ is at least } n \cdot k/2 \text{''} \tag{5.14}$$

because $|E(C_{min})| = k$ implies that every vertex of G has at least degree k. Obviously, the algorithm outputs C_{min} if and only if none of the edges of $E(C_{min})$ are contracted during the computation. Let us evaluate the probability of this event.

Let, for $i = 1, \ldots, n - 2$, $Event_i$ denote the event that no edge of $E(C_{min})$ has been contracted (randomly chosen) in the ith contraction step. The probability that no edge of $E(C_{min})$ was contracted during a complete execution of the algorithm is $Prob\left(\bigcap_{i=1}^{n-2} Event_i\right)$. Using conditional probabilities it can be expressed as

$$Prob\left(\bigcap_{i=1}^{n-2} Event_i\right) = Prob(Event_1) \cdot Prob(Event_2 | Event_1)$$

$$\cdot Prob(Event_3 | Event_1 \cap Event_2) \cdot \ldots \tag{5.15}$$

$$\cdot Prob\left(Event_{n-2} \,\middle|\, \bigcap_{j=1}^{n-3} Event_j\right)$$

To use (5.15) to get our result we need some lower bounds on

$$Prob\left(Event_i \,\middle|\, \bigcap_{j=1}^{i-1} Event_j\right)$$

for every $i = 1, 2, \ldots, n - 2$. Since the number of edges of G is at least $n \cdot k/2$ (see (5.14)),

$$Prob(Event_1) = \frac{|E| - |E(C_{min})|}{|E|} = 1 - \frac{k}{|E|} \underset{(5.14)}{\geq} 1 - \frac{k}{k \cdot n/2} = 1 - \frac{2}{n}. \tag{5.16}$$

In general, after $i - 1$ contractions the corresponding graph G/F_i has exactly $n - i + 1$ vertices. If $F_i \cap E(C_{min}) = \emptyset$[63], then C_{min} is a minimal cut of G/F_i, too. But this implies (see (5.14)) that every vertex of G/F_i has at

---

[62] $k \geq 1$ if and only if G is connected. But if G would be not connected, the contraction method would find the components of G and so all the minimal cuts, too.

[63] If none of the edges of C_{min} have been contracted in the first i steps, i.e., when $\bigcap_{j=1}^{i-1} Event_j$ happens.

least degree $k = |E(C_{min})|$ and so G/F_i has at least $k \cdot (n - i + 1)/2$ edges. Thus, for $i = 2, \ldots, n - 1$

$$Prob \left[Event_i \mid \bigcap_{j=1}^{i-1} Event_j \right] \geq \frac{|E(G/F_i)| - |E(C_{min})|}{|E(G/F_i)|} \tag{5.17}$$

$$\geq 1 - \frac{k}{k(n - i + 1)/2} = 1 - \frac{2}{(n - i + 1)}.$$

Inserting (5.16) and (5.17) into (5.15) we get

$$Prob \left(\bigcap_{i=1}^{n-2} Event_i \right) \geq \prod_{i=1}^{n-2} \left(1 - \frac{2}{n - i + 1} \right) =$$

$$\prod_{l=n}^{3} \left(\frac{l - 2}{l} \right) = \frac{1}{\binom{n}{2}} = \frac{2}{n \cdot (n - 1)}.$$

$\square$

Exercise 5.3.5.3. Modify the algorithm RANDOM CONTRACTION by the following change in the random choice. Instead of randomly choosing an edge, randomly choose two vertices x and y and identify them into one vertex. Prove that for some multigraphs, the probability that this modified algorithm finds a minimal cut is exponentially small in n.　$\square$

Theorem 5.3.5.2 shows that the probability of discovering a particular minimal cut is at least

$$\frac{2}{n(n - 1)} > \frac{2}{n^2}.$$

Following the idea of Section 5.2 to repeat the algorithm several times and to take the best output from all outputs produced, we obtain after $n^2/2$ independent runs of RANDOM CONTRACTION a minimal cut with a probability of at least

$$1 - \left(1 - \frac{2}{n^2} \right)^{n^2/2} > 1 - \frac{1}{e}.$$

The trouble is that the time complexity of $n^2/2$ repetitions of the algorithm RANDOM CONTRACTION is $O(n^4)$, which is not better than the $O(n^3)$ complexity of the best known deterministic algorithm. This is a reason to try to improve our randomized algorithm. Following the proof of Theorem 5.3.5.2 we see that the first contraction involves an edge from $E(C_{min})$ with small probabilities $\frac{2}{n}, \frac{2}{n-1}, \frac{2}{n-2}, \ldots$. The key observation is that these probabilities grow and may be very large (even $2/3$ in the last contraction). The first natural idea could be to use RANDOM CONTRACTION to reduce G to some G/F of l vertices and then to compute a minimal cut of G/F by the best known deterministic algorithm. In what follows we show that this approach beats the time complexity $O(n^3)$ of the best deterministic algorithm. Let $l : \mathbb{N} \to \mathbb{N}$ be a monotone function such that $1 \leq l(n) \leq n$ for every $n \in \mathbb{N}$.

Algorithm 5.3.5.4. l-COMB-CONTRACT

Input: a multigraph $G = (V, E)$ on n vertices, $n \in \mathbb{N}$.
Output: a cut (V_1, V_2) of G.
Step 1: Apply RANDOM CONTRACTION on G in order to get a multigraph G/F of $l(n)$ vertices.
Step 2: Apply a deterministic algorithm to compute a minimal cut of G/F. Output the corresponding cut of G.

Theorem 5.3.5.5. *For any function* $l : \mathbb{N} \to \mathbb{N}$, $1 \leq l(n) \leq n$, *the randomized algorithm* l-COMB-CONTRACT *works in time* $O(n^2 + (l(n))^3)$, *and it finds a minimal cut with a probability of at least*

$$\binom{l(n)}{2} \bigg/ \binom{n}{2}.$$

Proof. Step 1 consists of $n - l(n)$ contractions and so it can be realized in time $(n - l(n)) \cdot O(n) = O(n^2)$. Step 2 computes a minimal cut of a multigraph G/F of $l(n)$ vertices. Using the best known deterministic algorithm, this can be done in $O((l(n))^3)$ time.

Let us consider a fixed minimal cut C_{min} as in the proof of Theorem 5.3.5.2. Obviously, if G/F contains C_{min}, then a minimal cut of G is computed in Step 2. Thus, it is sufficient to give a lower bound on the probability that G/F contains the specific cut C_{min}. Following the proof of Theorem 5.3.5.2, we obtain that this is exactly $Prob\left(\bigcap_{i=1}^{n-l(n)} Event_i\right)$. Using (5.15), (5.16), and (5.17) we obtain

$$Prob\left(\bigcap_{i=1}^{n-l(n)} Event_i\right) \geq \prod_{i=1}^{n-l(n)} \left(1 - \frac{2}{n-i+1}\right)$$

$$= \prod_{i=1}^{n-2} \left(1 - \frac{2}{n-i+1}\right) \bigg/ \prod_{j=n-l(n)+1}^{n-2} \left(1 - \frac{2}{n-j+1}\right)$$

$$= \frac{1}{\binom{n}{2}} \bigg/ \frac{1}{\binom{l(n)}{2}} = \frac{\binom{l(n)}{2}}{\binom{n}{2}}.$$

□

Corollary 5.3.5.6. *For every* $l : \mathbb{N} \to \mathbb{N}$, $1 \leq l(n) \leq n$, $\frac{n^2}{(l(n))^2}$ *repetitions of* l-COMB-CONTRACT *provide an optimal cut with probability at least* $1 - 1/e$.

Since $\frac{n^2}{(l(n))^2}$ repetitions of l-COMB-CONTRACT lead to a time complexity of

$$O\left((n^2 + (l(n))^3) \cdot \frac{n^2}{(l(n))^2}\right) = O\left(\frac{n^4}{(l(n))^2} + n^2 \cdot l(n)\right),$$

the best choice for l is $l(n) = \lfloor n^{2/3} \rfloor$. In this case $\lfloor n^{2/3} \rfloor$-COMB-CONTRACT works in time $O\left(n^{8/3}\right)$ and provides a minimal cut with a probability greater than $1/2$. If one takes $\log_2 n$ repetitions of $\lfloor n^{2/3} \rfloor$-COMB-CONTRACT, then an optimal solution is provided with a probability of at least $1 - 1/n$ in time $O(n^{8/3} \log_2 n)$. So, this randomized algorithm has a high probability of success, and it is faster than the best known deterministic one.

Finally, we show that there is another possibility to improve the algorithm RANDOM CONTRACTION that leads to a randomized algorithm running in $O((n \log_2 n)^2)$ time. The basic idea is the same as for l-COMB-CONTRACT. The probability of randomly choosing an edge from $E(C_{min})$ grows with the number of contractions already realized. In l-COMB-CONTRACT we solved this problem by moving to a deterministic optimization after a number of random contractions. In our new algorithm RRC (recursive random contraction) we increase the number of independent runs of RANDOM CONTRACTION with the decrease of the size of the actual multigraph. More precisely, $O(n^2)$ independent runs of RANDOM CONTRACTION are needed to get a reasonable probability of its success.[64] But there is no reason to start with $O(n^2)$ copies of the input at the very beginning because the probability of choosing an edge from $E(C_{min})$ there is very small. So, we start with two copies of the input multigraph only and reduce it randomly to two multigraphs G/F_1 and G/F_2 of size roughly $n/\sqrt{2}$. Then we make two copies of G/F_1 and G/F_2 each and use the algorithm RANDOM CONTRACTION on each of the four multigraphs to reduce it to a multigraph of size roughly $n/2$. After that we again make two copies of each of the four multigraphs and randomly reduce each of the eight multigraphs to a size of roughly $n/(2 \cdot \sqrt{2})$, etc. Since we reduce the size by a factor of $1/\sqrt{2}$ in each stage, the number of stages before reaching size 2 is approximately $2 \cdot \log_2 n$. So, at the end we get $O(n^2)$ cuts and we take the best one of them. As we shall show later, since we work with a few large multigraphs and with many copies of very small multigraphs, the time complexity of our approach is asymptotically comparable with $O(\log_2 n)$ independent complete runs for the original algorithm RANDOM CONTRACTION.

In what follows we present our algorithm RRC as a recursive algorithm.

Algorithm 5.3.5.7. RRC(G)

Input: A multigraph $G = (V, E)$, $|V| = n$.
Output: A cut (V_1, V_2) of G.
Procedure: **if** $n \leq 6$ **then** compute a minimal cut of G by a deterministic method.
    else begin $h := \lceil 1 + n/\sqrt{2} \rceil$;
                realize two independent runs of RANDOM CONTRACTION on G in order to obtain multigraphs G/F_1 and G/F_2 of the size h;
                RRC(G/F_1);

---

[64] Obviously, this increases the time complexity by the multiplicative factor $O(n^2)$.

$$\mathrm{RRC}(G/F_2);$$

return the smaller of the two cuts produced by $\mathrm{RRC}(G/F_1)$
and $\mathrm{RRC}(G/F_2)$.

end.

Theorem 5.3.5.8. *The algorithm* RRC *works in* $O(n^2 \log_2 n)$ *time and finds a minimal cut with a probability of at least*

$$\frac{1}{\Omega (\log_2 n)}.$$

Proof. First, we analyze the time complexity $Time_{\mathrm{RRC}}$ of RRC. Since the size of a multigraph is reduced by the multiplicative factor $1/\sqrt{2}$ in each stage of the algorithm, the number of recursion calls is bounded by $\log_{\sqrt{2}} n = O(\log_2 n)$. Since the original algorithm RANDOM CONTRACTION works in time $O(n^2)$ on any multigraph of n vertices, we can roughly bound the time of reducing a multigraph of size m to a multigraph of size $\left\lceil 1 + \frac{m}{\sqrt{2}} \right\rceil$ by $O(m^2)$ too. So, we obtain the following recursion for $Time_{\mathrm{RRC}}$:

$$Time_{\mathrm{RRC}}(n) = 2 \cdot Time_{\mathrm{RRC}} \left(\left\lceil 1 + \frac{n}{\sqrt{2}} \right\rceil \right) + O(n^2).$$

One can easily check that $Time_{\mathrm{RRC}}(n) = O(n^2 \log_2 n)$.

Now we give a lower bound on the probability that RRC succeeded. Assume that C_{min} is a minimal cut of a given multigraph G, and $|E(C_{min})| = k$.

Now assume we have such a multigraph G/F of size l that G/F still contains C_{min}.[65] In the next stage two copies of G/F will be independently reduced to some multigraphs G/F_1 and G/F_2 of size $\lceil 1+l/\sqrt{2} \rceil$ each. What is the probability p_l that G/F_1 still contains C_{min}? Using the same consideration as in the proof of Theorem 5.3.5.2 we get

$$p_l \geq \frac{\binom{\lceil 1+l/\sqrt{2} \rceil}{2}}{\binom{l}{2}} = \frac{\left[1 + l/\sqrt{2}\right] \cdot \left(\lceil 1 + l/\sqrt{2} \rceil - 1\right)}{l \cdot (l-1)} \geq \frac{1}{2}.$$

Let $Pr(n)$ be the probability that RRC finds a fixed minimal cut of a given multigraph of n vertices. Then $p_l \cdot Pr(\lceil 1 + l/\sqrt{2} \rceil)$ is a lower bound on the conditional probability that RRC finds C_{min} by the reduction from G/F to G/F_1 and then by the following recursive call of RRC on G/F_1, if G/F has contained C_{min}. Thus,

$$\left(1 - p_l \cdot Pr\left(\lceil 1 + l/\sqrt{2} \rceil\right)\right)^2$$

is the upper bound on the probability that RRC does not compute C_{min} assuming G/F still contained C_{min}.[66] So, we obtain the following recursion for Pr:

---

[65] That is, no edge of $E(C_{min})$ has been contracted up till now.

[66] This is because RRC independently runs on two copies of G/F (via G/F_1 and G/F_2).

$$Pr(l) \geq 1 - \left(1 - p_l Pr\left(\left\lceil 1 + l/\sqrt{2}\right\rceil\right)\right)^2$$

$$\geq 1 - \left(1 - \frac{1}{2}Pr\left(\left\lceil 1 + l/\sqrt{2}\right\rceil\right)\right)^2. \tag{5.18}$$

One can easily verify that every function satisfying this recursion must be from $1/\Theta(\log_2 n)$. □

Exercise 5.3.5.9. Realize a detailed analysis of the recursion of (5.18) and show

$$\frac{1}{c_1 \cdot \log_2 n} \leq Pr(n) \leq \frac{1}{c_2 \cdot \log_2 n}$$

for some specific constants c_1, c_2 with a small difference $c_1 - c_2$. □

Exercise 5.3.5.10. Analyze the time complexity and the success probability of RRC if the reduction stages are changed to:

(i) l to $\lfloor l/2\rfloor$,
(ii) l to $\lfloor\sqrt{l}\rfloor$,
(iii) l to $l - \sqrt{l}$.

□

Obviously, $O(\log_2 n)$ independent runs of RRC on a multigraph of n vertices yield

(i) a time complexity $O(n^2 \cdot (\log_2 n)^2)$, and
(ii) a probability of computing a minimal cut of at least $1/2$.

So, this algorithm is essentially faster than the best deterministic algorithm for MIN-CUT and it is successful with a reasonable probability. $O((\log_2 n)^2)$ runs of RRC provide a minimal cut with a probability of at least $1 - 1/n$ and the time complexity grows to $O(n^2 \cdot (\log_2 n)^3)$ in this case only.

Summary of Section 5.3.5

One can design a randomized optimization algorithm by starting with a randomized polynomial time algorithm that finds an optimal solution with a small probability $1/p(n)$, where p is a polynomial. A naive possibility of increasing the probability of finding an optimum is to execute many independent runs of the algorithm. If the probability of computing an optimum decreases differently in different parts of the computation, one can consider one of the following two concepts in order to increase the probability of success more efficiently than by the naive repetition approach:

(i) Execute the parts of the randomized computation where the probability of success decreases drastically in a completely deterministic way. This approach may be successful if the "derandomization" of these parts can be efficiently done.

(ii) Do not execute a lot of complete independent runs of the algorithm on the same input. Prefer to execute many independent runs of the computation parts in which the probability of success drastically decreases, and only a few independent runs of the computation parts, where the decrease of the probability of success is almost negligible. This approach is especially efficient if the computation parts essentially decreasing the success probability are short.

For the MIN-CUT problem both approaches (i) and (ii) result in randomized optimization algorithms that are more efficient than the naive repetition approach.

5.3.6 MAX-SAT and Random Rounding

In this section we consider the MAX-SAT optimization problem. We consider two different randomized methods – random sampling and relaxation to linear programming with random rounding – for solving this problem. We shall see that both these methods yield randomized approximation algorithms for MAX-SAT, and that these methods are incomparable in the sense that, for some input instances of MAX-SAT, random sampling provides better results than random rounding and for another input instances random rounding is better than random sampling. So, the final idea is to let both methods run independently on any input instance and to take the best of the two outputs. We shall show that this combined algorithm is a randomized $(4/3)$-expected approximation algorithm for MAX-SAT.

First, remember Example 2.2.5.25, where we already showed that the following algorithm is a randomized 2-expected approximation algorithm for MAX-SAT.

Algorithm 5.3.6.1. RSMS (RANDOM SAMPLING FOR MAX-SAT)

> Input: A Boolean formula Φ over the set of variables $\{x_1, \ldots, x_n\}$, $n \in \mathbb{N}$.
> Step 1: Choose uniformly at random $\alpha_1, \ldots, \alpha_n \in \{0, 1\}$.
> Step 2: **output**$(\alpha_1, \ldots, \alpha_n)$.
> Output: an assignment to $\{x_1, \ldots, x_n\}$.

Let us repeat the argument why RSMS is a randomized 2-expected approximation algorithm for MAX-SAT.[67] Let Φ be a formula in CNF consisting of m clauses. Without loss of generality we may assume that no clause of Φ contains any constant or two occurrences of the same variable in its set of literals. For every clause F of k different literals, the probability that a random assignment satisfies F is $1 - 2^{-k}$. Since $1 - 2^{-k} \geq 1/2$ for every positive integer k, the expected number of satisfied clauses is at least $m/2$.

Since Φ has m clauses (i.e., at most m clauses can be satisfied due to an optimal assignment), RSMS is a randomized 2-expected approximation algorithm for MAX-SAT. If one considers only CNFs whose clauses consist of at least k different nonconstant literals, then the expected number of satisfied

---

[67] The thorough analysis of RSMS was given in Example 2.2.5.25.

clauses by a random assignment is $m \cdot \left(1 - 1/2^k\right)$ and so RSMS is a randomized $(2^k/(2^k - 1))$-expected approximation algorithm. We see that RSMS is very good for CNFs with long clauses, but probably not for CNFs with short clauses.

Our goal is to design a randomized approximation algorithm that would be good for CNFs having short clauses. To realize this goal we use the method of relaxation with random rounding. Roughly it means here that we formulate MAX-SAT as an integer linear program with Boolean variables, but we solve it by relaxing the variables to take the values from the real interval $[0, 1]$. If the computed optimal solution of the integer linear program assigns a value α_i to a variable x_i, then we finally assign 1 [0] to x_i with probability α_i $[1 - \alpha_i]$. So, the only difference between the algorithm RSMS and this approach is that the assignment to the variables of Φ is taken uniformly at random in RSMS and that the assignment to the variables of Φ in our new approach is taken randomly according to the probability precomputed by solving the corresponding linear programming problem.

The reduction of an input instance of MAX-SAT to an input instance of LIN-P can be described as follows. Let $\Phi = F_1 \wedge F_2 \wedge \cdots \wedge F_m$ be a formula over the set of variables $\{x_1, \ldots, x_n\}$ where F_i is a clause for $i = 1, \ldots, m$. Let $Set(F_i)$ be the set of literals of F_i. Let $Set^+(F_i)$ be the set of variables that appear uncomplemented in F_i, and let $Set^-(F_i)$ be the set of variables that appear complemented if F_i. Let $In^+(F_i)$ and $In^-(F_i)$, respectively, be the set of indices of variables in $Set^+(F_i)$ and $Set^-(F_i)$, respectively. Then the input instance $LP(\Phi)$ of LIN-P corresponding to Φ can be formulated as follows:

$$\text{maximize } \sum_{j=1}^{m} z_j$$

$$\text{subject to } \sum_{i \in In^+(F_j)} y_i + \sum_{i \in In^-(F_j)} (1 - y_i) \geq z_j \ \forall j \in \{1, \ldots, m\} \quad (5.19)$$

$$\text{where } y_i, z_j \in \{0, 1\} \text{ for all } i \in \{1, \ldots, n\}, j \in \{1, \ldots, m\}. \quad (5.20)$$

Observe that z_j may take the value 1 only if one of the variables in $Set^+(F_j)$ takes the value 1 or one of the variables in $Set^-(F_j)$ takes the value 0, i.e., only if F_j is satisfied. So, the objective function $\sum_{j=1}^{m} z_j$ counts the number of satisfied clauses.

Now consider the relaxed version of $LP(\Phi)$, where (5.20) is exchanged for

$$\text{where } y_i, z_j \in [0, 1] \text{ for all } i \in \{1, \ldots, n\}, j \in \{1, \ldots, m\}. \quad (5.21)$$

Let $\alpha(u)$ denote the value of u in an optimal solution of the relaxed $LP(\Phi)$ for every $u \in \{y_1, \ldots, y_n, z_1, \ldots, z_m\}$. We observe that

$$\sum_{j=1}^{m} \alpha(z_j) \text{ bounds from above the number of clauses that can be simultaneously satisfied in } \Phi \quad (5.22)$$

as the value of the objective function for the relaxed LP(Φ) bounds from above the value of the objective function for LP(Φ).

Now we are ready to present our algorithm based on the relaxation with random rounding.

Algorithm 5.3.6.2. RRRMS (RELAXATION WITH RANDOM ROUNDING FOR MAX-SAT)

Input: A formula $\Phi = F_1 \wedge F_2 \wedge \cdots \wedge F_m$ over $X = \{x_1, \ldots, x_n\}$ in CNF, $n, m \in \mathbb{N}$.

Step 1:   Formulate the MAX-SAT problem for Φ as the integer linear program LP(Φ) maximizing $\sum_{j=1}^{m} z_j$ by the constraints (5.19) and (5.20).

Step 2:   Solve the relaxed version of LP(Φ) according to (5.21). Let $\alpha(z_1)$, $\alpha(z_2)$, ..., $\alpha(z_m)$, $\alpha(y_1)$, ..., $\alpha(y_n) \in [0, 1]$ be an optimal solution of the relaxed LP(Φ).

Step 3:   Choose n values $\gamma_1, \ldots, \gamma_n$ uniformly at random from $[0, 1]$.
           for $i = 1$ **to** n **do**
                if $\gamma_i \in [0, \alpha(y_i)]$ **then** set $x_i = 1$
                else set $x_i = 0$
           {Observe that Step 3 realizes the random choice of the value 1 for x_i with the probability $\alpha(y_i)$.}

Output: An assignment to X.

Our first goal is now to prove that the expected number of clauses satisfied by RRRMS is at least $(1 - 1/e) \cdot \sum_{j=1}^{m} \alpha(z_j)$, i.e., $(1 - 1/e)$ times the maximum number of clauses that can be satisfied. To achieve this goal we present the following lemma, which provides even a stronger assertion than the lower bound $(1 - 1/e) \cdot \sum_{j=1}^{m} \alpha(z_j)$ on the number of satisfied clauses.

Lemma 5.3.6.3. *Let k be a positive integer, and let F_j be a clause of Φ with k literals. Let $\alpha(y_1), \ldots, \alpha(y_n), \alpha(z_1), \ldots, \alpha(z_m)$ be the solution of LP(Φ) by RRRMS. The probability that the assignment computed by the algorithm RRRMS satisfies F_j is at least*

$$\left(1 - \left(1 - \frac{1}{k} \right)^k \right) \cdot \alpha(z_j).$$

Proof. Since one considers the clause F_j independently from other clauses, one can assume without loss of generality that it contains only uncomplemented variables and that it is of the form $x_1 \vee x_2 \vee \cdots \vee x_k$. By the constraint (5.19) of LP(Φ) we have

$$y_1 + y_2 + \cdots + y_k \geq z_j. \tag{5.23}$$

The clause F_j remains unsatisfied if and only if all of the variables $y_1, y_2, \ldots, y_k$ are set to zero. Following Step 3 of RRRMS and the fact that each variable is rounded independently, this occurs with probability

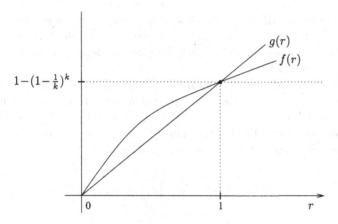

Fig. 5.4.

$$\prod_{i=1}^{k} (1 - \alpha(y_i)).$$

So, F_j is satisfied by the output of RRRMS with probability

$$1 - \prod_{i=1}^{k} (1 - \alpha(y_i)). \tag{5.24}$$

Under the constraint (5.23), (5.24) is minimized when $\alpha(y_i) = \alpha(z_j)/k$ for all $i = 1, \ldots, k$. Thus,

$$Prob(F_j \text{ is satisfied}) \geq 1 - \prod_{i=1}^{k} (1 - \alpha(z_j)/k). \tag{5.25}$$

To complete the proof it suffices to show, for every positive integer k, that

$$f(r) = 1 - (1 - r/k)^k \geq \left(1 - \left(1 - \frac{1}{k}\right)^k\right) \cdot r = g(r) \tag{5.26}$$

for every $r \in [0, 1]$ (and so for every $\alpha(z_j)$). Since f is a concave function in r, and g is a linear function in r (Fig. 5.4), it suffices to verify the inequality at the endpoints $r = 0$ and $r = 1$. Since $f(0) = 0 = g(0)$ and $f(1) = 1 - (1 - 1/k)^k = g(1)$, the inequality (5.26) holds. Setting $r = \alpha(z_j)$ in (5.26) and inserting (5.26) into (5.25) the proof is done. □

Applying Lemma 5.3.6.3 we obtain the following theorem.

Theorem 5.3.6.4. *The algorithm RRRMS is a polynomial-time randomized $(e/(e-1))$-expected approximation algorithm for* MAX-SAT *and a polynomial time randomized $(k^k/(k^k - (k-1)^k))$-expected approximation algorithm for* MAX-EkSAT.

Proof. The fact that RRRMS can be efficiently implemented is obvious. All steps but Step 2 can be realized in linear time. Step 2 corresponds to solving a problem instance of linear programming. This can be done by a polynomial time deterministic algorithm.

To prove that RRRMS is a randomized δ-expected approximation algorithm, it is sufficient to prove that the expected number of satisfied clauses by the output of RRRMS is $\sum_{i=1}^{m} \alpha(z_i)/\delta$. Let, for $i = 1, \ldots, m$, Z_i be the random variable that is equal to 1 if the clause F_i is satisfied and that is equal to 0 if the clause F_i is not satisfied. Lemma 5.3.6.3 claims that

$$E[Z_i] \geq \left(1 - \left(1 - \frac{1}{k}\right)^k\right) \alpha(z_i) \qquad (5.27)$$

if F_i consists of k different literals. Set $Z = \sum_{i=1}^{m} Z_i$. Obviously, Z is the random variable counting the number of satisfied clauses. If all clauses consist of k literals, then from the linearity of expectation we obtain

$$E[Z] = \sum_{i=1}^{m} E[Z_i] \underset{(5.27)}{\geq} \sum_{i=1}^{m} \left(1 - \left(1 - \frac{1}{k}\right)^k\right) \cdot \alpha(z_i)$$
$$\geq \left(1 - \left(1 - \frac{1}{k}\right)^k\right) \cdot \sum_{i=1}^{m} \alpha(z_i).$$

Since an optimal assignment cannot satisfy more than $\sum_{i=1}^{m} \alpha(z_i)$ clauses, the algorithm RRRMS is a $(1/(1-(1-1/k)^k))$-expected approximation algorithm. Since

$$\left(1 - \left(1 - \frac{1}{k}\right)^k\right)^{-1} = \frac{k^k}{k^k - (k-1)^k}$$

the part of the proof of MAXEkSAT is done.

Since $\left(1 - (1 - 1/k)^k\right) \geq (1 - 1/e)$ for every positive integer k, we obtain

$$E[Z] \geq \left(1 - \frac{1}{e}\right) \cdot \sum_{i=1}^{m} \alpha(z_i)$$

for every CNF Φ. So, the algorithm RRRMS is a randomized $(e/(e-1))$-expected approximation algorithm for MAX-SAT. $\square$

Observe that the algorithm RRRMS has the following nice property. One can increase the probability of getting an assignment to X with at least $E(Z)$ satisfied clauses without repeating several complete runs of RRRMS. If the computationally most involved Step 2 is once executed, then it suffices to take several independent random choices of $\gamma_1, \ldots, \gamma_n$ in Step 3. This is because Step 2 provides the probability distribution for the choice of $\gamma_1, \ldots, \gamma_n$ and we do not need to compute it again when we want to take another random choice.

In general the algorithm RRRMS is better for MAX-SAT than the algorithm RSMS because $2 > e/(e-1)$. But if one looks at the behavior of these algorithms more carefully, then one sees that the RSMS is better for problem instances with long clauses. For instance, for the MAX-EkSAT problems, RSMS provides $2^k/(2^k - 1)$ approximation ratio. Thus, a very natural idea is to combine both these algorithms in one and to take as output the best one from the two assignments computed.

Algorithm 5.3.6.5. COMB

Input: A formula Φ over X in CNF.
Step 1:   Compute an assignment β for X by the algorithm RSMS.
Step 2:   Compute an assignment γ for X by the algorithm RRRMS.
Step 3:   if the number of satisfied clauses by β is greater than the number of satisfied clauses by γ
             then output(β)
             else output(γ).

Theorem 5.3.6.6. *The algorithm* COMB *is a polynomial-time randomized* $(4/3)$-*expected approximation algorithm for* MAX-SAT.

Proof. Let U be the random variable that counts the number of clauses satisfied by the output of COMB. Let Y be the random variable counting the number of satisfied assignments by the output of the algorithm RSMS, and let Z be the random variable counting the number of satisfied assignments by the output of the algorithm RRRMS. We set $U = \max\{Y, Z\}$ and so $E[U] \geq \max\{E[Y], E[Z]\}$. Since any optimal solution cannot satisfy more than $\sum_{j=1}^{m} \alpha(z_j)$ clauses, it is sufficient to prove that

$$\max\{E[Y], E[Z]\} \geq \frac{3}{4} \sum_{j=1}^{m} \alpha(z_j).$$

We do this it by proving that

$$\frac{E[Y] + E[Z]}{2} \geq \frac{3}{4} \sum_{j=1}^{m} \alpha(z_j).$$

Let $C(k)$ be the set of clauses of Φ with exactly k literals for every $k \geq 1$. Applying Lemma 5.3.6.3 we have

$$E[Z] \geq \sum_{k \geq 1} \sum_{F_j \in C(k)} \left(1 - \left(1 - \frac{1}{k}\right)^k\right) \cdot \alpha(z_j). \tag{5.28}$$

From the analysis of the behavior of RSMS and from the fact $\alpha(z_j) \in [0, 1]$ we have

$$E[Y] = \sum_{k \geq 1} \sum_{F_j \in C(k)} \left(1 - \frac{1}{2^k}\right) \geq \sum_{k \geq 1} \sum_{F_j \in C(k)} \left(1 - \frac{1}{2^k}\right) \cdot \alpha(z_j). \quad (5.29)$$

Thus,

$$\frac{E[Y] + E[Z]}{2} \geq \frac{1}{2} \sum_{k \geq 1} \sum_{F_j \in C(k)} \left[\left(1 - \frac{1}{2^k}\right) + \left(1 - \left(1 - \frac{1}{k}\right)^k\right)\right] \cdot \alpha(z_j). \quad (5.30)$$

An easy calculation shows that

$$\left(1 - \frac{1}{2^k}\right) + \left(1 - \left(1 - \frac{1}{k}\right)^k\right) \geq \frac{3}{2} \quad (5.31)$$

for every positive integer k. Inserting (5.31) into (5.30) we finally obtain

$$E[U] \geq \frac{E[Y] + E[Z]}{2} \geq \frac{3}{4} \cdot \sum_{k \geq 1} \sum_{F_j \in C(k)} \alpha(z_j)$$

$$= \frac{3}{4} \cdot \sum_{j=1}^{m} \alpha(z_j).$$

$\square$

Summary of Section 5.3.6

The application of the approach of random sampling for optimization problems usually means a uniform random choice of the values of the variables (i.e., a random choice of a feasible solution) only. This is clearly very efficient and surprisingly this simple approach can provide randomized approximation algorithms with a reasonable approximation ratio. Examples of problems for which this approach works very well are MAX-CUT and MAX-SAT.

A nonuniform random choice of the values of the variables may be considered as a more elaborated version of random sampling. The probability distribution for this choice is precomputed by the relaxation method. We also call this special random sampling method the method of relaxation and random rounding. This method also works very well for MAX-SAT.

The approaches of random sampling and of relaxation with random rounding are incomparable for MAX-SAT. The best choice is to let both approaches run and then to take the best solution of the two solutions computed. This yields a randomized (4/3)-expected approximation algorithm for the MAX-SAT problem.

5.3.7 3SAT and Randomized Multistart Local Search

In Section 3.5 the concept of lowering worst case complexity of exponential algorithm was presented. Applying the divide-and-conquer technique we designed an $O(|F| \cdot 1.84^n)$ algorithm (Algorithm 3.5.2.1) for 3SAT. Here, we show that using randomization we can do still better. We shall present a one-sided-error Monte Carlo algorithm for 3SAT that runs in $O(|F| \cdot n^{3/2} \cdot (4/3)^n)$ time for any 3CNF formula F of n variables. Since the error probability approaches zero with a speed that is exponential in the number of repetitions,[68] this algorithm is more practical than Algorithm 3.5.2.1.

The design technique is based on a randomization of the local search technique from combinatorial optimization. Consider the neighborhood $Flip$[69] used to attack MAX-SAT by local search. The randomization that we use is twofold. First, one randomly generates $O(\sqrt{n} \cdot (4/3)^n)$ assignments to the variables of the given formula F. Similarly as in searching for an optimal solution to an instance of MAX-SAT, one consecutively performs the local search algorithm from each of the generated assignments. Secondly, a random decision is taken to generate a next neighbor. The main difference to the standard local search is that we do not require that the new assignment satisfies more clauses than the previous one. The only requirement is that the new assignment satisfies a clause that was not satisfied by the previous assignment. The second important difference is that we perform at most $3n$ steps of local search from any generated assignment because the probability to find an assignment satisfying F in the first $3n$ steps is larger than to find it later.[70] Thus, our algorithm can be described as follows.

Algorithm 5.3.7.1. SCHÖNING'S ALGORITHM

> Input: A formula F in 3CNF over a set of n Boolean variables.
>
> Step 1: $K := 0$;
> $UPPER := \lceil 20 \cdot \sqrt{3\pi n} \cdot \left(\frac{4}{3}\right)^n \rceil$
> $S := FALSE$.
>
> Step 2:   **while** $K < UPPER$ and $S := FALSE$ **do**
>   **begin** $K := K + 1$;
>   Generate uniformly at random an assignment $\alpha \in \{0, 1\}^n$;
>   **if** F is satisfied by α **then** $S := TRUE$;
> $M := 0$;
>   **while** $M < 3n$ and $S = FALSE$ **do**
>   **begin** $M := M + 1$;
>   Find a clause C that is not satisfied by α;
>   Pick one of the literals of C at random, and flip its value
>   in order to get a new assignment α;
>   **if** F is satisfied by α **then** $S := TRUE$
>   **end**
>   **end**

---

[68] Remember that every one-sided-error Monte Carlo algorithm possesses this excellent property.

[69] Two variable assignments are neighbors if they differ exactly in one bit, i.e., one can be obtained from the other one by flipping a bit.

[70] The argument for this claim will be given later in the analysis of the proposed algorithm.

Step 3:   **if** $S = TRUE$ **output** "F is satisfiable"
         else output "F is not satisfiable".

The general idea behind the above algorithm is that the small probabilities of finding a satisfying assignment in one local search attempt can be increased by a large number (but smaller than 2^n) of random attempts.

Theorem 5.3.7.2. SCHÖNING'S ALGORITHM *is a one-sided-error Monte Carlo algorithm for* 3SAT *that runs in time complexity* $O(|F| \cdot n^{3/2} \cdot (4/3)^n)$ *for any instance F of* 3SAT *over n Boolean variables.*

Proof. First, we analyze the worst case time complexity. SCHÖNING'S ALGORITHM performs at most $UPPER \in O(\sqrt{n} \cdot (4/3)^n)$ local searches from an assignment to the n variables of F. Any local search consists of at most $3n$ steps, and every step can be executed in time $O(|F|)$. Thus, the overall time complexity is in $O(|F| \cdot n^{3/2} \cdot (4/3)^n)$.

Now we analyze the failure probability of SCHÖNING'S ALGORITHM for a given formula F in 3CNF. If F is not satisfiable, then the algorithm outputs the right answer with certaincy.

Now consider that F is satisfiable. Let α^* be an assignment that satisfies F. Let p be the probability that one local search procedure that executes at most $3n$ local steps from a random assignment generates α^*. Obviously, p is a lower bound on finding an assignment that satisfies F in one run of the local search procedure (the inner cycle **while** in Step 2). The crucial point of this analysis is to show that

$$p \geq \frac{1}{2\sqrt{3 \cdot \pi \cdot n}} \cdot \left(\frac{3}{4}\right)^n. \tag{5.32}$$

The main idea behind is that the number $UPPER \gg p$ of independent attempts is sufficient to increase the probability of success to $1 - e^{-10}$.

In what follows we consider the distance between two assignments α and β as the number of bits in which α and β differ (i.e., the number of flips that the local search needs to move from α to β). Now, partition all assignments from $\{0,1\}^n$ into $n+1$ classes

$$Class(j) = \{\beta \in \{0,1\}^n \mid distance(\alpha^*, \beta) = j\}$$

according to their distance j to α^* for $j = 0, 1, \ldots, n$. Obviously, $|Class(j)| = \binom{n}{j}$, and the probability to uniformly generate an assignment from $Class(j)$ at random is exactly

$$p_j = \binom{n}{j} \bigg/ 2^n. \tag{5.33}$$

Now let us analyze the behavior of the local search. If $\alpha \in Class(j)$ does not satisfy F, then there exists at least one clause C that is not satisfied by α. Since α^* satisfies C, there exists a variable that occurs in C and whose flipping results in a $\beta \in Class(j-1)$. Thus, there exists a possibility to get an assignment β with a smaller distance to α than $distance(\alpha^*, \alpha)$. Since C consists of at most three literals and the algorithm chooses one of them randomly, we have a probability of at least $1/3$ "to move in the direction" to α^* (to decrease the distance to α^* by 1) in one step (i.e., we have a probability of at most $2/3$ to increase the distance to α^* by 1 in one step). Let, for all i, j, $i \leq j \leq n$, $q_{j,i}$ denote the probability to reach α^* from an $\alpha \in Class(j)$ by $j+i$ moves in the direction to α^* and i moves from the direction of α^* (i.e., in overall $j+2i$ steps). Then

$$q_{j,i} = \binom{j+2i}{i} \cdot \frac{j}{j+2i} \cdot \left(\frac{1}{3}\right)^{j+i} \cdot \left(\frac{2}{3}\right)^{i}$$

can be established by a short combinatorial calculation.[71] Obviously, the probability q_j to reach α^* from an $\alpha \in Class(j)$ is at least $\sum_{i=0}^{j} q_{j,i}$. Observe that SCHÖNING'S ALGORITHM allows $3n$ steps and so $j+2i$ steps can be executed for all $j \in \{0, 1, \ldots, n\}$ and $i \leq j$. Thus,

$$q_j \geq \sum_{i=0}^{j} \left[\binom{j+2i}{i} \cdot \frac{j}{j+2i} \cdot \left(\frac{1}{3}\right)^{j+i} \cdot \left(\frac{2}{3}\right)^{i} \right]$$

$$\geq \frac{1}{3} \sum_{i=0}^{j} \left[\binom{j+2i}{i} \cdot \left(\frac{1}{3}\right)^{j+i} \cdot \left(\frac{2}{3}\right)^{i} \right]$$

$$> \frac{1}{3} \binom{3j}{j} \cdot \left(\frac{1}{3}\right)^{2j} \cdot \left(\frac{2}{3}\right)^{j}.$$

Due to Stirling's formula

$$r! \sim \sqrt{2\pi r} \left(\frac{r}{e}\right)^{r}$$

we obtain

$$q_j \geq \frac{1}{3} \cdot \frac{(3j)!}{(2j)!j!} \cdot \left(\frac{1}{3}\right)^{2j} \cdot \left(\frac{2}{3}\right)^{j}$$

$$\sim \frac{1}{3} \cdot \frac{\sqrt{2\pi \cdot 3j} \cdot \left(\frac{3j}{e}\right)^{3j}}{\sqrt{2\pi \cdot 2j} \cdot \left(\frac{2j}{e}\right)^{2j} \cdot \sqrt{2\pi j} \cdot \left(\frac{j}{e}\right)^{j}} \cdot \left(\frac{1}{3}\right)^{2j} \cdot \left(\frac{2}{3}\right)^{j}$$

$$= \frac{1}{3} \cdot \frac{\sqrt{3}}{2 \cdot \sqrt{\pi j}} \cdot \frac{3^{3j}}{2^{2j}} \cdot \left(\frac{1}{3}\right)^{2j} \cdot \left(\frac{2}{3}\right)^{j}$$

$$= \frac{1}{2\sqrt{3\pi j}} \cdot \left(\frac{1}{2}\right)^{j}. \tag{5.34}$$

---

[71] Note that $\binom{j+2i}{i} \cdot \frac{j}{j+2i}$ is the number of different paths from $\alpha \in Class(j)$ to α^* where a path is determined by a word over the two-letter alphabet $\{+, -\}$ where $+$ means a movement in the direction to α^* and $-$ means a movement that increases the distance to α^*. Every such word must satisfy that each suffix of w contains more $+$ symbols than $-$ symbols.

Now we are ready to perform the final calculation for p. Clearly,

$$p \geq \sum_{j=0}^{n} p_j \cdot q_j. \tag{5.35}$$

Inserting (5.33) and (5.34) into (5.35) we obtain

$$p \geq \sum_{j=0}^{n} \left[\left(\frac{1}{2}\right)^n \cdot \binom{n}{j} \cdot \left(\frac{1}{2 \cdot \sqrt{3\pi j}} \cdot \left(\frac{1}{2}\right)^j \right) \right]$$

$$\geq \frac{1}{2 \cdot \sqrt{3\pi n}} \cdot \left(\frac{1}{2}\right)^n \cdot \sum_{j=0}^{n} \left[\binom{n}{j} \cdot \left(\frac{1}{2}\right)^j \right]$$

$$= \frac{1}{2 \cdot \sqrt{3\pi n}} \cdot \left(\frac{1}{2}\right)^n \cdot \left(1 + \frac{1}{2}\right)^n$$

$$= \frac{1}{2 \cdot \sqrt{3\pi n}} \cdot \left(\frac{3}{4}\right)^n = \tilde{p}.$$

(5.32) directly implies that the probability that we do not reach a satisfying assignment for F in one attempt is at most $(1 - \tilde{p})$. Therefore, the probability of failure after t attempts is at most

$$(1 - \tilde{p})^t \leq e^{-\tilde{p} t}. \tag{5.36}$$

Inserting

$$t = UPPER = 20 \cdot \sqrt{3\pi n} \cdot \left(\frac{4}{3}\right)^n$$

into (5.36) we obtain

$$(1 - \tilde{p})^{UPPER} \leq e^{-10} < 5 \cdot 10^{-5}.$$

Thus, SCHÖNING'S ALGORITHM is a one-sided-error Monte Carlo algorithm for 3SAT.   □

SCHÖNING'S ALGORITHM can be used for kSAT for any $k \geq 4$, too. The trouble is that the probability to move in the right direction by a local transformation (flip) may decrease to $1/k$, and so, a larger number of attempts is needed to sufficiently decrease the probability of failure. The exact analysis is left to the reader.

Exercise 5.3.7.3. [*] Prove, for every integer $k \geq 4$, that SCHÖNING'S ALGORITHM with an $UPPER \in \left(|F| \cdot h(n) \cdot \left(2 - \frac{2}{k}\right)^n \right)$ for a polynomial h is a one-sided-error Monte Carlo algorithm for kSAT.   □

Summary of Section 5.3.7

Combining the concepts of randomization and of lowering the worst case complexity of exponential algorithms can provide practical algorithms for hard problems. SCHÖNING'S ALGORITHM consists of exponentially many random attempts to find a satisfying assignment for a formula in 3CNF. Every attempt is performed by local search with a bounded number of local transformation steps. The crucial point is that the probability of success in one attempt is at least $1/Exp(n)$, where $Exp(n)$ is an exponential function that grows substantially slower than 2^n. Thus, performing $O(Exp(n))$ random attempts one can find a satisfying assignment with a probability almost 1 in time $O(|F| \cdot n \cdot Exp(n))$. SCHÖNING'S ALGORITHM is a $O(|F| \cdot (1.334)^n))$ one-sided-error Monte Carlo algorithm for 3SAT. From the algorithm design technique point of view SCHÖNING'S ALGORITHM is a randomized multistart local search. One randomly generates an assignment α and looks for a possibility to reach an satisfying assignment in a random short walk from α (i.e., by a short sequence of random local transformations).

5.4 Derandomization

5.4.1 Fundamental Ideas

In this chapter so far, we have seen that randomization is a powerful concept in algorithmics. The main reason for this claim is that we have several problems for which randomized algorithms are more efficient than the best known deterministic algorithms. Moreover, it is no negligible fact that randomized algorithms are often much simpler and therefore easier to implement than their deterministic counterparts. So, randomized algorithms may be more practical than deterministic ones for many information processing tasks.

On the other hand, it may be much more desirable to have deterministic algorithms that obtain a required approximation ratio or correct results efficiently for all input instances rather than merely with a high probability.[72] Thus, there is an interest in finding methods that can convert some randomized algorithms to deterministic ones without any essential increase in the amount of computational resources. Moreover, there is a theoretical reason to search for such methods. Up till now there is no known result claiming that randomized algorithms can do something essentially quicker than deterministic algorithms. For instance, we do not have even any polynomial-time randomized algorithm for an NP-complete problem. There is a hope of finding methods to simulate randomized computations without any exponential increase in the time complexity.

---

[72] Moreover, one removes the nontrivial problem of generating "good" pseudorandom sequences for the implementation of the randomized algorithm considered.

The goal of this section is to present two distinct methods that allow, in some cases, efficient transformations of randomized algorithms into deterministic ones. The application of these methods sometimes leads to the development of the best known deterministic algorithm for the considered computing problems.[73]

Informally, the concept of these two derandomization methods can be described as follows.

THE METHOD OF THE REDUCTION OF THE PROBABILITY SPACE.

We have already observed that if a randomized algorithm A uses only a small number of random bits $Random_A(n)$, then one can deterministically simulate all $2^{Random_A(n)}$ possible runs of A on the given input, and then select the right result. In fact $2^{Random_A(n)}$ is the size of the sample space from which the random sequences are chosen. Obviously, if A works in polynomial time and $2^{Random_A(n)}$ is bounded by a polynomial, a deterministic polynomial-time simulation is possible. But in general the size of the probability space is not bounded by a polynomial in the input size. The main idea of this concept is to try to reduce the size of the probability space of the work of A on a given input. Surprisingly, a reduction may be possible without any decrease in the number of random variables of A. It is sufficient to recognize that the total independence of the random choices is not necessary for the analysis of the probabilistic behavior of the randomized algorithms. If, for instance, one considers the random sampling for the problem instances of MAX-EkSAT with the set of variables $X = \{x_1, \ldots, x_n\}$, then it is not necessary that the n random choices for $x_1, \ldots, x_n$ are completely independent. To achieve that a random choice of the values for $x_1, x_2, \ldots, x_n$ implies a probability of at least $1 - 1/2^k$ to satisfy any particular clause, it is sufficient to require that, for any k variables from X, the random choices for these k variables are independent. As we shall see later this may be an essential difference to the requirement of the independence of all variables of X. The crucial point is that in such a case one can "approximate" the random choice of values of $x_1, x_2, \ldots, x_n$ by a random choice of values of some random variables $y_1, \ldots, y_k$.[74] If k is essentially smaller than n we are ready because the size of the probability space has been sufficiently reduced.

---

[73] It often happens that without using the way via the design of randomized algorithms and via their derandomization, it would be very hard to find an idea leading to the design of such an efficient deterministic algorithm.

[74] In fact, it means that one can efficiently determine (compute) the values for $x_1, \ldots, x_n$ from the values randomly chosen for $y_1, \ldots, y_k$.

THE METHOD OF CONDITIONAL PROBABILITIES.

This method is mainly applied for optimization problems, but applications for decision problems are not excluded. The usual situation with a randomized α-expected approximation algorithm is that one has a random variable Z measuring the quality of the output, and the expectation $E[Z]$ is known. To design a deterministic α-approximation algorithm it is sufficient to find the random sequence that leads to an output whose quality is at least $E[Z]$. The existence of such a random sequence is an obvious consequence of the definition of expectation. The method of conditional probabilities tries to deterministically compute such a random sequence, if possible. Surprisingly, for several problems this approach is successful despite of the fact that one does not see the possibility of a straightforward deterministic simulation if looking at the randomized algorithm.

We proceed in this section as follows. In Section 5.4.2 we explain the concept of the method of the reduction of probability space in its generality and show by which conditions this derandomization concept successfully works. Section 5.4.3 illustrates the method of the probability space reduction by applying it to derandomization of the random sampling algorithm for MAX-E3SAT. A detailed description of the method of conditional probabilities is given in Section 5.4.4 and the assumptions by which it can be successfully applied are discussed. The power of these derandomization methods is illustrated on some versions of the satisfiability problem in Section 5.4.5.

5.4.2 Derandomization by the Reduction of the Probability Space Size

As already mentioned above, the main idea is to decrease the size of the probability space when the total independence of all random choices is not ultimately required. In what follows we give a detailed description of this method in a general form.

Let A be a randomized algorithm. The run of A on a fixed input w can be viewed as a probabilistic experiment. The probability space $(\Omega, Prob)$ is defined by the set Ω of all random sequences used to proceed w. The typical case is that $\Omega = \{\alpha \in \{0,1\}^{Random_A(w)}\}$ and $Prob(\alpha) = \frac{1}{2^{Random_A(w)}}$ for every $\alpha \in \Omega$. All randomized algorithms presented here have either this probability space or it is possible to implement them in such a way that they would have such a probability space.

Obviously, if $Random_A(w)$ is polylogarithmic in $|w|$ for every input w, then $|\Omega|$ is polynomial in $|w|$ and we can deterministically simulate A by executing all $2^{Random_A(w)}$ computations[75] of A on w. But often $Random_A(w)$ is linear in the input length $|w|$. Thus, we would like to decrease $Random_A(w)$ if possible. Let $n = Random_A(w)$ in what follows.

---

[75] Remember, that if one fixes a random sequence, then the work of A on an input is completely deterministic.

One can consider random variables $X_1, X_2, \ldots, X_n$ to describe the random choices determining the values in the random sequences. Usually we had $Prob(X_i = 1) = \frac{1}{2} = Prob(X_i = 0)$ for all $i = 1, 2, \ldots, n$. In general we assume that $Prob(X_i = 1) = 1 - Prob(X_i = 0)$ and that $Prob(X_i = a)$ are rational numbers for $a \in \{0, 1\}$. Since it is often much easier to handle independent random variables than dependent random variables, we have always assumed that these random variables are independent. But sometimes, a weaker form of independence may suffice for a successful analysis of a randomized algorithm. The following definition presents such a kind of independence.

Definition 5.4.2.1. *Let $(\Omega, Prob)$ be a probability space and let $X_1, X_2, \ldots, X_n$ be some random variables. Let k be a positive integer, $2 \le k \le n$. $X_1, X_2, \ldots, X_n$ are called k-wise independent, if each subset consisting of k random variables of $\{X_1, \ldots, X_n\}$ is independent, i.e., if for all $i_1, \ldots, i_k \in \{1, \ldots, n\}$ $1 \le i_1 < \cdots < i_k \le n$ and all values $\alpha_1, \alpha_2, \ldots, \alpha_k$, $\alpha_j \in Dom(X_{i_j})$ for $j = 1, 2, \ldots, k$,*

$$Prob[X_{i_1} = \alpha_1, X_{i_2} = \alpha_2, \ldots, X_{i_k} = \alpha_k] = Prob[X_{i_1} = \alpha_1] \cdot \cdots \cdot Prob[X_{i_k} = \alpha_k].$$

Observe that this weak form of independence is sufficient for the analysis of the randomized algorithm RSMS based on random sampling for MAX-EkSAT, since the only independence we need there is the independence of the choice of the k random values for the variables in every particular clause. For instance, if

$$\Phi = (x_1 \vee \overline{x}_2 \vee x_5) \wedge (x_2 \vee \overline{x}_4 \vee \overline{x}_5) \wedge (\overline{x}_1 \vee x_3 \vee \overline{x}_4),$$

then we only need to have that the choices of random values for $x_1, x_2, \ldots, x_n$ are independent for each of the following sets of variables $\{x_1, x_2, x_5\}$, $\{x_2, x_4, x_5\}$, and $\{x_1, x_3, x_4\}$. Already this requirement assures us that each of the clauses is satisfied with the probability $1 - \frac{1}{2^3} = \frac{7}{8}$ and so the expected number of satisfied clauses is $3 \cdot \frac{7}{8} = \frac{21}{8}$.

Now, the question is: "How to use the knowledge that k-wise independence is sufficient for our purposes in order to decrease the size of the probability space?" For simplicity, we assume that the random variables $X_1, \ldots, X_n$ can take only values 0 or 1. The idea is to find a probability space (S, Pr) with elementary events $E_1, \ldots, E_m$ and an efficiently computable function $f : \{E_1, \ldots, E_m\} \times \{1, 2, \ldots, n\} \to \{0, 1\}$ such that

(i) for $i = 1, \ldots, n$, $X_i' = f(B, i)$ for $B \in \{E_1, \ldots, E_m\}$ is a random variable over (S, Pr) and X_i' **approximates** X_i in the sense that, for every $\alpha \in \{0, 1\}$,

$$|Pr[X_i' = \alpha] - Prob[X_i = \alpha]| \le \frac{1}{2n}, \text{ and}$$

(ii) $|S| = O(n^k)$ {Note that $|\Omega| = 2^n$}.

So, the exchange of the random variables $X_1, \ldots, X_n$ for the random variables $X_1', \ldots, X_n'$ corresponds to the move from the probability space $(\Omega, Prob)$

. to the probability space (S, Pr). The gain is that $|S|$ is polynomial in n while $|\Omega|$ is exponential in n. The loss is that by simulation of $(\Omega, Prob)$ in (S, Pr) $X_1', \ldots, X_n'$ are only k-wise independent in (S, Pr) while $X_1, \ldots, X_n$ are independent in $(\Omega, Prob)$, and that $X_1', \ldots, X_n'$ only approximate $X_1, \ldots, X_n$, i.e., do not have exactly the same probabilities of taking the same values. The approach presented below simplifies the general concept description a little bit. Instead of searching for a function f satisfying the property (i), we simply take the sample space as a space of functions $p : \{1, \ldots, n\} \to \{0, 1\}$ and set $X_i' := p(i)$. So, an elementary event in (S, Pr) is the choice of a function and this choice unambiguously determines the values of all $X_1', \ldots, X_n'$.

The next result shows how to find such (S, Pr) for every n and k. This theorem is very strong, because it guarantees the possibility of the reduction of the size of the sample space and a good approximation of the random choice if k-wise independence is sufficient for our purposes.

Theorem 5.4.2.2. *Let $(\Omega, Prob)$ be a probability space and $X_1, \ldots, X_n$ be random variables over $(\Omega, Prob)$ such that $\Omega = \{(X_1 = \alpha_1, \ldots, X_n = \alpha_n) \,|\, \alpha_i \in \{0, 1\} \text{ for } i = 1, \ldots, n\}$. Let k be a positive integer, $2 \le k < n$, and let q be a prime power such that $q \ge n$. Let $\mathrm{GF}(q) = \{r_1, r_2, \ldots, r_q\}$ be the finite field of q elements. Let*

$$A_{i,1} = \{r_1, r_2, \ldots, r_{d_i}\} \text{ and } A_{i,0} = \mathrm{GF}(q) - A_{i,1},$$

where $d_i = \lceil q \cdot \mathrm{Prob}\,[X_i = 1] - \frac{1}{2} \rceil$.
Then, the probability space (S, Pr), where $S = \{p \,|\, p \text{ is a polynomial over } \mathrm{GF}(q) \text{ with degree at most } k - 1\}$ and Pr is the uniform distribution over S, has the following properties:

(i) $|S| = q^k$,
(ii) for $i = 1, \ldots, n$, the random variables $X_i' : S \to \{0, 1\}$ defined by

$$X_i'(p) = 1 \text{ iff } p(r_i) \in A_{i,1}$$

and

$$X_i'(p) = 0 \text{ iff } p(r_i) \in A_{i,0}$$

satisfy the following properties:
a) $X_1', \ldots, X_n'$ are k-wise independent,
b) $|Pr[X_i' = \alpha] - Prob[X_i = \alpha]| \le \frac{1}{2q}$ for every $\alpha \in \{0, 1\}$.

Proof. Fact (i) is obvious because every polynomial of degree at most $k - 1$ is unambiguously determined by its k coefficients, and these coefficients are from $\{r_1, r_2, \ldots, r_q\}$.

Before starting the proof if (ii) it is important to observe, that for every $r_i \in \mathrm{GF}(q)$ and every $a, b \in \mathrm{GF}(q)$,

$$Pr(p(r_i) = a) = Pr(p(r_i) = b) = \frac{1}{q},$$

i.e.[76], that the number of occurrences of the value a in the sequence

$$p_1(r_i), p_2(r_i), \ldots, p_{q^k}(r_i)$$

where $S = \{p_1, p_2, \ldots, p_{q^k}\}$ is the same as the number of occurrences of the value b. This is the consequence of the fact that every polynomial $p \in S$ is uniquely determined by fixing its value at k different places.

To prove (ii.b) we observe that, for all $i = 1, \ldots, n$,

$$Pr[X_i' = 1] = \frac{|A_{i,1}|}{|GF(q)|}$$

$$= \frac{\left\lceil q \cdot Prob[X_i = 1] - \frac{1}{2} \right\rceil}{q}.$$

Thus, one can obtain $Pr[X_i' = 1]$ by rounding $Prob[X_i = 1]$ to the nearest multiple of $\frac{1}{q}$. In particular, this implies

$$|Pr[X_i' = 1] - Prob[X_i = 1]| \leq \frac{1}{2q}. \tag{5.37}$$

The upper bound

$$|Pr[X_i' = 0] - Prob[X_i = 0]| \leq \frac{1}{2q}$$

is a direct consequence of (5.37) and of the facts $Pr[X_i' = 0] = 1 - Pr[X_i' = 1]$, and $Prob[X_i = 0] = 1 - Prob[X_i = 1]$.

Now we prove (ii.a). Remember that, for $i = 1, 2, \ldots, n$,

$$A_{i,1} = \{r_1, r_2, \ldots, r_{d_i}\} \text{ and } A_{i,0} = \mathbb{Z}_q - A_{i,1}, \text{ where } d_i = \left\lceil q \cdot Prob[X_i = 1] - \frac{1}{2} \right\rceil.$$

Let $i_1, \ldots, i_k \in \{1, 2, \ldots, n\}$ be such that $1 \leq i_1 < \cdots < i_k \leq n$, and let $\alpha_l \in \{0, 1\}$ for $l = 1, \ldots, k$. Then

$$Pr[X_{i_1}' = \alpha_1, \ldots, X_{i_k}' = \alpha_k] = Pr[p(r_{i_1}) \in A_{i_1, \alpha_1}, \ldots, p(r_{i_k}) \in A_{i_k, \alpha_k}].$$

There are exactly

$$|A_{i_1, \alpha_1}| \cdot \cdots \cdot |A_{i_k, \alpha_k}|$$

distinct polynomials p from S fulfilling $p(r_{i_1}) \in A_{i_1, \alpha_1}, \ldots, p(r_{i_k}) \in A_{i_k, \alpha_k}$, because p is a polynomial of degree at most $k - 1$ and, hence, p is uniquely determined by fixing its value at k different places. Therefore,

$$Pr[p(r_{i_1}) \in A_{i_1, \alpha_1}, \ldots, p(r_{i_k}) \in A_{i_k, \alpha_k}] = \frac{|A_{i_1, \alpha_1}| \cdot \cdots \cdot |A_{i_k, \alpha_k}|}{|S|}$$

$$\underset{(i)}{=} \frac{|A_{i_1, \alpha_1}| \cdot \cdots \cdot |A_{i_k, \alpha_k}|}{q^k}$$

$$= \frac{|A_{i_1, \alpha_1}|}{q} \cdot \cdots \cdot \frac{|A_{i_k, \alpha_k}|}{q}$$

$$= Pr[X_{i_1}' = \alpha_1] \cdot \cdots \cdot Pr[X_{i_k}' = \alpha_k].$$

---

[76] i.e., that S over $GF(q)$ provides a universal hashing.

Thus $X_1', \ldots, X_n'$ are k-wise independent.   □

Exercise 5.4.2.3. Consider a special version of Theorem 5.4.2.2 when $Prob[X_i = 1] = Prob[X_i = 0] = \frac{1}{2}$. Find such a probability space (S, Pr) and such random variables $X_1', \ldots, X_n'$ over (S, Pr), that (i) and (ii.a) hold and additionally $Prob[X_i = \alpha] = Pr[X_i' = \alpha]$ for every $\alpha \in \{0, 1\}$.   □

Exercise 5.4.2.4. For which values of $Prob[X_i = 1]$ can one exactly "simulate" $X_1, \ldots, X_n$ by $X_1', \ldots, X_n'$?   □

Observe that one can use Theorem 5.4.2.2 not only as a powerful derandomization method. Also, in the case of implementing a randomized algorithm as a randomized algorithm, the number of random bits is an essential cost measure of the resulting program. Theorem 5.4.2.2 can be used to essentially decrease the number of random bits as described below.

Let A be a randomized algorithm with $Random_A(n) = n$ for every input length n, and let k-wise independence of the random variables $X_1, \ldots, X_n$ be sufficient for a successful work of A. A new algorithm $\text{RED}(A)$, reducing the number of random bits, can be described as follows:

Algorithm 5.4.2.5. $\text{RED}(A)$

Input: An input w as for A.

Step 1:   Choose uniformly at random an element p from the probability space (S, Pr) described in Theorem 5.4.2.2.
{Observe that this can be performed by $\lceil \log_2 |S| \rceil = O(k \cdot \log_2 q)$ random bits.}

Step 2:   Compute $X_1', X_2', \ldots, X_n'$ as described in (ii) of Theorem 5.4.2.2.

Step 3:   Run the algorithm A on w with the sequence of k-wise independent random bits $X_1', X_2', \ldots, X_n'$.

Output: The output of A computed in Step 3.

Obviously, if k is a constant according to the input length n, then we have reduced the number of random bits from n to $Random_{\text{RED}(A)} = O(\log_2 n)$.[77]

Thus, the general derandomization scheme based on the method of the probability space reduction can be described as follows:

Deterministic simulation of A by probability space reduction, PSR(A)

Input: An input w consistent as an input of A.

Step 1:   Create the probability space (S, Pr) as described in Theorem 5.4.2.2.

---

[77] Obviously, one does not need to assume that n is both the input length and the number of random bits. In general, for a constant k we get $Random_{\text{RED}(A)}(w) = O(\log_2 Random_A(w))$.

Step 2:   **for** every $p \in S$ **do**

simulate $\mathrm{RED}(A)$ on w with the random choice p and save the output $Result(p)$.

Step 3:   Estimate the "right" output from all outputs computed in Step 2. {Obviously, Step 3 depends on the kind of computing problem. If one considers an optimization problem then the output with the best cost is chosen. If A has been designed for a decidability problem, one has to look on the probabilities of the answers "accept" and "reject".}

The next section shows an application of this simulation scheme for MAX-E3SAT.

5.4.3 Probability Space Redutction and MAX-EkSAT

In this section we illustrate the method of the reduction of the probability space size by an example. Consider the algorithm RSMS for MAX-E3SAT. As we have already observed above, for the analysis of this random sampling method it is sufficient to require the independence of the variables occurring in the same clause. So, the 3-wise independence of random variables is sufficient to prove that the expected number of satisfied clauses of m given clauses is $\frac{7}{8} \cdot m$. If the input Φ consists of m clauses and contains n Boolean variables $x_1, \ldots, x_n$, then the number of random variables of the algorithm RSMS is exactly n. The probability space of RSMS for the input Φ is $(\Omega, Prob)$, where

(i) $\Omega = \{(\alpha_1, \alpha_2, \ldots, \alpha_n) \in \{0,1\}^n\}$, $(|\Omega| = 2^n)$, i.e., an elementary event corresponds to the choice of the Boolean values for n variables $X_1, \ldots, X_n$, and

(ii) for $i = 1, 2, \ldots, n$, $Prob(X_i = 1) = Prob(X_i = 0) = \frac{1}{2}$.

Remember that the meaning of a random choice α_i for x_i is in that the algorithm RSMS sets $x_i := \alpha_i$.

Following Theorem 5.4.2.2 we have the possibility of choosing $q = 2^r$ such that r is the smallest number with the property $q = 2^r \geq n$. We have decided to take this choice for q because we want to have an even q in order to achieve $Pr[X_i' = \alpha] = Prob[X_i = \alpha]$ for all $i \in \{1, \ldots, n\}$, and all $\alpha \in \{0,1\}$.

Now, the new sample space is

$$S = \{c_2 x^2 + c_1 x + c_0 \mid c_2, c_1, c_0 \in \mathrm{GF}(q)\}.$$

Since each such polynomial $c_2 x^2 + c_1 x_1 + c_0$ is unambiguously represented by the triple $(c_2, c_1, c_0) \in (\mathrm{GF}(q))^3$, we have $|S| = q^3$. Since q is even and $Prob[X_i = 1] = \frac{1}{2}$ we have

$$\lceil q \cdot Prob[X_i = 1] - \frac{1}{2} \rceil = \lceil 2^r \cdot \frac{1}{2} - \frac{1}{2} \rceil = \lceil 2^{r-1} - \frac{1}{2} \rceil = 2^{r-1} = \frac{q}{2},$$

i.e., $|A_{i,1}| = |A_{i,0}| = \frac{q}{2}$. Since $X_i'(p)$ takes the value 0 if and only if $p(i) \in A_{i,0}$, and $X_i'(p)$ takes the value 1 if and only if $p(i) \in A_{i,1}$, we obtain

$$Pr[X'_i = 0] = \frac{1}{2} = Pr[X'_i = 1].$$

So, $X'_1, \ldots, X'_n$ approximate $X_1, \ldots, X_n$ exactly and (ii) of Theorem 5.4.2.2 assures us that $X'_1, \ldots, X'_n$ are 3-wise independent. Applying the general simulation schemes $\text{RED}(A)$ and $\text{PSR}(A)$ presented in Section 5.4.2, we obtain the following deterministic $(8/7)$-approximation algorithm for MAX-E3SAT.

Algorithm 5.4.3.1. DERAND-RSMS-3

Input: A formula Φ in 3-CNF over a set of variables $\{x_1, \ldots, x_n\}$.

Step 1:   Find a positive integer r such that

$$q := 2^r \geq n$$

and r is the smallest integer with the property $2^r \geq n$.

Step 2:   **for** $c_0 = 0$ **to** $q - 1$ **do**
  for $c_1 = 0$ **to** $q - 1$ **do**
    for $c_2 = 0$ **to** $q - 1$ **do**
      begin
      for $i = 1$ **to** n **do**
        if $c_2 r_i^2 + c_1 r_i + c_0 \in \{r_1, r_2, \ldots, r_{q/2}\} \in \text{GF}(q)$
        then $x_i = 1$
        else $x_i = 0$;
      count the number of satisfied clauses of Φ by $x_1, \ldots, x_n$,
      and save the assignment $(\alpha_1, \ldots, \alpha_n)$ with the maximal
      number of satisfied clauses up till now.
      end

Output: $\alpha_1, \ldots, \alpha_n$.

We observe that the algorithm DERAND-RSMS-3 consists of $O(n^3)$ iterations of the algorithm RSMS. A run of RSMS on a formula Φ with m clauses and n variables has complexity in $O(n + m)$. So, the time complexity of DERAND-RSMS-3 is in $O(n^4 + n^3 \cdot m)$.

Since RSMS is a randomized $(8/7)$-expected approximation algorithm for the problem MAX-E3SAT, $\text{RED}(\text{RSMS})$ has also the expected approximation ratio of at most $8/7$. More precisely, the expected number of satisfied clauses by $\text{RED}(\text{RSMS})$ is at least $\frac{7}{8} \cdot m$. Thus, there must exist a choice of values c_0, c_1, c_2 such that the corresponding assignment $X'_1, \ldots, X'_n$ satisfies at least $\frac{7}{8}m$ clauses. Since DERAND-RSMS-3 looks on all choices of the values for c_0, c_1, c_2, DERAND-RSMS-3 must find an assignment that satisfies at least $\frac{7}{8}m$ clauses.

Exercise 5.4.3.2. Design and implement the algorithm $\text{RED}(\text{RSMS})$ for MAX-E3SAT. Which is the exact number of random variables for this implementation? $\square$

Exercise 5.4.3.3. Following the general concept and schemes of Section 5.4.2 design, for every integer $k > 3$,

(i) a randomized $(2^k/(2^k - 1))$-expected approximation algorithm for MAX-EkSAT that uses as few random bits (as a small sample space) as possible,

(ii) a deterministic $(2^k/(2^k - 1))$-approximation algorithm for MAX-EkSAT.

In both cases analyze the complexity of your algorithm.   □

Exercise 5.4.3.4. Is it possible to design an $O(n^4)$ deterministic δ-approximation algorithm for MAX-E6SAT by using the derandomization method of the reduction of the size of the probability space?   □

Finally, we observe that the derandomization method by the reduction of the size of the probability space is quite general and very powerful. But the complexity of the resulting deterministic algorithm may be too high. Already $O(n^4)$ for a formula in 3-CNF of n variables may be too large. Thus, from the practical point of view, the possibility of essentially reducing the number of random bits (choices) in a randomized algorithm may be the main current contribution of this method.

5.4.4 Derandomization by the Method of Conditional Probabilities

In this section we describe the general concept of the method of conditional probabilities. To simplify the presentation, we describe the method for maximization problems only. The adaptation to minimization problems is straightforward.[78] Some exemplary applications of this method are given in the next section.

Let A be a randomized algorithm for a maximization problem $MAX = (\Sigma_I, \Sigma_O, L, L_I, \mathcal{M}, cost, maximum)$. For any fixed input $w \in L_I$, we have a probability space $(\Omega, Prob)$ whose every elementary event can be determined by taking Boolean values for some random variables $X_1, X_2, \ldots, X_n$. In general, we allow $Prob[X_i = \alpha]$ for all $i = 1, \ldots, n$ and all $\alpha \in \{0, 1\}$ to be rational numbers.[79] For any output u of A we describe the cost of u ($cost(u)$) by a random variable Z. This means that $Prob[Z = \beta]$ is the probability that A, working on the input w, computes an output u with $cost(u) = \beta$. Our main goal is to convert A into a deterministic algorithm that computes an output whose cost is surely at least $E[Z]$.

Remember that the randomized algorithm can be viewed as a set of $2^{Random_A(w)}$ deterministic algorithms A_α where, for every $\alpha \in \{0, 1\}^n$, A_α works on the input w with the assignment α for $\{X_1, \ldots, X_n\}$. Since $E[Z]$ is a weighted average of the costs of all possible outputs of the work of A on w,

---

[78] Despite the fact that one usually applies this method for optimization problems, it is also possible to consider it for decision problems.

[79] Observe that we usually consider the uniform probability distribution.

there must exist an output v with $cost(v) \geq E[Z]$,[80] i.e., there must exist an assignment $\beta = \beta_1, \ldots, \beta_n \in \{0,1\}^n$ such that the deterministic algorithm A_β computes an output whose cost is at least $E[Z]$. The core of the method of conditional probabilities is to try to compute β deterministically and then to only simulate the work of A_β on the given input w. Obviously, β depends on the input w and so β has to be computed for every actual input again.[81] The efficiency of the resulting deterministic algorithm is then $Time_{A_\beta}(w)$ plus the complexity of computing β. Thus, the estimation of a right β is essential for the efficiency of this method (for the increase of the time complexity relative to $Time_A(n)$).

The idea presented here is to compute $\beta = \beta_1 \ldots \beta_n$ sequentially, bit after bit. The following method for computing β is called the **method of pessimistic estimators**. At the beginning the value 1 is chosen for β_1, if

$$E[Z|X_1 = 1] \geq E[Z|X_1 = 0]$$

and β_1 is set to 0 in the opposite case. In general, if $\beta_1, \beta_2, \ldots, \beta_i$ were already estimated, then one sets $\beta_{i+1} := 1$ if

$$E[Z|X_1 = \beta_1, X_2 = \beta_2, \ldots, X_i = \beta_i, X_{i+1} = 1] \geq$$
$$E[Z|X_1 = \beta_1, X_2 = \beta_2, \ldots, X_i = \beta_i, X_{i+1} = 0]$$

and one sets $\beta_{i+1} := 0$ in the opposite case. To show that this choice of $\beta = \beta_1, \ldots, \beta_n$ has the required property $cost(A_\beta(w)) \geq E[Z]$, we give the following technical lemma.

Lemma 5.4.4.1. *Let $(\Omega, Prob)$ be a probability space, and $X_1, \ldots, X_n$, and Z be random variables as described above. If, for a given input w, $\beta = \beta_1 \beta_2 \ldots \beta_n \in \{0,1\}^n$ is computed by the method of pessimistic estimators, then*

$$E[Z] \leq E[Z|X_1 = \beta_1] \leq E[Z|X_1 = \beta_1, X_2 = \beta_2] \leq \cdots$$
$$\leq E[Z|X_1 = \beta_1, \ldots, X_n = \beta_n] = cost(A_\beta(w)).$$

Proof. The fact that $E[Z|X_1 = \beta_1, \ldots, X_n = \beta_n] = cost(A_\beta(w))$ is obvious. In what follows we prove for every $i = 0, 1, \ldots, n-1$ that

$$E[Z|X_1 = \beta_1, \ldots, X_i = \beta_i] \leq E[Z|X_1 = \beta_1, \ldots, X_i = \beta_i, X_{i+1} = \beta_{i+1}]. \tag{5.38}$$

Since $X_1, X_2, \ldots, X_n$ are considered to be independent, it can be easily observed that

---

[80] It is impossible that all outputs would have costs smaller than $E[Z]$, because then $E[Z]$ cannot be a weighted average of them.

[81] Maybe it would be better to use the notation $\beta(w)$ instead of β to call attention to this fact, but we use the short notation β only because we consider w to be fixed during the whole Section 5.4.4.

$$E[Z|X_1 = \alpha_1, \ldots, X_i = \alpha_i] =$$
$$Prob[X_{i+1} = 1] \cdot E[Z|X_1 = \alpha_1, \ldots, X_i = \alpha_i, X_{i+1} = 1] +$$
$$Prob[X_{i+1} = 0] \cdot E[Z|X_1 = \alpha_1, \ldots, X_i = \alpha_i, X_{i+1} = 0]$$

for every $\alpha_1, \ldots, \alpha_i \in \{0,1\}^i$. Since $Prob[X_{i+1} = 1] = 1 - Prob[X_{i+1} = 0]$ and the weighted mean of two numbers cannot be larger than their maximum we obtain

$$E[Z|X_1 = \beta_1, \ldots, X_i = \beta_i] \leq$$
$$\max\{E[Z|X_1 = \beta_1, \ldots, X_i = \beta_i, X_{i+1} = 1], \quad\quad (5.39)$$
$$E[Z|X_1 = \beta_1, \ldots, X_i = \beta_i, X_{i+1} = 0]\}.$$

Since our choice for β_{i+1} corresponds to the choice of the maximum of the conditional probabilities in (5.39), (5.39) directly implies (5.38).   □

Thus, we can formulate a general derandomization scheme as follows.

Algorithm 5.4.4.2. COND-PROB(A)

Input: A consistent input w for A.
Step 1:   **for** $i := 1$ **to** n **do**
            if $E[Z|X_1 = \beta_1, \ldots, X_{i-1} = \beta_{i-1}, X_i = 1] \geq$
                 $E[Z|X_1 = \beta_1, \ldots, X_{i-1} = \beta_{i-1}, X_i = 0]$
            then $\beta_i := 1$
            else $\beta_i := 0$
Step 2:   Simulate the work of A_β on w, where $\beta = \beta_1\beta_2 \ldots \beta_n$.
Output: $A_\beta(w)$.

Lemma 5.4.4.1 directly implies that $cost(A_\beta(w)) \geq E[Z]$. The algorithm COND-PROB(A) is a polynomial-time algorithm if A is a polynomial-time algorithm and Step 1 can be implemented in polynomial time. Thus, the crucial point is whether the conditional probabilities $E[Z|X_1 = \alpha_1, \ldots, X_i = \alpha_i]$ are computable in polynomial time. In the following section we show examples where this is possible.

5.4.5 Conditional Probabilities and Satisfiability

We illustrate the usefulness of the method of conditional probabilities by applying it for some satisfiability problems. In fact, we show that all the randomized algorithms designed in Section 5.3.6 for MAX-E3SAT can be derandomized by this method. Moreover, the derandomization of the random sampling algorithm for MAX-E3SAT by the method of conditional probabilities results in a more efficient deterministic approximation algorithm than the algorithm DERAND-RSMS-3 obtained by the derandomization method for the probability space reduction in Section 5.4.3.

Let us first consider the problem MAX-EkSAT for a positive integer $k \geq 2$, and the algorithm RSMS based on random sampling. Let Φ over a set of Boolean variables $\{x_1, \ldots, x_n\}$ be an instance of MAX-EkSAT, and let $C = \{C_1, \ldots, C_m\}$ be the set of clauses of Φ, $m \in \mathbb{N}$. To analyze the work of RSMS on Φ we considered the probability space $(\Omega, Prob)$, where $\Omega = \{0, 1\}^n$ and every elementary event of Ω corresponds to the choice of the values of 0 and 1 for the random variables $X_1, \ldots, X_n$ with probabilities $Prob(X_i = \alpha) = 1/2$ for all $i = 1, \ldots, n$ and all $\alpha \in \{0, 1\}$. Moreover, we have for every $j \in \{1, \ldots, m\}$ the random variable Z_j indicating whether the clause C_j is satisfied. We proved

$$E[Z_j] = 1 - \frac{1}{2^k} = \frac{2^k - 1}{2^k}$$

for all $j = 1, \ldots, m$. The random variable $Z = \sum_{j=1}^{m} Z_j$ counts the number of satisfied clauses and because of the linearity of the expectation we obtained

$$E[Z] = \frac{2^k - 1}{2^k} \cdot m.$$

Now, to derandomize RSMS it is sufficient to show that, for any $i \in \{1, \ldots, n\}$ and for any $\alpha_1, \alpha_2, \ldots, \alpha_i \in \{0, 1\}^i$, one can efficiently compute the conditional probability $E[Z | X_1 = \alpha_1, X_2 = \alpha_2, \ldots, X_i = \alpha_i]$. Since

$$E[Z | X_1 = \alpha_1, \ldots, X_i = \alpha_i] = \sum_{j=1}^{m} E[Z_j | X_1 = \alpha_1, \ldots, X_i = \alpha_i],$$

we can do it successfully with the following algorithm.

Algorithm 5.4.5.1. CCP

Input: Φ and $\alpha_1, \ldots, \alpha_i \in \{0, 1\}^i$ for some positive integer i.

Step 1: **for** $j = 1$ **to** m **do**

      begin replace the variables $x_1, \ldots, x_i$ by the constants $\alpha_1, \ldots, \alpha_i$, respectively, in the clause C_j and denote by $C_j(\alpha_1, \ldots, \alpha_i)$ the resulting simplified clause;

      if $C_j \equiv 0$

      then set $E[Z_j | X_1 = \alpha_1, \ldots, X_i = \alpha_i] := 0$

      else if $C_j \equiv 1$

          then set $E[Z_j | X_1 = \alpha_1, \ldots, X_i = \alpha_i] := 1$

          else set $E[Z_j | X_1 = \alpha_1, \ldots, X_i = \alpha_i] := 1 - \frac{1}{2^l}$

          where l is the number of different variables appearing in $C_j(\alpha_1, \ldots, \alpha_i)$.

      end

Step 2: $E[Z | X_1 = \alpha_1, \ldots, X_i = \alpha_i] := \sum_{j=1}^{m} E[Z_j | X_1 = \alpha_1, \ldots, X_i = \alpha_i]$.

Output: $E[Z | X_1 = \alpha_1, \ldots, X_i = \alpha_i]$.

One can easily verify that CCP computes the conditional probability $E[Z|X_1 = \alpha_1, \ldots, X_i = \alpha_i]$ correctly. Step 1 can be implemented with $O(n+m)$ complexity and Step 2 can be performed with $O(m)$ complexity. CCP, then, works in linear time in the input length, i.e., it is very efficient.

Following the general scheme of the derandomization by conditional probabilities (Algorithm 5.4.4.2), we obtain the following result.

Theorem 5.4.5.2. *For every integer $k \geq 3$ the algorithm* COND-PROB(RSMS) *with the conditional probabilities computed by* CCP *is a* $(2^k/(2^k - 1))$-*approximation algorithm for* MAX-EkSAT *with* $Time_{\text{COND-PROB(RSMS)}}(N) = O(N^2)$, *where N is the input length.*

Proof. As proved in Lemma 5.4.4.1 the assignment computed by COND-PROB(RSMS) satisfies at least $E[Z]$ clauses. Since

$$E[Z] \geq \frac{2^k - 1}{2^k} \cdot m$$

for RSMS and any input Φ consisting of m clauses, the approximation ratio of COND-PROB(RSMS) is at most $2^k/(2^k - 1)$.

In Step 1 of COND-PROB(RSMS) we compute $2n$ conditional probabilities; each one in time $O(n+m)$ by CCP. Thus, Step 1 can be implemented in $O(N^2)$ time. Step 2 is the simulation of the algorithm RSMS for the given sequence β, which runs in $O(n+m) = O(N)$ time. $\qquad\square$

If one applies COND-PROB(RSMS) to any instance of MAX-SAT, then one obtains a 2-approximation algorithm for MAX-SAT. This is because RSMS is a 2-expected approximation algorithm for MAX-SAT. But we have also designed the randomized algorithm RRRMS for MAX-SAT by the method of relaxation and random rounding in Section 5.3.6. The difference with RSMS is that we have a probability space (Ω, Pr), where again $\Omega = \{0,1\}^n$ and every elementary event from Ω is given by taking the values 0 and 1 for the random variables $X_1, \ldots, X_n$, but the probability distribution Pr is given by the solution of the corresponding linear program for Φ. If $\alpha(x_1), \ldots, \alpha(x_n)$ is the solution of LP(Φ) for the variables $x_1, \ldots, x_n$, then $Pr[X_i = 1] = \alpha(x_i)$ and $Pr[X_i = 0] = 1 - \alpha(x_i)$. Lemma 5.3.6.3 claims that

$$E[Z_j] \geq \left(1 - \left(1 - \frac{1}{k}\right)^k\right) \cdot \alpha(z_j)$$

if the clause C_j contains k different variables and $\alpha(z_j)$ is the solution of LP(Φ) for the variable z_j. But the main point is the equality (5.24) claiming that

$$E[Z_j] = 1 - \prod_{i=1}^{k}(1 - \alpha(x_i)).$$

Using this statement, one can compute the conditional probability as follows:

Algorithm 5.4.5.3. CCP-LP

Input: Φ and $\alpha_j = \alpha(x_j)$ for $j = 1, \ldots, n$, where $\alpha(x_j)$ is the solution of LP(Φ) for the Boolean variable x_j, and $\beta_1 \ldots \beta_i \in \{0,1\}^i$ for some integer i, $1 \leq i \leq n$.

Step 1:   **for** $j = 1$ **to** m **do**
  begin replace the variables $x_1, \ldots, x_i$ by the constants $\beta_1, \ldots, \beta_i$, respectively, in the clause C_j and let $C_j(\beta_1, \ldots, \beta_i) = x_{l_1}^{\gamma_1} \vee x_{l_2}^{\gamma_2} \vee \cdots \vee x_{l_r}^{\gamma_r}$ be the resulting simplified clause;
  if $C_j \equiv \delta$ for some $\delta \in \{0,1\}$ {i.e., $r = 0$}
  then set $E[Z_j | X_1 = \beta_1, \ldots, X_i = \beta_i] := \delta$
  else set $E[Z_j | X_1 = \beta_1, \ldots, X_i = \beta_i] :=$
        $1 - \prod_{i=1}^{r} |\gamma_i - \alpha(x_{l_i})|$
  end

Step 2: $E[Z | X_1 = \beta_1, \ldots, X_i = \beta_i] := \sum_{j=1}^{m} E[Z_j | X_1 = \beta_1, \ldots, X_i = \beta_i]$.

Output: $E[Z | X_1 = \beta_1, \ldots, X_i = \beta_i]$.

The derandomization of the algorithm RRRMS can be realized as follows.

Algorithm 5.4.5.4. DER-RRRMS

Input: A formula Φ over $X = \{x_1, \ldots, x_n\}$ in CNF, $n \in \mathbb{N}$.

Step 1:   Formulate the instance Φ of MAX-SAT as the instance LP(Φ) of the problem of linear programming.

Step 2:   Solve the relaxed version of LP(Φ).

Step 3:   Compute $\beta_1, \ldots, \beta_n$ such that $E[Z] \leq E[Z | X_1 = \beta_1, \ldots, X_n = \beta_n]$ by the strategy described in COND-PROB() and using CCP-LP to compute the conditional probabilities.

Output: An assignment $\beta_1 \ldots \beta_n$ to X.

Theorem 5.4.5.5. *For every integer $k \geq 3$, the deterministic algorithm* DER-RRRMS *is a polynomial-time $\frac{k^k}{k^k - (k-1)^k}$-approximation algorithm for* MAX-EkSAT.

Proof. The fact that DER-RRRMS is a polynomial time algorithm is obvious. The expected approximation ratio of RRRMS is $\frac{k^k}{k^k - (k-1)^k}$ and so the approximation ratio of DER-RRRMS cannot be worse than $\frac{k^k}{k^k - (k-1)^k}$. $\square$

Observe that in Section 5.3.6 we have designed a randomized (4/3)-expected approximation algorithm for MAX-SAT. The idea was to take the better solution of the two solutions provided by independent runs of the algorithms RSMS and RRRMS. Since using the method of conditional probabilities we succeeded in derandomizing both RSMS and RRRMS, the following algorithm is a deterministic polynomial-time (4/3)-approximation algorithm for MAX-SAT.

Algorithm 5.4.5.6.

Input: A formula Φ over X in CNF.

Step 1:   Compute an assignment γ to X by COND-PROB(RSMS).
          Estimate $I(\gamma) :=$ the number of clauses of Φ satisfied by γ.

Step 2:   Compute an assignment δ to X by the algorithm DER-RRRMS.
          Estimate $I(\delta) :=$ the number of clauses of Φ satisfied by δ.

Step 3:   **if** $I(\gamma) \geq I(\delta)$ **then output**(γ)
          else output(δ).

Summary of Section 5.4

The aim of derandomization is to convert randomized algorithms into deterministic ones without any essential increase in the amount of computational resources. We have presented two fundamental derandomization methods; namely, the method of the reduction of the probability space and the method of conditional probabilities.

The method of the reduction of the probability space can be successful if one observes that the total independence of random variables is not necessary for the behavior of the algorithm. A transparent example is the random sampling algorithm for MAX-EkSAT. In the analysis of its behavior we only need the independence of every k-tuple of variables because one considers the probability as satisfying a particular clause separately from the other clauses. Theorem 5.4.2.2 provides a general strategy for reducing the size of the probability space in such a case and so for reducing the number of random variables, too. After reducing the size of the probability space, one simulates deterministically all random runs (according to the new probability space) of the randomized algorithms.

The method of conditional probabilities is mainly used for optimization problems. It is based on the idea of efficiently computing a specific random sequence. This random sequence determines a run of the randomized algorithm that computes a feasible solution whose cost is at least as good as the expected cost. This method works efficiently for the derandomization of the randomized approximation algorithms of Section 5.4.2 for MAX-SAT. In this way one obtains a (deterministic) polynomial-time (4/3)-approximation algorithm for MAX-SAT.

5.5 Bibliographical Remarks

The notion "Monte Carlo" was first presented by Metropolis and Ulam [MU49]. Randomized concepts were probably already used in physics, but they were secret because they were used in atomic research. At the beginning, the notion Monte Carlo algorithm was used for all randomized algorithms. The term Las Vegas algorithm was introduced by Babai [Bab79] in order to distinguish randomized algorithms that never err from those that err with a bounded probability.

The most comprehensive source of randomized algorithms is the excellent textbook of Motwani and Raghavan [MR95]. Involved overviews of paradigms for the design of randomized algorithm are given by Harel [Har87] and Karp [Kar91]. Modern chapters about randomized approaches can be also found in Brassard and Bratley [BB96] and Ausiello, Crescenzi, Gambosi, Kann, Marchetti-Spaccamela, and Protasi [ACG$^+$99]. An excellent source for the relation between randomized and approximation algorithms is given by Mayr, Prömel, and Steger [MPS98]. Bach and Shallit [BS96] is one of the best sources for the algorithmic number theory.

The Las Vegas algorithm QUADRATIC NONRESIDUE (Section 5.3.2) was used to efficiently compute the square root of a quadratic residue by Adleman, Manders, and Miller [AMM77].

Primality testing is one of the fundamental problems of computer science. Pratt [Pra75] proved that the primality problem is in NP. In the 17th century Fermat proved the so-called Fermat's (Little) Theorem, which has been the starting point for designing efficient primality testing algorithms. The primality test based on Fermat's Theorem does not work for all numbers, especially it does not work for the so-called Carmichael numbers. Carmichael numbers were defined by Carmichael [Car12], and the proof that there are infinitely many such numbers is due to Alford, Granville, and Pomerance [AGP92]. The first one-sided-error Monte Carlo algorithm for testing primality presented here is due to Solovay and Strassen [SS77]. A fascinating, transparent presentation of their concept can be found in Strassen [Str96]. Assuming that the Extended Riemann's Hypothesis holds, Miller [Mil76] designed a polynomial-time deterministic algorithm for primality testing. The test of Miller [Mil76] was modified by Rabin [Rab76, Rab80] to yield a one-sided-error Monte-Carlo algorithm for primality testing, which was presented as the MILLER-RABIN ALGORITHM here. The randomized algorithms described here have the feature that if the input is a prime, the output is always prime, while for composite inputs there is a small probability of making errors. This has motivated the effort to search for a Las Vegas algorithm for primality testing. First, Goldwasser and Kilian [GK86] gave such a Las Vegas algorithm that works correctly for all inputs but a small set of exceptional primes. Finally, an extremely complex concept of Adleman and Huang [AH87] provides a Las Vegas algorithm for primality testing. In 1983, Adleman, Pomerance and Rumely [APR83] achieved a breakthrough by giving a deterministic $(\log_2 n)^{O(\log \log \log n)}$ algorithm for primality testing. Currently, the major, fascinating breakthrough was achieved by Agrawal, Kayal and Saxena, who designed a polynomial-time algorithm for primality testing. This algorithm works in time $O((\log_2 n)^{12})$ and so it does not have the efficiency of randomized algorithms for primality testing. But to know that primality-testing is in P is one of the most exciting achievements of algorithmics. Good expository surveys of primality testing and algorithmic number theory can be found in Angluin [Ang82], Johnson [Joh86], Lenstra and Lenstra [LL90], Pomerance [Pom82], Zippel [Zip93], and Bach and Shallit [BS96].

The algorithm RANDOM CONTRACTION for MIN-CUT is due to Karger [Kar93]. Its concept is based on a deterministic algorithm for MIN-CUT designed by Nagamochi and Ibaraki [NI92]. The most efficient version RRC was designed by Karger and Stein [KS93].

The MAX-SAT problem is one of the fundamental optimization problems. Johnson [Joh74] designed a deterministic (3/2)-approximation algorithm for MAX-SAT already in 1974. This algorithm can be viewed as the derandomization of the random sampling algorithm for MAX-SAT by the method of conditional probabilities. The deterministic (4/3)-approximation algorithm for MAX-SAT is due to Yannakakis [Yan92]. The presentation in this book is based on Goemans and Williamson [GW94b]. The same authors [GW94a] improved the relaxation method by using semidefinite programming and they obtained an 1.139-approximation algorithm for problem instances of MAX-2SAT. A good source for the study of MAX-SAT is Mayr, Prömel, and Steger [MPS98].

There was a big effort in designing randomized practicable exponential algorithms for the satisfiability problems in the last decade. Paturi, Pudlák and Zane [PPZ97] proposed an $O(n^{(1-1/k)n})$ randomized algorithm for kSAT for $k \geq 3$. The SCHÖNING'S ALGORITHM presented in Section 5.3.7 for 3SAT is a special version of the randomized multistart local search algorithm proposed by Schöning in [Schö99] for kSAT. This algorithm runs in time $O(\frac{2k}{k+1})$ for kSAT for $k \geq 3$. A further development of the concept of Schöning can be found by Dantsin, Goerdt, Hirsch, and Schöning [DGHS00], Goerdt and Krivelevich [GoK01], and Friedman and Goerdt [FrG01].

One important issue that has not been addressed in Chapter 5 is the way in which computers can produce truly random bits. The possibility of generating them is unjustified, since a real computer is a totally deterministic entity, and hence, in principle, all of its actions can be predicted in detail. Therefore, a computer cannot generate truly random sequences of bits (numbers). There are two basic approaches to overcome this difficulty. The first one is to appeal to a physical source. The drawback of this method is not only its questionable true randomness from the physical point of view, but especially its cost. The second approach is based on the generation of **pseudorandom sequences**. A pseudorandom sequence is defined to be a sequence that cannot be distinguished from a truly random sequence in polynomial time. Knuth [Knu69] and Devroye [Dev86] are two encyclopedic sources of methods for generating pseudorandom sequences. General concepts are presented by Vazirani [Vaz86, Vaz87]. Further specific examples of pseudorandom generators can be found, for instance, by L'Écuyer [L'E88, L'E90], Blum and Micali [BMi84], Yao [Yao82], Blum, Blum, and Shub [BBS86], Brassard [Bra88], and Brassard and Bratley [BB96].

An excellent overview on derandomization methods is given by Siegling [Sie98]. The method of the reduction of the size of the probability space is due to Alon, Babai, and Itai [ABI86], where even a more general result than our Theorem 5.4.2.2 is presented. This concept is a generalization of the concept of pairwise independent random variables introduced by Lancaster [Lan65] (see also the survey [O'B80] by O'Brien), and developed by Karp and Wigderson [KW84] for converting certain randomized algorithms into deterministic ones. Our presentation of the method of conditional probabilities follows the works of Ragharan [Rag88] and Spencer [Spe94].

Probably the most fundamental research problem of theoretical computer science is the comparison of the computational powers of deterministic, randomized, and nondeterministic computations. Unfortunately, we have no separation between any two of this computation modes for polynomial-time computations. For space complexity $S(n) \geq \log_2 n$ only, we know that Las Vegas is as powerful as nondeterminism (see Gill [Gil77], and Macarie and Seiferas [MS97]). Since we conjecture that P $\neq$ NP, the concept of NP-completeness of Cook [Coo71] provides a method for the classification of problems according to computational hardness. Unfortunately, this concept does not help in the comparison between deterministic polynomial time and randomized polynomial time, because we do not have any efficient randomized algorithm for an NP-hard problem. The research in this area focuses on comparing the computational powers of determinism and randomization for restricted models of computation. The biggest success in this direction was achieved for the communication complexity of communication protocols that we have used to present simple, transparent examples illustrating the power of particular randomized computing modes. Example 5.2.2.6, showing an exponential gap between determinism and one-sided-error Monte Carlo randomization, is due to Freivalds [Fre77]. Ďuriš, Hromkovič, Rolim, and Schnitger [DHR+97] proved a linear relation between determinism and Las Vegas randomization for one-way protocols, and Mehlhorn and Schmidt [MS82] proved a polynomial[82] relation between Las Vegas and determinism for the general two-way protocols. Jájá, Prassanna Kumar, and Simon [JPS84] proved some exponential gaps relating Monte Carlo protocols to the deterministic and nondeterministic ones. For surveys on this topic one can look for Hromkovič [Hro97, Hro00] and Kushilevitz and Nisan [KN97]. Some of the results comparing different computation modes for communication protocols have also been extended to other computing models such as finite automata and distinct restricted kinds of branching programs (see, for instance, Ďuriš, Hromkovič, and Inoue [DHI00]. Freivalds [Fre77], Sauerhoff [Sau97, Sau99], Dietzfelbinger, Kutylowski, and Reischuk [DKR94], Hromkovič and Schnitger [HrS99, HrS00], and Hromkovič and Sauerhoff [HrSa00]).

---

[82] At most quadratic

6

Heuristics

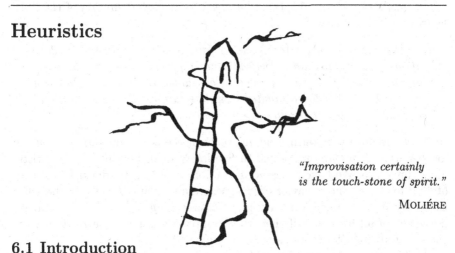

6.1 Introduction

This chapter is devoted to some algorithm design techniques which became known by the term heuristics. The term heuristic in the area of combinatorial optimization is not unambiguously specified and so it is used with different meanings. A heuristic algorithm in a very general sense is a consistent algorithm for an optimization problem that is based on some transparent (usually simple) strategy (idea) of searching in the set of all feasible solutions, and that does not guarantee finding any optimal solution. In this context people speak about local search heuristics, or a greedy heuristic, even when this heuristic technique results in an approximation algorithm. In a narrow sense a heuristic is a technique providing a consistent algorithm for which nobody is able to prove that it provides feasible solutions of a reasonable quality in a reasonable (for instance, polynomial) time, but the idea of the heuristic seems to promise good behavior for typical instances of the optimization problem considered. Thus, a polynomial-time approximation algorithm cannot be considered as a heuristic in this sense, independently of the simplicity of its design idea. Observe that the description of a heuristic in this narrow sense is a relative term because an algorithm can be considered to be a heuristic one while nobody is able to analyze its behavior. But after proving some reasonable bounds on its complexity and the quality of the produced solutions (even with an error-bounded probability in the randomized case), this algorithm becomes a (randomized) approximation algorithm and is not considered to be a heuristic any more.

Another general property of heuristics (independent of in which sense one specifies this term) is that they are very robust. This means that one can apply any heuristic technique to a large class of optimization problems, even if these problems have very different combinatorial structures. This is the main advantage of heuristics and the reason why they became so popular and widely used.

In this book we consider the following very narrow definition of the term heuristic:

> *A heuristic is a robust technique for the design of randomized algorithms for optimization problems, and it provides randomized algorithms for which one is not able to guarantee at once the efficiency and the quality of the computed feasible solutions, even not with any bounded constant probability $p > 0$.*

If one removes the term randomized from the above definition, then the local search algorithms presented in Section 3.6 can be viewed as heuristic algorithms, too. We have observed in Section 3.6.3 that there exist instances of TSP for which the behavior of local search algorithms is weak in the sense that they may produce arbitrary poor feasible solutions. This is a common drawback of all heuristics if their time complexity is not at least the size of the set of all feasible solutions.[1]

In this chapter we focus on two famous robust heuristics, namely simulated annealing and genetic algorithms. What they have in common is that both attempt to simulate some natural optimization processes. While simulated annealing is based on an analogy to thermodynamic processes, genetic algorithms look for optimization in population genetics in the process of evolution.

This chapter is organized as follows. Section 6.2 is devoted to simulated annealing. Simulated annealing is presented as an advanced local search algorithm, where it is allowed to move to a worse feasible solution (i.e., to also do deteriorations) with some probability. The idea behind this concept is to prevent getting stuck in a very poor local optimum by introducing the possibility to overcome "the hills" around local optima and so to find a better solution. The probability of deteriorating the quality of the solution decreases with the size of the deterioration and is exactly given by the laws of thermodynamics.

Section 6.3 is devoted to genetic algorithms, which are also based on the realization of a sequence of iterative improvements. But in contrast to local search, one starts with a set of feasible solutions called a population of individuals. Then, in one iterative step, a new population is generated. This is done by some operations that try to imitate the recombination of DNA sequences and by some random changes of local parts (genes). After that, the best solutions (individuals) are (randomly) chosen to be the members of the new population. The advantage of genetic algorithms is that one can take a collection of the best solutions of the population instead of a best solution

---

[1] Later, we shall see that the behavior may be still worse and that a time complexity that is essentially larger than $|\mathcal{M}(x)|$ is necessary to be able to guarantee a reasonable quality of feasible solutions, if any.

as the output. This can be especially useful if the optimization criteria are not sharp or complete and so the user can choose the best solution from the computed solutions with respect to some informal criterium.

Both Section 6.2 and 6.3 have the same structure. In the first parts of these sections the basic concept of the particular heuristic technique is presented and discussed. The second parts are devoted to the behavior of these heuristics and to the question of how to adjust the "free" parameters of the heuristics in order to get as good a behavior as possible.

Additionally, Section 6.2.3 presents randomized tabu search which can be viewed as a generalization of simulated annealing.

6.2 Simulated Annealing

6.2.1 Basic Concept

Simulated annealing may be viewed as a local search enriched by a kind of randomized decision that enables us to leave local optima in order to find better solutions. Let us carefully explain the concept of simulated annealing in what follows. Remember that, for a given neighborhood $Neigh$ of a minimization problem U, the scheme of local search is as follows.

LSS($Neigh$)

Input: An instance x of U.
Step 1: Find a feasible solution $\alpha \in \mathcal{M}(x)$.
Step 2: Find the best solution β from $Neigh_x(\alpha)$.
Step 3: **if** $cost(\beta) < cost(\alpha)$, **then** $\alpha := \beta$ and **goto Step 2 else stop**.

As we have shown in Section 3.6.3, there exist instances of TSP for which LSS($Neigh$) can get stuck in arbitrary poor local optima. There are several approaches in trying to overcome this main drawback of local search. In Section 3.6.2 we presented the Kernighan-Lin's variable search algorithm that allows us to search for an improvement of local optima α within a large distance from α using the greedy strategy. Another simple approach is to start LSS($Neigh$) several times with different randomly chosen initial solutions. The resulting algorithm is called **multistart local search**. It is true that the probability of reaching an optimal solution tends to 1 with the growth of the number of starts, but the number of starts necessary to guarantee a reasonable probability of success is usually much larger than the complexity of an exhaustive search of the set of all feasible solutions. Another possibility to try to leave local optima is the so-called **threshold local search**. The threshold local search algorithm allows moving to a weaker solution than the current one when the deterioration is not above a given threshold. The pathological input instances of TSP presented in Section 3.6.3 show that this approach also cannot prevent getting stuck in arbitrarily poor local optima.

The basic idea of simulated annealing is to add the possibility of leaving a local optimum (i.e., to move to a weaker solution than the current one) by some kind of coin tossing (random decision). The probability of accepting a deterioration in this iterative process depends on the size of deterioration as well as on the number of iterations executed up till now, and it is determined by the following physical analogy.

In condensed matter physics, **annealing** is a process for obtaining low-energy states of a solid in a heat bath. This process can be viewed as an optimization process in the following sense. At the beginning, one has a solid material with a number of imperfections in its crystal structure. The aim is to get the perfect structure of the solid, which corresponds to the state of the solid of minimal energy. The physical process of annealing consists of the following two steps:

(1) The temperature of the heat bath is increased to a maximum value at which the solid melts. This causes all the particles to arrange themselves randomly.

(2) The temperature of the heat bath is slowly decreased according to a given cooling schedule until a low-energy state of the solid (a perfect crystal structure) is achieved.

The crucial point is that this optimization process can be successfully modelled by the so-called METROPOLIS ALGORITHM that can be viewed as a randomized local search. Let, for a given state s of the solid, $E(s)$ be the energy of this state, and let k_B denote the Boltzmann constant.

METROPOLIS ALGORITHM

Step 1: Let s be the initial state of the solid with energy $E(s)$ and let T be the initial temperature of the heat bath.

Step 2: Generate a state q from s by applying a perturbation mechanism, which transfers s into q by a small random distortion (for instance, by a random displacement of a small particle).
if $E(q) \leq E(s)$ then $s := q$
else accept q as a new state with the probability

$$p(s \to q) = e^{-\frac{E(q)-E(s)}{k_B \cdot T}}$$

(i.e, remain in state s with the probability $1 - p(s \to q)$).

Step 3: Decrease T appropriately.
if T is not too close to 0 repeat Step 2,
else output(s).

First of all, we observe a strong similarity of the METROPOLIS ALGORITHM to the local search scheme. To move from the current state s one considers only a small, local change of the description of s in order to generate q. If the generated state q is at least as good as s (or even better), then q is considered as the new state. The main differences are:

(i) The METROPOLIS ALGORITHM may accept a deterioration with the probability

$$p(s \rightarrow q) = e^{-\frac{E(q)-E(s)}{k_B \cdot T}}.$$

(ii) The value of the required parameter T decides on the termination of the METROPOLIS ALGORITHM while the local optimality is the criterion for halting in the local search scheme.

The probability $p(s \rightarrow q)$ follows the laws of thermodynamics claiming that at temperature T, the probability $p(\Delta E)$ of an increase in energy of magnitude ΔE is given by

$$p(\Delta E) = e^{-\frac{\Delta E}{k_B \cdot T}}.$$

If the temperature is lowered sufficiently slowly, the solid can reach the so-called **thermal equilibration** at every temperature. Consider a fixed temperature T. Let $s_1, s_2, \ldots, s_m$ be all possible states of the solid, $m \in \mathbb{N}$. Let X be a random variable, where $X = i$ means that the solid is in the state s_i (at temperature T). The thermal equilibration is characterized by the **Boltzmann distribution** which relates the probability of the solid being in a state s_i to the temperature T, and is given by

$$Prob_T(X = i) = \frac{e^{-\frac{E(s_i)}{k_B \cdot T}}}{\sum\limits_{j=1}^{m} e^{-\frac{E(s_j)}{k_B \cdot T}}}.$$

The Boltzmann distribution is essential for proving the convergence of the METROPOLIS ALGORITHM to an optimal state. The most important properties of $p(s \rightarrow q)$ are:

- the probability $p(s \rightarrow q)$ of the movement from the state s to the state q decreases with $E(q) - E(s)$, i.e., large deteriorations are less probable than small deterioration, and
- the probability $p(s \rightarrow q)$ increases with T, i.e., large deteriorations are more probable at the beginning (when T is large) than later (when T becomes smaller and smaller).

Intuitively, this means that at the beginning it is possible to overcome "high hills" around very deep local optima in order to come to some deep valleys, where later only some not too deep local optima can be left. Very roughly, this optimization approach can be viewed as a recursive procedure in the following sense. First, one climbs on a top of a very high mountain and looks for the most promising areas (a deep valley). Then, one goes to this area and recursively continues to search for the minimum in this area only.

If one wants to use the strategy of the METROPOLIS ALGORITHM for randomized local search in combinatorial optimization, then one has to use the

following correspondence between the terms of thermodynamic optimization
and combinatorial optimization.

- the set of system states
- the energy of a state
- perturbation mechanism
- an optimal state
- temperature

- the set of feasible solutions
- the cost of a feasible solution
- a random choice from the neighborhood
- an optimal feasible solution
- a control parameter

The simulated annealing algorithm is a local search algorithm that is based
on the analogy to the METROPOLIS ALGORITHM. If one has fixed a neighbor-
hood *Neigh* for a minimization problem $U = (\Sigma_O, \Sigma_O, L, L_I, \mathcal{M}, cost, minimum)$
then the simulated annealing algorithm can be described as follows.

Simulated Annealing for U with respect to *Neigh*
SA(*Neigh*)

Input: An input instance $x \in L_I$.
Step 1:   Compute or select (randomly) an initial feasible solution $\alpha \in \mathcal{M}(x)$.
          Select an initial temperature (control parameter) T.
          Select a temperature reduction function f as a function of two pa-
          rameters T and time.
Step 2: $I := 0$;
          while $T > 0$ (or T is not too close to 0) **do**
              begin randomly select a $\beta \in Neigh_x(\alpha)$;
                  if $cost(\beta) \leq cost(\alpha)$ **then** $\alpha := \beta$
                  else begin generate a random number r uniformly in the
          range
                      $(0, 1)$;
                      if $r < e^{-\frac{cost(\beta) - cost(\alpha)}{T}}$
                      then $\alpha := \beta$
                  end;
                  $I := I + 1$;
                  $T := f(T, I)$
              end
Step 3:   **output**(α).

We see that there are two main free parameters to be chosen in the simulated annealing algorithm, namely, the neighborhood (as in any local search
algorithm) and the cooling scheme that determines the speed of the decrease
of the temperature. Observe that a slow decrease of T may result in an extremely large time complexity of simulated annealing, and we shall see later
that the increase of time complexity increases the probability of getting feasible solutions of high quality. The next section is devoted to the theoretical
and experimental knowledge of the tradeoffs between time complexity and
the quality of solutions produced by simulated annealing with respect to the
choice of neighborhoods and cooling schemes.

6.2.2 Theory and Experience

Simulated annealing has become a successful, widely used method in combinatorial optimization. The main reasons for this success are:

- the simulated annealing algorithm is based on a simple, transparent idea and it is easy to implement,[2]
- simulated annealing is very robust and can be applied for almost all (combinatorial) optimization problems,
- simulated annealing behaves better than local search (that also has the above formulated positive properties – simplicity and robustness) due to the replacement of the deterministic acceptance criterion of local search by a stochastic criterion.

Clearly, the success of the application of the simulated annealing algorithm to a particular optimization problem depends on the choice of the free parameters of the simulated annealing scheme – the neighborhood and the cooling schedule. Observe that the combinatorial structure of our optimization problems has nothing in common with the thermodynamic processes and so we cannot automatically take the parameters used in the physical annealing process.

In this section we present both theoretical and experimental results about the behavior of simulated annealing with respect to the choice of neighborhoods and cooling schedules. We omit the proofs of the theoretical results here because they are not essential for users of this method and their mathematical considerations are beyond the elementary level of this book.

The most important result says that some weak assumptions about the neighborhood and the cooling scheme are sufficient to assure an asymptotic convergence of simulated annealing to the set of optimal solutions.

Theorem 6.2.2.1. *Let U be a minimization problem and let Neigh be a neighborhood for U. The asymptotic convergence of the simulated annealing algorithm for an input x is guaranteed if the following conditions are satisfied.*

(i) Every feasible solution from $\mathcal{M}(x)$ is reachable from every feasible solution from $\mathcal{M}(x)$ (condition (iii) of the definition of a neighborhood in Definition 3.6.1.1), and

(ii) the initial temperature T is at least as large as the depth[3] of the deepest local, nonglobal minimum.

The asymptotic convergence means that one reaches a global optimum with probability tending to 1 with the number of iteration steps, i.e., one reaches an optimum after an infinite number of iterations. It is important

---

[2] This means that the cost of the development of a simulated annealing algorithm for a given optimization problem is low and the time of the development is short.

[3] The depth of a local minimum α is the size of the minimal deterioration that is sufficient for leaving α.

to note that the meaning of the asymptotic convergence result does not say that the limit of the sequence of feasible solutions generated by the simulated annealing algorithm is an optimal solution, as some people mistakenly believe. The right interpretation is that simulated annealing will find itself at a global optimum an increasing percentage of the time as time grows. This is good news from the convergence point of view because local search cannot guarantee any general result of this kind for polynomial-time searchable neighborhoods. On the other hand, this result is very far from the one we would like to have because we are interested in the quality of feasible solutions produced in a bounded time. But Theorem 6.2.2.1 does not guarantee anything for polynomial-time computations. Observe that our partially informal presentation of the convergence result for simulated annealing implicitly includes the assumption that the cooling is done slowly. There exist several further theoretical results saying under which assumption about the neighborhoods and cooling schemes the simulated annealing algorithm asymptotically converges to the set of optimal solutions. Typical requirements are the symmetricity of the neighborhoods (condition (ii) of the definition of a neighborhood in Definition 3.6.1.1), the uniformity of neighborhoods,

$$|Neigh_x(\alpha)| = |Neigh_x(\beta)| \text{ for all } \alpha, \beta \in \mathcal{M}(x),$$

and slow increase of T by specific cooling schedules.

The central question for simulated annealing is whether one can guarantee feasible solutions of high quality after a reasonable restricted number of iterative steps. Unfortunately, the answer is negative. To get a guarantee for an approximation of an optimal solution one needs the number of iterative steps to be at least quadratic in the size of the solution space. Thus, it is easier to execute the exhaustive search of $\mathcal{M}(x)$ than to run simulated annealing until it guarantees a good approximation. In the case of TSP, simulated annealing even needs

$$\Omega\left(n^{n^{2n-1}}\right)$$

iterative steps to reach a high quality feasible solution. Observe that $n^{n^{2n-1}}$ is essentially larger than $|\mathcal{M}(x)| \leq n!$ for $n = |x|$. The final observation is that the rate of convergence of simulated annealing is logarithmic in the number of iterative steps.

In Section 3.6 we exhaustively discussed the role of the size of neighborhoods for the success of local search and so we do not repeat this discussion here. Since, in contrast to local search, simulated annealing can leave local optima, one prefers small (simply definable) neighborhoods. This enables an efficient execution of one iterative step and so this strategy enabling the performance of many iterative steps usually provides better solutions than the strategy of using large neighborhoods and a small number of iterative steps. Theoretical results show that it is prudent to avoid neighborhoods with the following structural properties:

- spiky structure (topography),
- deep troughs,
- large plateau-like areas.

In general, it is not easy to assure nice structures of solution spaces, because for every neighborhood there may exist a pathological problem instance that has a very wild topology.

The rest of this section is devoted to cooling schedules. Experience shows that usually the largest improvement is achieved in the middle of the schedule. This is because the first part of the work of the simulated annealing algorithm is realized under a very high temperature that enables almost all changes (deterioration). Thus, this part can be viewed as a production of a sequence of almost random feasible solutions, i.e., an almost random search for an initial solution. On the other hand, if T is already small, large deteriorations are very improbable and so large changes are practically not possible.

If one speaks about cooling schedule it means that the following parameters have to be specified:

- an initial temperature T,
- a temperature reduction function[4] $f(T, t)$, and
- a termination condition (a value term such that simulated annealing stops for $T \leq term$).

CHOICE OF THE INITIAL TEMPERATURE.

The inital value T has to be large enough to allow all transitions to be accepted, because in the physical analogy it should correspond to heating up the solid until all particles are randomly arranged in the liquid phase. One possibility is to take T as the maximal difference in the cost between any two neighboring solutions. Since it may be hard to compute this value it is sufficient to efficiently find an estimation (upper bound) of it.

Another pragmatic possibility is to start with any value T and, after choosing a neighbor β of the initial situation α, to increase T in such a way that β will be accepted with a probability of almost 1. Doing so for a number of first iterative steps one can get a reasonable initial value T. Thus, these first steps can be viewed as the heating procedure in the physical analogy.

CHOOSING THE TEMPERATURE REDUCTION FUNCTION.

The typical choice of the temperature reduction function is to multiply T by some constant r, $0.8 \leq r \leq 0.99$. Using this reduction one works a constant number d of iterative steps with a fixed T, and after d steps

$$T := r \cdot T.$$

---

[4] Also called decrement function or cooling ratio

This means that the temperature T_k after k reductions is $r^k \cdot T$.

Another frequent choice[5] of the temperature reduction is

$$T_k := \frac{T}{\log_2(k+2)}.$$

The number d (the number of iterations for any fixed temperature) is usually chosen as the size of the neighborhood.

TERMINATION CONDITION.

One possible termination condition is independent of T and says that simulated annealing can stop when there is no change in the cost of the solutions for a time. Another possibility is to take a fixed constant term and to stop when $T \leq term$.

In thermodynamics one considers

$$term \leq \frac{\varepsilon}{\ln[(|\mathcal{M}(x)| - 1)/p]},$$

where p is the probability of getting a feasible solution with an ε-approximation ratio.

General empirical evidence supported by theoretical knowledge, is that the manner of cooling is not as important as the rate. So, one does not need to spend too much time with the choice between cooling schemes with the same cooling rate, but one should concentrate on the choice of the initial temperature T and on the choice of a sufficiently large number d of iterations with a fixed temperature.

Finally, we present some experience obtained by running simulated annealing in polynomial time. First, we list some general properties independent of particular applications and then we discuss the comparison of simulated annealing with other methods for specific optimization problems.

GENERAL OBSERVATIONS

- Simulated annealing has the potential to provide feasible solutions of high quality but at the cost of enormous computational efforts.
- The quality of the output does not essentially depend on the choice of the initial feasible solution (which is the case for local search).
- The essential parameters for the success of simulated annealing are the reduction rate and the neighborhood. A lot of experimental work is necessary to adjust the parameters for a particular problem well.

---

[5] Both reduction functions presented here are called static. There exist more complicated reduction functions where the speed of reduction varies in the time. Such reduction functions are called dynamic. They are based on some nontrivial stochastic analysis and so we omit the presentation of them in this introductory material.

- The average case time complexity of simulated annealing is close to the worst case time complexity.
- Considering the same neighborhood, simulated annealing behaves substantially better (higher quality in the same time) than local search or multistart local search.

Next, we observe that the success of simulated annealing may essentially depend on the class of optimization problems considered.

APPLICATIONS TO SPECIFIC OPTIMIZATION PROBLEMS

- For graph positioning problems (MAX-CUT, MIN-CUT, etc.), the simulated annealing performs better, with respect to both quality of the solutions and time complexity, than the Kernighan-Lin's variable-depth search algorithm (Section 3.6.2).
- For several engineering problems, for instance from the area of VLSI design, the simulated annealing method seems to be the best approach and it even beats nontrivial approximation algorithms.
- For TSP and similar problems, simulated annealing is weak and there are usually several different approaches that provide substantially better solutions in a shorter time.

6.2.3 Randomized Tabu Search

Tabu search is a heuristic based on local search. The main reason to present it here is that randomized tabu search may be viewed as a generalization of simulated annealing in the following sense. Local search algorithms and the simulated annealing algorithm are memory-less algorithms. Memory-less means here that the next step depends on the current feasible solution only (or cooling schedule in the case of simulated annealing) but does not depend on the history of the work of the algorithm. This is not the case for the Kernighan-Lin's variable-depth search algorithm,[6] where one generates a sequence of feasible solutions in such a way that no change is reversed during the generation of this solution sequence. Thus, the choice of the next neighboring feasible solution is not arbitrary, because some neighbors are forbidden (tabu). And this is the idea of the tabu search: to store some information about a sequence of the last feasible solutions generated and to use this information when generating the next feasible solution.

The general schema of tabu search for an optimization problem U may look like the following one. As usual for local search, we consider a fixed neighborhood $Neigh$ for U. Without loss of generality we assume that U is a minimization problem.

---

[6] Presented in Section 3.6.2

Tabu Search for U with respect to *Neigh*
TS(*Neigh*)

Input: An instance x of U.

Step 1: Choose an initial solution $\alpha \in \mathcal{M}(x)$;

Set $TABU := \{\alpha\}$; $STOP := FALSE$; $BEST := \alpha$.

Step 2: Take the best feasible solution $\beta \in Neigh_x(\alpha) - TABU$;

if $cost(\beta) < cost(\alpha)$ then $BEST := \beta$;

update $TABU$ and $STOP$;

$\alpha := \beta$.

Step 3: if $STOP = TRUE$ then output($BEST$) else goto Step 2.

Observe that TS(*Neigh*) exchanges α for β even if β is worse than α, which is the main difference between it and local search. The variable $BEST$ is used to store the best feasible solution that was generated by TS(*Neigh*) up till now and so $BEST$ is the output of TS(*Neigh*).

In the above scheme we did not specify how to update $STOP$ and $TABU$. The simplest possibility to update $STOP$ is the standard way used by local search. One sets $STOP := TRUE$ when no improvement was found (i.e., $cost(\beta) \geq cost(\alpha)$). Since $TABU$ may change this is not always the best possibility and one can decide to stop when several consecutive executions of Step 2 do not provide any improvement.

The specification of $TABU$ allows many different strategies. A basic one is to forbid any feasible solution generated in the last k steps for some positive integer k. This can avoid some repetitions or even short cyclic developments. Another possibility (similar to Kernighan-Lin) is to look at the representation of feasible solutions and to require that local transformations do not always change the same parts of the representation. A more advanced form of tabu search can modify the cost function when searching in the neighborhood $Neigh(\alpha)$. The solutions obtained by local transformations of parts of the representation that were changed recently will get some additional cost and the solutions obtainable from α by local transformations on parts of α that were not changed in the last steps become cheaper than their costs assigned by the original cost function. In this interpretation tabu does not mean forbidden, but it provides some additional preferences when choosing the next feasible solution. For instance, considering MAX-SAT and the Flip neighborhood, TS(*Flip*) may reduce the cost of a solution obtained by a flip of a variable that was frequently flipped in the last time and TS(*Flip*) increases the costs of solutions obtained by flips of variables whose values were not changed in the last k steps. Thus, the choice of the "best" solution from $Flip(\alpha)$ is done with respect to the cost functions and some penalties and preferences given by $TABU$.

Tabu search may be randomized in the same way as local search was randomized by simulated annealing. One takes a new, weaker solution with a probability that decreases with the size of the deterioration.

Randomized Tabu Search
RTS(*Neigh*)

Input: An instance x of a minimization problem U.
Step 1: Choose an initial solution $\alpha \in \mathcal{M}(x)$;
         Set $TABU := \{\alpha\}$; $STOP := FALSE$; $BEST := \alpha$;
         Choose a value T;
Step 2: Take the best feasible solution $\beta \in Neigh_x(\alpha) - TABU$;
         if $cost(\beta) < cost(\alpha)$ **then**
           begin
             $BEST := \beta$;
             $\alpha := \beta$
           end
         else accept β as a new solution $(\alpha := \beta)$ with the probability
           $prob(\alpha \to \beta) = e^{-\frac{cost(\beta) - cost(\alpha)}{T}}$;
         update $TABU$;
         update $STOP$;
Step 3: **if** $STOP = TRUE$ **then** output($BEST$)
         else goto Step 2.

Again, one can modify the above scheme by replacing the absolute application of $TABU$ with a relativized application without any prohibition, but with some penalties and preferences. An important point is that one can prove the asymptotic convergence of RTS(*Neigh*) even if RTS(*Neigh*) does not use the probability $prob(\alpha \to \beta)$ based on the Metropolis algorithm, but another probability reasonably decreasing with the difference $cost(\beta) - cost(\alpha)$.

Keywords introduced in Section 6.2.3

annealing, multistart local search, Metropolis algorithm, cooling schedule, thermal equilibrium, Boltzmann distribution, simulated annealing, tabu search, randomized tabu search

Summary of Section 6.2.3

Simulated annealing may be viewed as a local search enriched by a kind of randomized decisions that enable moving to a weaker feasible solution than the current one (i.e., that allows some cost deteriorations) and so to leave local optima in order to find better solutions. Simulated annealing is based on the analogy to annealing (a process for obtaining low-energy states of a solid in a heat bath) in condensed matter physics. The process of annealing can be described by the Metropolis algorithm which can be viewed as a local search algorithm for the minimization of the energy of the solid. If a generated state β of the solid has lower energy than the current state α, then α is always exchanged for β in the Metropolis algorithm. In the opposite case, α is exchanged for β with a probability (given by the laws of thermodynamics) that decreases with the size of the deterioration (the difference between the energies of β and α). If one considers feasible solutions instead of states of a solid and a cost function instead of energy, then the Metropolis algorithm can be used for solving minimization problems. The resulting algorithm is called simulated annealing.

Some mild assumptions about neighborhoods are sufficient to guarantee the asymptotic convergence of the simulated annealing algorithm. Unfortunately, this convergence is very slow and it is impossible to guarantee high-quality solutions in polynomial time. Anyway, simulated annealing is always better as a local search and, for some classes of optimization problems, simulated annealing is one of the few best methods for solving them. There exist also problems such as TSP, for which simulated annealing cannot compete with other methods.

Local search and simulated annealing are viewed as memory-less algorithms in the sense that the choice of the candidate for the next feasible solution does not depend on the history of the computation. Tabu search is a local search where one saves some information about a number of last steps (feasible solutions). This information is used either to forbid the generation of some neighboring feasible solutions or at least to give a penalty for the choice of some neighbors. Tabu search allows deteriorations (and so it can leave local optima) by always moving to the best neighboring solution that is not tabu. If a deterioration is accepted with a probability depending on the size of the deterioration, we speak about randomized tabu search, which can be viewed as a generalization of simulated annealing.

6.3 Genetic Algorithms

6.3.1 Basic Concept

Like simulated annealing, the concept of genetic algorithms is an optimization algorithm design technique that is inspired by some optimization processes in nature. Another important common property of both these algorithm design techniques is that they provide randomized algorithms and that the randomization is crucial for their successful behavior. While simulated annealing is based on the analogy to processes in physics, genetic algorithms are based on the analogy to optimization in the process of evolution of a population of individuals.

The main paradigms of the considered evolutionary process are the following ones:

(1) The individual of the population can be represented by strings (or vectors) over a finite alphabet, where a string is the code of the genetic structure of a chromosome (DNA sequence).

(2) To create a new individual one needs two parents on which the well-known genetic operation **crossover** is applied. The operation crossover performs an exchange of some sections of the parents' chromosomes, i.e., the "child" gets parts of the genetic information from both parents (for instance, the

first half of the chromosomes of the first parent and the second half from the second parent). For instance, consider two parents

$$\alpha = 01001101 \text{ and } \beta = 11100011.$$

If one has one crossover after point 5 only, then we obtain two children

$$\gamma = 01001011 \text{ and } \delta = 11100101.$$

Clearly, γ contains the first five bits of α and the last three bits of β. Symmetrically, δ contains the first five bits of β and the remaining three bits are the last three bits of α.

(3) There exists a **fitness value** for every individual, which judges the quality (evolutionary maturity) of the individuals. Parents are randomly chosen from the population, and the individuals with a high fitness value have higher probabilities of being chosen to be parents than the individuals with a low fitness value.

(4) The genetic operation of **mutation** is randomly applied to the individuals. This operation causes a random modification of a local part of the chromosomes.

(5) The individuals may pass away. Their lifetime is strongly connected to their fitness values.

If one wants to apply the above principles of modeling evolution processes in biology, one has to think about the following correspondence on the level of terminology.

- an individual
- a gene
- fitness value
- population
- mutation

- a feasible solution
- an item of the solution representation
- cost function
- a subset of the set of feasible solutions
- a random local transformation

There are two main differences between this scenario and simulated annealing. The first one is that one works with sets of feasible solutions instead of working with one feasible solution only. This may be viewed as an advantage of genetic algorithms because they may produce a collection of high-quality feasible solutions and there are applications that require the production of several different solutions. The second difference is connected with the first one. Genetic algorithms cannot be viewed as pure local search because of the creation of new individuals by the crossover operation. If the new individual gets the genetic code from both parents in a balanced way, then there is no small (local) neighborhood that enables its generation.

Following the genetic representation of individuals (paradigm 1 from above) the main assumption for applying genetic algorithms in combinatorial optimization is to have a convenient representation of feasible solutions by strings or vectors[7] in a similar manner as required for greedy algorithms. Observe that we can do so for almost all optimization problems considered here. A substantially harder requirement on the feasible solution representation is to find a representation that guarantees the generation of a representation of a feasible solution from every pair of feasible solutions by the crossover operation. Having a suitable representation of feasible solutions one can describe the general scheme of genetic algorithms as follows:

Genetic Algorithm Scheme (GAS)

Input: An instance x of an optimization problem $U = (\Sigma_I, \Sigma_O, L, L_I, \mathcal{M},$ $cost, goal)$.

Step 1:   Create (possibly randomly) an initial population $P = \{\alpha_1, \ldots, \alpha_k\}$ of size k;
$t := 0$;
{t denotes the number of created populations}.

Step 2:   Compute $fitness(\alpha_i)$, for $i = 1, \ldots, k$
{the fitness of every individual α_i may be $cost(\alpha_i)$}.
Use $fitness(\alpha_i)$ to estimate a probability distribution $Prob_P$ on P in such a way that feasible solutions with high fitnesses get assigned higher probabilities than feasible solutions with small fitnesses.

Step 3:   Use $Prob_P$ to randomly choose $k/2$ pairs of feasible solutions (β_1^1, β_1^2), $(\beta_2^1, \beta_2^2), \ldots, (\beta_{k/2}^1, \beta_{k/2}^2)$. Use the crossover operation on every pair of parents (β_i^1, β_i^2) for $i = 1, \ldots, k/2$ in order to create new individuals, and put them into P.

Step 4:   Apply randomly the mutation operation to each individual of P.

Step 5:   Compute the fitness $fitness(\gamma)$ of all individuals γ in P and use it to choose $P' \subseteq P$ of cardinality k.
Possibly improve every individual of P' by local search with respect to a neighborhood.

Step 6: $t := t + 1$;
$P := P'$;
if the stop criterion is not fulfilled goto Step 2
else give the best individuals of P as the output.

There are many free parameters to be adjusted in the above genetic algorithm scheme and numerous possibilities for their choice. We move the discussion on this topic, except the representation of feasible solutions (individuals), to the next section and focus on the knowledge of the general properties and behavior of genetic algorithms.

The first thing that has to be done before applying the concept of genetic algorithms is to choose a suitable representation of feasible solutions as well as to determine the crossover operation on this representation in such a way that

---

[7] Note that sometimes people consider nonlinear data representations like trees or other special structures but then it is hard to search for a biological interpretation of the substitute for the crossover operation.

it maps two feasible solutions into two feasible solutions. Examples of problems with straightforward representations are cut problems and maximum satisfiability problems. For instances of cut problems with $V = \{v_1, v_2, \ldots, v_n\}$, the feasible solutions are strings $\alpha = \alpha_1 \alpha_2 \ldots \alpha_n \in \{0, 1\}^n$, where α determines a cut (V_1, V_2) in such a way that $V_1 = \{v_i \in V \mid \alpha_i = 1\}$ and $V_2 = \{v_j \in V \mid \alpha_j = 0\}$. Obviously, any crossover of two strings $\alpha, \beta \in \{0, 1\}^n$ results in strings over $\{0, 1\}^n$. Since every string over $\{0, 1\}^n$ is a representation of a cut (of a feasible solution) the crossover operation is well-defined and the scheme of genetic algorithms can be applied in a straightforward way. Similar situations occur for maximum satisfiability problems when a string over $\{0, 1\}^n$ represents the assignment of Boolean values to the n Boolean variables of a given formula. Considering optimization problems like TSP, knapsack problem, bin packing, and minimum set cover problem one can easily see that the representations of feasible solutions used in the previous chapters do not allow any simple crossover operation as introduced above. For instance, if $(\alpha_1, \alpha_2, \ldots, \alpha_n)$ and $(\beta_1, \beta_2, \ldots, \beta_n)$ represent the characteristic vectors of two set covers of an input instance $(X, \{S_1, S_2, \ldots, S_n\})$, a vector $\alpha_1 \alpha_2 \ldots \alpha_i \beta_{i+1} \ldots \beta_n$ can represent a $\varphi \subseteq Pot(X)$ that does not cover X (i.e., the constraints are not satisfied). How to proceed in such cases will be discussed in the next section.

Similarly as for simulated annealing, the basic question is what can be proved about the behavior of genetic algorithms. The effort goes in the direction of trying to give a guarantee on the quality of the best generated feasible solutions. Unfortunately, no reasonable assumptions assuring a fast convergence to global optima have been found and it is questionable whether they even exist. Concerning theoretical analysis of the behavior of genetic algorithms, only two substantial results were achieved which we roughly present in what follows.

Consider a restricted genetic algorithm without the crossover operation, which creates a new generation by applying the mutation operation only. If the mutation is done on every item (gene) of an individual with the same small probability, and the probability of the choice of an individual for the membership in the new generation depends on its fitness, then we see a strong similarity between simulated annealing and this restricted genetic algorithm. The mutation operation can be viewed as a random choice from a probably small neighborhood and the randomized choice of individuals for the next generation can be viewed as a randomized acceptance criterion on the new individuals. By suitable choices of all these probabilities there is a possibility to prove similar asymptotic convergence results as for simulated annealing.

The second result analyzing the behavior of genetic algorithms is the so-called **Schema Theorem** that tries to find practically formal arguments for explaining the intuition of a reasonable behavior of genetic algorithms. In what follows we present a version of the Schema Theorem for problems whose feasible solutions can be represented by binary strings.

Definition 6.3.1.1. *Let* $U = (\Sigma_I, \Sigma_O, L, L_I, \mathcal{M}, cost, goal)$ *be an optimization problem. Let, for a given instance* $x \in L_I$ *of size* n, *every* $\alpha \in \{0,1\}^n$ *represent a feasible solution to* x *(i.e.,* $\mathcal{M}(x) = \{0,1\}^n$*). A* **schema for** $\mathcal{M}(x)$ *is any vector* $s = (s_1, s_2, \ldots, s_n) \in \{0,1,*\}^n$. *A* **set of feasible solutions of a schema** $s = (s_1, s_2, \ldots, s_n)$ *is*

$$Schema(s_1, s_2, \ldots, s_n) = \{\gamma_1, \gamma_2, \ldots, \gamma_n \in \mathcal{M}(x) \mid \gamma_i = s_i$$
$$\text{for all } i \in \{1, \ldots, n\} \text{ such that } s_i \in \{0,1\},$$
$$\text{and } \gamma_j \in \{0,1\} \text{ for all } j \in \{1, 2, \ldots, n\}$$
$$\text{such that } s_i = *\}.$$

The **length of a schema** s, *denoted by* **length**(s), *is the distance between the first and the last non-* position in* s. *The* **order of a schema** s, *denoted by* **order**(s), *is the number of non-* positions. The* **fitness of a schema** $s = (s_1, s_2, \ldots, s_n)$ *in a population* P *is the average fitness of feasible solutions in* $Schema(s)$, *i.e.,*

$$Fitness(s, P) = \frac{1}{|Schema(s) \cap P|} \cdot \sum_{\gamma \in Schema(s) \cap P} cost(\gamma).$$

The **fitness ratio of a schema** s *in a population* P *is*

$$Fit\text{-}ratio(s, P) = \frac{Fitness(s, P)}{\frac{1}{|P|} \sum_{\beta \in P} cost(\beta)}.$$

Aver-Fit$(P) = \frac{1}{|P|} \sum_{\beta \in P} cost(\beta)$ *is called the* **average fitness of the population** P.

We see that a schema is nothing else than fixing values of a few items (genes) in the representation of feasible solutions. The positions labeled by $*$ are free and so $|Schema(s)|$ is exactly 2 to the number of $*$ in s. A schema with a $Fitness(s, P)$ that is essentially higher than the average fitness of the population P seems to be good genetic information that should be spread in the evolution process. Our aim is now to show that the cardinality of $Schema(s) \cap P$ grows relative to $|P|$ if s has a high (at least greater than 1) fitness ratio. In what follows we consider a simple version of genetic algorithms, where all parents die and the new generation consists of mutated children only.

First, we prove three simple technical lemmata. Let P_0 be the initial population, and let P_t, for $t = 1, 2, \ldots$, be the population created in the tth iteration step of a genetic algorithm. Let, for a fixed schema s and a fixed population P_t, $t = 1, 2, \ldots, X_{t+1}(s)$ be a random variable counting the number of parents chosen from P_t in a random way and being in $Schema(s)$. The probability of choosing $\alpha \in P_t$ is considered to be

$$Pr_{par}(\alpha) = \frac{cost(\alpha)}{\sum_{\beta \in P_t} cost(\beta)} = \frac{cost(\alpha)}{|P_t| \cdot Aver\text{-}Fit(P_t)}. \tag{6.1}$$

Lemma 6.3.1.2. *For every $t \in \mathbb{N}$ and every schema s,*

$$E[X_{t+1}(s)] = \text{Fit-ratio}(s, P_t) \cdot |P_t \cap \text{Schema}(s)|.$$

Proof. The procedure of choosing parents consists exactly of $|P_t|$ independent random choices of one individual from P_t. Let $X_{t+1}^i(s)$ be the indication random variable that is 1 if the chosen individual in the ith choice is in $\text{Schema}(s)$ and 0 else. For every $i = 1, 2, \ldots, |P_t|$,

$$
\begin{aligned}
E[X_{t+1}^i(s)] \;&=\; \sum_{\alpha \in \text{Schema}(s) \cap P} Pr_{par}(\alpha) \\[2mm]
&\underset{(6.1)}{=}\; \sum_{\alpha \in \text{Schema}(s) \cap P} \frac{\text{cost}(\alpha)}{|P_t| \cdot \text{Aver-Fit}(P_t)} \\[2mm]
&=\; \frac{\sum_{\alpha \in \text{Schema}(s) \cap P} \text{cost}(\alpha)}{|P_t| \cdot \text{Aver-Fit}(P_t)} \\[2mm]
&=\; \frac{|\text{Schema}(s) \cap P_t|}{|P_t| \cdot \text{Aver-Fit}(P_t)} \cdot \left(\frac{1}{|\text{Schema}(s) \cap P_t|} \cdot \sum_{\alpha \in \text{Schema}(s) \cap P} \text{cost}(\alpha) \right) \\[2mm]
&=\; |\text{Schema}(s) \cap P_t| \cdot \frac{\text{Fitness}(s, P_t)}{|P_t| \cdot \text{Aver-Fit}(P_t)} \\[2mm]
&=\; \frac{1}{|P_t|} \cdot |\text{Schema}(s) \cap P_t| \cdot \text{Fit-ratio}(s, P).
\end{aligned}
$$

Since $X_{t+1} = \sum_{i=1}^{|P_t|} X_{t+1}^i(s)$ and because of the linearity of expectation we have

$$
\begin{aligned}
E[X_{t+1}(s)] \;&=\; \sum_{i=1}^{|P_t|} E[X_{t+1}^i(s)] \\[2mm]
&=\; \sum_{i=1}^{|P_t|} \frac{1}{|P_t|} \cdot |\text{Schema}(s) \cap P_t| \cdot \text{Fit-ratio}(s, P) \\[2mm]
&=\; |\text{Schema}(s) \cap P_t| \cdot \text{Fit-ratio}(s, P).
\end{aligned}
$$

$\square$

Let Parents_{t+1} be the set of parents chosen from P_t as described above. We know that $E[X_{t+1}(s)]$ is the expected number of individuals from $\text{Schema}(s)$ in Parents_{t+1}, and that $|P_t| = |\text{Parents}_{t+1}|$. Now consider that the next step of a genetic algorithm works as follows. The individuals from Parents_{t+1} are uniformly paired at random and so $\frac{1}{2} \cdot |P_t|$ pairs of parents are created. Every pair of parents produces two children by a simple crossover with one crossover position. The crossover position is uniformly chosen at random from the n positions (i.e., every position is chosen with probability $\frac{1}{n}$). Let Children_{t+1} be the set of $|P_t|$ children generated in the above-described way.

Let $Y_{t+1}(s)$ be the random variable that counts the number of individuals from $Schema(s)$ in $Children_{t+1}$. Let $Y_{t+1}^i(s)$ be the random variable that counts the number of children from $Schema(s)$ created by the ith pair of parents.[8]

Lemma 6.3.1.3. *For every schema s and every $t = 1, 2, \ldots,$*

$$E[Y_{t+1}(s)] \geq \frac{|P_t|}{2} \cdot \left[2 \cdot \left(\frac{E[X_{t+1}(s)]}{|P_t|} \right)^2 \right.$$

$$\left. +2 \cdot \frac{n - length(s)}{n} \cdot \frac{E[X_{t+1}(s)]}{|P_t|} \cdot \left(1 - \frac{E[X_{t+1}(s)]}{|P_t|} \right) \right].$$

Proof. Because of $Y_{t+1}(s) = \sum_{i=1}^{|P_t|/2} Y_{t+1}^i(s)$, $E[Y_{t+1}^i(s)] = E[Y_{t+1}^j(s)]$ for all $i, j \in \{1, \ldots, |P_t|/2\},$[9] and the linearity of expectation we have

$$E[Y_{t+1}(s)] = \sum_{i=1}^{|P_t|/2} E[Y_{t+1}^i(s)] = \frac{|P_t|}{2} \cdot E[Y_{t+1}^1(s)].$$

It suffices to prove a lower bound on $Y_{t+1}^1(s)$. The probability that both parents are from $Schema(s)$ is exactly

$$\left(\frac{E[X_{t+1}(s)]}{|P_t|} \right)^2.$$

Obviously, in this case both children are from $Schema(s)$ for every choice of the crossover position.

Now, we consider the situation when one parent is from $Schema(s)$ and the second one is not. The probability of getting such a parent is

$$2 \cdot \frac{E[X_{t+1}(s)]}{|P_t|} \left(1 - \frac{E[X_{t+1}(s)]}{|P_t|} \right).$$

If the crossover position does not lie between the minimal non-$*$ position and the maximal non-$*$ position of s (i.e., the crossover operation does not break the schema s), then one child of these parents is in $Schema(s)$. The probability of this event is $(n - length(s))/n$. Thus,

$$E[Y_{t+1}^1(s)] \geq$$

$$2 \left(\frac{E[X_{t+1}(s)]}{|P_t|} \right)^2 + \frac{n - length(s)}{n} \cdot 2 \cdot \frac{E[X_{t+1}(s)]}{|P_t|} \cdot \left(1 - \frac{E[X_{t+1}(s)]}{|P_t|} \right).$$

In this lower bound on $E[Y_{t+1}(s)]$ we neglect[10] the probability of getting a child from $Schema(s)$ when the crossover operation breaks s in the parents of $Schema(s)$. $\square$

---

[8] Observe that $Y_{t+1}^i(s)$ is from $\{0, 1, 2\}$.

[9] This is because of the uniform probability distribution used to get from $Parents_{t+1}$ to $Children_{t+1}$.

[10] Observe that we also neglect the possibility of getting a child from $Schema(s)$ by a crossover operation over two parents not in $Schema(s)$.

Corollary 6.3.1.4. *For every schema s and every $t = 1, 2, \ldots$,*

$$E[Y_{t+1}(s)] \geq \frac{n - length(s)}{n} \cdot E[X_{t+1}(s)].$$

Proof. Following Lemma 6.3.1.3 we obtain

$$
\begin{aligned}
E[Y_{t+1}(s)] &\geq |P_t| \cdot \left(\frac{E[X_{t+1}(s)]}{|P_t|} \right) \\
&\quad \cdot \left[\frac{E[X_{t+1}(s)]}{|P_t|} + \frac{n - length(s)}{n} \cdot \left(1 - \frac{E[X_{t+1}(s)]}{|P_t|} \right) \right] \\
&\geq E[X_{t+1}(s)] \cdot \frac{n - length(s)}{n} \cdot \left[\frac{E[X_{t+1}(s)]}{|P_t|} + 1 - \frac{E[X_{t+1}(s)]}{|P_t|} \right] \\
&= E[X_{t+1}(s)] \cdot \frac{n - length(s)}{n}.
\end{aligned}
$$

$\square$

The final part of the genetic algorithm considered in this analysis is to take every child α from $Children_{t+1}$ and to realize a mutation on every item (gene) of α with probability pr_m that is usually very close to 0. Let P_{t+1} consist of children mutated in the above described way and let $Z_{t+1}(s)$ be the random variable that counts the number of individuals from $Schema(s)$ in P_{t+1}. The following assertion is obvious.

Lemma 6.3.1.5. *For every schema s and every $t = 1, 2, \ldots$,*

$$
\begin{aligned}
E[Z_{t+1}(s)] &\geq (1 - pr_m)^{order(s)} \cdot E[Y_{t+1}(s)] \\
&\geq (1 - order(s) \cdot pr_m) \cdot E[Y_{t+1}(s)].
\end{aligned}
$$

Thus, combining all lemmata above we get the following result.

Theorem 6.3.1.6 (The Schema Theorem for GAS). *For every schema s and every $t = 1, 2, \ldots$, the expected number of individuals from $Schema(s)$ in the $(t+1)$-st population P_{t+1} is*

$$E[Z_{t+1}] \geq Fit\text{-}ratio(s, P_t) \cdot \frac{n - length(s)}{n} \cdot (1 - order(s) \cdot pr_m) \cdot |P_t \cap Schema(s)|.$$

Proof.

$$
\begin{aligned}
E[Z_{t+1}] &\underset{L.6.3.1.5}{\geq} (1 - order(s) \cdot pr_m) \cdot E[Y_{t+1}(s)] \\[2mm]
&\underset{Cor.6.3.1.4}{\geq} (1 - order(s) \cdot pr_m) \cdot \frac{n - length(s)}{n} \cdot E[X_{t+1}(s)] \\[2mm]
&\underset{L.6.3.1.2}{\geq} (1 - order(s) \cdot pr_m) \cdot \frac{n - length(s)}{n} \\
&\qquad \cdot Fit\text{-}ratio(s, P_t) \cdot |P_t \cap Schema(s)|.
\end{aligned}
$$

$\square$

Since the probability pr_m is usually chosen to be very small, the multiplicative factor $(1 - order(s) \cdot pr_m)$ is almost 1 if $order(s)$ is not too large. Also $\frac{n - length(s)}{n}$ is close to 1 if $length(s)$ is not very large. This means that short schemata with large fitness $Fit\text{-}ratio(s, P_t)$ will increase their representation in the population during the evolution.

Thus, the ideal situations for a genetic algorithm are those where short schemata combine to form better and better solutions. The assumption that this corresponds to reality is known as the so-called **building-block hypothesis**.

Note that the growth of the number of individuals from $Schema(s)$ in the population by a slightly lowered factor $Fit\text{-}ratio(s, P_t)$ needs not necessarily correspond to a good development because preferring the schema s (or still more other ones) may converge to some local optimum.

6.3.2 Adjustment of Free Parameters

Following the scheme of genetic algorithms presented in the previous section we see that there are several parameters that have to be adjusted in concrete implementations. Some of them are simply numbers, but some of them are connected with a choice of a strategy from a variety of possibilities. In what follows we discuss the following free parameters of a genetic algorithm:

- population size,
- selection[11] of the initial population,
- fitness estimation and selection mechanism for parents,
- representation of individuals and the crossover operation,
- probability of mutation,
- selection mechanism for a new population,
- stop criterion.

Observe that in contrast to simulated annealing we have many possibilities for choosing a reasonable implementation of the genetic algorithm concept, even if the representation of individuals was fixed before.

POPULATION SIZE.

It is obvious (like in the case of multistart simulated annealing) that

(i) a small population size increases the risk of converging to a local optimum whose fitness may be essentially smaller than the fitness of a global optimum, and

(ii) large population sizes increase the probability to reach a global optimum.

---

[11] Called also *seeding* in the literature, when one tries to precompute high-quality feasible solutions for the initial population.

On the other hand, the amount of computer work grows with the size of the population. Again, one has to find a reasonable compromise between the time complexity and probability of finding a high-quality solution. There is no theory providing any clear advice in the choice of the population size. Several practitioners believe that the size of 30 individuals is quite adequate for several problems, while other experimental works suggest taking the population size from the interval $[n, 2n]$ for problem instances of size n. In general, practitioners prefer small population sizes in order to be competitive with other approaches with respect to the time complexity.

SELECTION OF THE INITIAL POPULATION.

One frequently used possibility is to take a random selection, where every individual (feasible solution) has the same probability of being chosen. Experience shows that it can be helpful to take some precomputed high-quality feasible solutions into the initial population, because they can speed up the convergence of genetic algorithms. This precomputation can be done by local search or any other fast heuristic. But if the selection is done completely deterministically, one risks a fast convergence to some local optimum. Probably the best possibility is to combine random choice of individuals with a precomputation of some individuals with high fitness.

FITNESS ESTIMATION AND SELECTION MECHANISM FOR THE PARENT ROLE.

The simplest way of choosing fitness for a maximization problem is to say that a fitness of an individual (a feasible solution) α is exactly $cost(\alpha)$, i.e., to identify the fitness with the cost function. In this case one can assign the probability

$$p(\alpha) = \frac{cost(\alpha)}{\sum_{\beta \in P} cost(\beta)}$$

to every α from the population P and then choose the individuals for the parent role with respect to this probabilistic distribution.

Let $\max(P) = \max\{cost(\alpha) \mid \alpha \in P\}$ and $\min(P) = \min\{cost(\beta) \mid \beta \in P\}$. If $\frac{\max(P) - \min(P)}{\min(P)}$ is small (i.e., $\max(P)/\min(P)$ is close to 1) then the differences between $p(\alpha)$ and $p(\beta)$ are small even if $cost(\alpha)$ essentially differs from $cost(\beta)$. In such a case the probabilistic distribution p over P is very close to the uniform distribution, and so there is almost no preference for individuals with higher fitnesses. A standard way to overcome this difficulty is to set fitness as

$$fitness(\alpha) = cost(\alpha) - C$$

for some constant $C < \min(P)$. The choice of C close to $\min(P)$ means strong preferences for individuals with high costs.

One can also consider a partially deterministic choice of parents. The subset of individuals of P with the highest fitnesses is deterministically moved to the set of parents and the rest of the parents are chosen in a random way. For instance, one takes the $k/2$ best individuals $\alpha_1, \ldots, \alpha_{k/2}$ and randomly chooses a β_i to mate α_i, for all $i = 1, \ldots, k/2$.

A completely different approach for a random choice of the parents is based on the so-called **ranking**. Let $\alpha_1, \ldots, \alpha_n$ be the sorted sequence of the population P with $cost(\alpha_1) \leq cost(\alpha_2) \leq \cdots \leq cost(\alpha_n)$. Then, one sets

$$prob(\alpha_i) = \frac{2i}{n(n+1)}$$

for $i = 1, \ldots, n$. In the probability distribution $prob$ the probability of the choice of the best individual is roughly twice the probability of the choice of the individual $\alpha_{\lfloor n/2 \rfloor}$.

REPRESENTATION OF INDIVIDUALS AND CROSSOVER.

The choice of a representation of individuals is of the same importance for genetic algorithms as the choice of a neighborhood for local search. Moreover, not every representation is suitable because of the crossover operation. Thus, to simply represent the individuals by binary strings as for the cut problems and the maximum satisfiability problems is impossible for several other problems.

Some problems, like TSP, consider a permutation of $1, 2, \ldots, n$ as the representation of feasible solutions. Obviously, this representation is not suitable for the crossover operation when crossover is interpreted in a very narrow way. For instance, for two parents

$$P_1 = 2\ 1\ 8\ 7\ 4\ 3\ 6\ 5 \quad \text{and} \quad P_2 = 7\ 8\ 3\ 5\ 1\ 4\ 2\ 6$$

and the simple crossover operation after the 4th position, the resulting strings

$$I_1 = 2\ 1\ 8\ 7\ 1\ 4\ 2\ 6 \quad \text{and} \quad I_2 = 7\ 8\ 3\ 5\ 4\ 3\ 6\ 5$$

do not represent any permutation of $1, 2, 3, 4, 5, 6, 7, 8$. A possibility to overcome this difficulty is to take copies of the prefix parts before the crossover position and then to complete the rest in such a way that it preserves the relative order for the elements of the second parent. In this way we get for P_1 and P_2 and the crossover position 4 the following children

$$C_1 = 2\ 1\ 8\ 7\ 3\ 5\ 4\ 6 \quad \text{and} \quad I_2 = 7\ 8\ 3\ 5\ 2\ 1\ 4\ 6,$$

which are permutations of $1, 2, 3, 4, 5, 6, 7, 8$.

Some practitioners solve the problem with the crossover operation by allowing individuals that do not represent any feasible solution, but drastically decreasing the fitness of such individuals. The risk of this approach is that some representations may be so bad for the crossover operation that the number of individuals that do not represent any feasible solution grows fast.

Another general strategy may be not to apply the crossover operation on the actual representations of the parents, but rather on a representation of some characteristics (parameters, properties) of the parents, and then to search for children who have the mixed properties obtained by this crossover.

In all cases above we have considered the simple crossover with exactly one crossover position. One can also combine these concepts with more complicated crossover operations. A straightforward extension is to allow k crossing positions for some fixed constant k. The extreme case is to consider a binary vector $(x_1, x_2, \ldots, x_n) \in \{0, 1\}^n$ as a mask and to take the i-th gene from the first parent if and only if $x_i = 1$ for $i = 1, 2, \ldots, n$. The possible advantage of this crossover is that the growth of the representation of schemata of high fitness does not depend on their length anymore, but on their order[12] only.

PROBABILITY OF MUTATIONS.

The role of mutation in genetic algorithms is similar to the role of randomized deteriorations in simulated annealing. The mutation preserves a reasonable level of population diversity, and so it enables the optimization process to escape from local optima with respect to the crossover operation. Usually, the probability of randomly changing one gene in the representation of an individual is adjusted to be smaller than $\frac{1}{100}$. Some practitioners propose $\frac{1}{n}$ or $\frac{1}{k^{0.93}\sqrt{n}}$, where n is the number of genes and k is the size of the population. Another possibility is to dynamically change the probability of the mutation during the computation. Thus, if the individuals of the population are too similar (one has a small diversity of the population), then the probability of mutation has to increase in order to prevent a premature convergence to a poor local optimum.

SELECTION MECHANISM FOR A NEW POPULATION.

There are the following two principal possibilities for the choice of a new population.

(i) The children completely replace their parents, i.e., the new population consists of children only. This strategy is called **en block** strategy.

(ii) There is some mechanism for the choice of a new population from the individuals of the old population and the created children. What has to be adjusted in this mechanism is the degree of the preference of children for parents. The **élitist** model suggests deterministically taking a few individuals with the highest fitness from the old population and allowing the majority of the new population to consist of children, i.e., the élitist

---

[12] The number of genes

model strongly prefers children to parents. Another extreme is to take the individuals of the old population with all children together, to estimate the fitness of everybody, and then to choose deterministically or randomly the individuals of a new population with respect to their fitness. A lot of possibilities lie between these two extremes.

After choosing the individuals of a new population one can immediately start to generate a next population or try to still improve the average fitness of the new population. A possibility is to choose a small neighborhood and to apply local search or simulated annealing to every individual in order to get individuals as local optima. Then the new population consists of the individuals resulting in these local search procedures.

STOP CRITERION.

If one is asked to provide feasible solutions in a restricted amount of time,[13] then one can fix the number of generated populations at the beginning. Another possibility is to measure the average fitness of the populations and the differences between individuals. If the average fitness did not essentially change in the last few generations or the individuals seem to be very similar each to each other, then the genetic algorithm can stop.

Above, we have discussed a variety of possibilities how to specify an individual implementation of the genetic algorithm scheme. Despite of this large number of combinations presented, we are far from an exhaustive survey on this topic. To see this, we close this section by presenting a model of genetic algorithms called the **island** model that differs essentially from the versions of genetic algorithms considered above. One starts with r populations $P_1, P_2, \ldots, P_r$, and applies h iterative steps of a version of genetic algorithms to each of them. Informally, one assumes that every population develops on an isolated island. Then, the best individuals of each of the r generated populations are broadcasted to all other populations (islands). After that, each of the r populations again develops in h generation steps independently from other populations. Next, the best individuals of the population are exchanged between the islands, etc. Obviously, there is an enormous variability of specifications of individual implementations of this concept. The idea behind this island model is to prevent a premature convergence of a population. The populations on islands can develop in different ways. Before an isolated generation converges to some local optimum, new high-quality individuals with possibly different structures are added to the population and so getting stuck in a poor local optimum can be overcome.

Keywords introduced in Section 6.3

genetic algorithm, population, crossover operation, mutation, fitness value, gene, parents, children, scheme, length of a scheme, order of a scheme, fitness of a scheme,

---

[13] For instance, in a real-time application

average fitness of a population, en block selection mechanism, litist model, island model

Summary of Section 6.3

The concept of genetic algorithms for solving optimization problems is based on an analogy to evolutionary biology. The feasible solutions (called individuals) are represented as strings which are viewed as chromosomes consisting of genes. One starts with a set of feasible solutions called a population. New individuals (called children) are created by the crossover operation over two individuals (parents) in a way simulating the well-known crossover operation over chromosomes from genetics. The created children are randomly mutated. A genetic algorithm is usually highly randomized, because the choice of parents, the choice of the crossover position, as well as the choice of the individuals for the membership in a new population are recommended to be realized in a random way with respect to the quality (fitness) of individuals. The random mutation is crucial for preventing a premature convergence and so for the success of genetic algorithms in many applications.

Similarly as for simulated annealing, there is no result guaranteeing that genetic algorithms provide high-quality feasible solutions in polynomial time. On the other hand, it can be shown that the randomized mutation can be used to somehow simulate the randomized deteriorations of simulated annealing, and so some versions of genetic algorithms assure the asymptotic convergence to a global optimum. The Schema Theorem proves that the number of individuals containing the same small groups (combinations) of genes of high local fitness grows from one generation to the next generation.

Probably the main advantage of genetic algorithms over simulated annealing is that the output of a genetic algorithm is a set of feasible solutions. This is of special interest if not all constraints or all optimization criteria of the particular practical problem can be exactly formulated, and so the intuition and the experience of the user may be applied in the final step in order to choose a suitable feasible solution. This is the typical case in evaluating experiments in macromolecular biology, where the objective functions are rather guessed with respect to some experience than estimated with respect to some theoretical knowledge.

There is a large number of free parameters of genetic algorithms. They need to be individually adjusted with respect to the optimization problem considered. As usual one tries to get a reasonable tradeoff between the computational complexity and the quality of produced solutions. Most of the considerations related to the parameter adjustment concentrate on the prevention of a premature convergence to a poor local optimum.

6.4 Bibliographical Remarks

The Metropolis algorithm for the simulation of the annealing process in condensed matter physics was discovered by Metropolis, A.W. Rosenbluth, M.N.

Rosenbluth, A.H. Teller, and E. Teller, [MRR+53] in 1953. Černý [Čer85] and Kirkpatrick, Gellat, and Vecchi [KGV83] independently showed that the Metropolis algorithm could be used in combinatorial optimization by mapping the states of the solid to the feasible solutions of optimization problems as described in Section 6.2.1.

The best sources for both theoretical and experimental study of the behavior of simulated annealing are the exhaustive monographs of van Laarhoven and Aarts [LA88], Aarts and Korst [AK89], and Otten and van Ginneken [OvG89]. In discussing the asymptotic convergence of simulated annealing in Section 6.2.2 we followed Hájek [Háj85, Háj88]. Excellent surveys on distinct aspects of simulated annealing are given by Johnson and McGeoch [JG97], Aarts, Korst, and van Laarhoven [AKL97], and Dowsland [Dow95].

The roots of the idea of tabu search go back to the 1970s, because the Kernighan-Lin's variable depth search algorithm [KL70] already used ideas to forbid the move to some neighbors (see Section 3.6.2). The modern form followed here is due to Glover [Glo86] and Hansen [Han86]. Some further steps in the formalization of this algorithm design technique were realized by Glover [Glo89, Glo90] and de Werra and Hertz [dWH89]. Faigle and Kern [FaK92] have proposed a version of randomized tabu search and they also proved that randomized tabu search asymptotically converges to a global optimum. An important result of [FaK92] is that one can choose the probability of moving from a current feasible solution to a neighboring feasible solution in a wide range without losing the probabilistic convergence property. Excellent surveys on tabu search are presented in Glover and Laguna [GLa95] and Hertz, Taillard, and de Werra [HTW97].

The fundamental principle of natural selection as the main evolutionary principle was formulated by Darwin long before the discovery of the genetic mechanism. The basic laws of the transference of hereditary factors from parents to offspring were discovered by Mendel in 1865. The experimental proof of the fact that chromosomes are the main carriers of hereditary information by Morgan was the starting point of modern genetics. Genetic algorithms use the terminology borrowed from genetics and try to model an optimization process as a biological evolution. Genetic algorithms were initially developed in the 1960s and the 1970s by Holland, who was mainly interested in adaptation. The first systematic treatment including also a theoretical analysis of their behavior was presented in Holland [Hol75]. Independently, Rechenberg [Rec73] and Schwefel [Sch81] investigated the evolutionary strategies in combinatorial optimization. Genetic algorithms presented here are the result of the combination and the development of all these roots.

There is a lot of literature about genetic algorithms. General, exhaustive surveys on this topic are presented, for instance, in Goldberg [Gol89], Davis [Dav87, Dav91], Michalewicz [Mic92], Michalewicz and Fogel [MiF98], and Schwefel [Sch95]. A nice, comprehensive overview was given by Reeves [Ree95a]. In order to make a theoretical analysis of the behavior of genetic algorithms it is reasonable to look for the fundamental research in population genetic theory (see, for instance, Crow and Kimura [CK70] and Crow [Cr86]). An excellent survey of the theoretical analysis of the convergence of genetic algorithms was presented by Mühlenbein [Müh97] in [AL97]. The idea to use genetic algorithms for the evolution of some optimization characteristics (control parameters) instead of a direct optimization of the individuals (feasible solutions) follows the works of Branke, Kohlmorgen, Schmeck, and Veith [BKS+95], Branke, Middendorf, and Schneider [BMS98], and Julstrom and Raidl [JR00]. More details about the island model are given by Tanese [Ta89], Whitley and Starkweather [WhS90], and Schmeck, Branke, and Kohlmorgen [SBK00].

There are only a few papers devoted to the theoretical analysis of the complexity of genetic algorithms and simulated annealing for solving specific problems. Droste, Jansen, and Wegener [DJW98, DJW98a, DJW99, DJW01], Jansen and Wegener, [JWe99], and Wegener [Weg00] provide the first results comparing the convergence speed of some simple versions of genetic algorithms and simulated annealing for some artificial optimization problems. The conclusion is that they are incomparable [Weg00], i.e., simulated annealing may be better than genetic algorithms for a specific problem, and genetic algorithms may be better than simulated annealing for another problem.

7

A Guide to Solving Hard Problems

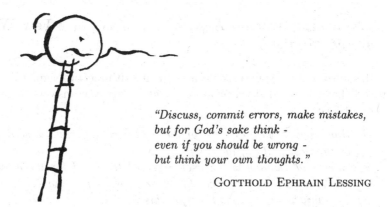

"Discuss, commit errors, make mistakes,
but for God's sake think -
even if you should be wrong -
but think your own thoughts."

GOTTHOLD EPHRAIN LESSING

7.1 Introduction

In the previous chapters we have introduced a variety of algorithm design techniques and algorithmic concepts in order to attack hard problems. A lot of specific algorithms were presented in order to illustrate the power and possible success of their applications. Nevertheless, we did not consider the question

"What needs to be done when one is asked to solve a hard problem?"

up till now. We only listed a number of possible approaches and discussed what they can do and what they probably cannot do. The aim of this chapter is to discuss the search for a suitable method for solving a given problem depending on the requirements and constraints prescribed by the user. We present this discussion in a systematic way as follows.

Section 7.2 is devoted to the initial decisions made with respect to the (economical) importance of solving the given problem, and with respect to the investment provided for the design and the implementation of an algorithmic solution. Section 7.3 presents the most common combinations of the approaches presented in the previous chapters. Section 7.4 is devoted to the problem of the theoretical and experimental comparison of different algorithms designed[1] for a specific algorithmic problem. Some experimental results from the literature illustrate the strengths and weaknesses of particular approaches. Section 7.5 discusses the possibility of speeding up the algorithmic design techniques presented in the previous chapters using parallel processing.

---

[1] Possibly by distinct algorithm design techniques

Section 7.6 provides a concise, informal introduction to the concepts of DNA computing and quantum computing as potential future technologies for solving hard problems. We close this chapter with a glossary of selected keywords in this book. The glossary is presented on the level of informal concepts and ideas with pointers to formal definitions and results presented in this book.

7.2 Taking over an Algorithmic Task or a Few Words about Money

If someone asks you to solve a hard algorithmic problem, the first things that have to be made clear are connected with the answers to the following questions:

(i) *How high is the influence of the quality of algorithmic solutions on the economical profit of the user?*

(ii) *How much money is the user prepared to invest for the design and the implementation of an algorithm that solves her/his task?*

(iii) *How much time do you have for this job?*

You especially have to take care of a reasonably balanced tradeoff between the economical importance of the algorithmic job and the investment for the algorithm design. If the quality of produced feasible solutions of an optimization problem is a decisive factor in saving of millions of dollars in each of the next few years and the user is willing to pay the salary for one or two co-workers for a few months for the algorithm design, then you must tell the user that she/he has to forget about it. It is important to tell everybody that the quality of your software product strongly depends upon the amount of intellectual work invested in it and the overall time that you have for this project. The user has to learn from you how much her/his investment can influence the quality of the output of your algorithmic project in order to be able to decide which investment is the most profitable.

When these first questions have been answered and the financial framework is more or less clear, then you can start to deal with initial decisions concerning algorithmics. Assuming that the given task is not a standard one (for which you already have algorithms in your library), the fundamental strategy can be roughly described as follows:

(i) If you have a small investment and not much time for the project, then you have to consider robust algorithm design techniques like local search methods, simulated annealing, and genetic algorithms. Thus, you reduce the work with the specifics of the problem to the adjustment of the free parameters of these robust methods. But even in this framework the amount of experimental work you do in adjusting the parameters depends on time and resources for this work.

(ii) If the quality of the algorithmic solutions is very important in the considered application, qualified experts should be asked to carefully analyze
the given problem and to try to use all its specifics in order to find an
algorithm that perfectly fits the specific features of the given problem.

Thus, the above-described strategy says that the deepness of the analysis of
the specifics of a given problem should be strongly correlated with the degree
of the importance of finding a high-quality solution. This strategy is based on
the experience that a deep problem-oriented analysis usually provides better
algorithms than any straightforward application of a robust heuristic (Aarts
and Korst [AK89], Černý [Čer98], Wegener [Weg00]). The recent development in combinatorial optimization and operations research shows especially
that involved mathematical instruments are more powerful than any heuristics based on physical or biological analogies (see the preface in Aarts and
Lenstra [AL97a] and Wegener [Weg00]). On the other hand, one should not
consider this strategy as an absolute principle. For some optimization problems (like the layout problem in VLSI circuit design [AK89]), robust heuristics
may result in excellent solutions if the particular heuristic fits the typical input instances of the considered problem. A reasonable approach to solving
important problems is to try to apply several different approaches and then
to choose the best one of the designed algorithms in some competition.[2]

One also has to take into account that focusing on a deep problem analysis
and randomized, approximation, or other mathematical approaches requires
having high-qualified experts in the team, while the mathematical and algorithmic knowledge sufficient to apply simple heuristics is on an elementary
level.[3]

After making rough, initial decisions one has to think about concrete concepts and techniques, their possible combinations, and their comparison in
an experiment that is relevant to the specific application. These topics are
covered in the subsequent sections.

7.3 Combining Different Concepts and Techniques

We have learned in the previous chapters that the main art of solving hard
problems consists in saving a huge amount of computer work (complexity)
by making as small changes as possible in the formulation (specification) of
the problem or in the requirements on the solutions. Fascinating examples
are provided by randomization and approximation, where for some specific
algorithmic problems, an intractable amount of complex work of the best
deterministic algorithms can be reduced to a matter of few minutes by moving

---

[2] Section 7.4 is devoted to the topic of the theoretical and experimental comparison
of different methods in a particular application.

[3] This, together with the costs, is probably the main reason why heuristics become
so popular.

from the requirement of getting the exact (optimal) solution with a 100% guarantee to the assurance of getting a high-quality approximation of the optimal solutions with a high probability. As we have seen, similar effects can be achieved by reasonably restricting the set of feasible input instances. The aim of this section is to discuss some standard profitable combinations of the approaches presented in Chapters 3, 4, 5, and 6.

Remember that we have presented the following concepts:

- pseudo-polynomial-time algorithms
- parameterized complexity
- lowering the worst case complexity of exponential algorithms
- approximation algorithms
- dual approximation algorithms
- stability of approximation
- randomized algorithms (foiling an adversary, abundance of witnesses, random sampling, random rounding, fingerprinting)
- derandomization

and the following algorithm design techniques:

- local search
- Kernighan-Lin variable depth search
- branch-and-bound
- relaxation to linear programming
- simulated annealing
- tabu search
- genetic algorithms

in order to attack hard problems. To realize any of the above-mentioned concepts one can use the standard robust algorithm design techniques[4] such as divide-and-conquer, dynamic programming, greedy algorithms, and local search.

First, let us consider the situation when one unconditionally wants to find optimal solutions to input instances of an optimization problem, whatever the costs should be. We know that this is impossible for every input instance of a large size n, but this may be realistic for most of the typical input instances of size n in the considered application. The best method in this case seems to be branch-and-bound with a precomputation of some bounds on the cost of the optimal solutions. Any of the above-listed concepts and algorithm design techniques may be used in the precomputation. Currently, the most successful method for getting a good estimation on the optimal cost is the relaxation to linear programming or other combinations with some more advanced methods of operations research. If the considered optimization problem

---

[4] Remember that we distinguish between design techniques that give a clear framework for the algorithm design and concepts that provide ideas and goals but no specification or method for how to reach them.

is well approximable, then the use of an efficient approximation algorithm in the precomputation is recommended.

The idea to use one method as a precomputation for another method is of importance for genetic algorithms, too. A precomputation of a few high-quality individuals for the initial population may speed up the convergence of genetic algorithms.

Another standard combination in combinatorial optimization is mixing randomization with the concept of approximation. This resulted in the design of efficient randomized approximation algorithms as presented in Chapter 5. Applying the derandomization techniques can lead to discovering completely new (deterministic) approximation algorithms. The combination of the techniques of the relaxation to linear programming with the technique of random rounding is one of the standard examples of the development of randomized approximation algorithms.

Randomization is a universal concept that can be combined with almost all other approaches. For instance, combining randomization with the concepts of lowering the worst case complexity of exponential algorithms leads to practical algorithms for satisfiability problems and the combination of randomization and tabu search provides a powerful heuristic. Randomization is a substantial part of simulated annealing[5] and genetic algorithms. Due to the methods of foiling an adversary and abundance of witnesses, randomization seems to be the best method for attacking the hardest problem instances. We conjecture that randomization can bring some further breakthrough in the still-not-investigated combinations involving concepts of parameterized complexity and stability of approximation.

Classical general algorithm design techniques such as divide-and-conquer, dynamic programming, local search, and greedy algorithms may provide approximation algorithms with a reasonable approximation ratio. There is a special connection between pseudo-polynomial-time algorithms and approximation algorithms. The idea is to reduce the size of all input values of an input instance of an integer-valued problem by dividing all items by the same large number. This division is usually connected with some rounding. Any pseudo-polynomial-time algorithm runs efficiently on the reduced input instance. If the solution to the reduced input instance is a good approximation of an optimal solution to the original input instance, then one obtains an efficient approximation algorithm or even a PTAS. Since the reduction between input instances may be viewed as an approximation of an input instance by another one, there are good chances to get approximation algorithms for several problems in this way.

Finally, observe that Kernighan-Lin's variable depth search is nothing else than a combination of local search and greedy technique.

---

[5] Simulated annealing can be viewed as randomized local search.

7.4 Comparing Different Approaches

Assume that we have to solve a hard algorithmic problem and that this problem is very important from the user's point of view. In such a case one usually follows a variety of ideas and approaches that result in the design of several different algorithms for the given problem. We have to choose the best one for the actual application. The possibility of making this decision by a pure theoretical analysis is very rare. This can be done only if you are able to prove that one of the algorithms computes the right results or high-quality approximations of optimal solutions with a reasonably high probability and the other algorithms cannot give this kind of assurance. But your decision for this algorithm is right only when additionally this algorithm is competitive with the other ones with respect to performance. And to compare the performances of different algorithms by experiment is often reasonable because we are usually not very interested in the worst case time complexity but rather in the "average" time complexity on inputs that are typical for the considered application.

Thus, the standard job to be done is to carry out an experimental comparison of different algorithms. Let us consider a more general case when the hard problem is an optimization problem. This means that we have to compare both crucial parameters – the quality of computed solutions as well as the time complexity (amount of executed computer work). Having two parameters, it can obviously happen that none from our collection of algorithms dominates all others, i.e., there is no algorithm that would be the best one for both parameters. In such a case one should follow the user's preferences because in some applications the performance may be more important than small deteriorations in the solution quality and in other applications the quality of approximation is strongly preferred over the performance.

In preparing an experimental comparison one should deal with the following matters:

(i)   All algorithms should be implemented in the same programming language and have to be run on the same computer.
(ii) The test data have to be representative for the application considered.
(iii) How can the quality of the outputs of the algorithms be measured when the costs of optimal solutions are not available?

In what follows we discuss the points (i), (ii), and (iii).

UNIFORMITY OF IMPLEMENTATION.

The condition (i) is important for the comparison of the performance of different algorithms. Both requirements, the same programming language and the same computer (the same processor and the same system software), are necessary for a fair competition. Because of this one should be very careful in claiming that an algorithm is better than another one on the basis of experimental results presented in different papers.[6]

THE CHOICE OF TEST DATA.

Choosing good test data is a nontrivial problem because we would like to have data that are representative for our particular application. One possibility is to generate the data randomly. This may be suitable for some applications, but you should be careful with this approach in general. There are many applications for which randomly generated data are very far from that which is typical. Obviously, if one is able to fix a formal specification of the set of typical input instances, a random choice from this set is possible. But to get such a specification is usually a very nontrivial task.

An alternative way to choose test data is to take a collection of data that actually appeared in the considered application. One can try to collect input instances appearing in some period of time in the particular application. For some optimization problems there are public sets of test data available. For instance, Reinelt [Rei91, Rei94] collected a database of instances of TSP. This database (called TSPLIB) is available via anonymous ftp from soft-lib.rice.edu. TSPLIB contains problem instances of different sizes (even instances with more than 80,000 vertices) from different application areas (for instance, VLSI circuit design, or real geographical instances).

MEASUREMENT OF SOLUTION QUALITY.

The simplest way to perform the competition is to compare the costs of the outputs with each other. In this case one can obtain an ordering of the algorithms with respect to the average order of the quality of the solutions over all instances in the test set.

If one wants to have a more involved comparison, one could prefer to take the differences in the costs of feasible solutions produced by different algorithms into account. But the values of the cost differences may be misleading because they have to be relativized with respect to the cost of an optimal solution. The trouble is that usually we do not know the costs of the optimal solutions and we are not able to compute them. One possibility to overcome this difficulty is to generate test problem instances in such a way that the optimal solutions are known. But it is questionable whether such input instances may be considered to be representatives of the particular application. Another frequently used possibility is to compare the costs of the results with the so-called **Held-Karp bound** (Held and Karp [HeK80, HeK71]) that is the bound obtained by the method of relaxation to linear programming. This

---

[6] Unfortunately, this is a frequent mistake in the literature on experimental algorithmics.

method may provide surprisingly good bounds. Johnson and McGeoch [JG97] reported that the bounds obtained in this way usually do not differ from the optimum by more that 0.01% for their test set of input instances of the metric TSP. Wolsey [Wol80] and Shmoys and Williamson [SW90] even showed that the lower bounds from linear programming cannot be smaller than 2/3 times the optimal value in the worst case for the metric TSP.

Currently, there is a large amount of empirical knowledge of algorithmic methods for solving specific hard problems. An excellent and detailed report on a comparison of different methods for solving TSP is given by Johnson and McGeoch [JG97]. They documented that the Kernighan-Lin algorithm is the best one for random instances of the metric TSP with respect to the tradeoff between the quality and time complexity. A higher solution quality can only be achieved by simulated annealing and genetic algorithms if an essential increase of time complexity is acceptable. The algorithms considered in this huge experimental study were different local search algorithms, constructive algorithms such as greedy and the Christofides algorithm, simulated annealing, genetic algorithms and several of their modifications and combinations. On the other hand, Aarts and Lenstra report in the preface of [AL97a] that linear-programming-based optimization techniques dominate the local search and heuristic approaches.

As a sampling of the huge number of reports on experiments with algorithms on combinatorial optimization we mention Bentley [Ben90], Reinelt [Rei92], Fredman, Johnson, McGeoch, and Ostheimer [FJGO95], van Laarhoven, Aarts and Lenstra [LAL92], Vaessens, Aarts, and Lenstra [VAL96], Reiter and Sherman [ReS65], Johnson [Joh90a], Johnson, Aragon, McGeoch, and Scheron [JAGS89, JAGS91], Gendreau, Laporte, and Potvin [GLP97], Anderson, Glass, and Potts [AGP97], and Aarts, van Laarhoven, Lin, and Pan [ALLP97].

7.5 Speedup by Parallelization

All algorithms considered up till now were sequential algorithms, which means that they are implemented by executing a sequence of operations over data. The idea of parallel computing is to use several independent processors to cooperate in order to compute a given task. For our purposes it does not matter whether we model parallel computations as

(i) a collection of processors that share a common memory and realize the cooperation (communication) by direct read/write access to this memory, or

(ii) as an interconnection network of processors where the communication links between pairs of processors are used for their cooperation.

The main point is that the time complexity $Time_A(n)$ of an algorithm is nothing else than the number of operations to be executed (i.e., the amount of computer work). So, if one uses $p(n)$ processors instead of one to execute a run of A on an input of size n, then the result cannot be computed faster than in time

$$\frac{Time_A(n)}{p(n)}.$$

This is obvious, because the amount of work (the number of executed operations) done by a parallel algorithm is at most the number of processors times the parallel time and so

$$sequential\ time \leq (number\ of\ processors) \cdot (parallel\ time). \qquad (7.1)$$

An immediate consequence of this fact is that parallelism itself cannot be considered as an independent approach for solving hard problems. This is clear, because if one has to execute an intractable number of operations by a parallel algorithm, then the number of processors must be intractable or the parallel time is intractable.

The art of parallelization is to find such a parallel implementation of an algorithm that parallel time is as close as possible to sequential time divided by the number of processors. Thus, the main contribution of parallelism is the speedup of the execution of algorithms. The importance of such a speedup can be illustrated by the following example. Let A be an excellent approximation algorithm that is able to compute high-quality solutions to typical input instances of a hard optimization problem in one hour. But the application is a real-time one, where it is necessary to produce feasible solutions in at most 5 minutes. We do not see any possibility to design a faster (sequential) algorithm that would guarantee the same solution quality. Now, parallelism can decide the success in the considered application. Taking a parallel computer of 20 processors (or a network of 20 PCs) one could have a good chance of achieving an execution speedup of the sequential realization by 10 if the algorithm has some reasonable properties.[7] In general, for many optimization problems like scheduling problems, the minimization of time in waiting for a solution is at least as important as the solution quality. Sometimes, parallelism may be the only way to fit both these requirements.

This section does not aim to present any fundamentals of modeling parallel computation or of designing parallel algorithms. Readers interested in this topic can look, for instance, at

(i) Jájá [Já92], Leighton [Lei92], and Reif [Rei93] for excellent general sources,
(ii) Karp and Ramachandran [KR90], Kindervater, Lenstra, and Shmoys [KLS89] and Díaz, Serna, Spirakis, and Torán [DSST97] for the techniques for the design of parallel algorithms,
(iii) Greenlaw, Hoover, and Ruzzo [GHR95] for the complexity theory of parallel computation,

---

[7] It is not inherently sequential, for instance.

(iv) Kindervater and Lenstra [KiL85, KiL85a] and Kindervater and Trienekens [KiT85] for extensive bibliographies of papers about parallel algorithms (also with an emphasis on combinatorial optimization), and

(v) Leighton [Lei92] and Hromkovič, Klasing, Monien, and Peine [HKM+96] for methods for solving communication tasks in networks of processors (parallel architectures).

The aim of this section is to briefly discuss the suitability of parallel implementations of the following algorithm design techniques for hard problems:

(i) local search,
(ii) branch-and-bound,
(iii) random sampling,
(iv) abundance of witnesses,
(v) Kernighan-Lin's variable depth search and tabu search,
(vi) simulated annealing, and
(vii) genetic algorithms.

Observe that we are speaking about design techniques and not about concepts such as pseudo-polynomial-time algorithms. This is because the given framework of the presented concepts is too rough to make any conclusion about possible parallelization. On the other hand, the specifications of algorithm design techniques provide enough details for specifying the possibilities of their parallelization.

We start with some general considerations as illustrated in Figure 7.1. When thinking about the possibility to parallelize a given sequential algorithm A we mainly follow the following two parameters:

• the amount of work to be done (time complexity), and
• the necessary amount of information exchange (communication) between processors that execute the work of A in parallel.

The reason for taking the amount of communication into account is that the communication is costly. To communicate between processors may be time consuming because of the necessary synchronization, the determination of message routing, the solution of conflicts appearing on communication links when too many processors want to speak each with each other, etc. Moreover, if the dependencies between the parallel processes are too strong, the processors spend more time waiting for data (results of other processors) than executing the primary computer work. In such cases the achieved speedup in the comparison with a sequential implementation may be considerably smaller than the number of processors used for the parallel implementation. Note that the term "amount of communication" should not simply be interpreted as the number of bits exchanged between the processors during the parallel computation. If you have many messages that can be communicated in a few communication actions (i.e., if one made the effort to organize a physical connection between two processors, then this connection is used to exchange a lot of information), then the parallel algorithm is usually efficient because there are long periods of time in which the processors can independently work without any interruption. Thus, the worst case is, when one has frequent requirements to communicate many short messages and the processors cannot continue to compute until their

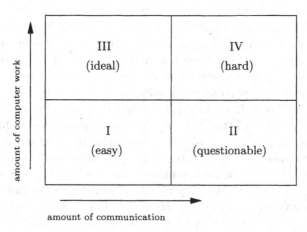

Fig. 7.1.

requirements for data from other processors are satisfied. Thus, the term amount of communication means the frequency of communication rather than the number of bits communicated in what follows.

Let us look at the partition[8] of the class of algorithms according to their complexity and the requirements for communication of their parallel execution as depicted in Figure 7.1. Square I represents efficient algorithms that may simply be parallelized because the cooperation between processes running in parallel does not require any intensive information transfer. Anyway, one considers using parallelism for such algorithms for real-time applications only, because the overall amount of computer work to be done is not large enough not to be quickly executable by any sequential implementation.

The algorithms of type II are efficient (the amount of computer work to be done is small) and any parallelization of these algorithms forces a large amount of communication activity. This means that one has to make a huge effort (to use many processors) to get a small speedup. In this case it is really questionable whether parallelism is the right way to speed up the algorithms. In such cases one should use parallelism only if even small speedups are of importance and one does not see any possibility to achieve such speedups with a good choice of data structures or some other ideas of sequential implementations.

---

[8] Clearly, this partition is a slightly simplified representation of the parallelization problem.

The algorithm class III represents the ideal situation from the parallelization point of view. We have to execute a lot of work and so there is usually a need to speedup the computations. Further, the parallelization can be performed efficiently, because the work to be done can be partitioned into several jobs that can be executed almost independently, i.e., a small amount of communication actions is sufficient for the coordination of the work of the processors running in parallel.

The parallelization of algorithms in case IV is usually the hardest task. We have to execute a lot of computer work and so the requirement for the speedup is very natural, but there is no natural possibility of partitioning the computations of the given algorithms into parts that could be executed with a small number of dependencies. A solution may be to design circuits that quickly execute some frequently appearing subcomputations (i.e., to think .about parallelization on the bit level). Another possibility is to forget about the parallelization of the given algorithm and to try to design a completely new algorithm that allows an efficient parallel implementation. The parallelization becomes really hard when the computing task is inherently sequential in the sense that there does not exist any parallel algorithm for the task that would be remarkable faster than the best sequential algorithm. An excellent presentation of the complexity theory of the classification of algorithmic problems according to the existence of fast parallel algorithms is given by Greenlaw, Hoover, and Ruzzo [GHR95].

Next we look at a possible classification of the algorithm design techniques considered in this book according to the four classes introduced in Figure 7.1.

LOCAL SEARCH.

Local search can be viewed as an iterative improvement of an initial feasible solution. Since this iteration[9] is inherently sequential, parallelism can speed up local search only by decreasing the time complexity of every iteration. Thus, one can use parallelism for

- determining whether a given solution is a local optimum with respect to the given neighborhood, and
- computing a better neighbor of a given solution that is not a local optimum.

How large the speedups are that can be achieved depends on the size of the neighborhoods as well as on their structures, and on the complexity of the local transformation execution. Visiting a feasible solution α, the simplest approach is to partition $Neigh(\alpha)$ into p disjoint parts (if possible) and to use p processors to search independently in these parts. If one searches for a best $\beta \in Neigh(\alpha)$ or tries to prove the local optimality of α, the only communication part of this parallel computation is executed at the end, after all processors have finished the search in their parts of the neighborhood. To

---

[9] That is, visiting feasible solutions one after the other.

agree on the best solution found by all p processors is a standard task that can be quickly performed in parallel. If the p processors want to find only a feasible solution that is better than the current one, every processor that succeeded immediately informs all other ones and distributes the new feasible solution[10] in order to start the work on the next iteration step.

In general, the parallelization of one iteration step of local search can be viewed as a problem of type III, but parallelism cannot help to achieve a speedup below the number of iterations necessary to find a local optimum. An excellent survey on the complexity of parallel local search is presented by Yannakakis in Section 6 of [Yan97].

A highly parallelizable version of local search is the multistart local search, where one performs a number p of independent local searches starting from p different initial solutions.

BRANCH-AND-BOUND.

Backtracking and the branch-and-bound technique are very suitable for a parallel implementation. There is a lot of work to be done and the communication requirements are low. The simplest case is the parallel execution of backtracking. One simply partitions the set of all feasible solutions into subsets and searches inside these subsets in parallel. If some processor has finished its work and other ones are still working, a repartition of the still unvisited part of the solution set among the processors can be considered. Besides this, the only communication needed is for the task of finding the best feasible solution from the best solutions of particular processors. A little bit more communication is required when performing branch-and-bound in parallel. Each time a processor finds a solution better than the best known one (better than the precomputed estimation), it broadcasts the cost of this solution to all other processors. If the precomputed estimation on the cost of the optimal solutions is close to $Opt_U(x)$, then the number of improvements and so the number of executed broadcastings will be small. The situation may change if the precomputed bound[11] on $Opt_U(x)$ is far from $Opt_U(x)$. But experience shows that this rarely causes too frequent communication activity.

RANDOM SAMPLING.

This very simple randomized technique consists of choosing a feasible solution at random and considering it as the output. This is usually very efficient and one can ask where the need for parallelism is. But the typical case is that one needs a number of independent choices in order to get a good feasible

---

[10] We omit the discussion on how to organize an agreement when several processors find better solutions at the same time.

[11] We omit discussing the parallel complexity of the precomputation, because it substantially depends on the particular method used.

solution with a high probability. If the probability of getting a high-quality solution in some random attempt is small, then the number of independent random choices has to be large in order to sufficiently increase the probability of success. This is an ideal situation for the concept of parallelism, because the random choices can be done in parallel by distinct processors. Observe that the idea of

- executing a number of independent runs of a randomized algorithm on a given input in parallel, and
- computing the output from the outputs of all executed runs

is a general concept that is applicable for many randomized algorithms.

ABUNDANCE OF WITNESSES.

To get a witness with a high probability one can do several independent random choices from the set of all witness candidates. This is exactly the same situation as for random sampling and so one can simply implement the random choices in parallel. Now we present an idea that is a little bit more involved. Let us consider a situation when one random choice of a witness candidate is sufficient because there is a real abundance of witnesses in the set of witness candidates. For instance, for every $x, y \in \{1, 2, \ldots, 2^n\}$, $x \neq y$, the set of approximately $n^4/(4 \cdot \ln n)$ primes from $\{2, 3, \ldots, n^4\}$ contains at least $n^4/(4 \cdot \ln n) - n$ witnesses of the fact $x \neq y$ (i.e., $x \bmod b \neq y \bmod b$). Thus, the probability of choosing a candidate that is not a witness of $x \neq y$ is at most $4 \ln n / n^3$. For large enough n, one does not see any reason to make several random attempts. Thus, this approach seems to be inherently sequential, because one has to choose b at random, to compute $x \bmod b$ and $y \bmod b$, and to compare the results. In the framework of communication protocols one has to perform the submission of b and $x \bmod b$, i.e., the submission of a message of length $\lceil 8 \cdot \log_2 n \rceil$. The general idea in such situations is to look for the possibility of using witnesses of a smaller representation size than the size of the original witnesses, and then to use several small candidates for witnesses instead of a large one. In our example one can consider primes from the interval $[2, n^{1+\varepsilon}]$ (instead of the interval $[1, n^4]$), for some $0 < \varepsilon < 1$. Then, the probability of choosing a nonwitness is at most $\frac{\ln(1+\varepsilon)}{n^\varepsilon}$. Taking k such random choices in parallel decreases the probability of error to

$$\left(\frac{\ln(1 + \varepsilon)}{n^\varepsilon} \right)^k.$$

KERNIGHAN-LIN'S VARIABLE DEPTH SEARCH AND TABU SEARCH.

For these techniques one has almost the same situation as for local search. The only non-negligible difference is that the subsets of forbidden neighbors vary during the computation and so considering a fixed (static) partition of the neighborhoods among the processors does not need to be the most efficient approach. To overcome this difficulty (an unbalanced partition of the search to be done in the "available" neighborhood[12]) one should consider a dynamic partition of the neighborhoods among the processors (working in parallel) or a kind of load balancing during the search in the neighborhood of a particular feasible solution.

SIMULATED ANNEALING.

The nature of the work of simulated annealing is still more inherently sequential than the nature of local search and so to design a parallel algorithm for simulated annealing is a hard task. This is because not only the sequence of iterative improvements, but also the random choice of a neighbor (instead of the exhaustive search in the area of all neighbors by local search) seems to be a typically sequential task. If one wants to design a parallel simulated annealing algorithm, then one has to take into account that, in order to achieve a reasonable speedup, a lot of additional computer work may be necessary. One can consider the following two independent approaches to implement simulated annealing as a parallel algorithm.

(i) *Pipelining.* One tries to perform the random generation of a neighbor to the current solution, the calculation of its cost, and the decision whether or not the new solution will be accepted in parallel. So, one has to work on the level of problem instance representation and on the level of the elementary operations over the items of this representation. Whether this approach can be successful depends on the particular optimization problem and the representation of its feasible solutions. For instance, no successful parallelization on this level is known for TSP. On the other hand, Casotto, Romeo, and Sangiovanni-Vincentelli [CRS87] and Kravitz and Rutenbar [KRu87] present successful implementations of this approach for the placement problem.

(ii) *Parallel search for a new solution.* Each of the available processors works independently by randomly generating a feasible solution and (randomly) deciding whether this new solution has to be accepted. If none of the processors succeeds, all processors continue in the search for a next solution. If exactly one of the processors accepts its new solution, this solution is taken as the next one. If there are more processors that accept their generated solutions at the same time, a (randomized) competition is used to select the next solution.

A completely different strategy is to consider multistart simulated annealing, which consists of a number of independent simulated annealing algorithms

---

[12] This happens when most neighbors in a part of the neighborhood assigned to a processor are forbidden, but all feasible solutions in a part of the neighborhood of another processor should be visited.

starting from random initial solutions. Obviously, the multistart simulated annealing is highly parallelizable, but this cannot be viewed as a parallel implementation of simulated annealing. An excellent and involved overview about the design of parallel simulated annealing algorithms is given in Aarts and Korst (chapter 6 of [AK89]).

A final observation is that simulated annealing is of sequential nature and it is not so easy to design parallel algorithms that would essentially speed up its sequential execution. The way out is to change the thinking on the level of the proper simulated annealing. This led to the introduction of the Boltzmann machine (see Hinton and Sejnowski [HiS83], Hinton, Sejnowski, and Ackley [HSA84], and Ackley, Hinton, and Sejnowski [AHS85]) that can be viewed as a combination of parallelism and of the idea of optimizing by physical laws. An excellent and involved explanation of this topic is given by Aarts and Korst [AK89].

GENETIC ALGORITHMS.

Genetic algorithms are almost ideal for a parallel implementation. This is because the genetic operators work independently on different individuals of a population and so the following steps:

- generation of the initial population,
- fitness estimation of the individuals, and
- application of the crossover operation and the mutation

can be implemented in parallel in a straightforward way. Communication actions between processors are necessary for the selection procedures (the selection of parents and the selection of a new population) because to estimate the probabilities of the random choice of particular individuals from the population requires global knowledge that is distributed among the processors.[13] Such global knowledge is, for instance, the average fitness of the population or the order of the individuals according to their fitnesses. But this global knowledge can be efficiently computed by standard parallel algorithms with an acceptable amount of communication activity.[14]

The island model of genetic algorithms simplifies this problem of computing global information. The evolution on particular islands is performed independently for a period of time and the exchange of the best individuals after a period of time does not require any high communication activity. Thus, already a straightforward parallel implementation of the island model of genetic algorithms provides an essential speedup of sequential genetic algorithms. Tanese [Ta89] and Whitley and Starkweather [WhS90] are the original

---

[13] Assuming every processor has the knowledge of one individual or of a small group of individuals.

[14] The communication effort corresponds approximately to the effort of broadcasting in an interconnection network.

sources of the island model and its parallel implementation. An excellent and involved overview on this topic is given by Schmeck, Branke, and Kohlmorgen [SBK00].

GENERAL REMARKS.

As already mentioned, we do not have any detailed enough specification to be able to deal with a parallel implementation of concepts such as pseudo-polynomial-time algorithms, lowering the worst case complexity of exponential algorithms, dual approximation algorithms, approximation algorithms, or parameterized complexity. If one wants to consider this problem, one has to look for the possibility of a parallel implementation of the standard algorithm design techniques used to realize these concepts. In this book we have frequently used divide-and-conquer, dynamic programming, and greedy algorithms. Greedy algorithms are inherently sequential because they create a feasible solution by sequentially specifying one item of the solution representation after another. The only possibility is to implement the choice of the value of the next item in parallel. Since the sequential implementations of greedy algorithms are usually very efficient, one rarely makes the effort to speed them up in parallel. Divide-and-conquer and dynamic programming are very suitable for parallel implementations, because the created subproblems can usually be solved independently.[15] Additionally, some problems allow an efficient parallel computation of a solution to a problem instance from the solutions to some of its problem subinstances. Thus, the achieved speedup of a parallel implementation may be reasonably high.

7.6 New Technologies

7.6.1 Introduction

All algorithms considered in this book so far are based on the model of standard electronic computers (Turing machines in the theory). The crucial point is that the definition of a computing model is given by a set of executable elementary operations and by some specifications of the memory organization and of the access to the memory. The standard model considered here allows operations over computer words (symbols of a fixed alphabet) like arithmetic operations and comparisons. We always assumed having a direct access memory which may be essential for an efficient implementation, but which does not matter from the computability and tractability points of view. The first motivation for another computing model is connected to the question whether nature possesses other possibilities of storing and processing information that cannot be simply simulated by the Turing machine model.

---

[15] Provided that the data needed are available in a global shared memory.

The second motivation is related to the development of technologies. The development of computers has a long history, starting with the abacus and continuing via mechanical devices to electronic computers. The capacity and the speedup grew enormously in this development. Moore observed that the computer capacity doubles every 18 months. These growth and speedup are the consequence of miniaturization. But there are known limits to miniaturization with current computer technologies, and so this development cannot continue forever on any macroscopic level. Thus, to obtain an essential improvement in the abilities of computers one has to go to the microscopic world (to the molecular level).

The above ideas resulted in the development of two completely different concepts – *DNA computing* and *quantum computing*. While DNA computing is based on chemical operations over DNA strands, quantum computing is based on the laws of quantum physics. Currently, the DNA-based technology is already available for small input sizes, but still not really competitive with classical algorithmics presented in this book. The mathematical models of DNA computations are equivalent to the model of Turing machines from the computability point of view and one can sequentially simulate DNA computations with the increase of time complexity that is proportional to the number and lengths of DNA strands under the chemical operations. Thus, if one knows that the execution of an exponential number of "classical" operations is necessary in order to solve a given arithmetic problem, it is clear that this high number of operations must be executed on the molecular level of DNA computations, too. But the power of DNA computing lies in the fact that one can do a chemical operation over 10^{16} DNA strands at once, i.e., in the possibility of using massive parallelism of a size that is not achievable by any electronic parallel computer or any interconnection network. Since we cannot sequentially execute 10^{16} operations on any electronic computer in hundreds of years, the advantage of DNA computing over electronic computing is the ability to "efficiently" execute exponential algorithms for input sizes, for which classical computing does not have any chance.

The situation with quantum computing is completely different. There is no available quantum computer and experimental physics is trying to build a 7-bit quantum computer under nontrivial physical difficulties. Because of this, it is not probable that we will have quantum computers in the near future. The crucial points are that the best known simulation of the quantum computing model by Turing machines requires an exponential increase of time complexity and that there exists a polynomial-time quantum algorithm for computing the factors of a given number (the factorization problem that is considered to be hard for classical polynomial-time computation). Thus, in contrast to DNA computing, there is a hope that P is a proper subset of the class of problems solvable by polynomial-time quantum computations. The physical construction of quantum computers would have an enormous impact on algorithmics, and especially on cryptography.[16]

Despite the fact that DNA computing and quantum computing are not competitive with respect to the presented algorithmic approaches for solving hard problems in the current practice, the situation may change in the future. This is the reason to provide at least an informal introduction to these two new computing technologies. Note that there are already involved monographs devoted to these topics (see, for instance, Păun, Rozenberg, and Salomaa [PRS98], Gruska [Gru99], Hirvensalo [Hi00], and Preskill [Pre00]) and we do not have the chance to go deeply in detail here. This is especially the case for quantum computing, where a nontrivial knowledge of quantum theory is a prerequisite of an involved study. The goal of this section is to provide an introduction to fundamental concept and ideas of DNA computing and quantum computing, and to discuss their possible impact on algorithmics in the future. Section 7.6 is organized as follows. First, the concepts of DNA computing is explained and illustrated using some DNA algorithms in Section 7.6.2. Section 7.6.3 provides a rough and informal description of the kind of the work done by quantum computers, and contains a discussion about the power of quantum computation.

7.6.2 DNA Computing

The fundamental idea behind DNA computing is to represent data as DNA strands (chains) and to use chemical manipulations (operations) over DNA strands to compute. To represent data as DNA strands is a natural idea because we usually represent data as words over an alphabet (as sequences of symbols) and DNA strands can be represented by sequences over the alphabet $\{A$ (adenine), G (guanine), C (cytosine), T (thymine)$\}$.

The idea to use chemical operations over DNA strands is connected with the concept of massive parallelism because one can execute a chemical operation over a multiset of DNA strands in a test tube all at once, and the cardinality of this multiset may be huge (for instance, 10^{16}).

The idea of using massive parallelism on the level of molecules in order to speed up computations was proposed already in 1961 by Feynman [Fey61], but more than thirty years were needed until an experimental demonstration of this idea became reality. In 1994 Adleman [Adl94] successfully realized a molecular computation in order to solve a small instance (7 vertices) of the NP-hard Hamiltonian path problem. This major experimental milestone was the start of the interdisciplinary research area called DNA computing. The method of Adleman was immediately modified and improved by Gifford [Gif94], Kaplan, Thaler, and Libchaber [KTL97], and Morimoto and Suyama [MoS97].

---

[16] Observe that the security of the RSA algorithm in public-key cryptography is based on the assumption that the problem of factorization is hard.

Lipton [Lip94, Lip95] defined the first abstract model of molecular computation. The elements of the model (its memory) are a set of test tubes, and each test tube contains a multiset of DNA strands. This model allows operations on the contents of the test tubes, which can be executed in one lab step. A simplification of this model was introduced by Adleman [Adl95]. Several further abstract models were introduced and investigated in recent years (see, for instance, Csuhaj-Varju, Freund, Kari, and Păun [CFK+96, CKP96], Liu, Guo, Coudon, Corn, Lagally, and Smith [LGC+96], Kurtz, Mahaney, Royer, and Simon [KMR+96], Reif [Rei95], and Yokomori and Kabayashi [YoK97]).

There are many papers showing that the models of DNA computing are universal, i.e., equivalent to the Turing machine model (for some excellent surveys see Păun, Rozenberg, and Salomaa [PRS98] and Salomaa [Sal98]). For some of these models Beaver [Bea95], Reif [Rei98], and Papadimitriou [Pap95] independently proved that they can achieve an exponential speedup of linear-space-bounded Turing machines working in exponential time. Perhaps, the most interesting theoretical result was established by Pudlák [Pud94], who introduced the so-called **genetic Turing machines**, and showed that these machines can recognize all languages from *PSPACE* (i.e., all NP-hard languages, too) in polynomial time. See Gifford [Gif94], Smith and Schweitzer [SmS95], Rubin [Rub96], Delcher, Hood, and Karp [DHK96], and Reif [Rei98] for some overviews on different aspects of DNA complexity.

The aim of Section 7.6.2 is to provide an elementary introduction to DNA computing. The rest of this section is organized as follows. First, we describe a simple model of DNA computation, and then we present two examples of DNA algorithms for hard problems. Finally, we discuss the theoretical and technological limits of the applicability of DNA computing. We do not provide any introduction to molecular biology. The book by Drlica [Drl92] is a beautiful introduction to molecular biology and genetic engineering for readers without any background on this topic.

A MODEL OF DNA COMPUTING.

The memory of this model consists of a constant number of test tubes, each one containing a multiset of DNA strands (a multiset of strings over $\{A, G, C, T\}$). We consider the following operations over the tubes. Let $MulS(T)$ denote the multiset of strings in the test tube T.

Copy	$Copy(T_1, T_2)$, for the test tubes T_1 and T_2, copies the content $MulS(T_1)$ of the test tube T_1 to the (originally empty) test tube T_2.
Merge	$Merge(T_1, T_2, T)$, for given test tubes T_1, T_2, and an empty test tube T, causes $MulS(T) = MulS(T_1) \cup MulS(T_2)$.
Detect	$Detect(T) \in \{yes, no\}$ for every test tube T. $Detect(T) = yes$ iff $MulS(T) \neq \emptyset$.

Separate	$Separate(T, w)$, for a test tube T and a string w, removes all strings from T that do not contain w as a substring.
Extract	$Extract(T, w)$, for a test tube T and a string w, removes all strings from T that contain w as a substring.
Separate-Pref	$Separate\text{-}Pref(T, w)$, for a test tube T and a string w, removes all strings from T that do not begin with the prefix w.
Separate-Suff	$Separate\text{-}Suff(T, w)$, for a test tube T and a string w, removes all strings from T that do not end with the suffix w.
Amplify	$Amplify(T)$, for a test tube T, causes the number of DNA strands in T to be doubled.
Length-Separate	$Length\text{-}Separate(T, l)$, for a test tube T, and a length l, removes all strings from T whose lengths are different from l.
Concatenate	$Concatenate(T)$, for a test tube T, is a random operation that results in a multiset of DNA strands that are random concatenations of the DNA strands in T. If T contains a huge number of copies of each of the strings in T, then after $Concatenate(T)$, T contains all possible concatenations of strings in T up to some length with a very high probability.
Cut	$Cut(T)$, for a test tube T, is an operation that randomly cuts the strands from $MulS(T)$ into shorter DNA strands.
Select	$Select(T)$, for a test tube T, randomly selects a DNA strand from T and puts it in a new test tube.

To explain how these operations can be executed in a chemical lab is beyond the scope of this book. For more technological details one should consult Adleman [Adl94, Adl95] and partially the introduction to the monograph [PRS98] by Păun, Rozenberg, and Salomaa. Excellent books on molecular biology are Drlica [Drl92], Alberts, Bray, Lewis, Raff, Roberts, and Watson [ABL⁺94], and Walker and Rapley [WR97].

The key complexity measures (computational measures) of the abstract models of DNA computation are:

- the number of operations, and
- the size of the test tubes (the size of a test tube T is the cardinality of $MulS(T)$, i.e., replications are included in counting).

The importance of the first complexity measure is two-fold. A lab operation is no matter of microseconds as it is in the case of an electronic computer, but its execution may be a matter of several hours. Secondly, the correct execution of most of the operations cannot be guaranteed with certainty by the current technology, i.e., an error probability must be accepted. Thus, the probability of getting a correct output with a DNA algorithm fast decreases with the number of its operations.[17]

---

[17] We shall later discuss this problem in detail, because the competitiveness of DNA computing with respect to classical algorithmic techniques depends on the development of biochemical technologies with a small probability of errors.

To design a DNA algorithm one has to find a suitable representation of data using existing DNA strands and then to design the proper algorithm as a sequence of operations from the above-mentioned list. We do not deal with the data representation because this requires learning the fundamentals of biochemistry. A typical algorithmic concept in DNA computing is straightforward. For a decision problem, it simply corresponds to an exhaustive search in the space of all solution candidates. The operations *Amplify* and *Concatenate* are usually used to obtain a set of all candidates for a solution in a test tube. The operations *Extract, Separate, Separate-Pref, Separate-Suff,* and *Length-Separate* are then used to remove all nonsolutions from the test tube. Finally, the operation *Detect* provides the output of the algorithm. We illustrate this concept with two examples from the literature.

ADLEMAN'S EXPERIMENT AND HAMILTONIAN PATH PROBLEM.

We sketch the concept of the Adleman's DNA algorithm for the labeled directed Hamiltonian path problem, LDHP, which is a decision problem. An input instance of the LDHP problem is a graph $G = (V, E)$ with two labeled vertices v_{in} and v_{out} in G. The idea of the Adleman's algorithm is to construct all possible sequences of vertices in a test tube and then to extract those which are not Hamiltonian paths from v_{in} to v_{out}.

ADLEMAN'S ALGORITHM

Input: A directed graph $G = (V, E)$ of n vertices $v_1, v_2, \ldots, v_n$ and two designated vertices $v_1 = v_{in}$, $v_n = v_{out} \in V$.

Preparation: Take n "convenient" DNA strands of an equal length l to represent the n vertices of G, and put these n DNA strands into a test tube T.

Step 1:   Use $2n \log_2 n$ times the operation *Amplify* to get at least $2^{2n \cdot \log_2 n} = n^{2n}$ copies of each DNA strand in T.

Step 2:   Apply the operation $Concatenate(T)$ to get a set of sequences of vertices of G in T that contains all sequences up to the length n with high probability.

Step 3:   Apply $Separate\text{-}Pref(T, v_{in})$ to remove all sequences that do not start with v_{in}.

Step 4:   Apply $Separate\text{-}Suff(T, v_{out})$ to remove all sequences that do no end with v_{out}.

Step 5:   Apply $Length\text{-}Separate(T, l \cdot n)$ to remove all sequences that do not correspond to paths of n vertices.

Step 6:   Apply consecutively $n - 2$ operations

$$Separate(T, v_2), Separate(T, v_3), \ldots, Separate(T, v_{n-1})$$

in order to remove all paths that do not contain all n vertices of G.

Output: $Detect(T)$.

It is obvious that ADLEMAN'S ALGORITHM decides the LDHP problem. Observe that this algorithm uses $2n \log_2 n + n + 3$ operations, i.e., the number of operations grows with n. This makes the ADLEMAN'S ALGORITHM unreliable for currently available technologies. Note that the operation *Concatenation* in step 2 is more complicated than presented here. In fact, one has to have not only DNA strands for vertices in the test tube, but also DNA strands for edges. A DNA strand for an edge (u, v) must have the first half the suffix[18] of the strand for u, and the second half has to be the prefix of the DNA strand for v. Then, building double-strands as sequences of vertices can work.

3-COLORABILITY.

A graph G is k-colorable if there is a possibility to color the vertices of G by k colors in such a way that no pair of vertices connected by an edge has the same color. The 3-colorability problem is to decide, for a given graph G, whether G is 3-colorable. This problem is known to be NP-complete. In what follows we give a DNA algorithm for this problem. The idea is again to generate a huge number of DNA strands that represent all possible assignments of three colors to the vertices of the given graph, and then to extract all such color assignments that are not feasible 3-colorings.

Algorithm 3-COL

Input: A graph $G = (V, E)$, $V = \{v_1, v_2, \ldots, v_n\}$.

Preparation: Take n DNA strands $strand(v_1), \ldots, strand(v_n)$ of the same length l to represent the vertices of G, and three strands $strand(c_1)$, $strand(c_2)$, $strand(c_3)$ of a length h to represent the colors. Put the strands $strand(v_i)strand(c_j)$ in a test tube U for $i = 1, \ldots, n$, $j = 1, 2, 3$.

Step 1: Use $3n$ times the operation $Amplify(U)$ in order to get 2^{3n} copies of every strand $strand(v_i)strand(c_j)$ in U.

Step 2: Apply the operation $Concatenate(U)$ in order to get strands of the form $u_1u_2 \ldots u_m$, $m \in \mathbb{N} - \{0\}$, where
$$u_k \in \{strand(v_i)strand(c_j) \mid i \in \{1, \ldots, n\}, j \in \{1, 2, 3\}\}$$
for $k = 1, \ldots, m$.

Step 3: Apply $Length\text{-}Separate(U, n \cdot (l+h))$ in order to remove all sequences that do not correspond to the assignment of colors to n vertices. {Note that some vertices may occur several times in this sequence, and so this sequence does not need to represent any consistent assignment of colors to all vertices}.

Step 4: **for** $i = 1$ **to** n **do** $Separate(U, v_i)$
{After this, every sequence contains exactly n vertices and each of the vertices of G occurs exactly once in the sequence}.

---

[18] Note that in biological reality we do not speak about a suffix and a prefix, but about the "biological complements" of this suffix and prefix.

Step 5: **for** $i = 1$ **to** $n - 1$ **do**
                    do begin $Copy(U, T_1)$; $Copy(U, T_2)$; $Copy(U, T_3)$
                    for $j = 1$ **to** 3 **do in parallel**
                    do begin $Extract(T_j, strand(v_i)strand(c_{(j+1) \bmod 3}))$;
                    $Extract(T_j, strand(v_i)strand(c_{(j+2) \bmod 3}))$;
                    {After this, T_j contains only colorings, where the vertex v_i is
                    colored by the jth color}.
                    for every $\{v_i, v_k\} \in E$ **do**
                         $Extract(T_j, strand(v_k)strand(c_j))$;
                    {After this, for $j = 1, 2, 3$, T_j contains only colorings where the
                    vertex v_i is colored by the jth color and none of its neighbors
                    is colored by the jth color.};
                    $Merge(T_1, T_2, T)$; $Merge(T, T_3, U)$
                    end
                    {After the rth loop of the main cycle each vertex $v \in \{v_1, v_2, \ldots, v_r\}$
                    has a color that is different from the colors of all its neighbors.}
                    end
Step 6: $Detect(U)$.

Observe that Steps 1, 2, 3, and 4 of 3-COL are similar to the work of the ADLEMAN'S ALGORITHM. The crucial step of 3-COL is Step 5, where one removes all inconsistent colorings. Observe that we really need more than one test tube to realize this step. We used three test tubes to separately handle the cases when v_i has the colors c_1, c_2, and c_3. After removing all coloring attempts that fail on edges incident to v_i, we joined the contents of these three tubes into the original one. Since the number of edges of G can be quadratic in n, the time complexity of Step 5 seems to be in $\Theta(n^2)$. The quadratic growth is unrealistic from the practical point of view, even if the current biochemical technologies are essentially improved. But the crucial point is that 3-COL can be implemented in $O(n)$ steps. This is because the current technology allows the operation $Extract(T, S)$, which, for a test tube T and a set of strings S, removes all strings from T that contain at least one substring $w \in S$. Thus, the inner cycle

"**for** every $\{v_i, v_k\} \in E$ **do** $Extract(T_j, strand(v_k)strand(c_j))$"

in Step 5 can be executed by $Extract(T_j, S_{ij})$, where $S_{ij} = \{strand(v_k) strand(c_j) \mid \{v_i, v_k\} \in E\}$.

LIMITS OF APPLICABILITY AND PERSPECTIVES.

As already mentioned, a DNA-based computer may be viewed as a parallel machine that may use a huge amount (10^{16}, for instance) of processors. This massive parallelism can cause an enormous speedup of sequential computations. On the other hand, if one needs to execute 10^{100} operations in order to solve a problem instance, then test tubes of the size of the known universe would not suffice to execute them in the time of its existence. Thus, DNA algorithms are able to simulate the work of exponential algorithms for inputs that are essentially larger than the inputs that could be handled by an electronic computer, but the input size acceptable for DNA computing is also bounded by a physical constant. This means that it is not completely fair to consider polynomial-time DNA computations as tractable ones. This is not only because of the potential exponential growth of the number and the size of the molecules in the test tubes with the number of lab operations, but also because of the time needed to execute the chemical operations.

The most serious problem for the competitiveness of DNA computations is the reliability of the lab operations that differs from operation to operation. One defines the **error rate** of a specific operation as the fraction of molecules that commits an error during the execution of this operation. If one has to execute a sequence of n operations $\sigma_1, \sigma_2, \ldots, \sigma_n$ with error rates $e_1, e_2, \ldots, e_n$, respectively, then the error rate of the final result may be roughly calculated as

$$1 - (1 - e_1) \cdot (1 - e_2) \cdot \cdots \cdot (1 - e_n).$$

Thus, if some current technologies have an error rate $5 \cdot 10^{-2}$, the error rate grows too fast with the number of executed operations. One can try to increase the probability of getting correct solutions by increasing the number of copies of DNA strands in the test cube.[19] An interesting calculation was done by Adleman, Rothemund, Roweiss, and Winfree [ARR+96] who considered a special DNA algorithm for breaking Data Encryption Standard (DES). To get a 63% chance of getting the key in the final tube:

(i) 1.4 gram of DNA strands is sufficient if an error rate of 10^{-4} is attainable,
(ii) less than 1 kg of DNA strands is sufficient, if the error rate can be bounded by 10^{-3}, and
(iii) approximately 23 Earth masses of DNA strands are necessary if the error rate is 10^{-2}.

We recommend the work of Boneh, Dunworth, Lipton, and Sgall [BDL+96] for the study of the tradeoffs between the number of lab operation steps and the number of DNA strands for various problems.

We have observed that the future success of DNA computing mainly depends on the results of the research in molecular biology and chemistry. The achievable degree of precision and the reliability of lab methods will be crucial for the competitiveness of DNA computation with respect to the classical computation. On the other hand, there is a need for development on the algorithmic level, too. The current DNA algorithms are based on an exhaustive search in exponentially large spaces, i.e., they execute an exponential number of classical operations. This does not always need to be the case. A combination of some of the algorithmic concepts presented in this book with DNA

---

[19] Note that this can be viewed as the execution of a number of random attempts of the experiment.

computing can lead to practical DNA algorithms. The first step in this direction was done by Bach, Condon, Glaser, and Tangway [BCG+98], who merged the concept of lowering the worst case exponential complexity with DNA computing. This combination, additionally connected with the concept of randomization, can be found by Chen and Ramachandran [ChR00] and by Díaz, Esteban, and Ogihara [DEO00].

Whether DNA computing will become a standard computing technology or not is still open. But the experience with the successful development of lab methods in molecular biology and chemistry gives a reason for optimism. We conclude this section with the following text of Adleman [Adl94] who has great expectations for DNA computing.

> "For the long term, one can speculate about the prospect for molecular computation. It seems likely that a single molecule of DNA can be used to encode the instantaneous description of a Turing machine and that currently available protocols and enzymes could (at least under idealized conditions) be used to induce successive sequence modifications, which would correspond to the execution of the machine. In the future, research in molecular biology may provide improved techniques for manipulating macromolecules. Research in chemistry may allow for the development of synthetic designer enzymes. One can imagine the eventual emergence of a general purpose computer consisting of nothing more than a single macromolecule conjugated to a ribosome collection of enzymes that act on it."

7.6.3 Quantum Computing

When Turing [Tu36] introduced the model of a Turing machine, he argued that his model can simulate the mental work of a mathematician who calculates on the symbolic level by following some method (algorithm). The computer science pioneers such as Turing, Church, Post, and Gödel considered Turing machines and other equivalent formal systems as a formal model of computation, or, in other words, as a mathematical formalization of the intuitive notion of algorithms. This means that they postulate the formalization of the notion of computation (algorithm) as a mathematical axiom. Remember that an axiom is something that people believe, but that cannot be verified by mathematical proof. This is the kernel of the criticism of this corner stone of computer science from physicists. Physicists claim that information is physical, and that it is stored, transmitted, and processed always by physical means. Thus, the model of computation should be a physical model that corresponds to the laws of physics and that can be experimentally verified. Why is this difference between mathematical modeling and physical modeling so important? There is no doubt that the Turing machine model can be realized as a physical device (electronic computer) and so any elementary operation of any reasonable mathematical model of computation can be simply executed in the physical world. Moreover, Benioff [Be80] proved that Turing machines can be built from quantum components, i.e., Turing machines can be considered as realistic objects on the level of quantum mechanics. But the crucial point is that the elementary actions (operations, step development) of quantum mechanical systems cannot be seen as simple and natural from the Turing machine point of view. To explain this a little bit more carefully, let us present a brief history of quantum computing.

In the beginning of the 1980s, Feynman [Fey82] asked whether a quantum mechanical system could be simulated by an ordinary computer without any exponential slowdown.[20] He personally conjectured that an efficient simulation is impossible and asked for the opportunity to build a computer that runs according to the laws of quantum physics, and so could efficiently simulate quantum mechanical systems.

In 1985, Deutsch [Deu85] introduced a quantum model of computation and described the universal quantum computer. This paper of Deutsch [Deu85] was seminal for the development of the theory of quantum computing because his model provided a solid ground for the theory of quantum computation as well as for the complexity theory of quantum computation. Thus, algorithmic problems that are solvable by polynomial-time quantum computers can be assumed to be tractable from the quantum mechanical point of view. The introduction of a quantum computer did not violate the Church thesis because the universal quantum computer can be simulated by a Turing machine (Bernstein and Vazirani [BV93]). Since all known simulations of quantum computation by a classical computation are connected with an exponential increase in time complexity, the question whether these are problems solvable in quantum polynomial time but not in classical polynomial time became the central question in the area of quantum computation.

The development of quantum algorithms in Deutsch and Josza [DJ92], Bernstein and Vazirani [BV93], and Simon [Sim94] culminated in the work of Shor [Sho94], who designed an efficient quantum algorithm for factorization. Remember that we neither have a polynomial-time algorithm for factorization, nor do we know whether this problem is NP-hard. But the factorization problem is considered to be hard and one conjectures that it is not in P. Thus, this result of Shor supports the hypothesis that quantum computation can efficiently solve problems that cannot be solved by classical (randomized) computation in polynomial time.

To explain the model of quantum computation requires learning the fundamentals of quantum mechanics, which is beyond the scope of this book. To provide at least an approximate idea of how quantum computation works and why it may be powerful, we give a rough sketch of a quantum model of computation. To do it, we draw an analogy to randomized computation.

---

[20] Note that any classical physical system can be efficiently (with a polynomial slowdown) simulated by standard computer models.

Let $s_1, s_2, \ldots, s_m$ be all possible states (configurations[21]) of a randomized machine A. Then, for every non-negative integer t, there exist probabilities $p_{t,1}, p_{t,2}, \ldots, p_{t,m}$, where

(i) for all $i \in \{1, \ldots, m\}$, $1 \geq p_{t,i} \geq 0$ is the probability of being in state s_i after t computation steps, and

(ii) $\sum_{i=1}^{m} p_{t,i} = 1$, i.e., $p_{t,1}, \ldots, p_{t,m}$ determine a probability distribution over the states of A.

The computation of A can be described by a matrix $M_A = [q_{ij}]_{i,j=1,\ldots,m}$, where q_{ij} is the probability of moving from state s_j to state s_i in one step of the computation of A. Obviously, every randomized algorithm has the property

$$\sum_{i=1}^{m} q_{ij} = 1 \qquad (7.2)$$

for each $j \in \{1, \ldots, m\}$, because $q_{1j}, q_{2j}, \ldots, q_{mj}$ is a probability distribution over the set of states of A with respect to the reachability from state s_j. We observe that

$$\begin{pmatrix} p_{t,1} \\ p_{t,2} \\ \vdots \\ p_{t,m} \end{pmatrix} = \begin{pmatrix} q_{11} & q_{12} & \cdots & q_{1m} \\ q_{21} & q_{22} & \cdots & q_{2m} \\ \vdots & \vdots & \ddots & \vdots \\ q_{m1} & q_{m2} & \cdots & q_{mm} \end{pmatrix} \cdot \begin{pmatrix} p_{t-1,1} \\ p_{t-1,2} \\ \vdots \\ p_{t-1,m} \end{pmatrix}$$

for every positive integer t. Due to the property (7.2) we have

$$\sum_{i=1}^{m} p_{t,i} = \sum_{i=1}^{m} p_{t-1,i}. \qquad (7.3)$$

Thus, starting with a probability distribution $p_{0,1}, p_{0,2}, \ldots, p_{0,m}$ (i.e., with the assumption $\sum_{i=1}^{m} p_{0,i} = 1$), (7.3) implies that $p_{t,1}, p_{t,2}, \ldots, p_{t,m}$ is a probability distribution for every t, too.

Now we outline the work of a quantum computer by explaining some similarities and some differences between randomized computation and quantum computation. The description of quantum mechanical systems is connected to Hilbert spaces, which are special vector spaces. For simplicity, consider the n-dimensional vector space over complex numbers with the orthonormal basis $\{X_1, X_2, \ldots, X_n\}$, where $X_1 = (1, 0, \ldots, 0)$, $X_2 = (0, 1, 0, \ldots, 0)$, $\ldots$, $X_n = (0, 0, \ldots, 0, 1)$. Any quantum state of this system can be written as

$$S = \alpha_1 X_1 + \alpha_2 X_2 + \cdots + \alpha_n X_n,$$

---

[21] Here, a state (configuration) is considered to be a complete description of the machine in a given time moment of the computation.

where α_i are complex numbers called **amplitudes**,[22] and

$$\sum |\alpha_i|^2 = 1.$$

S should not be viewed as a vector, but as a **superposition** of vectors (basic states) X_1, X_2, ..., and X_n. The similarity to randomized computation is that $|\alpha_1|^2, |\alpha_2|^2, \ldots, |\alpha_n|^2$ can be viewed as a probability distribution over the set of basic states $\{X_1, X_2, \ldots, X_n\}$. The difference between this and the randomized case is in the interpretation. While in randomized computation we have a probability of reaching a (basic) state in a time unit t, in quantum computation we assume that the quantum system is in all basic states $X_1, X_2, \ldots, X_n$ at once[23] and $|\alpha_i|^2$ is the probability of seeing the basic state X_i if one looks at $S = \alpha_1 X_1 + \alpha_2 X_2 + \cdots + \alpha_n X_n$. The crucial point is that there is no opportunity to look at a whole superposition S and to learn all $\alpha_1, \ldots, \alpha_n$ (all information involved in S). Any physical measurement of a superposition S results in a basic state and $|\alpha_i|^2$ is the probability that X_i is the output of this measurement. We also say that $|\alpha_i|^2$ is the probability that X_i is **observed**. Moreover, any measurement of S with a result X_i destroys the computation (the evolution) of the quantum system in the sense that the system collapses to the state

$$X_i = 0 \cdot X_1 + \cdots + 0 \cdot X_{i-1} + 1 \cdot X_i + 0 \cdot X_{i+1} + \cdots + 0 \cdot X_n.$$

Taking superpositions as a description of quantum systems it remains to be explained how the system evolves, i.e., which are the operations over superpositions. Analogously to randomized computation, the evolution of the system can be described by matrices. Let $A = [a_{ij}]_{i,j=1,\ldots,n}$ be an $n \times n$ matrix. Applying A to a superposition $S = \alpha_1 X_1 + \cdots + \alpha_n X_n$ we obtain a superposition $S' = \alpha'_1 X_1 + \cdots + \alpha'_n X_n$ by

$$\begin{pmatrix} \alpha'_1 \\ \alpha'_2 \\ \vdots \\ \alpha'_n \end{pmatrix} = \begin{pmatrix} a_{11} & a_{12} & \ldots & a_{1n} \\ a_{21} & a_{22} & \ldots & a_{2n} \\ \vdots & \vdots & \ddots & \vdots \\ a_{n1} & a_{n2} & \ldots & a_{nn} \end{pmatrix} \cdot \begin{pmatrix} \alpha_1 \\ \alpha_2 \\ \vdots \\ \alpha_n \end{pmatrix}. \tag{7.4}$$

All matrices that preserve the property

$$\sum_{i=1}^{n} |\alpha'_i|^2 = \sum_{i=1}^{n} |\alpha_i|^2 \tag{7.5}$$

for any vector $(\alpha_1, \alpha_2, \ldots, \alpha_n)^{\mathsf{T}}$ are allowed. The matrices satisfying the requirement (7.5) are called **unitary matrices**.[24]

---

[22] With respect to the base considered

[23] This is something that is not possible in our macroworld.

[24] A unitary matrix A is a matrix whose transpose complex conjugate is the inverse matrix A^{-1} to A.

A quantum computation starts in a known superposition S. Its own computation can be viewed as an application of a sequence of unitary matrices $A_1, A_2, \ldots, A_t$, i.e., as computing $S' = A_t \cdot A_{t-1} \cdots A_1 \cdot S$. The computation finishes with a measurement of S' that provides a basic state of the considered Hilbert space.

To illustrate the concept of quantum computation we present a simple example of a quantum system that realizes a fair coin toss (a generation of a truly random bit). Note that no abstract deterministic machine can compute a truly random bit.[25] Consider the 2-dimensional vector space over complex numbers with basic states $(1,0)^\mathsf{T}$, and $(0,1)^\mathsf{T}$. Obviously, $\{(1,0)^\mathsf{T}, (0,1)^\mathsf{T}\}$ is an orthonormal basis of this Hilbert space. We use the basic state $(1,0)^\mathsf{T}$ to represent the classical bit 1, and the basic state $(0,1)^\mathsf{T}$ to represent the classical bit 0. To compute we use the unitary matrix

$$H_2 = \begin{pmatrix} \frac{1}{\sqrt{2}} & \frac{1}{\sqrt{2}} \\ \frac{1}{\sqrt{2}} & -\frac{1}{\sqrt{2}} \end{pmatrix}.$$

Obviously,

$$H_2 \cdot (1,0)^\mathsf{T} = \left(\frac{1}{\sqrt{2}}, \frac{1}{\sqrt{2}}\right)^\mathsf{T} \text{ and } H_2 \cdot (0,1)^\mathsf{T} = \left(\frac{1}{\sqrt{2}}, -\frac{1}{\sqrt{2}}\right)^\mathsf{T}.$$

Looking at the superposition

$$\frac{1}{\sqrt{2}} \cdot (1,0)^\mathsf{T} + \frac{1}{\sqrt{2}} \cdot (0,1)^\mathsf{T}$$

one obtains the output $(1,0)^\mathsf{T}$ with probability $1/2 = \left(1/\sqrt{2}\right)^2$ and the output $(0,1)^\mathsf{T}$ with probability $1/2$, too. Since $(1,0)^\mathsf{T}$ represents the classical bit 1 and $(0,1)^\mathsf{T}$ represents the classical bit 0, the fair coin toss was performed. The same situation happens when measuring the superposition

$$\frac{1}{\sqrt{2}} \cdot (1,0)^\mathsf{T} + \left(-\frac{1}{\sqrt{2}}\right) \cdot (0,1)^\mathsf{T},$$

i.e., it does not matter whether we start from $(1,0)^\mathsf{T}$ or from $(0,1)^\mathsf{T}$. Another interesting point is that

$$H_2 \cdot (H_2 \cdot (1,0)^\mathsf{T}) = H_2 \cdot \left(\frac{1}{\sqrt{2}}, \frac{1}{\sqrt{2}}\right)^\mathsf{T} = (1,0)^\mathsf{T}, \text{ and}$$

$$H_2 \cdot (H_2 \cdot (0,1)^\mathsf{T}) = H_2 \cdot \left(\frac{1}{\sqrt{2}}, -\frac{1}{\sqrt{2}}\right)^\mathsf{T} = (0,1)^\mathsf{T},$$

i.e., after two applications of H_2 we always obtain the initial basic state. This means that any measurement after an odd number of computation steps

---

[25] One has to connect the machine to a physical source in order to get random bits.

provides a random bit, and any measurement after an even number of steps results in the initial basic state with probability 1.

To finish this section, we briefly discuss the power of quantum computation. Considering the above description of quantum computation we observe the following possible sources of its high computational power.

(i) **massive parallelism**

One transforms a superposition with a unitary transformation in one computation step of a quantum computer independent of the dimension of the vector space considered.

(ii) **infinite number of superpositions**

It is true that we can observe only one of the finitely many basis vectors, but the computation works over infinitely many superpositions.

(iii) **free choice of a basis**

If one wants to perform an observation of a superposition, then one has a free choice of an orthonormal basis of the system state space for this observation.

(iv) **inference**

If $\alpha_1, \alpha_2, \ldots, \alpha_k$ are amplitudes of a superposition and

$$1 = \sum_{i=1}^{k} |\alpha_i|^2 \neq \left| \sum_{i=1}^{k} \alpha_i \right|^2 ,$$

we speak about inference. The following example of an extreme destructive inference shows why we are unable to find an efficient simulation of quantum computation by a random computation. Let a quantum computation reach a basic vector X twice, once with the amplitude $\frac{1}{2}$ and once with the amplitude $-\frac{1}{2}$. Then the probability of being in X is

$$\left(\frac{1}{2} + \left(-\frac{1}{2} \right) \right)^2 = 0^2 = 0.$$

This contrasts with the solution when X is reached by exactly one of these two possibilities, and when the probability of being in X is $\left(\frac{1}{2} \right)^2 = \left(-\frac{1}{2} \right)^2 = \frac{1}{4}$. Obviously, such a behavior is impossible for classical randomized computation.

Currently, it is very hard to estimate the computational power of quantum computation. Due to Shor [Sho94] we have an efficient quantum algorithm for factorization and we do not believe that there exists a randomized polynomial-time algorithm for this task.[26] The negative experience of physicists in trying to efficiently simulate quantum systems with classical computers makes us believe that the realization of a quantum computer could essentially enrich our computational facilities.

---

[26] Remember that the hardness of factorization is unknown and we do not know whether this problem is NP-hard.

An excellent and exhaustive overview on quantum computing is given in the textbook [NC00] by Nielsen and Chuang. An involved, nice survey on this topic is given by Buhrman and Röhrig [BR03].

7.7 Glossary of Basic Terms

The goal of this section is to provide a concise glossary of a few basic terms of algorithmics for hard problems. The terms listed here are strongly connected with the concepts and techniques for the design of algorithms for hard problems, and should be in the vocabulary of every computer scientist. Here, we prefer to explain the informal meaning of every keyword rather than to provide its formal description. Pointers to the formal definitions and the applications in this book are included, too.

- **Adleman's experiment**
  Adleman's experiment is the first demonstration of the possibility to use DNA technologies for computational tasks. An input instance of the Hamiltonian path problem of 7 vertices was solved in this experiment (Section 7.6.2).
- **approximation algorithm**
  An approximation algorithm for an optimization problem is an algorithm that provides a feasible solution whose quality does not differ too much from the quality of an optimal solution. The quality of an approximation algorithm is measured by the approximation ratio of the algorithm (Section 4.2.1).
- **approximation-preserving reduction**
  An approximation-preserving reduction is a polynomial-time reduction (between two optimization problems) that preserves the approximation ratio of computed feasible solutions to some extent (Section 4.4.3).
- **approximation ratio**
  The approximation ratio of an algorithm A on an input x of a minimization [maximization] problem U is the optimal cost for x divided by the cost of the feasible solution to x provided by the algorithm A [the cost of the computed feasible solution divided by the optimal cost] (Definition 4.2.1.1).
- **backtracking**
  Backtracking is a method for solving optimization problems by a systematic exhaustive search for the optimal solution in the set of all feasible solutions or for determining an optimal strategy in a finite game by a search in the set of all configurations of the game. Backtracking is the basis for branch-and-bound (Section 2.3.4).
- **bin-packing problem**
  The bin-packing problem is an optimization problem. An input instance consists of a sequence $w_1, \ldots, w_n$ of rational numbers from $[0, 1]$ and the objective is to distribute them among bins of unit size 1 in such a way that a minimal number of bins is used (Section 2.3.2).

- **branch-and-bound**
  Branch-and-bound is an algorithm design technique for solving optimization problems. It makes backtracking more efficient by omitting the generation of some subtrees of the backtrack tree (by cutting), because it is possible to recognize that these parts do not contain any optimal solution. Branch-and-bound usually starts with computing a bound on the cost of an optimal solution, and then this bound is used to make backtracking more efficient. Different methods can be used for bounding the optimal cost. The typical method used for this purpose is relaxation to linear programming (Section 3.4).

- **Carmichael number**
  Carmichael numbers are composite numbers n with the property $a^{n-1} \equiv 1 \pmod{n}$ for all $n \in \{1, 2, \ldots, n-1\}$ (which is satisfied for all primes, too). Because of their existence (they are rare, but infinitely many) one cannot use Fermat's Theorem to obtain an equivalent definition of primality (which makes primality testing more complex) (Section 5.3.3).

- **Chinese Remainder Theorem**
  The Chinese Remainder Theorem is a theorem about the structure of $\mathbb{Z}_n$ that can be expressed as a cartesian product of finite fields. It provides an important idea for the design of a randomized algorithm for primality testing (Theorems 2.2.4.32 and 2.2.4.33).

- **Christofides algorithm**
  Christofides algorithm is a 1.5-approximation algorithm for the metric traveling salesperson problem (Section 4.3.5).

- **Church thesis**
  The Church thesis, also called the Church-Turing thesis, is the first axiom of computer science. It claims that a Turing machine is a mathematical formalization of the intuitive notion "algorithm" (Section 7.6).

- **clique problem**
  The clique problem is a decision problem. The task is, for a given graph G and a positive integer k, to decide whether G contains a clique of k vertices (Section 2.3.2).

- **crossover**
  The operation crossover is used in genetic algorithms. It expresses the analogy between genetic algorithms and population genetics by simulating the way how characteristics are passed from one generation to another by means of genes (Section 6.3).

- **decision problem**
  A decision problem is to indicate whether a given input has the prescribed property or not (Section 2.3.2).

- **derandomization**
  Derandomization is a concept of designing efficient deterministic algorithms by converting randomized algorithms to deterministic ones (Section 5.4).

- **divide-and-conquer**
  Divide-and-conquer is an algorithm design technique. It is based on breaking the given problem into several subinstances in such a way that one can easily compute a solution to the given problem instance from the solutions to the desired subinstance. Since the solution to the subinstances are computed in the same way, divide-and-conquer may be viewed as a recursive procedure (Section 2.3.4).

- **DNA computing**
  DNA computing is a new computing technology. Data are stored in DNA strands and the computation is executed by performing chemical operations over DNA strands in a tube (Section 7.6.2).

- **dual approximation algorithms**
  Dual approximation algorithms approximate the set of feasible solutions (feasibility) rather than the optimality. The cost of the computed solutions is at least as good as the optimal cost, but the solutions do not need to be feasible because they may break the constraints a little bit. Dual approximation algorithms are suitable for situations where one cannot exactly determine the constraints on the feasibility of solutions. The concept of dual approximation algorithms can also be used to design approximation algorithms (Section 4.2.4 and 4.3.6).

- **dynamic programming**
  Dynamic programming is an algorithm design technique. The similarity to divide-and-conquer is that both these techniques solve problems by combining the solutions to problem subinstances. The difference between them is in the fact that divide-and-conquer does it recursively, while dynamic programming works in a bottom-up fashion by starting with computing solutions to the smallest subinstances, and continuing to larger and larger subinstances until the original problem instance is solved. The main advantage of dynamic programming over divide-and-conquer is that a dynamic programming algorithm solves every problem subinstance at most once, while divide-and-conquer may compute a solution to the same subinstance many times (Section 2.3.4).

- **Euclid's algorithm**
  Euclid's algorithm is an efficient algorithm for computing the greatest common divisor of two integers (Section 2.2.4).

- **Euler's criterion**
  Euler's criterion provides an equivalent definition of primality that can be efficiently verified by a randomized algorithm. It is the basis of the Solovay-Strassen algorithm for primality.

- **feasible solution**
  A feasible solution to an instance of an optimization problem is a solution that satisfies all constraints of the given problem instance.

- **fingerprinting**
  Fingerprinting is a technique for designing efficient randomized algorithms for equivalence problems. It can be viewed as a special case of the method

of abundance of witnesses. To compare two representations of some objects one randomly chooses a mapping from a special class of mappings and applies this mapping to both representations. The images of this mapping are called fingerprints. Fingerprints are substantially shorter than the original representations and one solves the equivalence problem by comparing the fingerprints (Sections 5.2.3 and 5.3.4).

- **foiling an adversary**
  Foiling an adversary is a fundamental paradigm of the design of randomized algorithms. It is based on the fact that a randomized algorithm can be viewed as a probability distribution over a set of deterministic algorithms. While an adversary can be able to find hard input instances that foil a small fraction of the deterministic algorithms in the set, it is difficult to devise a single input that is likely to defeat a randomly chosen algorithm (Section 5.2.3).

- **Fundamental Theorem of Arithmetics**
  The Fundamental Theorem of Arithmetics says that every integer can be expressed as a product of nontrivial powers of primes, and that up to a rearrangement of the factors, this prime factorization is unique (Theorem 2.2.4.14).

- **genetic algorithms**
  The concept of genetic algorithms is a heuristic for solving optimization problems. It is based on an analogy to the optimization in the process of evolution of a population of individuals (Section 6.3).

- **greedy method**
  The greedy method is a classical algorithm design technique for solving optimization problems. One consecutively specifies a feasible solution in a step-by-step manner. In each of the specification steps the greedy method takes the (locally) best possible choice for the parameter of the solution (Section 2.3.4).

- **Hamiltonian cycle problem**
  The Hamiltonian cycle problem (HC) is a decision problem. For a given graph G, one has to decide whether G contains a Hamiltonian cycle (Sections 2.3.2 and 3.6.3).

- **Held-Karp bound**
  If one designs an algorithm for an optimization problem, then it is not so easy to judge its quality in an experimental analysis because one needs to compare the costs of the outputs of the algorithm with the optimal costs which are usually not known. In such cases one can take the cost of solutions computed by the technique of relaxation to linear programming (called Held-Karp bounds on the optimal costs) as substitutes for the unknown optimal costs (Section 7.4).

- **heuristics**
  Heuristics are algorithms (or algorithm design techniques) for which we are not able to guarantee the computation of correct results in a reasonable time. Their basic advantages are simplicity and robustness, and so

they usually provide the cheapest way for developing algorithms for given problems. Simulated annealing, genetic algorithms, and tabu search are well-known representatives of heuristics (Chapter 6).

- **inference**
  Inference is an important feature of quantum computation. If there are two different ways to reach a configuration (basic state) and both have positive probabilities, the inference can cause the probability of reaching this configuration to be 0. The classical randomized computations are unable to simulate such behavior (Section 7.6.3).

- **integer programming**
  Integer programming is an NP-hard optimization problem. The objective is to compute an optimal integer solution to a system of linear equations (Sections 2.3.2 and 3.7).

- **island model**
  The island model is a special version of genetic algorithms. Here, we take several populations that develop in an isolated way for a time. Then, the best individuals of each population are broadcasted to all populations, and one continues with the isolated development for a time, etc. The idea behind the island model is to prevent a premature convergence of a population. Another advantage of the island model over standard genetic algorithms is its suitability for efficient parallel implementations (Section 6.3.2 and 7.5).

- **Kernighan-Lin's variable depth search**
  Kernighan-Lin's variable depth search is an algorithm design technique for optimization problems. It combines local search with the greedy method (Section 3.6.2).

- **knapsack problem**
  The knapsack problem (KP) is an NP-hard optimization problem. An input instance consists of a number of objects (items) with given weights and costs and an upper bound on the weight of the knapsack. The aim is to maximize the common costs of objects in the knapsack under the constraints given by the upper bound on the weight of the knapsack. There is an FPTAS for KP, and so this problem is considered easy from the approximation point of view (Sections 2.3.4 and 4.3.4).

- **Las Vegas**
  Las Vegas algorithms are randomized algorithms that never err. There are two possible definitions of the concept of Las Vegas randomization. The first one requires a correct output for every possible run of the algorithm and measures complexity as expected complexity. The second definition allows the answer "I don't know" with a bounded probability and measures complexity in the worst case manner (Sections 5.2 and 5.3.2).

- **linear programming**
  Linear programming is an optimization problem that can be solved in polynomial time. One minimizes a linear function under constraints that are given by a system of linear equations (Sections 2.2.1 and 3.7).

- **local optimum**
  One can define a neighborhood on any set of all feasible solutions to an instance of an optimization problem by assigning a set of neighbors to each feasible solution. Then, a solution is a local optimum if none of its neighbors has a better cost (Section 3.6).

- **local search**
  Local search is a simple and robust algorithm design technique for optimization problems. One starts with a feasible solution and tries to improve it by searching in its neighborhood. A local search algorithm halts if it reaches a local optimum (Section 2.3.4 and 3.6).

- **lowering worst case complexity of exponential algorithms**
  This concept aims to design exponential algorithms of worst case complexity in $O(c^n)$ for some $c < 2$ or, more generally, of worst case complexity that is essentially smaller than the number of solution candidates[27] (Sections 3.5 and 5.3.7).

- **makespan scheduling**
  Makespan scheduling is an NP-hard optimization problem. The objective is to minimize the execution time of a given set of jobs on a given number of machines (Sections 2.3.2, 4.2.1, and 4.3.6).

- **maximum clique problem**
  The objective of the maximum clique problem is to find a clique of the maximal size in a given graph. This NP-hard optimization problem is one of the hardest problems with respect to polynomial-time approximability (Sections 2.3.2 and 4.4).

- **maximum cut problem**
  The objective of the maximum cut problem is searching for a cut of the maximal size in a given graph. This NP-hard maximization problem is in APX (Sections 2.3.2 and 4.3.3).

- **maximum satisfiability (MAX-SAT)**
  The objective of the maximum satisfiability problem is to find an assignment that satisfies the maximal number of clauses of a given formula in the conjunctive normal form. MAX-SAT is one of the paradigmatic NP-hard problems and it is in APX.

- **method of conditional probabilities**
  The method of conditional probabilities is a derandomization method for converting randomized optimization algorithm into deterministic ones. The idea is to (deterministically) compute a random sequence that leads to an output whose approximation ratio is at least as good as the expected approximation ratio (Sections 5.4.1, 5.4.4, and 5.4.5).

- **method of reduction of the probability space**
  The method of the reduction of the probability space is a derandomization method for converting randomized algorithms into deterministic ones. It is based on reducing the number of random bits to such a small number that

---

[27] Feasible solutions in the case of optimization problems

one is able to consecutively simulate all possible runs of the randomized algorithm in polynomial time. This reduction is not necessary for the analysis of the probabilistic behavior of the randomized algorithms (Sections 5.4.1, 5.4.2, and 5.4.3).

- **Metropolis algorithm**
  The Metropolis algorithm is a model of the optimization process of annealing in condensed matter physics. It can be viewed as a randomized local search that may also accept deteriorations of feasible solutions with some probability. The Metropolis algorithm is the base for simulated annealing (Section 6.2.1).

- **Monte Carlo**
  Monte Carlo algorithms are randomized algorithms that may err (Chapter 5). We distinguish one-sided-error, two-sided-error, and unbounded-error Monte Carlo algorithms with respect to their error probability.

- **mutation**
  The operation of mutation is used in genetic algorithms to perform random changes on the individuals of a given population.

- **neighborhood**
  For every instance I of an optimization problem one defines a neighborhood as a mapping that assigns a set of feasible solutions (neighbors) to each feasible solution of the set of all feasible solutions to I. A neighborhood is called **exact** if every local optimum is a global optimum, too. A neighborhood is called **polynomial-time searchable** if there exists a polynomial-time algorithm that, for every feasible solution α, decides whether α is a local optimum (Section 3.6).

- **NP-complete**
  A language (decision problem) is NP-complete if it is in NP and is NP-hard (Section 2.3.3).

- **NP-hard**
  A language L (decision problem) is NP-hard if all languages from NP are polynomial-time reducible to L. An optimization problem is NP-hard if its threshold language is NP-hard (Section 2.3.3).

- **parallel computation**
  A parallel computation is a computation mode in which several processors cooperate in order to solve a problem.

- **parameterization**
  A parameterization for a given algorithmic problem is such a partition of the set of all input instances of this problem into classes that, for a given input instance, one can efficiently assign it to its class (Section 3.3.1).

- **parameterized complexity**
  Parameterized complexity is a concept for the design of algorithms for hard problems. The idea is to use a parameterization in order to classify the input instances of a given problem with respect to their individual hardness (Section 3.3).

- **PCP-Theorem**
  The PCP-Theorem is a fundamental result of theoretical computer science.
  It says that every proof of a polynomial length in the length of its theorem
  can be transformed into a proof that can be probabilistically checked by
  looking at only 11 bits of this proof. This surprising result is the core of
  the theory for proving lower bounds on polynomial-time approximability
  (Section 4.4).

- **polynomial-time approximation scheme**
  A polynomial-time approximation scheme (PTAS) is a polynomial-time
  algorithm which for an input instance I and a positive real number ε
  computes a feasible solution to I with an approximation ratio at most ε.
  Such an algorithm is called a fully PTAS if it works in polynomial time
  with respect to both input size and ε^{-1} (Definition 4.2.1.6 and Sections
  4.3.4 and 4.3.6).

- **primality testing**
  The problem of primality testing is to decide, for a given positive integer,
  whether it is a prime. This problem is fundamental for modern cryptography. It is not known to be in P, but there is a Las Vegas polynomial-time
  algorithm that solves it (Section 5.3.3).

- **Prime Number Theorem**
  The Prime Number Theorem is a fundamental result of number theory.
  It says that the number of primes among the integers $1, 2, 3, \ldots, n$ tends
  to $n/\ln n$ as n increases. It has important applications for the design of
  approximation algorithms, especially for the method of abundance of witnesses and the fingerprinting technique (Theorem 2.2.4.16 and Section 5.2).

- **pseudo-polynomial-time algorithm**
  Pseudo-polynomial-time algorithms are algorithms that are efficient for the
  input instances of integer-valued problems with integer values bounded by
  a polynomial in the length of the input. Pseudo-polynomial-time algorithms can be viewed as a special version of the concept of parameterized
  complexity (Section 3.2).

- **quantum computer**
  The quantum computer is a model of computations based on the principles
  of quantum mechanics (Section 7.6.3).

- **random rounding**
  Random rounding is a technique for the design of randomized algorithms.
  It is usually connected with the method of relaxation to linear programming. One relaxes an optimization problem in order to get an optimal
  solution to the relaxed problem in polynomial time, and then this solution
  is randomly rounded in order to obtain a feasible solution to the original
  problem (Sections 5.2.3 and 5.3.6).

- **random sampling**
  A random sample from a population is often a representative of the whole
  population. Random sampling can efficiently provide solutions to prob-

lems of searching for objects with prescribed properties (Sections 5.2.3 and 5.3.6).

- **randomized computation**
  Randomized computation is a computation mode where the next action does not only depend on the correct state but also on the result of a coin tossing (Chapter 5).

- **relative error**
  The relative error of an approximation algorithm measures the quality of feasible solutions computed by the algorithm. The relative error of an algorithm A and an input x is the ratio of the absolute difference between the optimal cost and the cost of $A(x)$ (the solution to x computed by A) and the optimal cost (Definition 4.2.1.1).

- **relaxation to linear programming**
  Relaxation to linear programming is an optimization algorithm design technique that consists of three steps. In the first step an input instance U of the considered optimization problem is expressed as an input instance I' of linear integer programming. The second step is relaxation which means that one computes an optimal solution to I' as an instance of linear programming. In the third step this solution is used to compute feasible solutions to the instance I of the original problem (Sections 3.7, 4.3.2, and 5.3.6).

- **satisfiability problem**
  The satisfiability problem is one of the paradigmatic NP-complete problems. The task is to decide whether a given Boolean formula in the conjunctive normal form is satisfiable (Sections 2.3.2, 3.5, 5.3.7, and 5.4).

- **Schema Theorem**
  The assertion of the Schema Theorem is the best known analytic result (unfortunately, often misinterpreted) about the behavior of genetic algorithms. It says that the number of individuals with some short representation subparts of a high fitness grows in the population under some reasonable assumptions (Section 6.3).

- **simplex algorithm**
  The simplex algorithm is an algorithm for solving instances of linear programming. It is nothing else than a local search algorithm that moves between vertices of a polytope via edges of the polytope. It has an exponential worst case complexity, but is very fast on average (Section 3.7.3).

- **simulated annealing**
  Simulated annealing may be viewed as a local search enriched by a kind of randomized decision that enables leaving local optima in order to find a better solution. It is based on the Metropolis algorithm that models the annealing process as an optimization process of the state of a solid (Section 6.2).

- **Solovay-Strassen algorithm**
  The Solovay-Strassen algorithm is a polynomial-time one-sided-error Monte Carlo algorithm for primality testing. It is designed by the method

of abundance of witnesses and is based on the characterization of primes by Jacobi symbols (Section 5.3.3).

- **stability of approximation**
  Stability of approximation is a concept for attacking hard optimization problems. The main idea is to find some crucial characteristics of the input instances that partition the set of all input instances into infinitely many classes with respect to the achievable polynomial-time approximability (Sections 4.2.3, 4.3.4, and 4.3.5).

- **tabu search**
  Tabu search is a heuristic based on local search. In contrast to classical local search and simulated annealing it is not a memory-less algorithm. It dynamically restricts the actual neighborhoods dependent on the development in a number of the last steps (Section 6.2.3).

- **traveling salesperson problem**
  The traveling salesperson problem (TSP) is one of the paradigmatic NP-hard optimization problems. The task is to find a Hamiltonian cycle of minimal cost in a given weighted complete graph. It is one of the hardest optimization problems in NPO with respect to polynomial-time approximability (Sections 2.3.2, 3.6.3, and 4.3.5).

- **vertex cover**
  The vertex cover problem is, for a given graph G and a number k, to decide whether G contains a vertex cover of size k.

References

*"Unhappy are those
to whom everything is clear."*

LOUIS PASTEUR

[AB95]   Andreae, T., Bandelt, H.-J.: Performance guarantees for approximation
          algorithms depending on parameterized triangle inequalities. *SIAM Jour-
          nal on Discrete Mathematics* 8 (1995), 1–16.

[ABI86]   Alon, N., Babai, L., Itai, A.: A fast and simple randomized parallel algo-
          rithm for the maximal independent set problem. *Journal of Algorithms*
          7 (1986), 567–583.

[ABL+94] Alberts, B., Bray, D., Lewis, J., Raff, M., Roberts, K., Watson, J. D.:
          Molecular Biology of the Cell. 3rd edition, Garland Publishing, New York,
          1994.

[ACG+99] Ausiello, G., Crescenzi, P., Gambosi, G., Kann, V., Marchetti-
          Spaccamela, A., Protasi, M.: *Complexity and Approximation (Combi-
          natorial Optimization Problems and Their Approximability Properties),*
          Springer-Verlag 1999.

[Adl78]   Adleman, L.: Two theorems on random polynomial time. In: *Proc. 19th
          IEEE FOCS,* IEEE 1978, pp. 75–83.

[Adl94]   Adleman, L. M.: Molecular computation of solution to combinatorial
          problems. *Science* 266 (1994), 1021–1024.

[Adl95]   Adleman, L. M.: On constructing a molecular computer. In: [LiB96], pp.
          1–22.

[ADP77]   Ausiello, G., D'Atri, A., Protasi, M.: On the structure of combinatorial
          problems and structure preserving reductions. In: *Proc. 4th ICALP '77,
          Lecture Notes in Computer Science* 52, Springer-Verlag 1977, pp. 45–60.

[ADP80]   Ausiello, G., D'Atri, A., Protasi, M.: Structure preserving reductions
          among convex optimization problems. *Journal of Computer and System
          Sciences* 21 (1980), 136–153.

[AGP92]   Alford, W. R., Granville, A., Pomerance, C.: There are infinitely many
          Carmichael numbers. *University of Georgia Mathematics Preprint Series,*
          1992.

[AGP97]   Anderson, E. J., Glass, C. A., Potts, C. N.: Machine scheduling. In:
          [AL97a], pp. 361–414.

504   References

[AH87]   Adleman, L., Huang, M.: Recognizing primes in random polynomial time.
            In: *Proc. 19th ACM STOC*, ACM, 1987, pp. 482–469.
[AHS85]   Ackley, D. H., Hinton, G. E., Sejnowski, T. J.: A learning algorithm for
            Boltzmann machines. *Cognitive Science* 9 (1985), 147–169.
[AHU74]   Aho, A. J., Hopcroft, J. E., Ullman, J. D.: *The Design and Analysis of
            Computer Algorithms*. Addison-Wesley, 1974.
[AHU83]   Aho, A. J., Hopcroft, J. E., Ullman, J. D.: *Data Structures and Algo-
            rithms*. Addison-Wesley, 1983.
[AHO$^+$96] Asano, T., Hori, K., Ono, T., Hirata, T.: Approximation algorithms for
            MAXSAT: Semidefinite programming and network flows approach. Tech-
            nical Report, 1996.
[AK89]   Aarts, E. H. L., Korst, J. H. M.: *Simulated Annealing and Boltzmann ma-
            chines (A Stochastic Approach to Combinatorial Optimization and Neural
            Computing)*. John Wiley & Sons, Chichester, 1989.
[AKL97]   Aarst, E. H. L., Korst, J. H. M., van Laarhoven, P. J. M.: Simulated
            annealing. In: [AL97a], pp. 91–120.
[AKS02]   Agrawal, M., Kayal, N., Saxena, N.: Primes in P. Unpublished
            manuscript.
[AL97]   Arora, S., Lund, C.: Hardness of approximation. In: *Approximation Al-
            gorithms for NP-hard Problems* (D.S. Hochbaum, Ed.), PWS Publishing
            Company, Boston, 1997.
[AL97a]   Aarts, E., Lenstra, J. K. (Eds.): *Local Search in Combinatorial Opti-
            mization*. Wiley-Interscience Series in Discrete Mathematics and Opti-
            mization, John Wiley & Sons, 1977.
[AL97b]   Aarts, E., Lenstra, J. K.: Introduction. In:[AL97a], pp. 1–18.
[Ali95]   Alizadeh, F.: Interior point method in semidefinite programming with ap-
            plications to combinatorial optimization. *SIAM Journal on Optimization*
            5, no. 1 (1995), 13–51.
[ALLP97]   Aarts, E. H. L., van Laarhoven, P. J. M., Lin, C. L., Pan, P.: VLSI layout
            synthesis. In [AL97a], pp. 415–440.
[ALM92]   Arora, S., Lund, C., Motwani, R., Sudan, M., Szegedy, M.: Proof veri-
            fication and hardness of approximation problems. In: *Proc. 33rd IEEE
            FOCS*, IEEE, 1992, pp. 14–23.
[AMM77]   Adleman L., Manders, K., Miller, G. L.: On taking roots in finite fields.
            In: *Proc. 18th IEEE FOCS*, IEEE, 1977, pp. 151–163.
[AMS$^+$80] Ausiello, G., Marchetti-Spaccamela, A., Protasi, M.: Towards a unified
            approach for the classification of NP-complete optimization problems.
            Theoretical Computer Science 12 (1980), 83–96.
[Ang82]   Angluin, D.: Lecture notes on the complexity on some problems in num-
            ber theory. Technical Report 243, Department of Computer Science, Yale
            University, 1982.
[AOH96]   Asano, T., Ono, T., Hirata, T.: Approximation algorithms for the maxi-
            mum satisfiability problem. *Nordic Journal of Computing* 3 (1996), 388–
            404.
[APR83]   Adleman, L., Pomerance, C., Rumely, R.: On distinguishing prime num-
            bers from composite numbers. *Annals of Mathematics* 117 (1983), 173–
            206.
[Aro94]   Arora, S.: Probabilistic checking of proofs and hardness of approximation
            problems, Ph.D. thesis, Department of Computer Science, Berkeley, 1994.

[Aro96]   Arora, S.: Polynomial time approximation shemes for Euclidean TSP and other geometric problems. In: *Proc. 37th IEEE FOCS*, IEEE, 1996, pp. 2–11.

[Aro97]   Arora, S.: Nearly linear time approximation schemes for Euclidean TSP and other geometric problems. In: *Proc. 38th IEEE FOCS*, IEEE, 1997, pp. 554–563.

[ARR+96] Adleman, L. M., Rothemund, P. W. K., Roweiss, S., Winfree, E.: On applying molecular computation to Data Encryption Standard. In: [BBK+96], pp. 28–48.

[AS92]   Arora, S., Safra, S.: Probabilistic checking proofs: A new characterization of NP. In: *Proc. 38th IEEE FOCS*, IEEE, 1997, pp. 2–13.

[Asa92]   Asano, T.: Approximation algorithms for MaxSat: Yannakakis vs. Goemanns-Williamson. In: *Proc. 33rd IEEE FOCS*, IEEE, 1992, pp. 2–13.

[Asa97]   Asano, T.: Approximation algorithms for MAXSAT: Yannakakis vs. Goemans-Williamson. In: Proc. *5th Israel Symposium on the Theory of Computing and Systems*, 1997, pp. 24–37.

[AV79]   Angluin, D., Valiant, L.: Fast probabilistic algorithms for Hamiltonian circuits and matchings. *Journal of Computer and System Sciences* 18 (1979), 155–193.

[Baa88]   Baase, S.: *Computer Algorithms – Introduction to Design and Analysis.* 2nd edition, Addison-Wesley, 1988.

[Bab79]   Babai, L.: Monte Carlo algorithms in graph isomorphism techniques. In: *Research Report no. 79-10*, Département de mathématiques et de statistique, Université de Montréal, Montréal, 1979.

[Bab85]   Babai, L.: Trading group theory for randomness. In: *Proc. 17th ACM STOC*, ACM, 1985, pp. 421–429.

[Bab90]   Babai, L.: E-mail and the unexpected power of interaction. In: *Proc. 5th Annual Conference on Structure in Complexity Theory*, 1990, pp. 30–44.

[BB88]   Brassard, G., Bratley, P.: *Algorithms: Theory and Practice.* Prentice-Hall, 1988.

[BB96]   Brassard, G., Bratley, P.: *Fundamentals of Algorithmics.* Prentice-Hall, Englewood Cliffs, 1996.

[BBK+96] Baum, E., Boneh, D., Kaplan, P, Lipton, R., Reif, J., Soeman, N. (Eds.): *DNA Based Computers.* Proc. 2nd Annual Meeting, Princeton, 1996.

[BBS86]   Blum, L., Blum, M., Shub, M.: A simple unpredictable pseudorandom number generation. *SIAM Journal on Computing* 15 (1986), 364–383.

[BC99]   Bender, M. A., Chekuri, C.: Performance guarantees for the TSP with a parametrized triangle inequality. In: *Proc. 6th WADS'99, Lecture Notes in Computer Science* 1663, Springer-Verlag, 1999, pp. 80–85.

[BC00]   Bender, M. A., Chekuri, C.: Performance guarantees for the TSP with a parametrized triangle inequality. *Information Processing Letters* 73, no. 1–2 (2000), 17–21.

[BCG+98] Bach, E., Condon, A., Glaser, E., Tangway, C.: DNA models and algorithms for NP-complete problems. *Journal of Computer and System Sciences* 57 (1988), 172–186.

[BDF+92] Balasubramanian, R., Downey, R., Fellows, M., Raman, V.: Unpublished manuscript, 1992.

[BDL+96] Boneh, D., Dunworth, C., Lipton, R. J., Sgall, J.: On the computational power of DNA. *Discrete Applied Mathematics* 71 (1996), 79–94.

[Be80]   Benioff, J. A.: The computer as a physical system: A microsopic quantum mechanical Hamiltonian model of computers as represented by Turing machines. *Journal of Statistical Physics* 22 (1980), 563–591.

[BE95]   Beigel, R., Eppstein, D.: 3-coloring in time $O(1.344^n)$: A No-MIS algorithm. In *Proc. 36th IEEE FOCS*, IEEE, 1995, pp. 444–453.

[Bea95]   Beaver, D.: Computing with DNA. *Journal of Computational Biology* 2 (1995), 1–7.

[Ben90]   Bentley, J. L.: Experiments on traveling salesman heuristics. In: *Proc. 1st ACM-SIAM Symposium on Discrete Algorithms*, ACM, New York 1990, and SIAM, Philadelphia, 1990, pp. 91–99.

[BFL91]   Babai, L., Fortnow, L., Lund, C.: Non-deterministic exponential time has two-prover interactive protocols. In: *Computational Complexity* 1 (1991), pp. 3–40.

[BFLS91] Babai, L., Fortnow, L., Levin, L., Stegedy, M.: Checking computations on polylogarithmic time. In: *Proc. 23rd ACM STOC*, ACM, 1991, pp. 21–31.

[BFR00]   Balasubramanian, R., Fellows, M., Raman, V.: An improved fixed parameter algorithm for vertex cover. *Information Processing Letters*, to appear.

[BHK$^+$99] Böckenhauer, H.-J., Hromkovič, J., Klasing, R., Seibert, S., Unger, W.: Towards the notion of stability of approximation for hard optimization tasks and the traveling salesman problem. *Electronic Colloqium on Computational Complexity*, Report No. 31 (1999). Extended abstract in: *Proc. CIAC 2000, Lecture Notes in Computer Science* 1767, Springer-Verlag, 2000, pp. 72–86.

[BHK$^+$00] Böckenhauer, H.-J., Hromkovič, J., Klasing, R., Seibert, S., Unger, W.: Approximation algorithms for the TSP with sharpened triangle inequality. *Information Processing Letters* 75 (2000), 133–138.

[BKS$^+$95] Branke, J., Kohlmorgen, U., Schmeck, H., Veith, H.: Steuerung einer Heuristik zur Losgrößenplanung unter Kapazitätsrestriktionen mit Hilfe eines parallelen genetischen Algorithmus. In: *Proc. Workshop Evolutionäre Algorithmen*, Göttingen, 1995 (in German).

[BM88]   Babai, L., Moran, S.: Arthur-Merlin games: A randomized proof system, and a hierarchy of comlexity classes. *Journal of Computer and System Sciences* 36 (1988), 254–276.

[BM95]   Bacik, R., Mahajan, S.: Semidefinite programming and its applications to NP problems. In: *Proc. 1st COCOON, Lecture Notes in Computer Science* 959, Springer-Verlag, 1995.

[BMi84]   Blum, M., Micali, S.: How to generate cryptographically strong sequences of pseudo-random bits. *SIAM Journal on Computing* 13 (1984), 850–864.

[BMS98]   Branke, J., Middendorf, M., Schneider, F.: Improved heuristics and a genetic algorithm for finding short supersequences. *OR Spektrum* 20 (1998), 39–45.

[BN68]   Bellmore, M., Nemhauser, G.: The traveling salesperson problem: A survey. *Operations Research* 16 (1968), 538–558.

[Boc58a] Bock, F.: An algorithm for solving 'traveling-salesman' and related network optimization problems: abstract. *Bulletin Fourteenth National Meeting of the Operations Research Society of America*, 1958, p. 897.

[BoC93]   Bovet, D. P., Crescenzi, C.: *Introduction to the Theory of Complexity*. Prentice-Hall, 1993.

[BR03]   Buhrman, H., Röhrig, H.; Distributed quantum computing. In: *Proc. 28th MFCS'03, Lecture Notes in Computer Science 2747*, Springer-Verlag 2003, pp. 1–20.

[Bra88]   Brassard, G.: *Modern Cryptology: A Tutorial. Lecture Notes in Computer Science 325*, Springer-Verlag, 1988.

[Bra94]   Brassard, G.: Cryptology column – Quantum computing: The end of classical cryptograpy? *ACM Sigact News* 25, no. 4 (1994), 15–21.

[Bre89]   Bressoud, D. M.: *Factorization and Primality Testing*. Springer-Verlag, 1989.

[BS94]   Bellare, M., Sudan, M.: The complexity of decision versus search. In: *Proc. 26th ACM STOC*, ACM, 1994, pp. 184–193.

[BS96]   Bach, E., Shallit, J.: *Algorithmic Number Theory, Vol. 1*, MIT Press, 1996.

[BS00]   Böckenhauer, H.-J.: Seibert S.: Improved lower bounds on the approximability of the traveling salesman problem. *Theoretical Informatics and Applications*, 34, no. 3 (2000), pp. 213–255.

[BSW+98]   Bonet, M., Steel, M. S., Warnow, T., Yooseph, S.: Better methods for solving parsimony and compatibility. Unpublished manuscript, 1998.

[Bu89]   Buss, S.: Unpublished manuscript, 1989.

[BV93]   Bernstein, E., Vazirani, U. V.: Quantum complexity theory. In: *Proc. 25th ACM STOC*, ACM, 1993, pp. 11–20 (also: *SIAM Journal of Computing* 26 (1997), 1411–1473).

[Car12]   Carmichael, R.: On composite numbers p which satisfy the Fermat congruence $a^{p-1} \equiv 1$. *American Mathematical Monthly* 19 (1912), 22–27.

[CCD98]   Calude, C. S., Casti, J., Dinneen, M. J. (Eds.): *Unconventional Models of Computation*. Springer-Verlag 1998.

[Čer85]   Černý, V.: A thermodynamical approach to the traveling salesman problem: An efficient simulation algorithm. *Journal of Optimization Theory and Applications* 45 (1985), 41–55.

[Čer98]   Černý, V.: personal communication.

[CFK+96]   Csuhaj-Varju, E., Freund, R., Kari, L., Păun, Gh.: DNA computing based on splicing: Universality results. In: *Proc. 1st Annual Pacific Symposium on Biocomputing*, Hawai 1996 (L. Hunter, T. E. Klein, Eds.), World Scientific, 1996, pp. 179–190.

[Cha79]   Chachian, L. G.: A polynomial algorithm for linear programming. *Doklady Akad. Nauk USSR* 224, no. 5 (1979), 1093–1096 (in Russian), translated in *Soviet Math. Doklady* 20, 191–194.

[Chr76]   Christofides, N.: Worst case analysis of a new heuristic for the travelling salesman problem. Technical Report 388, Graduate School of Industrial Administration, Carnegie-Mellon University, Pittsbourgh, 1976.

[ChR00]   Chen, K., Ramachandran, V.: A space-efficient randomized DNA algorithm for k-SAT. In: [CoR00], pp. 171–180.

[Chv79]   Chvátal, V.: A greedy heuristic for the set-covering problem. *Mathematics on Operations Research* 4 (1979), 233–235.

[Cob64]   Cobham, A.: The intrinsic computation difficulty of functions. In: *Proc. 1964 Int. Congress of Logic Methodology and Philosophie of Science*, (Y. Bar-Hillel, Ed.), North-Holland, Amsterdam, 1964, pp. 24–30.

[Coo71]   Cook, S. A.: The complexity of theorem proving procedures. In: *Proc. 3rd ACM STOC*, ACM, 1971, pp. 151–158.

[CoR00]   Condon, A., Rozenberg, G. (Eds.): *DNA Based Computing. Proc. 6th Int. Meeting*, Leiden Center for Natural Computing, 2000.

[Cr86]   Crow, J. F.: *Basic Concepts in Population, Quantitive and Evolutionary Genetics*. Freeman, New York, 1986.

[Cro58]   Croes, G. A.: A method for solving traveling salesman person. *Operations Research* 6 (1958), 791–812.

[CRS87]   Casotto, A., Romeo, F., Sangiovanni-Vincentelli, A. L.: A parallel simulated annealing algorithm for the placement of macro-cells. *IEEE Transactions on Computer-Aided Design* 6 (1987.), 838–847.

[CK70]   Crow, J. F., Kimura, M.: *An Introduction to Population Genetic Theory*. Harper and Row, New York, 1970.

[CK99]   Crescenzi, P., Kann, V.: A Compendium of NP Optimization Problems. http://www.nada.kth.se/theory/compendium/

[CKP96]   Csuhaj-Varju, E., Kari, L., Păun, Gh.: Test tube distributed system based on splicing. *Computers and Artificial Intelligence* 15 (1996), 211–232.

[CKS⁺95] Crescenzi, P., Kann, V., Silvestri, R., Trevisan, L.: Structure in approximation classes. In: *Proc. 1st Computing and Combinatorics Conference (CONCOON), Lecture Notes in Computer Science* 959, Springer-Verlag, 1995, pp. 539–548.

[CLR90]   Cormen, T., Leiserson, C., Rivest, R.: *Introduction to Algorithms*, MIT Press, and McGraw-Hill, 1990.

[Cro86]   Crow, J. F.: *Basic Concepts in Population Quantitative and Evolutionary Genetics*, Freeman, New York, 1986.

[Dan49]   Dantzig, G. B.: Programming of independent activities, II, Mathematical model. *Econometrics* 17 (1949), 200–211.

[Dan63]   Dantzig, G. B.: *Linear Programming and Extensions*. Princeton University Press, 1963.

[Dan82]   Dantsin, E.: Tautology proof system based on the splitting method. Ph.D. dissertation, Steklov Institute of Mathematics (LOMI), Leningrad, 1982 (in Russian).

[Dav87]   Davis, L. (Ed.): *Genetic Algorithms and Simulated Annealing*. Morgan Kauffmann, Los Altos, 1987.

[Dav91]   Davis, L. (Ed.): *Handbook of Genetic Algorithms*. Van Nostrand Reinhold, New York, 1991.

[Deu85]   Deutsch, D.: Quantum theory, the Church-Turing principle and the universal quantum computer. In: *Proceedings of the Royal Society*, London, 1992, vol. A439, pp. 553-558.

[Dev86]   Devroye, L.: *Non-Uniform Random Variate Generation*. Springer-Verlag, 1986.

[DEO00]   Díaz, S., Esteban, J. L., Ogihara, M.: A DNA-based random walk method for solving k-SAT. In: [CoR00], pp. 181–191.

[DF92]   Downey, R. G., Fellows, M. R.: Fixed-parameter tractability and completeness. *Congressus Numerantium* 87 (1992), 161–187.

[DF95a]   Downey, R. G., Fellows, M. R.: Fixed-parameter tractability and completeness I: Basic results. *SIAM Journal of Computing* 24 (1995), 873–921.

[DF95b]   Downey, R. G., Fellows, M. R.: Fixed-parameter tractability and completeness II: On completeness for $W[1]$. *Theoretical Computer Science* 141 (1995), 109–131.

[DF99]   Downey, R. G., Fellows, M. R.: *Parameterized Complexity*. Monographs in Computer Science, Springer-Verlag 1999.

[DFF56]   Dantzig, G. B., Ford, L. R., Fulkerson, D. R.: A primal-dual algorithm for linear programming. In: *Linear Inequalities and Related Systems*, (H.W. Kuhn, A. W. Tucker, Eds.), Princeton University Press, 1956.

[DFF+61]   Dunham, B., Fridshal, D., Fridahal, R., North, J. H.: Design by natural selection. IBM Res. Dept. RC-476, June 2000, 1961.

[DGHS00]   Dantsin, E., Goerdt, A., Hirsch, E.A., Schöning, U.: Deterministic algorithms for k-SATbased on covering codes and local search. In: *Proc. ICALP '00, Lexture Notes in Computer Science*, Springer-Verlag 2000, pp. 236-247.

[DHI00]   Ďuriš, P., Hromkovič, J., Inoue, K.: A separation of determinism, Las Vegas and nondeterminism for picture recognition. In: *Proc. IEEE Conference on Computation Complexity*, IEEE, 2000, pp. 214-228.

[DHK96]   Delcher, Al. L., Hood, L., Karp, R. M.: Report on the DNA/Biomolecular Computing Workshop, 1996.

[DHR+97]   Ďuriš, P., Hromkovič, J., Rolim, J.D.P., Schnitger, G.: Las Vegas versus determinism for one-way communication complexity, finite automata and polynomial-time computations. In: *Proc. STACS '97, Lecture Notes in Computer Science* 1200, Springer-Verlag, 1997, pp. 117-128.

[Dij59]   Dijkstra, E. W.: A note on two problems in connexion with graphs. *Numerische Mathematik* 1 (1959), 269-271.

[DJ92]   Deutsch, D., Josza, R.: Rapid solution of problems by quantum computation. *Proceedings of the Royal Society, London A* 439 (1992), 553-558.

[DJW98]   Droste, S., Janses, T., Wegener, I.: A rigorous complexity analysis of the $(1+1)$ evolutionary algorithm for separable functions with Boolean inputs. *Evolutionary Computation* 6 (1998), 185-196.

[DJW98a]   Droste, S., Janses, T., Wegener, I.: On the optimization of unimodal functions with the $(1+1)$ evolutionary algorithm. In: *Proc. 5th Parallel Problem Solving from Nature, Lecture Notes in Computer Science* 1998, Springer-Verlag, Cambridge University Press, 1998, pp. 47-56.

[DJW99]   Droste, S., Janses, T., Wegener, I.: Perhaps not a free lunch but at least a free appetizer. In: *Proc. 1st Genetic and Evolutionary Computation Conference* (W. Banzaf, J. Daida, A. E. Eiben, M. H. Garzon, V. Honarir, M. Jakiela, and R. E. Smith, Eds.), Morgan Kaufmann, San Francisco, 1999, pp. 833-839.

[DJW01]   Droste, S., Janses, T., Wegener, I.: On the analysis of the $(1+1)$ evolutionary algorithm. *Theoretical Computer Science*, to appear.

[DKR94]   Dietzfelbinger, M., Kutylowski, M., Reischuk, R.: Exact lower bounds for computing Boolean functions on CREW PRAMs. *Journal of Computer and System Sciences* 48 (1994), pp. 231-254.

[Dow95]   Dowsland, K. A.: Simulated annealing. In: [Ree95], pp. 20-69.

[Drl92]   Drlica, K.: *Understanding DNA and Gene Cloning. A Guide for the CURIOUS*. John Wiley & Sons, New York, 1992.

[DSST97]   Díaz, J., Serna, M., Spirakis, P. Torán, J.: *Paradigms for Fast Parallel Approximability*, 1997.

[dWH89]   de Werra, D., Hertz, A.: Tabu search techniques: a tutorial and an application to neural networks. *OR Spektrum* 11 (1989), 131-141.

510 References

[Eas58]   Eastman, W. L.: Linear programming with pattern constraints. Ph.D. Thesis, Report No. Bl.20, The Computation Laboratory, Harvard University, 1958.

[Edm65]   Edmonds, J.: Paths, Trees, and Flowers. *Canadian International Mathematics* 17 (1965), 449–467.

[ELR85]   O'hEigeartaigh, M., Lenstra, J. K., Rinnovy Kan, A. H. (Eds.): *Combinatorial Optimization: Annotated Bibliographies*. John Wiley & Sons, Chichester, 1985.

[Eng99]   Engebretsen, L.: An explicit lower bound for TSP with distances one and two. In: *Proc. 16th STACS, Lecture Notes in Computer Science* 1563, Springer-Verlag, 1999, pp. 373–382. (full version: *Electronic Colloquium on Computational Complexity* TR99-046, 1999).

[Eve80]   Even, S.: *Graph Algorithms*. Computer Science Press, 1980.

[FaK92]   Faigle, U., Kern, W.: Some convergence results for probabilistic tabu search. *ORSA Journal on Computing* 4 (1992), 32–37.

[Fei96]   Feige, U.: A threshold of $\ln n$ for approximation set cover. In: *Proc. 28th ACM STOC*, ACM, 1996, pp. 314–318.

[Fel88]   Fellows, M. R.: On the complexity of vertex set problems. Technical report, Computer Science Department, University of New Mexico, 1988.

[Fey61]   Feynman, R. P.: In: *Miniaturization* (D. H. Gilbert, Ed.), Reinhold, New York, 1961, pp. 282 - 296.

[Fey82]   Feynman, R.: Simulating physics with computers. *International Journal of Theoretical Physics* 21, nos. 6/7 (1982), 467–488.

[Fey86]   Feynman, R.: Quantum mechanical computers. *Foundations of Physics* 16, no. 6 (1986), 507–531; originally appeared in *Optics News*, February 1985.

[FF62]   Ford, L. R., Fulkerson, D. R.: *Flows in Networks*. Princeton University Press 1962.

[FG95]   Feige, U., Goemans, M.: Approximating the value of two prover proof systems, with application to Max2Sat and MaxDiCut. In: *Proc 3rd Israel Symposium on the Theory of Computing and Systems*, 1995, pp. 182–189.

[FGL91]   Feige, U., Goldwasser, S., Lovász, L., Safra, S, Szegedy M.: Approximating clique is almost NP-complete. In: *Proc 32nd IEEE FOCS*, IEEE, 1991, pp. 2–12.

[FJGO95]   Fredman, M. L., Johnson, D. S., McGeoch, L. A., Ostheimer, G.: Data structures for traveling salesman. *Journal of Algorithms* 18 (1995), 432–479.

[FK94]   Feige, U., Kilian, J.: Two-prover one-round proof systems: Their power and their problems. In: *Proc. 24th ACM STOC*, ACM, 1992, pp. 733–744.

[FL88]   Fellows, M. R., Langston, M. A.: Nonconstructive advances in polynomial-time complexity. *Information Processing Letters* 28 (1988), 157–162.

[FRS88]   Fortnow, L., Rompel, J., Sipser, M.: On the power of multi-prover interactive protocols. In: *Proc. 3rd IEEE Symposium on Structure in Complexity Theory*, IEEE, 1988, pp. 156–161.

[Fre77]   Freivalds, R.: Probabilistic machines can use less running time. In: *Information Processing 1977*, IFIP, North-Holland, 1977, pp. 839–842.

[FrG01]   Friedman, J., Goerdt, A.: Recognizing more unsatisfiable random 3-SATinstances efficiently. In: *ICALP '01, Lecture Notes in Computer Science*, Springer-Verlag 2001, pp. 310-321.

[Gar73]   Garfinkel, R.: On partitioning the feasible set in a branch-and-bound algorithm for the asymetric traveling salesperson problem. *Operations Research* 21 (1973), 340–342.

[GB65]   Colomb, S. W., Baumert, L. D.: Backtrack programming. *Journal of the ACM* 12, no. 4 (1965), 516 –524.

[GHR95]   Greenlaw, R. Hoover, H., Ruzzo, W.: *Limits to Parrallel Computation: P-completness Theory*. Oxford University Press, 1995.

[Gif94]   Gifford, D.: On the path to computing with DNA. *Science* 266 (1994), 993–994.

[Gil77]   Gill, J.: Computational complexity of probabilistic Turing machines. *SIAM Journal on Computing* 6 (1977), 675–695.

[GJ78]   Garey, M. R., Johnson, D. S.: Strong NP-completeness results: Motivations, examples and applications. *Journal of the ACM* 25 (1978), 499–508.

[GJ76]   Garey, M. R., Johnson, D. S.: Approximation algorithms for combinatorial problems: An annotated bibliography. In: *Algorithms and Complexity: Recent Results and New Directions* (J.F. Traub, Ed.), Academic Press, 1976, pp. 41–52.

[GJ79]   Garey, M. R., Johnson, D. S.: *Computers and Intractability. A Guide to the Theory on NP-Completeness*. W. H. Freeman and Company, 1979.

[GK86]   Goldwasser, S., Kilian, J.: Almost all primes can be quickly certified. In: *Proc. 18th ACM STOC*, ACM, 1986, pp. 316–329.

[GLa95]   Glover, F., Laguna, M.: Tabu search. In: [Ree95], pp. 70–150.

[Glo86]   Glover, F.: Future paths for integer programming and links to artificial intelligence. *Computers & Operations Research* 5 (1986), 533–549.

[Glo89]   Glover, F.: Tabu search: Part I. *ORSA Journal on Computing* 1 (1989), 190–206.

[Glo90]   Glover, F.: Tabu search: Part II. *ORSA Journal on Computing* 2 (1990), 4–32.

[GLP97]   Gendreau, M., Laporte, G., Potvin, J.-Y.: Vehicle routing: handling edge exchanges. In: [AL97a], pp. 311–336.

[Gol89]   Goldberg, D. E.: *Genetic Algorithms in Search, Optimization, and Machine Learning*. Addison-Wesley, Reading, 1989.

[GLS81]   Grötschel, M., Lovász, L., Schrijver, A.: The Elipsoid method and its consequences in combinatorial optimization. *Combinatorica* 1 (1981), 169–197.

[GMR89]   Goldwasser, S., Micali, S., Rackoff, C.: The knowledge complexity of interactive proof-systems. *SIAM Journal of Computing* 18 (1989), 186–208.

[GMW91]   Goldreich, O., Micali, S., Wigderson, A.: Proofs that yield nothing but their validity, or all languages in NP have zero-knowledge proof system. *Journal of the ACM* 38 (1991), 691–729.

[GN00]   Gramm, J., Niedermeier, R.: Faster exact solutions for MAX2SAT. In: *Proc. CIAC'2000, Lecture Notes in Computer Science* 1767, Springer-Verlag, 2000, pp. 174–186.

[Goe97]   Goemans, M.: Semidefinite programming in combinatorial optimization. In: *Proc. 16th International Symposium on Mathematical Programming*, 1997.

[GoK01]   Goerdt, A., Krivelevich, M.: Efficient recognition of random unsatisfiable
           k-SATinstances by spectral methods. In: *Proc. STACS '01, Lecture Notes
           in Computer Science*, Springer-Verlag 2001, pp. 294-304.

[Gol95]   Goldreich, O.: Probabilistic proof systems (survey). Technical report, De-
           partment of Computer Science and Applied Mathematics, Weizmann In-
           stitute of Science, 1995.

[Gra66]   Graham, R.: Bounds for certain multiprocessor anomalics. *Bell System
           Technical Journal* 45 (1966), pp. 1563–1581.

[GrKP94]   Graham, R., Knuth, D. E., Patashnik, O.: *Concrete Mathematics: A
           Foundation for Computer Science*. Addison-Wesley, 1994.

[Gru99]   Gruska, J.: *Quantum Computing*. McGraw-Hill, 1999.

[GW94a]   Goemans, M., Williamson, D.: .878-Approximation algorithms for MAX
           CUT and MAX 2SAT. In: Proc *26th ACM STOC*, ACM, 1994, pp. 422–
           431.

[GW94b]   Goemans, M., Williamson, D.: New 3/4-approximation algorithms for the
           maximum satisfiability problem. *SIAM Journal on Discrete Mathematics*
           7 (1994), 656–666.

[GW95]   Goemans, M., Williamson, D.: Improved approximation algorithms for
           maximum cut and satisfiability problems using semidefinite program-
           ming. *Journal of the ACM* 42 (1995), 1115–1145.

[Háj85]   Hájek, B.: A tutorial survey of the theory and applications of simulated
           annealing. In: *Proc. 24th IEEE Conference on Decision and Control*,
           IEEE, 1985, pp. 755–760.

[Háj88]   Hájek, B.: Cooling schedules for optimal annealing. *Mathematics of Op-
           erations Research* 13 (1988), 311–329.

[Han86]   Hansen, P.: The steepest ascent mildest descent heuristic for combinato-
           rial programming. Talk presented in: *Congress on Numerical Methods in
           Combinatorial Optimization*, Capri, 1986.

[Har87]   Harel, D.: *Algorithmics: The Spirit of Computing*. Addison-Wesley, 1987;
           2nd edition 1992.

[HáS89]   Hájek, B., Sasaki, G.: Simulated annealing: To cool it or not. *Systems
           Control Letters* 12, 443–447.

[Hås96a]   Håstad, J.: Clique is hard to approximate within $n^{1-\varepsilon}$. In: *Proc. 37th
           IEEE FOCS*, IEEE, 1996, pp. 627–636.

[Hås96b]   Håstad, J.: Testing of the long code and hardness for clique. In: *Proc.
           28th ACM STOC*, ACM, 1996, pp. 11–19.

[Hås97a]   Håstad, J.: Clique is hard to approximate within $n^{1-\varepsilon}$. Technical Report
           TR97-038, *Electronic Colloquium on Computational Complexity*, 1997.
           An earlier version was presented in [Hås96a].

[Hås97b]   Håstad, J.: Some optimal inapproximability results. In: *Proc. 29th ACM
           STOC*, ACM, 1997, pp. 1–10. Also appeared as Technical Report TR97-
           037, *Electronic Colloquium on Computational Complexity*.

[HeK71]   Held, M., Karp, R. M.: The traveling-salesman problem and minimum
           spanning trees, part II. *Mathematical Programming* 1 (1971), 6–25.

[HeK80]   Held, M., Karp, R. M.: The traveling-salesman problem and minimum
           spanning trees. *Operations Research* 18 (1970), 1138–1162.

[HH98]   Hofmeister, T., Hühne, M.: Semidefinite programming and its application
           to approximation algorithms. In: [MPS98], pp. 263-298.

[Hir98]   Hirsch, E. A.: The new upper bounds for SAT. In: *Proc. 9th ACM-SIAM Symposium on Discrete Algorithms*, 1998, pp 521–530; extended version – to appear in *Journal of Automated Reasoning*.

[Hi00]   Hirvensalo, M.: *Quantum Computation. Springer Series on Natural Computing*. Springer-Verlag, 2000, to appear.

[Hir00]   Hirsch, E. A.: A new algorithm for MAX-2-SAT. In: *Proc. 17th STACS' 2000, Lecture Notes in Computer Science* 1700, Springer-Verlag, 2000, pp. 65–73.

[HiS83]   Hinton, G. E., Sejnowski, T. J.: Optimal perceptual inference. In: *Proc. IEEE Conference on Computer Vision and Pattern Recognition*, IEEE, Washington, 1983, pp. 448–453.

[HK65]   Hall, M., Knuth, D. E.: Combinatorial analysis and computers. *American Mathematical Monthly* 72 (1965), 21–28.

[HK70]   Held, M., Karp, R.: The traveling salesperson problem and minimum spanning trees. *Operations Research* 18 (1970), 1138–1162.

[HK71]   Held, M., Karp, R.: The traveling salesperson problem and minimum spanning trees: Part II. *Mathematical Programming* 1 (1971), 6–25.

[HKM⁺96]   Hromkovič, J., Klasing, R., Monien, B., Peine, R.: Dissemination of information in interconnection networks (Broadcasting and Gossiping). In: *Combinatorial Network Theory* (Ding-Zhu Du, Frank Hau, Eds.), Kluwer Academic Publishers, 1996, pp. 125 – 212.

[HL96]   Hofmeister, T., Lefman, H.: A combinatorial design approach to MaxCut. *Random Structures and Algorithms* 9 (1996), 163–175.

[Hoa92]   Hoare, C. A. R.: Quicksort. *Computer Journal* 5, no. 1 (1962), 10–15.

[Hoc97]   Hochbaum, D. S. (Ed.): *Approximation Algorithm for NP-hard Problems*. PWS Publishing Company, Boston, 1997.

[Hol75]   Holland, J. H.: *Adaptation in Natural and Artificial Systems*. University of Michigan Press, Ann Arbor, 1975.

[HPS94]   Hougardy, S., Prömel, H., Steger, A.: Probabilistically checkable proofs and their consequences for approximation algorithms. *Discrete Mathematics* 136, (1994), 175–223.

[Hro97]   Hromkovič, J.: *Communication Complexity and Parallel Computing*. EATCS Monographs, Springer-Verlag, 1997.

[Hro98]   Hromkovič, J.: Towards the notion of stability of approximation algorithms for hard optimization problems. Department of Computer Science I, RWTH Aachen, May 1998.

[Hro99a]   Hromkovič, J.: Stability of approximation algorithms and the knapsack problem. In: *Jewels Are Forever* (J. Karhumäki, M. Mauer, G. Pann, G. Rozenberg, Eds.), Springer-Verlag, 1999, pp. 29–46.

[Hro99b]   Hromkovič, J.: Stability of approximation algorithms for hard optimization problems. In: *Proc. SOFSEM '99, Lecture Notes in Computer Science* 1725, Springer-Verlag, 1999, pp. 29–46.

[Hro00]   Hromkovič, J.: Communication protocols: An exemplary study of the power of randomness. In: *Handbook of Randomized Computing* (P. Pardalos, S. Rajasekaran, J. Reif, J. Rolim, Eds.), Kluwer Academic Publishers.

[Hro03]   Hromkovič, J.: *Theoretical Computer Science. Introduction to Automata, Computability, Complexity, Algorithmics, Randomization, Communication, and Cryprography*. EATCS Series, Springer-Verlag, 2003.

514   References

[HrS99]   Hromkovič, J., Schnitger, G.: On the power of Las Vegas II: Two-way
           finite automata. In: *Proc. ICALP '99, Lecture Notes in Computer Sci-
           ence* 1644, Springer-Verlag, 1999, pp. 433–442 (full version–to appear in
           Theoretical Computer Science).

[HrS00]   Hromkovič, J., Schnitger, G.: On the power of Las Vegas for one-way
           communication complexity, OBDDs, and finite automata. *Information
           and Computation*, to appear.

[HrSa00]   Hromkovič, J., Sauerhoff, M.: Tradeoffs between nondeterminism and
           complexity for communication protocols and branching programs. In
           Proc. STACS '2000, Lecture Notes in Computer Science 1770, Springer-
           Verlag 2000, pp. 145–156.

[HS76]   Horowitz, E., Sahni, S.: *Fundamentals of Data Structures*. Computer Sci-
           ence Press, 1976.

[HS78]   Horowitz, E., Sahni, S.: *Fundamentals of Computer Algorithms*. Com-
           puter Science Press, 1978.

[HS85]   Hochbaum, D. S., Shmoys, D.: A best possible heuristic for the k-center
           problem. *Mathematics of Operations Research* 10 (1975), 180–184.

[HS87]   Hochbaum, D. S., Shmoys, D. B.: Using dual approximation algoritms for
           scheduling problems: practical and theoretical results. *Journal of ACM*
           34 (1987), pp. 144–162.

[HSA84]   Hinton, G. E., Sejnowski, T. J., Ackley, D. H.: Boltzmann machines con-
           straint satisfaction networks that learn. Technical report CMU-CS-84-119
           Carnegie-Mellon University, 1984.

[HT74]   Hopcroft, J. E., Tarjan, R. E.: Efficient algorithms for graph manipula-
           tion. *Journal of the ACM* 21, no. 4 (1974), 294–303.

[HTW97]   Hertz, A., Taillard, E., de Werra, D.: Tabu search. In: [AL97a], pp. 121–
           136.

[HU79]   Hopcroft, J. E., Ullman, J. D.: *Introduction to Automata Theory, Lan-
           guages and Computation*. Addison-Wesley, 1979.

[Hu82]   Hu, T. C.: *Combinatorial Algorithms*. Addison-Wesley, Reading, 1982.

[IK75]   Ibarra, O. H., Kim, C. E.: Fast approximation algorithms for the knap-
           sack and sum of subsets problem. *Journal of the ACM* 22 (1975), 463–468.

[IS65]   Ignall, E., Schrage, L.: Application of the branch-and-bound technique
           to some flow-shop scheduling problems. *Operations Research* 13 (1965),
           400–412.

[JAGS89]   Johnson, D. S., Aragon, C. R., McGeoch, L. A, Scheron, C.: Optimiza-
           tion by simulated annealing; and experimental evaluation: Part I, graph
           partitioning. *Operations Research* 37 (1989), 865–892.

[JAGS91]   Johnson, D. S., Aragon, C. R., McGeoch, L. A, Scheron, C.: Optimization
           by simulated annealing: An experimental evaluation; part II, graph col-
           oring and number partitioning. *Operations Research* 39 (1991) 378–406.

[Já92]   Jájá, J.: *An Introduction to Parallel Algorithms*. Addison-Wesley, 1992.

[Jer73]   Jeroslow, R. J.: The simplex algorithm with the pivot rule of maximizing
           criterion improvement. *Discrete Mathematics* 4 (1973), 367–378.

[JG97]   Johnson, D. S., McGeoch, L. A.: The traveling salesman problem: A case
           study. In [AL97a], pp. 215–310.

[Joh73]   Johnson, D. S.: Near-optimal Bin-packing algorithms. Doctoral disser-
           tion, Massechusetts Institute of Technologie, MIT Report MAC TR-109,
           1973.

[Joh74]   Johnson, D. S.: Approximation algorithms for combinatorial problems. *Journal of Computer and System Sciences* 9 (1974), 256–289.

[Joh86]   Johnson, D. S.: Computing in the Math Department: Part I (The NP-completeness column: An ongoing guide). *Journal of Algorithms* 7 (1986), 584–601.

[Joh87]   Johnson, D. S.: The NP-completeness column: An ongoing guide. *Journal of Algorithms* 8, no. 2 (1987), 285–303, and no. 3 (1987), 438–448.

[Joh90]   Johnson, D. S.: A catalog of complexity classes. In: *Handbook of Theoretical Computer Science, Volume A*, (J. van Leeuwen, Ed.), Elsevier and MIT Press, 1990, pp. 67–161.

[Joh90a] Johnson, D. S.: Local optimization and the traveling salesman problem. In: *Proc. ICALP '90, Lecture Notes in Computer Science* 443, Springer-Verlag 1990, pp. 446–461.

[JPS84]   Jájá, J., Prassanna Kumar, V.K., Simon, J.: Information transfer under different sets of protocols. *SIAM Journal of Computing* 13 (1984), 840–849.

[JPY88]   Johnson, D. S., Papadimitriou, Ch., Yannakakis, M.: How easy is local search? *Journal of Computer and System Sciences* 37 (1988), 79–100.

[JR00]   Julstrom, B., Raidl, G. R.: A weighted coding in a genetic algorithm for the degree-constrained minimum spanning tree problem. In: *Proc. ACM Symposium of Applied Computing*, ACM, 2000.

[JWe99]   Jansen, T., Wegener, I.: On the analysis of evolutionary algorithms – a proof that crossover really can help. In: *Proc. 7th ESA, Lecture Notes in Computer Science* 1643, Springer-Verlag, 1999, pp. 184–193.

[Kar72]   Karp, R. M.: Reducibility among combinatorial problems. In: *Complexity of Computer Computations* (R. E. Miller, J. W. Thatcher, Eds.). Plenum Press, 1972, pp. 85–103.

[Kar84]   Karmarkar, N.: A new polynomial-time algorithm for linear programming. *Combinatorica* 4 (1984), 373–395.

[Kar91]   Karp, R.: An introduction to randomized algorithms. *Discrete Applied Mathematics* 34 (1991), 165–201.

[Kar96]   Karloff, H.: How good is the Goemans-Williamson MaxCut algorithm? In: *Proc. 28th ACM STOC*, ACM, 1996, pp. 427–434.

[Kar93]   Karger, D. K.: Global min-cuts in RNC, and other ramifications of a simple min-cut algorithm. In: *Proc. 4th ACM-SIAM Symposium on Discrete Algorithms*, 1993, pp. 21–30.

[KGV83]   Kirkpatrick, S., Gellat, P. D., Vecchi, M. P.: Optimization by simulated annealing. *Science* 220 (1983), 671–680.

[Kha79]   Khachian, L.: A polynomial algorithm in linear programmng. *Soviet Mathematics Doklady* 20 (1979), 191–194.

[KiL85]   Kindervater, G. A. P., Lenstra, J. K.: An introduction to parallelism in combinatorial optimization. In: [LeL85], pp. 163–184.

[KiL85a] Kindervater, G. A. P., Lenstra, J. K.: Parallel algorithms. In: [ELR85], pp. 106–128.

[KiT85]   Kindervater, G. A. P., Trienekens, H. W. J. M.: Experiments with parallel algorithms for combinatorial problems. Technical Report 8550/A, Erasmus University, Rotterdam, Econometric Institute, 1985.

[KK82]   Karmarkar, N., Karp, R.: An efficient approximation scheme for the one-dimensional bin packing problem. In: *Proc. 23rd IEEE FOCS*, IEEE, 1982, pp. 312–320.

[KL70] Kernighan, B. W., Lin, S.: An efficient heuristic procedure for partition-ing graphs. *Bell System Technical Journal* 49 (1970), 291–307.

[KL96] Klein, P., Lu, H.: Efficient approximation algorithms for semidefinite pro-grams arising from MaxCut and Coloring. In: *Proc. 28th ACM STOC*, ACM, 1996, pp. 338–347.

[KLS89] Kindervater, G. A. P., Lenstra, J. K., Shmoys, D. B.: The parallel com-plexity of TSP heuristics. *Journal of Algorithms* 10 (1989), 249–270.

[KTL97] Kaplan, P., Thaler, D., Libchaber, A.: Parallel overlap assembly of paths through a directed graph. In: *3rd. DIMACS Meeting on DNA Based Com-puters*, University of Pennsylvania, 1997.

[KM72] Klee, V., Minty, G. J.: How good is the simplex algorithm? In: *Inequalities III* (O. Shesha, Ed.), Academic Press, New York, 1972, pp. 159–175.

[KMR+96] Kurtz, S. A., Mahaney, S. R., Royer, J. S., Simon, J.: Active transport in biological computing. In: *Proc. 2nd Annual DIMACS Meeting on DNA-Based Computers*, 1996.

[KMS94] Karger, D., Motwani, R., Sudan, M.: Approximate graph coloring by semidefinite programming. In: *Proc. 35th IEEE FOCS*, IEEE, 1994, pp. 2–13.

[KMS+94] Khanna, S., Motwani, R., Sudan, M., Vazirani, U.: On syntactic ver-sus computational views of approximability. In *Proc. 26th ACM STOC*, ACM, 1994, pp. 819–830.

[KN97] Kushilevitz, E., Nisan, N.: *Communication Complexity*, Cambridge Uni-versity Press, 1997.

[KW84] Karp, R.M., Wigderson, A.: A fast parallel algorithm for the maximal independent set problem. In: *Proc. 16th ACM STOC*, ACM, 1984, pp. 266–272.

[Knu68] Knuth, D. E.: *The Art of Computer Programming; Volume 1: Fundamen-tal Algorithms*. Addison-Wesley, 1968; 2nd edition, 1973.

[Knu69] Knuth, D. E.: *The Art of Computer Programming; Volume 2: Seminu-merical Algorithms*. Addison-Wesley, 1969; 2nd edition, 1981.

[Knu73] Knuth, D. E.: *The Art of Computer Programming; Volume 3: Sorting and Searching*. Addison-Wesley, 1973.

[Knu75] Knuth, D. E.: Estimating the efficiency of backtrack programms. *Math-ematics of Computation* 29 (1975), 121–136.

[Ko92] Kozen, D. C.: *The Design and Analysis of Algorithms*. Springer-Verlag, 1992.

[KP80] Karp, R. M., Papadimitriou, Ch.: On linear characterizations of combi-natorial optimization problems. In: *Proc. 21st IEEE FOCS*, IEEE, 1980, pp. 1–9.

[KP89] Kum, S. H., Pomerance, C.: The probability that a random probable prime is composite. *Mathematics of Computation* 53, no. 188 (1989), 721–741.

[KR90] Karp, R. M., Ramachandran, V.: Parallel algorithms for shared-memory machines. In [Lee90], pp. 869–941.

[Kru56] Kruskal, J. B., Jr.: Parallel computation and conflicts in memory access. *Information Processing Letters* 14, no. 2 (1956), 93–96.

[KRu87] Kravitz, S. A., Rutenbar, R.: Placement by simulated annealing on a multiprocessor. *IEEE Transactions on Computer-Aided Design* 6 (1987), 534–549.

[Kuh55]   Kuhn, H. W.: The Hungarian method for the assignment problem. *Naval Research Logistics Quartely* 2 (1955), 83–97.

[Kul99]   Kullman, O.: New methods for 3-SAT decision and worst case analysis. *Theoretical Computer Science* 223 (1999), 1–71.

[KS93]   Karger, D. R., Stein, C.: An $\Theta(n^2)$ algorithm for minimum cuts. In: *Proc. ACM STOC*, ACM, 1993, pp. 757–765.

[LAL92]   van Laarhoven, P. J. M., Aarts, E. H. L., Lenstra, J. K.: Job shop scheduling by simluated annealing. *Operations Research* 40 (1992), 113–125.

[Lap92]   Laporte, G.: The travelling salesman problem: An overview of exact and approximate algorithms. *European Journal of Operations Research* 59 (1992), 231–247.

[Law76]   Lawler, E. L.: *Combinatorial Optimization: Networks and Matroids.* Holt, Rinehart and Winston, 1976.

[Law79]   Lawler, E. L.: Fast approximation algorithms for knapsack problems. *Mathematics of Operations Research* 4 (1979), 339–356.

[L'E88]   L'Écuyer, P.: Efficient and portable combined random number generators. *Communications of the ACM* 31, no. 6 (1988), 742–749.

[L'E90]   L'Écuyer, P.: Random numbers for simulation. *Communication of the ACM* 33, no. 10 (1990), 85–97.

[Lei92]   Leighton, F. T.: *Introduction to Parallel Algorithms and Architectures: Arrays, Trees, Hypercubes.* Morgan Kaufmann 1992.

[LeL85]   van Leeuwen, J., Lenstra, J. K.: *Parallel Computers and Computation, CWI Syllabus,* Vol. 9, Center for Mathematics and Computer Science, Amsterdam, 1985.

[LD91]   Lweis, H. R., Denenberg, L.: *Data Structures & Their Algorithms.* Harper Collins Publishers, 1991.

[Lee90]   van Leeuwen, J. (Ed.): *Handbook of Theoretical Computer Science; Volume A: Algorithms and Complexity.* Elsevier and MIT Press, 1990.

[LFK+92] Lund, C., Fortnow, L., Karloff, H., Nisan, N.: Algebraic methods for interactive proof systems. *Journal of the ACM* 39, no. 4 (1992), 859–868.

[LGC+96] Liu, Q., Guo, Z., Condon, A. E., Corn, R. M., Lagally, M. G., Smith, L. M.: A surface-based approach to DNA computation. In: *Proc. 2nd Annual Princeton Meeting on DNA-Based Computing,* 1996.

[Lan65]   Lancaster, H.O.: Pairwise statistical independence. *Ann. Mathematical Statistics,* 36 (1965) 1313–1317.

[LiB96]   Lipton, R. J., Baum, E. B. (Eds.): *DNA Based Computers, Proc. DIMACS Workshop,* Princeton 1995, AMS Press, 1996.

[Lin65]   Lin, S.: Computer solutions of the traveling salesman problem. *Bell System Technical Journal* 44 (1965), 2245–2269.

[Lip94]   Lipton, R. J.: Speeding up computations via molecular biology. Princeton University Draft, 1994.

[Lip95]   Lipton, R. J.: DNA solution of hard computational problems. *Science* 268 (1995), 542–545.

[LK73]   Lin, S, Kernighan, B. W.: An effective heuristic algorithm for the traveling-salesman problem. *Operations Research* 21 (1973), 498–516.

[LA88]   van Laarhoven, P. J. M., Aarts, E. H. L.: *Simulated Annealing: Theory and Applications.* Kluwer, Dordrecht, 1988.

[Llo93]   Lloyd, S.: A potentially realizable quantum computer. *Science* 261 (1991), 1569–1571.

[LL85]   Lawler, E., Lenstra, J., Rinnovy Kan, A., Shmoys, D.: *The Traveling Salesman Problem.* John Wiley & Sons, 1985.

[LL90]   Lenstra, A. K., Lenstra, M. W. Jr.: Algorithms in number theory. In: *Handbook of Theoretical Computer Science* (J. van Leeuwen, Ed.), Vol. A, Elsevier, Amsterdam, 1990, pp. 673–715.

[LMS⁺63]   Little, J. D. C., Murtly, K. G., Sweeny, D. W., Karel, C.: An algorithm for the Traveling-Salesman Problem. *OR* 11 (1963), 972–989.

[LP78]   Lewis, H. R., Papadimitriou, Ch.: The efficiency of algorithms. *Scientific American* 238, no. 1 (1978).

[Lov75]   Lovász, L.: On the ratio of the optimal integral and functional covers. *Discrete Mathematics* 13 (1975), 383–390.

[Lov79]   Lovász, L.: Graph theory and integer programming. *Annals of Discrete Mathematics* 4 (1979), 146–158.

[Lov80]   Lovász, L.: A new linear programming algorithm – better or worse than the simplex method? *The Mathematical Intelligence* 2 (1980), 141–146.

[Lue75]   Lueker, G.: Two polynomial complete problems in non-negative integer programming. Manuscript TR-178, Department of Computer Science, Princeton University, Princeton, 1975.

[LW66]   Lawler, E. L., Wood, D. W.: Branch-and-bound methods: A survey. *Operations Research* 14, no. 4 (1966), 699–719.

[LY93]   Lund, C., Yannakakis, M.: On the hardness of approximating minimization problems. In: *Proc. 25th ACM STOC*, ACM, 1993, pp. 286–293.

[Man89a]   Manber, U.: *Introduction to Algorithms: A Creative Approach.* Addison-Wesley, Reading, 1989.

[Man89b]   Manber, U.: Memo functions, the graph traverser, and a simple control situation. In: *Machine Intelligence* 5 (B. Meltzer, D. Michie, Eds.), American Elsevier and Edinburgh University Press 1989, pp. 281–300.

[Meh84a]   Mehlhorn, K.: *Data Structures and Algorithms 1: Sorting and Searching.* EATCS Monographs, Springer-Verlag, 1984.

[Meh84b]   Mehlhorn, K.: *Data Structures and Algorithms 2: Graph Algorithms and NP-Completeness.* EATCS Monographs, Springer-Verlag, 1984.

[Meh84c]   Mehlhorn, K.: *Data Structures and Algorithms 3: Multi-Dimensional Searching and Computational Geometry.* EATCS Monographs, Springer-Verlag, 1984.

[MC80]   Mead, C. A., Conway, L. C.: *Introduction to VLSI Design.* Addison-Wesley, Reading, 1980.

[Mic92]   Michalewicz, Z.: *Genetic Algorithms + Data Structures = Evolution Programs.* Springer-Verlag, 1992.

[MiF98]   Michalewicz, Z., Fogel, D. B.: *How to Solve It: Modern Heuristics.* Springer-Verlag 1998.

[Mil76]   Miller, G.: Riemann's hypothesis and tests for primality. *Journal of Computer and System Sciences* 13 (1976), 300–317.

[Mit96]   Mitchell, I. S. B.: Guillotine subdivisions approximate polygonal subdivisions: Part II – a simple polynomial-time approximation scheme for geometric k-MST, TSP and related problems. Technical Report, Department of Applied Mathematics and Statistics, Stony Brook, 1996.

[MNR96]   Motwani, R., Naor, J., Raghavan, P.: Randomized approximation algorithms in combinatorial optimization. In: *Approximation Algorithms for NP-hard Problems* (D. S. Hochbaum, Ed.), PWS Publishing Company, 1997, pp. 447–481.

[Mon80]   Monier, L.: Evaluation and comparison of two efficient probabilistic primality testing algorithms. *Theoretical Computer Science* 12 (1980), 97–108.

[MoS97]   Morimoto, N, Suyama, M. A. A.: Solid phase DNA solution to the Hamiltonian path problem. In: *3rd DIMACS Meeting on DNA Based Computers*, University of Pennsylvania 1997.

[MS79]   Monien, B., Speckenmeyer, E.: 3-satisfiability is testable in $O(1.62^r)$ steps. Bericht Nr. 3/1979, Reihe Theoretische Informatik, Universität Paderborn, 1979.

[MS82]   Mehlhorn, K., Schmidt, E.: Las Vegas is better than determinism in VLSI and ditributed computing. In: *Proc 14th ACM STOC*, ACM, 1982, pp. 330–337.

[MS85]   Monien, B., Speckenmeyer, E.: Solving satisfiability in less than 2^n steps. *Discrete Applied Mathematics* 10 (1985), 287–295.

[MS97]   Macarie, I. I., Seiferas, J. I.: Strong equivalence of nondeterministic and randomized space-bounded computations. Manuscript, 1997.

[MPS98]   Mayr, E. W., Prömel, H. J., Steger, A. (Eds.): *Lecture on Proof Verification and Approximation Algorithms*. Lecture Notes in Computer Science 1967, Springer-Verlag, 1998.

[MR95]   Motwani, R., Raghavan, P.: *Randomized Algorithms*. Cambridge University Press, 1995.

[Müh97]   Mühlenberg, H.: Genetic algorithms. In: [AL97], pp. 137–171.

[MU49]   Metropolis, I. N., Ulam, S.: The Monte Carlo method. *Journal of the American Statistical Assosiation* 44, no. 247 (1949), 335–341.

[MRR+53] Metropolis, N., Rosenbluth, A. W., Rosenbluth, M. N., Teller A. H., Teller, E.: Equation of state calculation by fast computing machines. *Journal of Chemical Physics* 21 (1953), 1087–1091.

[NI92]   Nagamochi, H., Ibarski, T.: Computing edge connectivity in multigraphs and capacitated graphs. *SIAM Journal on Discrete Mathematics* 5 (1992), 54–66.

[Nis96]   Nisan, N.: Extracting randomness: How and why–A survey. In: *Proc. IEEE Symposium on Structure in Complexity Theory*, IEEE, 1996.

[NC00]   Nielsen, M.A., Chuang,I.L.: *Quantum Computation and Quantum Information*. Cambridge University Press 2000.

[NP80]   Nešetřil, J., Poljak, S.: Geometrical and algebraic correspondences of combinatorial optimization. In: *Proc. SOFSEM '80*, Computer Research Center Press, Bratislava, 1980, pp. 35–77 (in Czek).

[NR99]   Niedermeier, R., Rossmanith, P.: New upper bounds for MaxSat. In: *Proc. 26th ICALP'99*, Lecture Notes in Computer Science 1644, Springer-Verlag, 1999, pp. 575–584.

[OtW96]   Ottmann, T., Widmayer, P.: *Algorithmen und Datenstrukturen*. Spektrum Akademischer Verlag, 1996 (in German).

[OvG89]   Otten, R. H. J. M., van Ginneken, L. P. P. P.: *The Annealing Algorithm*. Kluwer Academic Publishers, 1989.

[O'B80]   O'Brien, G.L.: Pairwise independent random variables. *Ann. Probability* 8 (1980), 170–175.

[Pap77]   Papadimitriou, Ch.: The Euclidean travelling salesman problem is NP-complete. *Theoretical Computer Science* 4 (1977), 237–244.

[Pap94]   Papadimitriou, Ch.: *Computational Complexity*. Addison-Wesley, 1994.

[Pap95]   Papadimitriou, Ch.: personal communication to J. Reif.

[Ple82]   Plesník, J.: Complexity of decomposing graphs into factors with given diameters or radii. *Mathematica Slovaca* 32, 379–388.

[Ple83]   Plesník, J.: *Graph Algorithms.* VEDA, Bratislava, 1983 (in Slovak).

[Ple86]   Plesník, J.: Bad examples of the metric traveling salesman problem for the 2-change heuristic. *Acta Mathematica Universitatis Comenianae* 55 (1986), 203–207.

[PM81]   Paz, A., Moran, S.: Nondeterministic polynomial optimization problems and their approximation. *Theoretical Computer Science* 15 (1981), 251–277.

[Pol75]   Pollard, J. M.: A Monte Carlo method of factorization. *BIT* 15, (1975), 331–334.

[Pom81]   Pomerance, C.: On the distribution of pseudoprimes. *Mathematics of Computation* 37, no. 156 (1981), 587–593.

[Pom82]   Pomerance, C.: The search for prime numbers. *Scientific American* 2476 (1982).

[Pom87]   Pomerance, C.: Very short primality proofs. *Mathematics of Computation* 48, no. 177 (1987), 315–322.

[PPZ97]   Paturi, R., Pudlák, P., Zane, F.: Satisfiability coding lemma. In: *Proc. 38th IEEE FOCS*, IEEE, 1997, pp. 566–574.

[PPSZ98]   Paturi, R., Pudlák, P., Saks, E., Zane, F.: An improved exponential-time algorithm for k-SAT. In: *Proc. IEEE FOCS*, IEEE, 1998, pp. 628–637.

[Pra75]   Pratt, V.: Every prime has a succint certificate. *SIAM Journal on Computing* 4, no. 3 (1975), 214–220.

[Pre00]   Preskill, J.: *Lecture Notes on Quantum Information and Quantum Computation.* Web address:
www.theory.caltech.edu/people/preskill/ph229.

[PRS98]   Păun, G., Rozenberg, G., Salomaa, A.: *DNA Computing (New Computing Paradigms).* Springer-Verlag, 1998.

[PS77]   Papadimitriou, Ch., Steiglitz, K.: On the complexity of local search for the traveling salesman problem. *SIAM Journal of Computing* 6 (1977), 76–83.

[PS78]   Papadimitriou, Ch., Steiglitz, K.: Some examples of difficult traveling salesman problems. *Operations Research* 26 (1978), 434–443.

[PS82]   Papadimitriou, Ch., Steiglitz, K.: *Combinatorial Optimization: Algorithms and Complexity.* Prentice-Hall, Englewood Cliffs, 1982.

[PS84]   Papadimitriou, Ch., Sipser, M.: Communication complexity. *Journal of Computer and System Sciences* 28 (1984), pp. 260–269.

[Pud94]   Pudlák, P.: Complexity theory and genetics. In: *Proc. 9th Conference on Structure in Complexity Theory*, 1994, pp. 183–195.

[PVe00]   Papadimitriou, Ch., Vempala, S.: On the approximability of the travling salesperson problem. In: *Proc. 32nd ACM STOC*, ACM, 2000.

[PY91]   Papadimitriou, Ch., Yannakakis, M.: Optimization, approximation, and complexity classes. *Journal of Computer and System Sciences* 43, 3 (1991), 425–440.

[PY93]   Papadimitriou, Ch., Yannakakis, M.: On limited nondeterminism and the complexity of the V-C dimension. In: *Proc. 8th Conference on the Structure in Complexity Theory*, 1993, pp. 12–18.

[Rab76]   Rabin, M. O.: Probabilistic algorithms. In: *Algorithms and Complexity: Recent Results and New Directions.* (J. F. Traub, Ed.), Academic Press, 1976, pp. 21–39.

[Rab80]   Rabin, M. O.: Probabilistic algorithm for primality testing. *Journal of Number Theory* 12 (1980), 128–138.

[Rag88]   Raghavan, P.: Probabilistic construction of deterministic algorithms: Approximating packing integer programs. *Journal of Computer and System Sciences* 37 (1988), 130–143.

[Rec73]   Rechenberg, I.: Evolutionsstrategie: Optimierung technischer Systeme nach Prinzipien der biologischen Information. Fromman, Freiburg, 1973, (in German).

[Ree95]   Reeves, C. R. (Ed.): *Modern Heuristic Techniques for Combinatorial Problems.* McGraw-Hill, London, 1995.

[Ree95a] Reeves, C. R.: Genetic algorithms. In: [Ree95], pp. 151–196.

[Rei91]   Reinelt, G.: TSPLIB: A traveling salesman problem library. *ORSA Journal on Computing* 3 (1991), 376–384.

[Rei92]   Reinelt, G.: Fast heuristics for large geometric traveling salesman problems. *ORSA Journal on Computing* 4 (1992), 206–217.

[Rei93]   Reif, J. H. (Ed.): *Synthesis of Parallel Algorithms.* Morgan Kaufmann, 1993.

[Rei94]   Reinelt, G.: *The Traveling Salesman: Computational Solutions for TSP Applications. Lecture Notes in Computer Science* 840, Springer-Verlag, 1994.

[Rei95]   Reif, J.: Parallel molecular computation: Models and simulations. In: *Proc. 17th ACM Symp. on Parallel Algorithms and Architectures*, ACM, 1995, pp. 213–223.

[Rei98]   Reif, J.: Paradigms for biomolecular computation. In: [CCD98], pp. 72–93.

[ReS65]   Reiter, S., Sherman, G.: Discrete optimizing. *Jorunal of the Society for Industrial and Applied Mathematics* 13, (1965), 864–889.

[RND82]   Reingold, E. M., Nievergelt, J., Deo, N.: *Combinatorial Algorithms: Theory and Practice.* Prentice-Hall, Englewood Cliffs, 1982.

[Ros00]   Rosen, K. M. (Ed.): *Handbook of Discrete and Combinatorial Mathematics.* CRC Press LLC, 2000.

[RS65]   Reiter, S., Sherman, G.: Discrete optimizing. *Journal of the Society for Industrial and Applied Mathematics* 13 (1965), 864–889.

[RSL77]   Rosenkrantz, D. J., Stearns, R. E., Lewis, P. M.: An analysis of several heuristics for the traveling salesman problem. *SIAM Journal on Computing* 6 (1977), 563–581.

[RT87]   Rödl, V., Tovey, C. A.: Multiple optima in local search. *Journal of Algorithms* 8 (1987), 250–259.

[Rub96]   Rubin, H.: Looking for the DNA killer app. *Nature* 3 (1996), 656–658.

[Sah75]   Sahni, S.: Approximate algorithms for the 0/1 knapsack problem. *Journal of the ACM* 22 (1975), 115–124.

[Sal98]   Salomaa, A.: Turing, Watson-Crik and Lindemeyer. Aspects of DNA complementarity. In: [CCD98], pp. 94–107.

[Sau97]   Sauerhoff, M.: On nondeterminism versus randomness for read-once branching programs. *Electronic Colloquium on Computational Complexity* TR 97 - 030 (1997).

522   References

[Sau99]   Sauerhoff, M.: On the size of randomized OBDDs and read-once branch-
           ing programs for k-stable functions. In: Proc. *STACS '99, Lecture Notes
           in Computer Science* 1563, Springer-Verlag, 1999, pp. 488–499.

[SBK00]   Schmeck, H., Branke, J., Kohlmorgen, U.: Parallel implementations of
           evolutionary algorithms. In: *Solutions to Parallel and Distributed Com-
           puting Problems* (A. Zomaya, F. Ercal, S. Olariu, Eds.), John Wiley &
           Sons, 2000, to appear.

[SC79]   Stockmeyer, L. J., Chandra, A. K.: Intrinsically difficult problems. In:
           Scientific American 240, no. 5 (1979), 140–159.

[Sch81]   Schwefel, H.-P.: *Numerical Optimization of Computer Models*. John Wi-
           ley & Sons, Chichester, 1981.

[Sch95]   Schwefel, H.-P.: *Evolution and Optimum Seeking*. John Wiley & Sons,
           Chichester, 1995.

[Schö99]   Schöning, U.: A probabilistic algorithm for k-SAT and constraint satis-
           faction problems. In: *Proc. 40th IEEE FOCS*, IEEE, 1999, pp. 410–414.

[Sed89]   Sedgewick, R.: *Algorithms*. 2nd edition, Addison-Wesley, 1989.

[SG74]   Sahni, S., Gonzales, T.: P-complete problems and approximate solutions.
           Compututer Science Technical Report 74-5, University of Minnesota,
           Minneapolis, Minn., 1974.

[SG76]   Sahni, S., Gonzales, T.: P-complete approximation problems. *Journal of
           the ACM* 23 (1976), 555–565.

[SH78]   Sahni, S., Horowitz, E.: Combinatorial problems: Reducibility and ap-
           proximation. *Operations Research* 26, no. 4 (1978), 718–759.

[Sha79]   Shamir, A.: Factoring numbers in $O(\log n)$ arithmetic steps. *Information
           Processing Letters* 8, no. 1 (1979), 28–31.

[Sh92]   Shallit, J.: Randomized algorithms in 'primitive' cultures, or what is the
           oracle complexity of a dead chicken. *ACM Sigact News* 23, no. 4 (1979),
           77–80; see also *ibid.* 24, no. 1 (1993), 1–2.

[Sha92]   Shamir, A.: IP=PSPACE. *Journal of the ACM* 39, no. 4 (1992), 869–877.

[She92]   Shen, A.: IP=PSPACE: Simplified proof. *Journal of the ACM* 39, no. 4
           (1992), 878–880.

[Sho94]   Shor, P. W.: Algorithms for quantum computation: Discrete logarithmics
           and factoring. In: *35th IEEE FOCS*, IEEE, 1994, pp. 124–134.

[Sie98]   Siegling, D.: Derandomization. In: [MPS98], pp. 41–61.

[Sip92]   Sipser, M.: The history and status of the P versus NP question. In: *Proc.
           24th ACM STOC*, ACM, 1992, pp. 603–618.

[Sim94]   Simon, D. R.: On the power of quantum computation. In: *Proc. 35th
           IEEE FOCS*, IEEE, 1994, pp. 116–123, (also: *SIAM Journal on Comput-
           ing* 26 (1997), 1484–1509).

[Spe94]   Spencer, J.: Randomization, derandomization and antirandomization:
           three games. *Theoretical Computer Science* 131 (1994), 415–429.

[SS77]   Solovay, R., Strassen, V.: A fast Monte Carlo test for primality. *SIAM
           Journal on Computing* 6, no. 1 (1977), 84–85; erratum, *ibid.* 7, no. 1
           (1978), 118.

[SmS95]   Smith, W., Schweitzer, A.: DNA computers in vitro and vivo. NEC Re-
           search Institute Technical Report 95 - 057 - 3 - 0058 - 3, 1995.

[Sti85]   Stinson, D. R.: *An Introduction to the Design and Analysis of Algorithms*.
           The Charles Babbage Research Centre, St. Pierre, Manitoba, 1985; 2nd
           edition, 1987.

[Str96]   Strassen, V.: Zufalls-Primzahlen und Kryptographie. In: *Highlights aus der Informatik* (I. Wegener, Ed.), Springer-Verlag, 1996, pp. 253–266 (in German).

[Sud92]   Sudan, M.: Efficient checking of polynomials and proofs and the hardness of approximation problems. Ph.D. thesis, Department of Computer Science, Berkeley, 1992.

[SW90]   Shmoys, D. B., Williamson, D. P.: Analysing the Held-Karp TSP bound: A monotonicity property with applications. *Information Processing Letters* 35 (1990), 281–285.

[Ta89]   Tanese, R.: Distributed genetic algorithms. In: *Proc. 3rd Int. Conference on Genetic Algorithms* (J. D. Schaffer, Ed.), Morgan Kaufmann, San Mateo, 1989, pp. 434–439.

[Tre97]   Trevisan, L.: Reductions and (Non)-Approximability. Ph.D. thesis, Computer Science Department, University of Rome "La Sapienze", 1997.

[TSS⁺96]   Trevisan, L., Sorkin, G., Sudan, M, Williamson, D.: Gadgets, approximation, and linear programming. In: *Proc. 37th IEEE FOCS*, IEEE, 1996, pp. 617–626.

[Tu36]   Turing, A. M.: On computable numbers, with an application to the Entscheidungsproblem. *Proceedings London Mathematical Society*, Ser. 2, 42 (1936), 230–265; a correction, 43 (1936), 544–546.

[Tu50]   Turing, A. M.: Computing machinery and intelligence. *Mind* 59 (1950), 433–460.

[UAH74]   Ullman, J., Aho, A., Hopcroft, J.: *The Design and Analysis of Computer Algorithms.* Addison-Wesley, 1974.

[VAL96]   Vaessens, R. J. M., Aarts, E. H. L, Lenstra, J. K.: A local search template. In: *Parallel Problem Solving from Nature 2* (R. Manner, B. Manderick, Eds.), North-Holland, Amsterdam, 1996, pp. 65–74.

[Vaz86]   Vazirani, U. V.: Randomness, adversaries, and computation. Doctoral dissertation, Deptartment of Computer Science, University of California, Berkeley, 1986.

[Vaz87]   Vazirani, U. V.: Efficiency considerations in using semi-random sources. In: *Proc. 19th ACM STOC*, ACM, 1987, pp. 160–168.

[Ver94]   Verma, R. M.: A general method and a master theorem for divide-and-conquer recurrences with applications. *Journal of Algorithms* 16 (1994), 67–79.

[vNe53]   von Neumann, J.: A certain zero-sum two-person game equivalent to the optimal assignment problem. In: *Contributions to the Theory of Games II* (H. W. Kuhn, A. W. Tucker, Eds.), Princeton University Press 1953.

[Wal60]   Walker, R., J.: An enumerative technique for a class of combinatorial problems. In: *Proc. of Symposia of Applied Mathematics*, Vol. X, AMS 1960.

[Weg93]   Wegener, I.: *Theoretische Informatik: eine algorithmenorientierte Einführung.* B.G. Teubner, 1993 (in German).

[Weg00]   Wegener, I.: On the expected runtime and the success probability of evolutionary algorithms. In: *Preproceedings of 26th WG 2000*, University of Konstanz 2000, pp. 229–240 (also: *Proc. 26th WG 2000, Lecture Notes in Computer Science* 1928, Springer-Verlag, 2000, pp. 1–10).

[Wel71]   Wells, M. B.: *Elements of Combinatorial Computing.* Pergamon Press, Oxford, 1971.

524   References

[WhS90] Whitley, L. D., Starkweather, T.: Genitor II: A distributed genetic algorithm. *Expt. Theor. Artif. Intelligence* 2 (1990), 189–214.

[Wi78] Williams, H.: Primality testing on a computer. *Ars Combinatoria* 5, no. 6 (1979), 347–348.

[Win87] Winter, P.: Steiner problem in networks: A survey. *Networks* 17, no. 1 (1987), 129–167.

[Wo93] Wood, D.: *Data Structures, Algorithms, and Performance*. Addison-Wesley, 1993.

[Wol80] Wolsey, L. A.: Heuristic analysis, linear programming, and branch-and-bound. *Mathematical Programming Studies* 13 (1980), 121–134.

[WR97] Walker, M. R., Rapley, R.: *Route Maps in Gene Technology*. Blackwell Science, Oxford, 1997.

[Yan79] Yannakakis, M.: The effect of a connectivity requirement on the complexity of maximum subgraph problems. *Journal of the ACM* 26 (1979), 618–630.

[Yao82] Yao, A. C.-C.: Theory and applications of trapdoor functions. In: *Proc. 23rd IEEE FOCS*, IEEE, 1982, pp. 80–91.

[Yan92] Yannakakis, M.: On the approximation of maximum satisfiability. In: *Proc. 3rd ACM-SIAM Symposium on Discrete Algorithms*, 1992, pp. 1–9.

[Yan97] Yannakakis, M.: Computational complexity. In: [AL97a], pp. 19–58.

[YoK97] Yokomori, T., Kabayashi, S.: DNA-EC, a model of DNA-computing based on equality checking. In: *3rd. DIMACS Meeting on DNA-Based Computing*, University of Pennsylvania, 1997.

[Zad73] Zadeh, N.: A bad network problem for the simplex method and other minimum cost flow algorithms. *Mathematical Programming* 5 (1973), 255–266.

[Zad73a] Zadeh, N.: More pathological examples for network flow problems. *Mathematical Programming* 5 (1973), 217–224.

[Zip93] Zippel, R. E.: *Efficient Polynomial Computations*. Kluwer Academic Publishers, Boston, 1993.

Index

Monographs in Theoretical Computer Science · An EATCS Series

K. Jensen
Coloured Petri Nets
Basic Concepts, Analysis Methods
and Practical Use, Vol. 1
2nd ed.

K. Jensen
Coloured Petri Nets
Basic Concepts, *Analysis Methods*
and Practical Use, Vol. 2

K. Jensen
Coloured Petri Nets
Basic Concepts, Analysis Methods
and *Practical Use,* Vol. 3

A. Nait Abdallah
The Logic of Partial Information

Z. Fülöp, H. Vogler
Syntax-Directed Semantics
Formal Models Based on Tree Transducers

A. de Luca, S. Varricchio
**Finiteness and Regularity
in Semigroups and Formal Languages**

E. Best, R. Devillers, M. Koutny
Petri Net Algebra

S.P. Demri, E. S. Orłowska
**Incomplete Information:
Structure, Inference, Complexity**

J.C.M. Baeten, C.A. Middelburg
Process Algebra with Timing

L. A. Hemaspaandra, L. Torenvliet
Theory of Semi-Feasible Algorithms

E. Fink, D. Wood
Restricted-Orientation Convexity

Zhou Chaochen, M. R. Hansen
Duration Calculus

M. Große-Rhode
**Semantic Integration of Heterogeneous
Software Specifications**

Texts in Theoretical Computer Science · An EATCS Series

J. L. Balcázar, J. Díaz, J. Gabarró
Structural Complexity I
2nd ed. (see also overleaf, Vol. 22)

M. Garzon
Models of Massive Parallelism
Analysis of Cellular Automata
and Neural Networks

J. Hromkovič
**Communication Complexity
and Parallel Computing**

A. Leitsch
The Resolution Calculus

G. Păun, G. Rozenberg, A. Salomaa
DNA Computing
New Computing Paradigms

A. Salomaa
Public-Key Cryptography
2nd ed.

K. Sikkel
Parsing Schemata
A Framework for Specification
and Analysis of Parsing Algorithms

H. Vollmer
Introduction to Circuit Complexity
A Uniform Approach

W. Fokkink
Introduction to Process Algebra

K. Weihrauch
Computable Analysis
An Introduction

J. Hromkovič
Algorithmics for Hard Problems
Introduction to Combinatorial Optimization,
Randomization, Approximation, and Heuristics
2nd ed.

S. Jukna
Extremal Combinatorics
With Applications in Computer Science

P. Clote, E. Kranakis
**Boolean Functions and Computation
Models**

L. A. Hemaspaandra, M. Ogihara
The Complexity Theory Companion

C.S. Calude
Information and Randomness.
An Algorithmic Perspective
2nd ed.

J. Hromkovič
Theoretical Computer Science
Introduction to Automata, Computability,
Complexity, Algorithmics, Randomization,
Communication and Cryptography

A. Schneider
Verification of Reactive Systems
Formal Methods and Algorithms

Former volumes appeared as
EATCS Monographs on Theoretical Computer Science